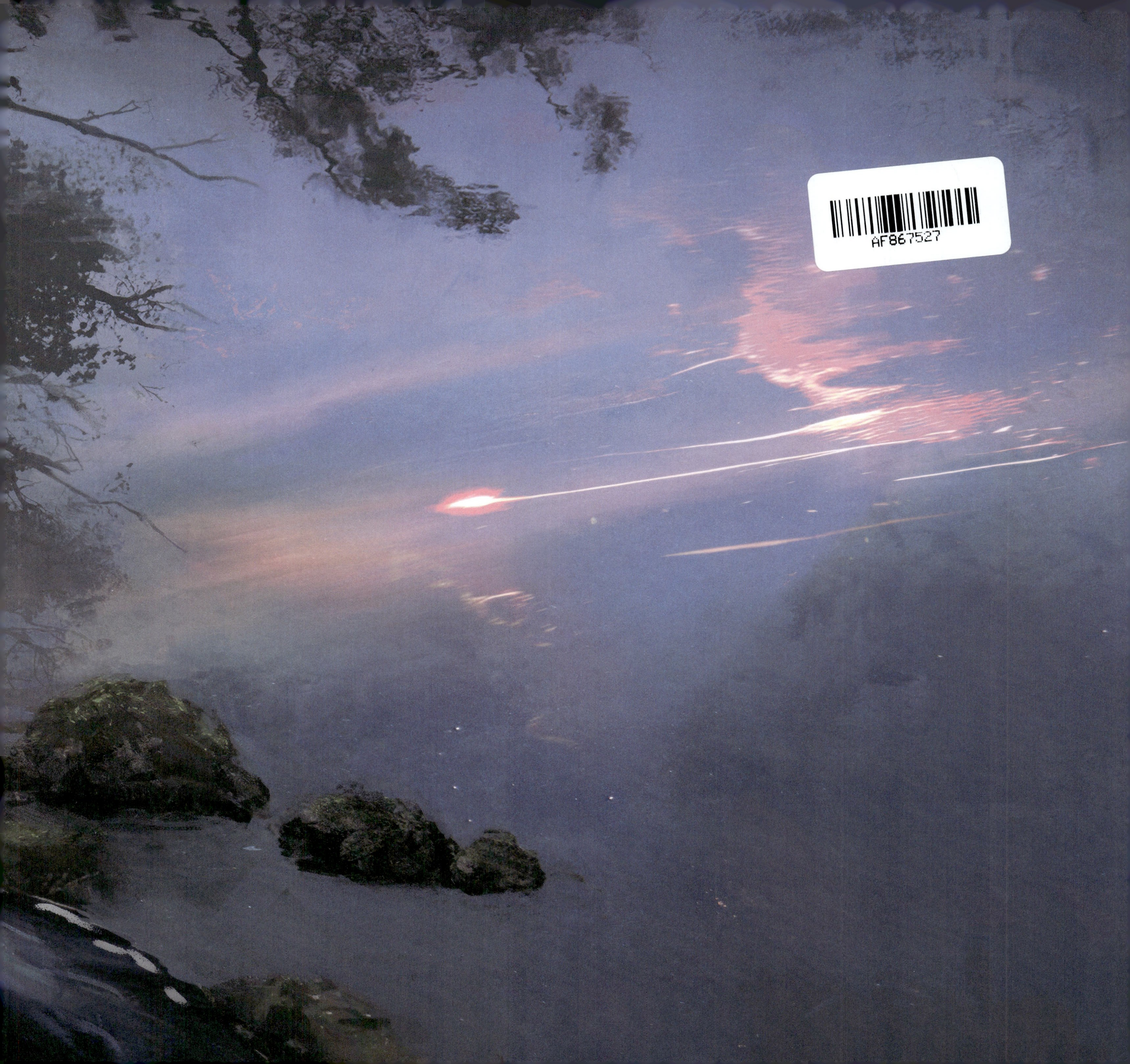

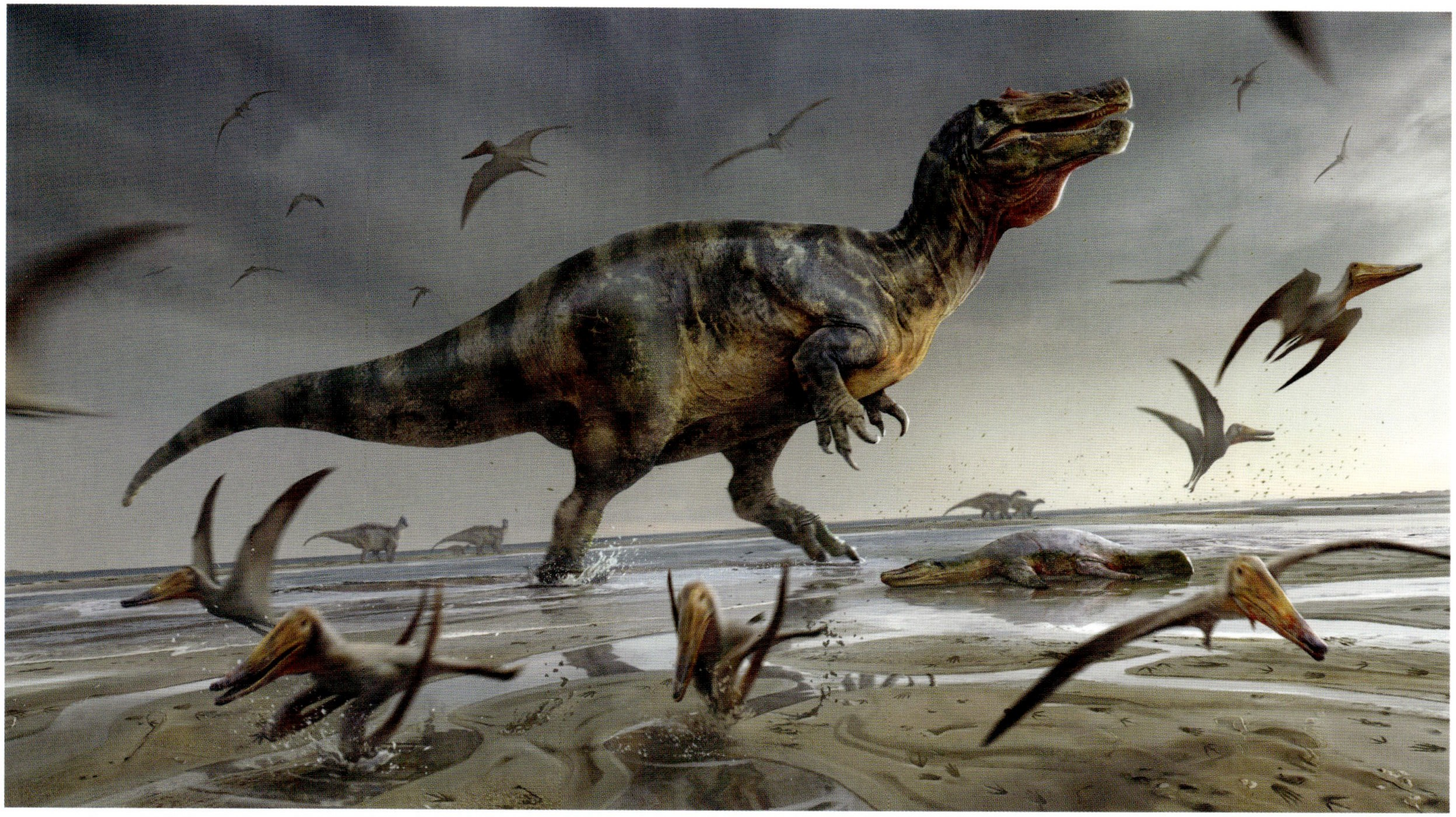

MESOZOIC ART II

DINOSAURS AND OTHER ANCIENT ANIMALS IN ART

EDITED BY STEVE WHITE AND DARREN NAISH

FOREWORD BY CHRIS PACKHAM

BLOOMSBURY WILDLIFE

LONDON • OXFORD • NEW YORK • NEW DELHI • SYDNEY

DEDICATED TO JENNY AND ROSE

BLOOMSBURY WILDLIFE
Bloomsbury Publishing Plc
50 Bedford Square, London, WC1B 3DP, UK
Bloomsbury Publishing Ireland Limited
29 Earlsfort Terrace, Dublin 2, D02 AY28, Ireland

BLOOMSBURY, BLOOMSBURY WILDLIFE and the Diana logo are trademarks of Bloomsbury Publishing Plc

First published in the United Kingdom 2025

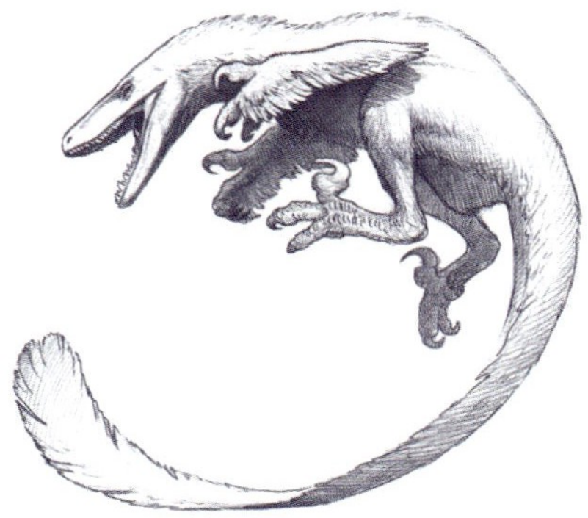

A catalogue record for this book is available from the British Library.

Library of Congress Cataloguing-in-Publication data has been applied for.

ISBN: HB: 978-1-3994-1221-6 ; ePub: 978-1-3994-1220-9; ePDF: 978-1-3994-1222-3

10 9 8 7 6 5 4 3 2 1

Designed by Austin Taylor

Printed and bound in China by RR Donnelley Asia

MIX
Paper | Supporting responsible forestry
FSC www.fsc.org FSC® C144853

To find out more about our authors and books visit www.bloomsbury.com and sign up for our newsletters. For product safety related questions contact productsafety@bloomsbury.com

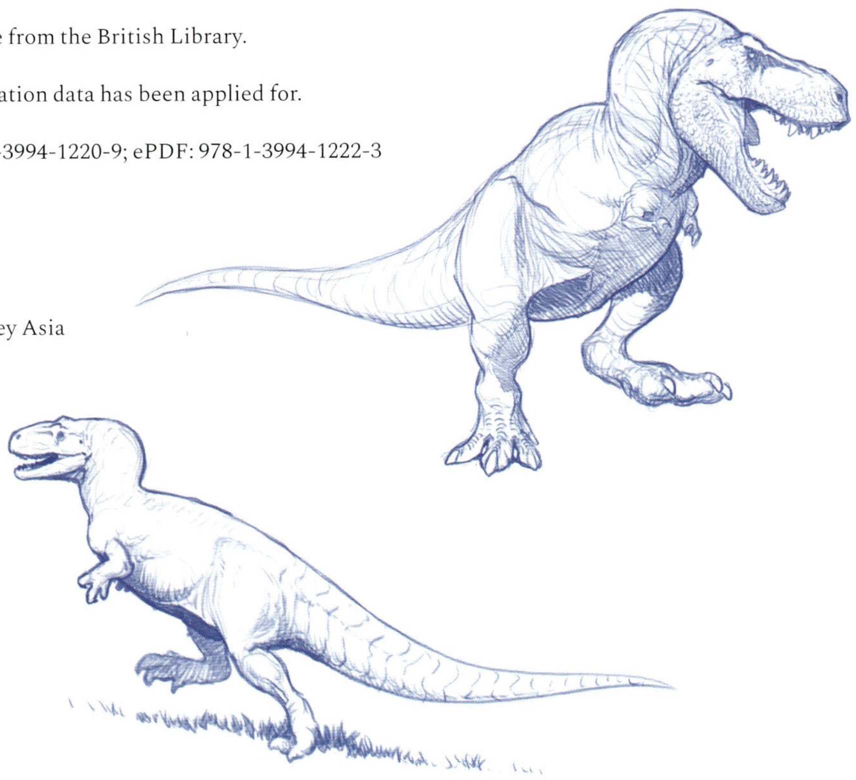

[Front endpaper]
BALAUR AND SKY
From its arboreal vantage point, a *Balaur* is surprised by colourful events in the sky. The animal cannot know it, but this is extraterrestrial debris raining through the atmosphere, the harbinger of the devastating extinction event about to occur far to the west in Mexico. The dinosaur's colour scheme reflects adaptation to forest life and echoes that of modern bird groups like turacos and magpies.
Gaëlle Seguillon • Digital | 2023

[Back endpaper]
A SPRING NIGHT DURING THE K/PG EVENT
Uncertainty remains over the exact look of the few seconds when a giant rock from space raced through our atmosphere and crashed into our planet 66 million years ago, but here's one possibility. This scene makes deliberate reference to research showing that the impact happened during the northern spring.
Gaëlle Seguillon • Digital | 2023

[Hardback cover]
QIJIANGLONG
Qijianglong – its name means 'dragon of Qijiang' – is a giant sauropod from the Early Cretaceous of western China. This image speculates on the resting and sleeping habits of sauropods, and with the idea that the moon was larger in the sky during the geological past than it is today.
Rebecca Dart • Digital | 2020

[Jacket, front cover]
PSITTACOSAUR RIVALRY
Fossils show that *Psittacosaurus* lived in groups for at least part of their lives. Cheek horns, a short, powerful beak and bipedal abilities suggest that they fought face to face while standing tall. This image perhaps suggests that squabbles for dominance occurred within their social groups.
Anthony J Hutchings • Digital | 2020

[Previous page]
WHITE ROCK CHONKER
Anthony J Hutchings • Digital | 2022

[This page, top]
DEINONYCHUS
Frederic Wierum • Digital | 2022

[This page, bottom]
T. REX AND *ALLOSAURUS*
Frederic Wierum • Digital | 2022

CONTENTS

FOREWORD

THEY'RE GONE... but to say they will never be forgotten is an understatement of sublime absurdity. Nothing, nothing, has the enduring fame of dinosaurs. Humans can compose symphonies, paint masterpieces, discover new planets, pose as movie stars, rock stars and sports stars, but although they sparkle, they will never achieve the celebrity afforded to that remarkable fauna of extinct animals we too broadly call 'dinosaurs'. The allure of these ancient creatures will outlive da Vinci, Mozart, Monroe, Mercury and Messi. In fact, I'll bet you that more people – especially children – have heard of *Tyrannosaurus rex* than all these human pretenders! It's bigger, fiercer, critically far-longer 'extincter' – and therefore will always be better.

You see, being all but invisible for 66 million years has an emphatic attraction. I know through paintings what da Vinci and Mozart looked like. I'd recognise these figures in a crowd of portraits, and there are probably more photographs of Monroe, Mercury and Messi than every Cretaceous fossil ever found. We all know them intimately...and obviously they don't change. I remember Marilyn Monroe from when I first saw her in the 1960s and she is still sensational...but static. But the *T. rex* I met back then, well, that's grown into a whole new organism! Everything is different! Everything! Gone is the tail-dragging, cold-blooded, unintelligent, scaly, lumbering, roaring giant lizard. *T. rex* has been reborn. And as this simply fabulous book illustrates, it, and all its mates, enjoy constant re-imagination. Their re-evolution continues...so much more exciting!

For me, science is the art of understanding truth and beauty. Art and science are not alienated fields and in this magnificent book, it's the art which leads – but it is informed by the science and the results are truly beautiful. They are also brilliantly diverse, from the dynamic 'graphic novel strips', the photographically inspired images with shallow focus and cascading perspectives, to the classic giant dioramas which have defined how we see 'dinosaurs' for more than one hundred years. Ultimately, though, what is most important is that this 'Mesozoic art' communicates: it makes us think, it excites us, it connects us ever more accurately with the real animals we will never meet.

Now, a genie, a lamp, a wish. A date with Marilyn or *T. rex*? No contest!

CHRIS PACKHAM

NEW FOREST 2024

Chris Packham is a renowned wildlife TV presenter, conservationist and photographer. He lives in the UK.

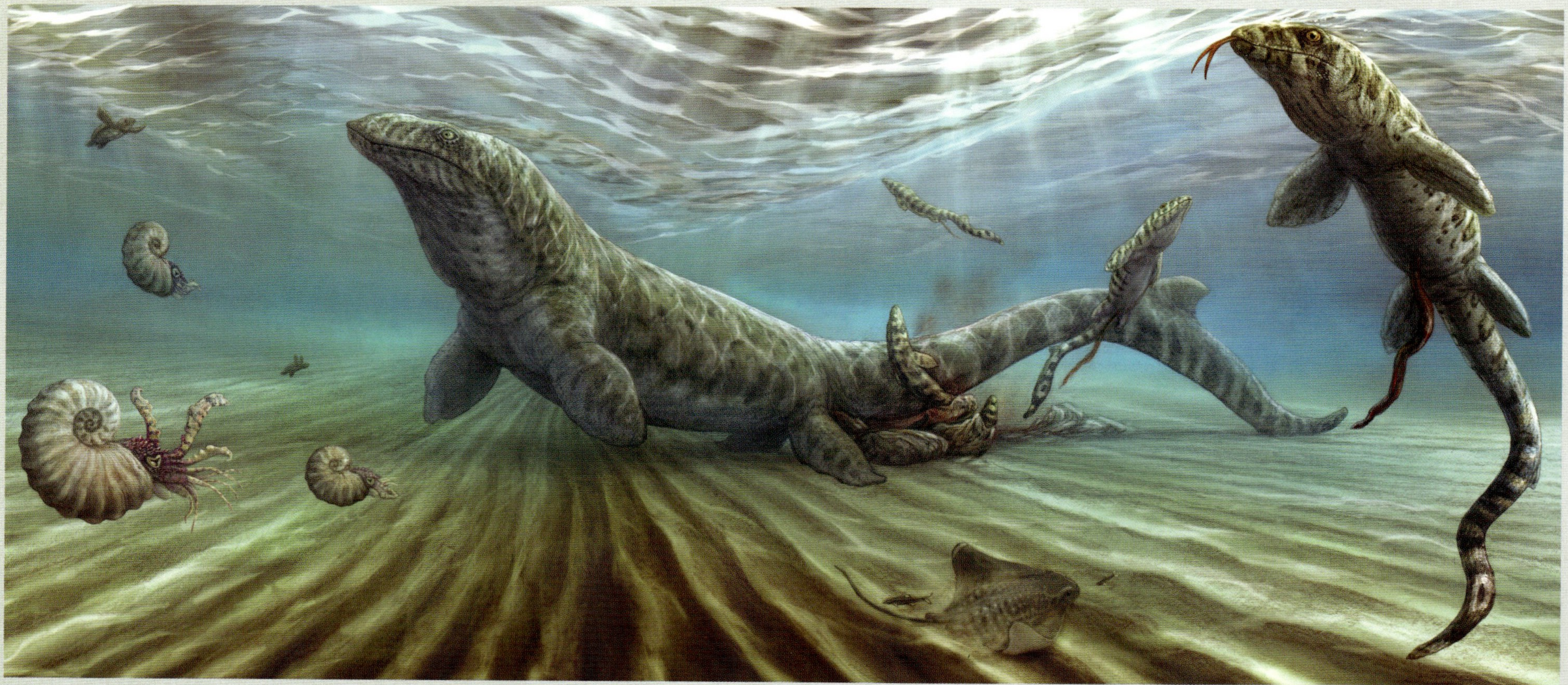

BIRTHING *PLATYCARPUS*

Brian Engh • Graphite on paper, digital colouring | 2019

INTRODUCTION – POP ART

WELCOME TO *MESOZOIC ART II*, the successor to our previous volume, published in late 2022. Many comments made about the first volume of *Mesozoic Art* were from readers who enjoyed the unexpectedly wide range of styles we included. The new volume you hold in your hands presents an equally diverse selection of approaches to palaeoart, but in some ways goes even further.

Most people's exposure to palaeoart usually comes via an illustration that accompanies a press release or news report, this most usually pertaining to a newly discovered dinosaur or the announcement of some novel insight into a significant moment in Earth's history. Such artwork invariably exists at the more 'realistic' end of the genre. That makes sense, as the image is portraying, as accurately as possible, the new animal and, often, the environment it inhabited. Many of the artists whose work is included in this volume (and in the previous one) have produced artwork created to accompany announcements of this sort.

A question we, this book's editors and writers, are regularly asked is how a person might break into palaeoart as a career. The sad fact is that any chance of success is exceedingly slim. Success in the field is generally measured via the production of artwork for the aforementioned press releases, or work for museums, magazines such as *National Geographic*, or high-end book illustrations. Those artists who provide this sort of artwork are few and far between, and indeed the opportunities to produce it are scarce.

Those artists that do succeed – those who can be considered professional palaeoartists – generally provide imagery at the realistic, even photo-realistic end of the palaeoart spectrum. Yet even a superficial cruise across social media will show that palaeoart as a whole is far more catholic.

There are many artists – let's take Natalia Jagielska and DJ Washington as two excellent examples, both featuring in this volume – whose work functions as a synergy of scientific accuracy and a particular graphic style. There are those within the palaeoart community who might even have a hard time recognising it as palaeoart at all, regarding it instead as 'palaeontologically themed' art. Fair enough. Yet both artists are as disciplined in their portrayal of ancient animals as those artists at the 'realistic' end of the spectrum.

That said, it's likely that most artists featured in this book have influences outside of palaeontology. Natalia's work, for example, borrows from Japanese art, while Lewis LaRosa is a first in a palaeoart book in that he's a highly regarded comic artist.

Comic books form a driving force in popular culture – 'pop culture' – a genre that was quick to embrace dinosaurs long before pop culture's significance was recognised. Dinosaurs and their literary ilk – the pterodactyl (a colloquial term for the genus *Pterodactylus* used to represent *all* pterosaurs), the mammoth, the sabre-toothed tiger – were soon staples of the burgeoning pulp fantasy genre and comic books. 1941 saw the first appearance of the Savage Land in a very early incarnation of Marvel Comics. This dinosaur-infested tropical wilderness in Antarctica featured regularly in the Marvel Universe. Its wannabe ruler was Sauron, once Dr. Karl Lykos until he was bitten by, er, mutant pterodactyls, turning him into a superpowered mutant *Pteranodon* who enjoyed transforming people into dinosaurs.

Whilst embracing dinosaurs as thoroughly as Sauron, pop culture's very popularity could also be detrimental to the public's general perception of what these animals were like. As the 20th century drew to its close, the Dinosaur Renaissance overturned the image of dinosaurs as giant monstrous reptiles dull of wit, slow of foot and cold of blood. An interest in them as warm-blooded, semi-avian, living animals had risen considerably, but it was a spectacularly popular movie – 1993's *Jurassic Park* – that took dinophilia to a new zenith, further enhanced by its sequels. The downside to this is that the movies have crystallised, or perhaps fossilised, the public's perception of dinosaurs: it has become hard for people to accept dinosaurs as animals, not movie monsters. Scientists ever since have had to fight hard to explain that *T. rex* could not, in fact, outrun a jeep and that *Velociraptor* was not a scaly-skinned Cretaceous Einstein as big as an adult human.

Decades on, it's a battle that rages still, ironically fuelled by an unstoppable love for dinosaurs and driven by pop culture's embrace of a Sure Thing. Within this cultural setting, *Mesozoic Art II* is, we submit, important in helping balance out these sometimes downright incorrect perceptions. Despite the variety of styles, every artist whose work features here has endeavoured to ensure that the animals and environments that they've portrayed are at the cutting edge of current palaeontological thinking and knowledge. To those readers for whom this is in any way new, we hope that you're enthralled and excited by visions of the past that are often beautiful, sublime and showcase environments just as complex as those of the modern world. And to those who are familiar with thoughts about the ancient world, it should remain obvious that these are exciting, fast-moving times, and that more people – worldwide, from many backgrounds – are contributing anew. And, finally, that the tough work involved in being a palaeoartist can combine connections with the natural world, a deep understanding of science *and* a love of art, and with visual imagery of many sorts. This is despite the hardships and state of opportunity within the field, such as it is.

DARREN NAISH & STEVE WHITE

A NOTE ON TERMINOLOGY

Readers may notice some apparent inconsistency when reference is made to the geological age of an organism. 'Early' or 'Late' are used in reference to the chronological age of an organism (e.g., '*T. rex* lived in Late Cretaceous times'). 'Lower, and 'Upper' refers to the age of the geological formation (e.g. the Morrison Formation or Hell Creek Formation) – literally, the rocks from which a fossil is found in (e.g., '*T. rex* was found in Upper Cretaceous sediments').

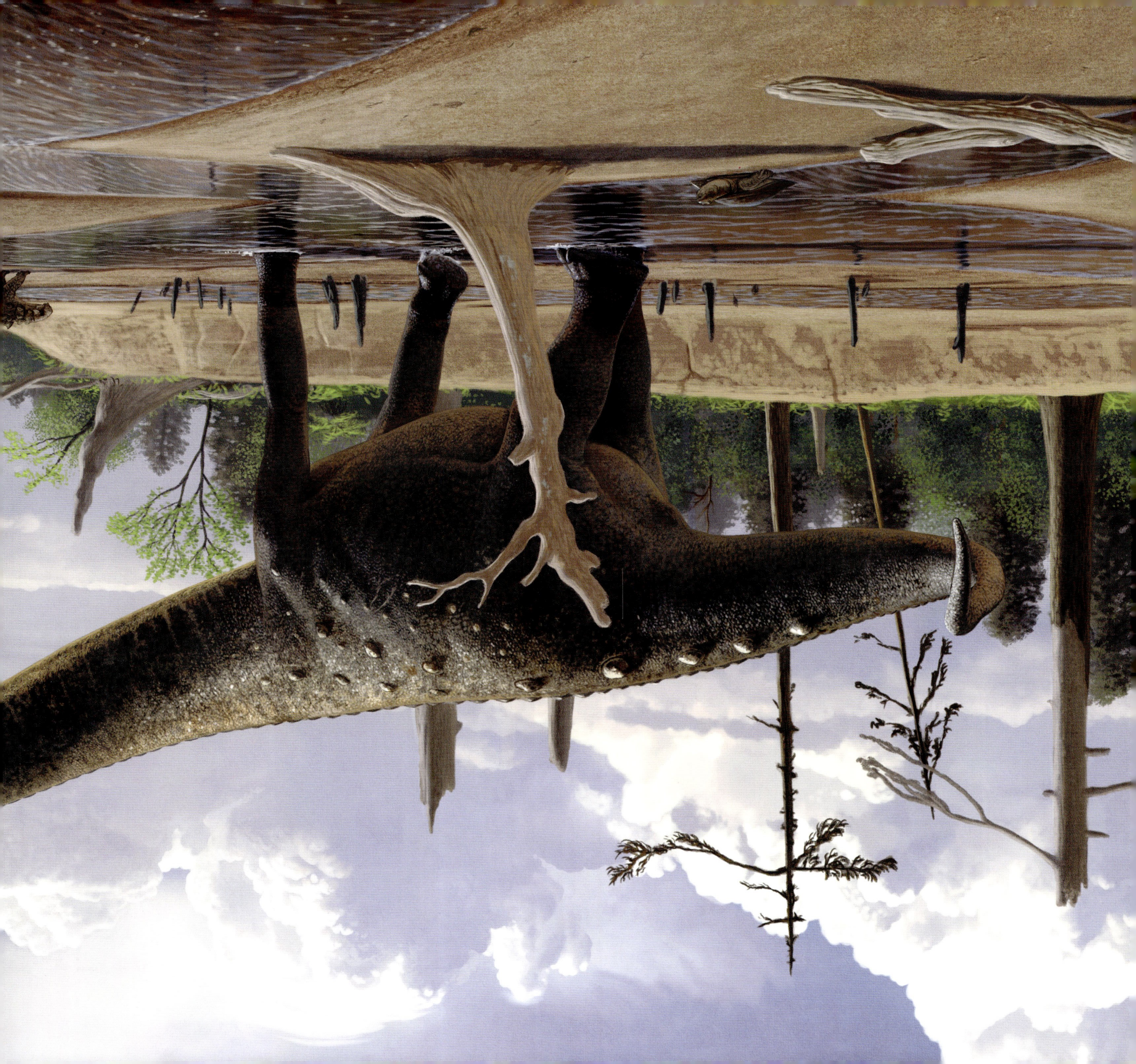

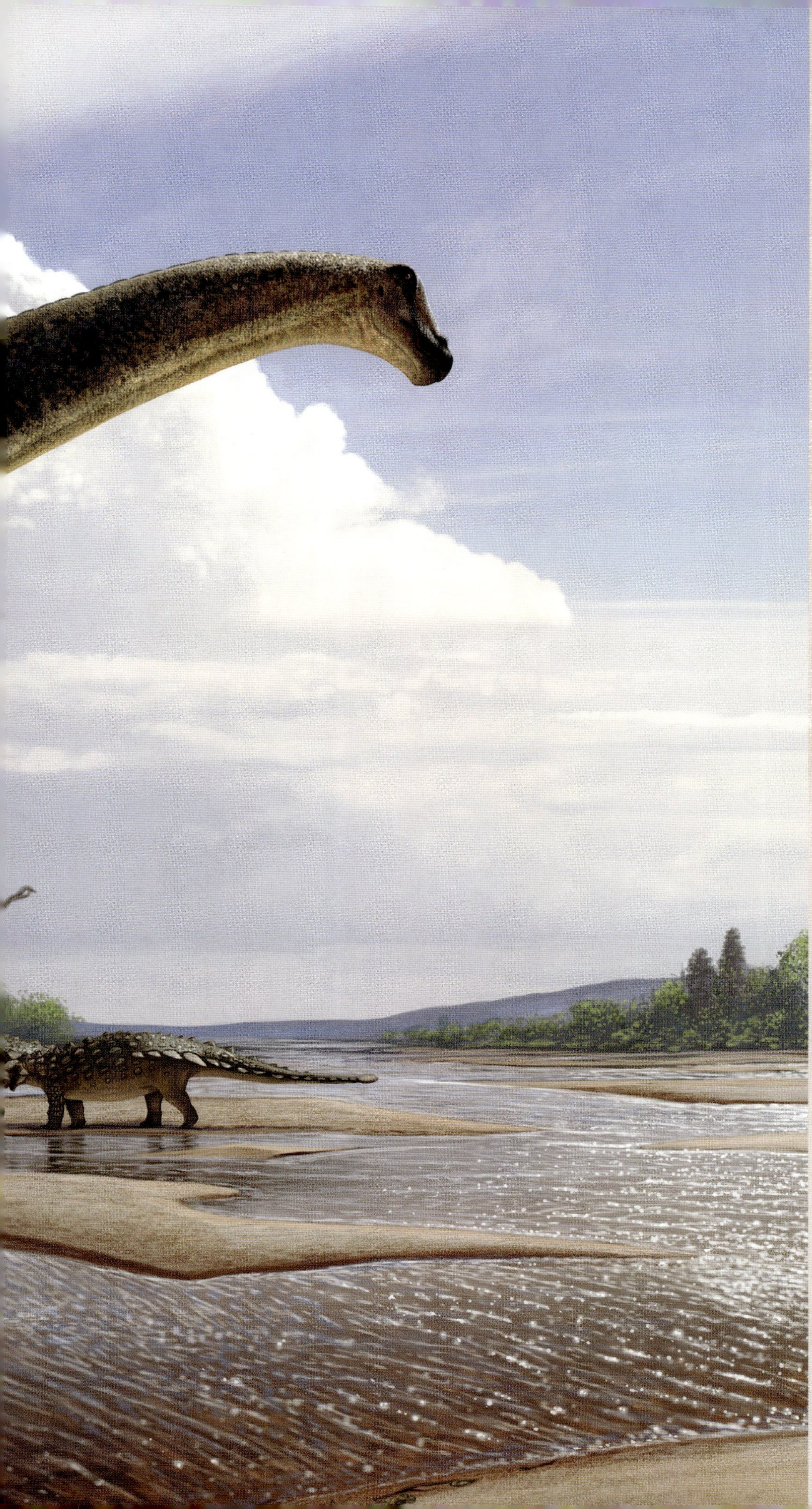

ANDREY ATUCHIN

ANDREY ATUCHIN is a Russian-born palaeoartist and natural science illustrator, now based in Calgary, Canada. Born in 1980 in Siberia, Andrey developed a deep interest in nature, wildlife and prehistoric life from a young age. This passion grew into a career as an illustrator, specializing in palaeontology, natural history, and wildlife.

Known for his meticulous attention to anatomical detail and artistic flair, his illustrations have contributed to academic papers, museum exhibits, popular science books and magazines. He has collaborated with palaeontologists worldwide, helping visualise new fossil discoveries, including dinosaurs and other prehistoric species from North America to Australia.

Largely self-taught, Andrey honed his craft through years of experience and partnerships with leading researchers. His work merges scientific accuracy with artistic expression, aiming to educate and inspire. In recognition of his exceptional skill, he was awarded the Society of Vertebrate Paleontology's Lanzendorf-National Geographic PaleoArt Prize for Scientific Illustration in 2020.

Since relocating to Canada, Andrey has continued his passion for scientific illustration. He quickly adapted to Canadian culture, especially Alberta's renowned palaeontological history, and enjoys exploring the province's landscapes as an avid hiker and wildlife photographer.

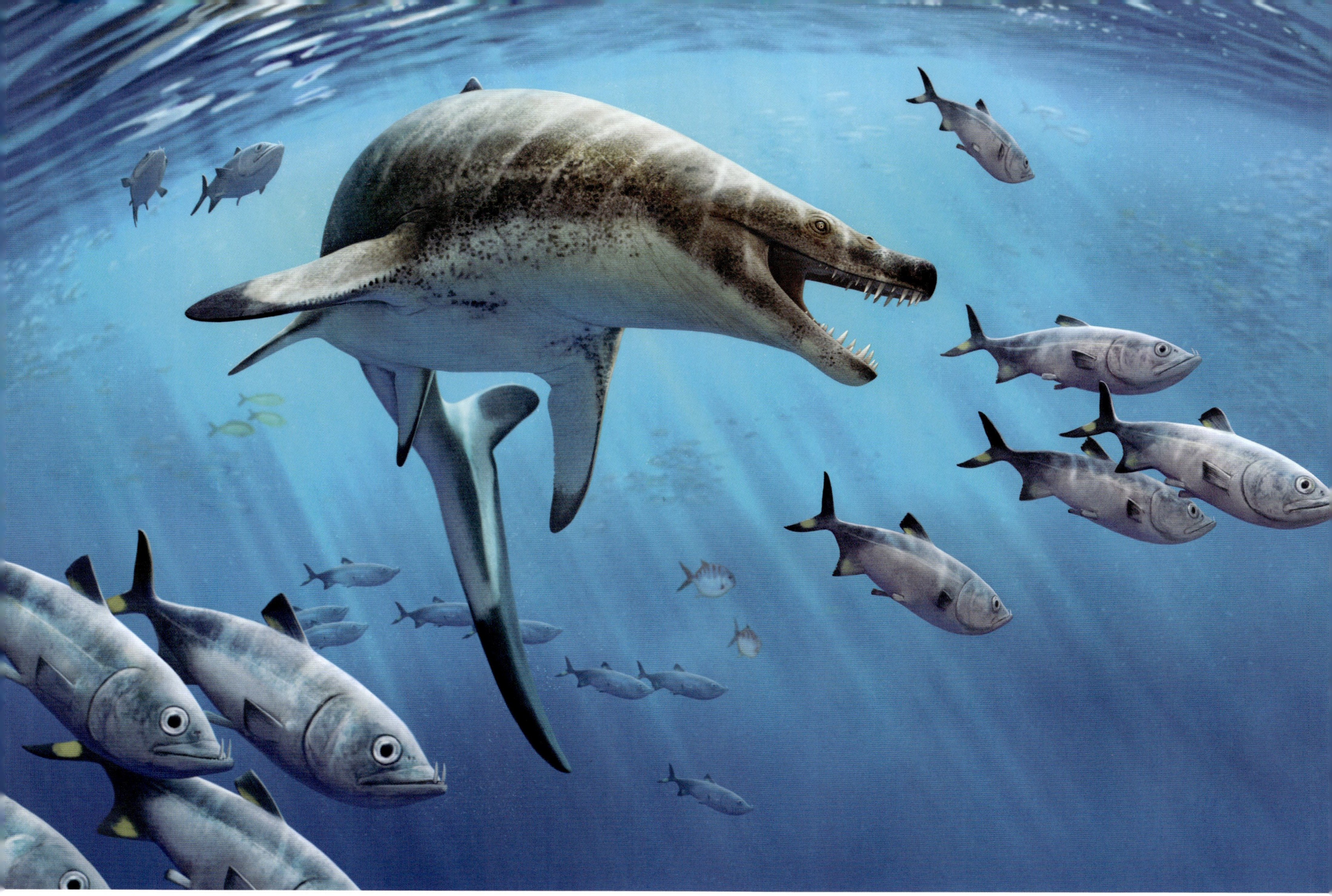

[Previous pages]

ABDARAINURUS

This reconstruction shows Late Cretaceous animals of the Mongolian Alagteeg Formation, particularly the sauropod *Abdarainurus*, named in 2020. *Abdarainurus* was a mid-sized titanosaur about 12m (39.3ft) long. Armour plates and nodules were probably scattered across its back and sides. Also shown is *Pinacosaurus*, an armoured dinosaur, and softshell turtles.

• Digital | 2020

▲ 'DAGGER' THE MOSASAUR

We are currently in a golden age of mosasaur science, mosasaurs being the great sea-going lizards of the Cretaceous. This reconstruction shows *Khinjaria*, a Moroccan mosasaur named in 2024 and noted for its long, interlocking, proportionally large, knife-shaped teeth.

• Digital | 2024

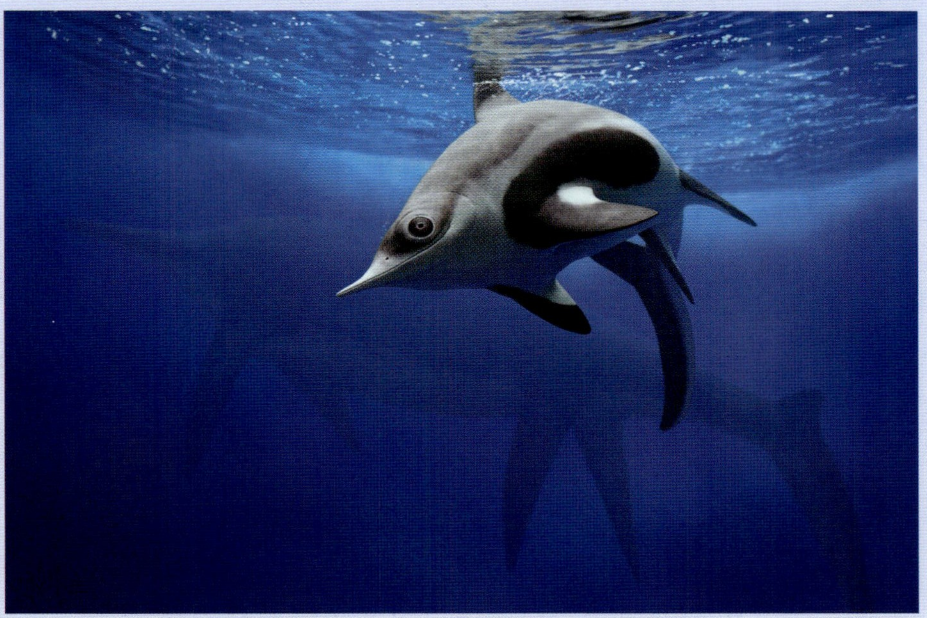

◂ AURORORBOREALIA

Auroroborealia – a fish-shaped reptile belonging to the group known as ichthyosaurs – is one of the oldest members of the ichthyosaur subgroup Parvipelvia. This was the only ichthyosaur group that persisted beyond the Triassic and into the Jurassic. *Auroroborealia* lived alongside more archaic ichthyosaurs like the giant shastasaur shown in the background.

• Digital | 2021

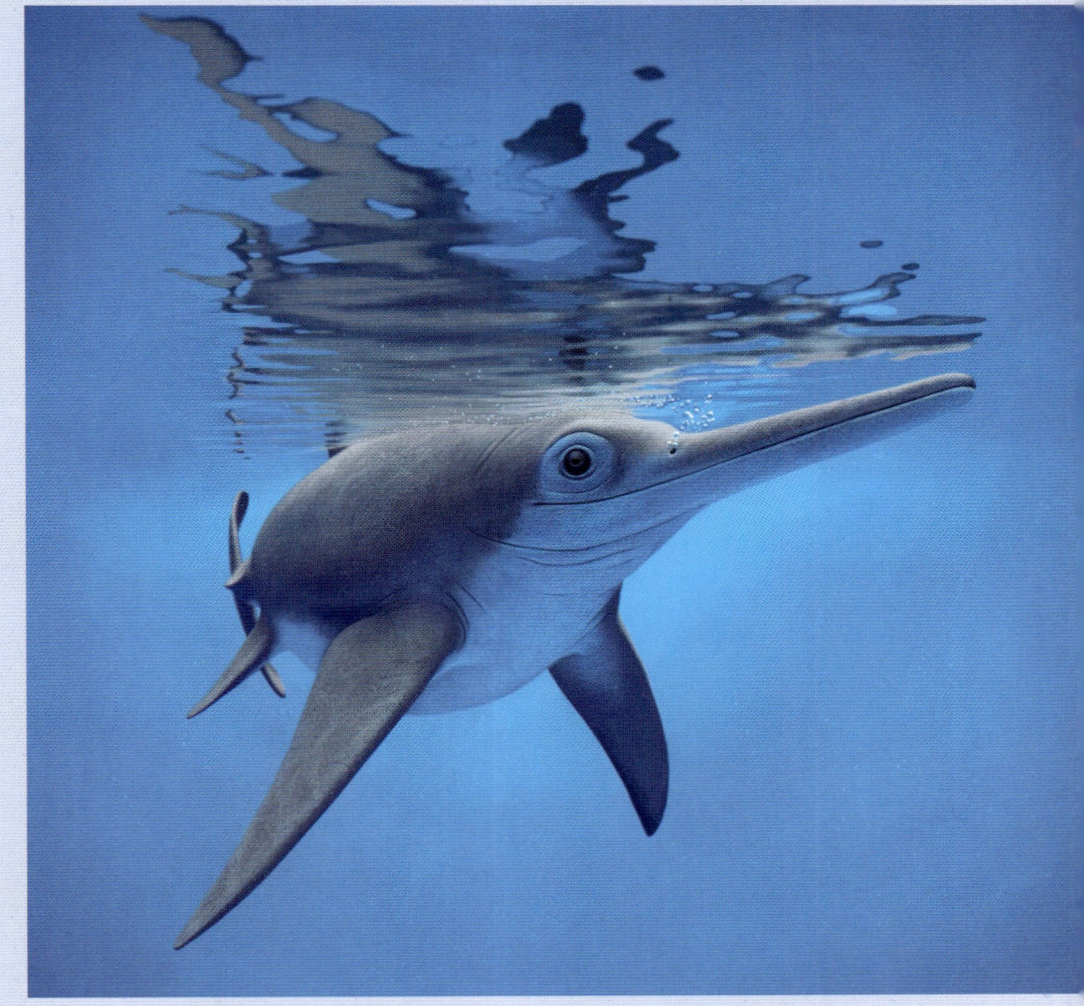

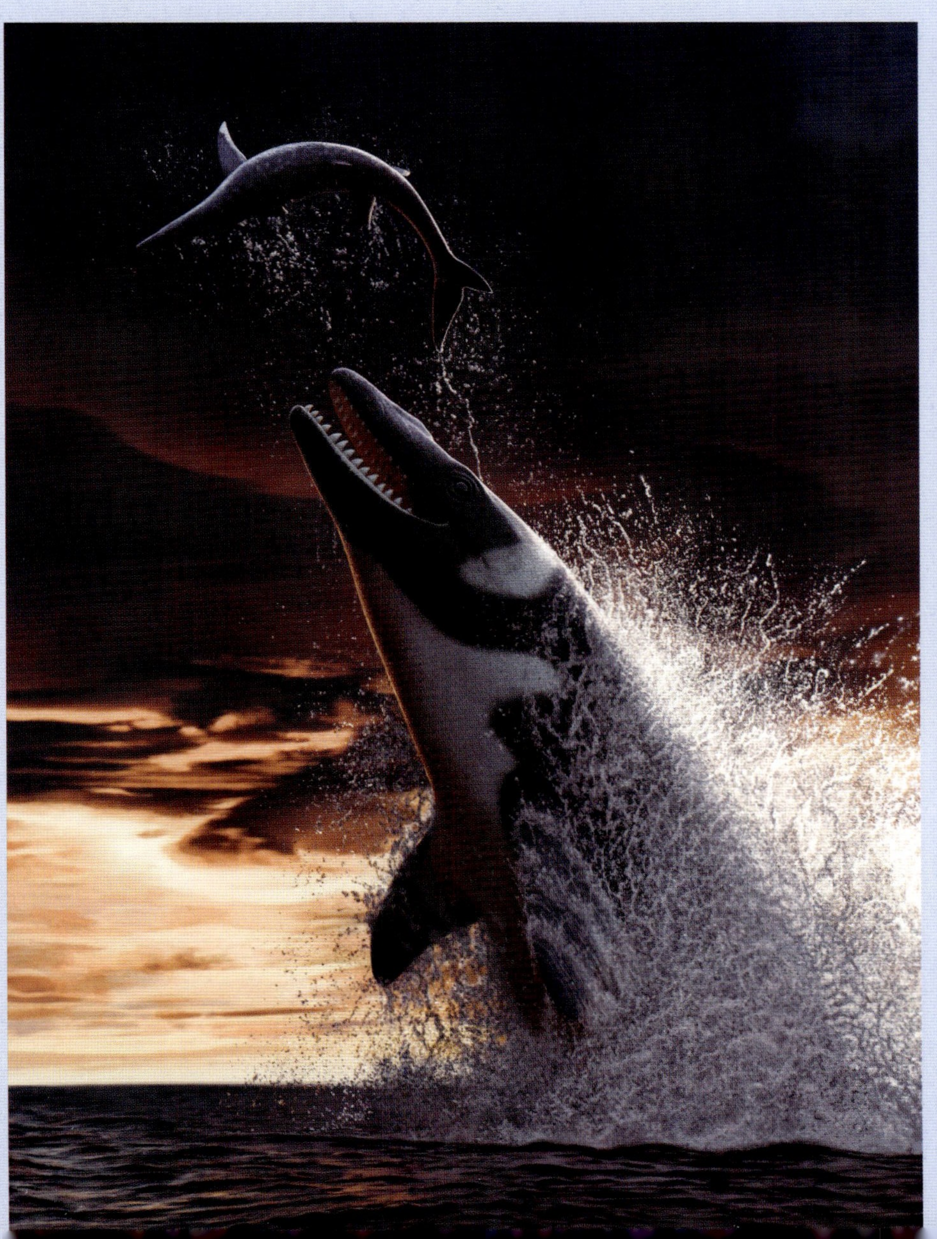

◂ THALASSOTITAN

It is difficult to resist the idea that ancient marine reptiles indulged in behaviours like those practised by marine mammals today. This image shows the Moroccan mosasaur *Thalassotitan* throwing the carcass of an unfortunate *Halisaurus*, a smaller and very different kind of mosasaur. Killer whales today do this with marine mammals and other prey.

• Digital | 2023

▴ MAIASPONDYLUS

Recent decades have seen a huge increase in the number of ichthyosaur species from the Cretaceous, revealing that their diversity was higher than traditionally thought. Among these new types is *Maiaspondylus*, named in 2006 from Canada. Its name – which means 'caring mother vertebra' – refers to the two embryos preserved within the body cavity of the original specimen.

• Digital | 2020

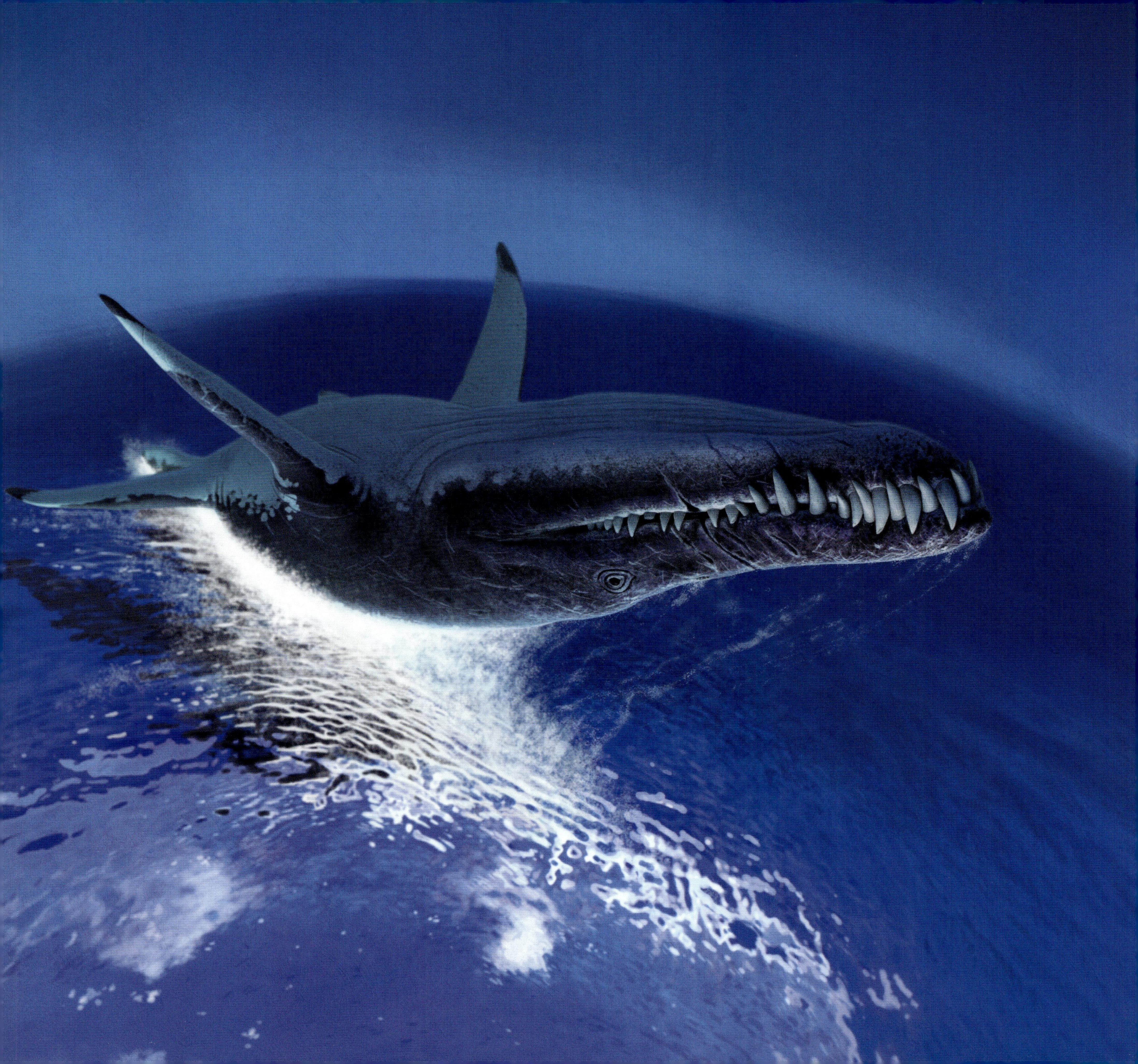

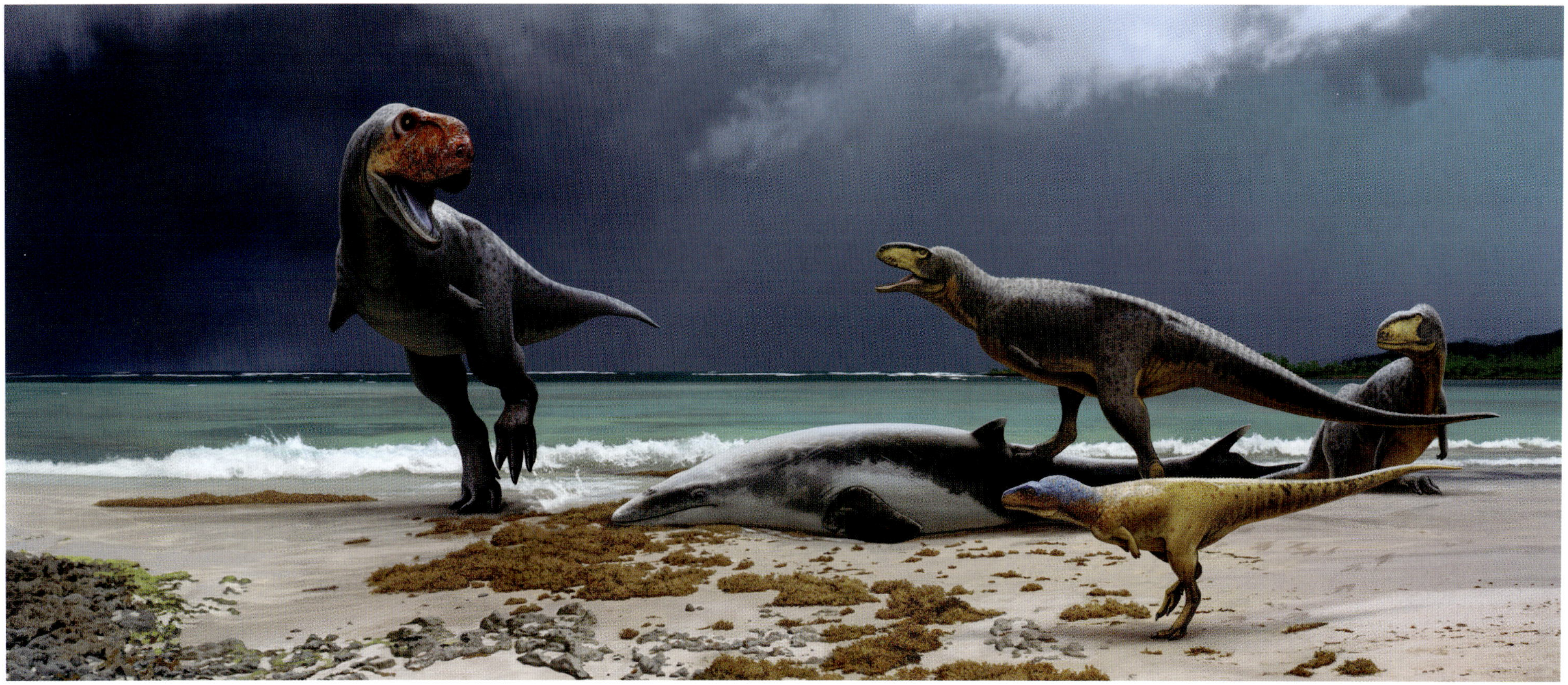

▲ ABELISAUR BEACH TIME

Most dinosaur communities, like the Moroccan one shown here, included several large predator species. These dinosaurs might have usually avoided one another, but there were occasions where they surely interacted, perhaps aggressively, such as scavenging a beached marine reptile. Those shown here are as-yet-unnamed types of abelisaurs.

• Digital | 2023

DZHARATITANIS ▶

Dzharatitanis from Uzbekistan, named in 2021, is one of many recently named Central Asian dinosaurs. It was initially identified as belonging to the short-necked group Rebbachisauridae, as emphasised in this reconstruction. However, recent studies suggest a titanosaur identity.

• Digital | 2021

◀ VOLGA PLIOSAURID

Pliosaurids – a group of mostly short-necked plesiosaurs – lived in seas worldwide during the Jurassic and Cretaceous. This one is an as-yet-unnamed species from Russia's Volga region. The teeth in these animals varied in size and shape. At the tip of the jaw, these were long and interlocking. Further back, a group of big, canine-like teeth projected downwards from the upper jaw's edge.

• Digital | 2020

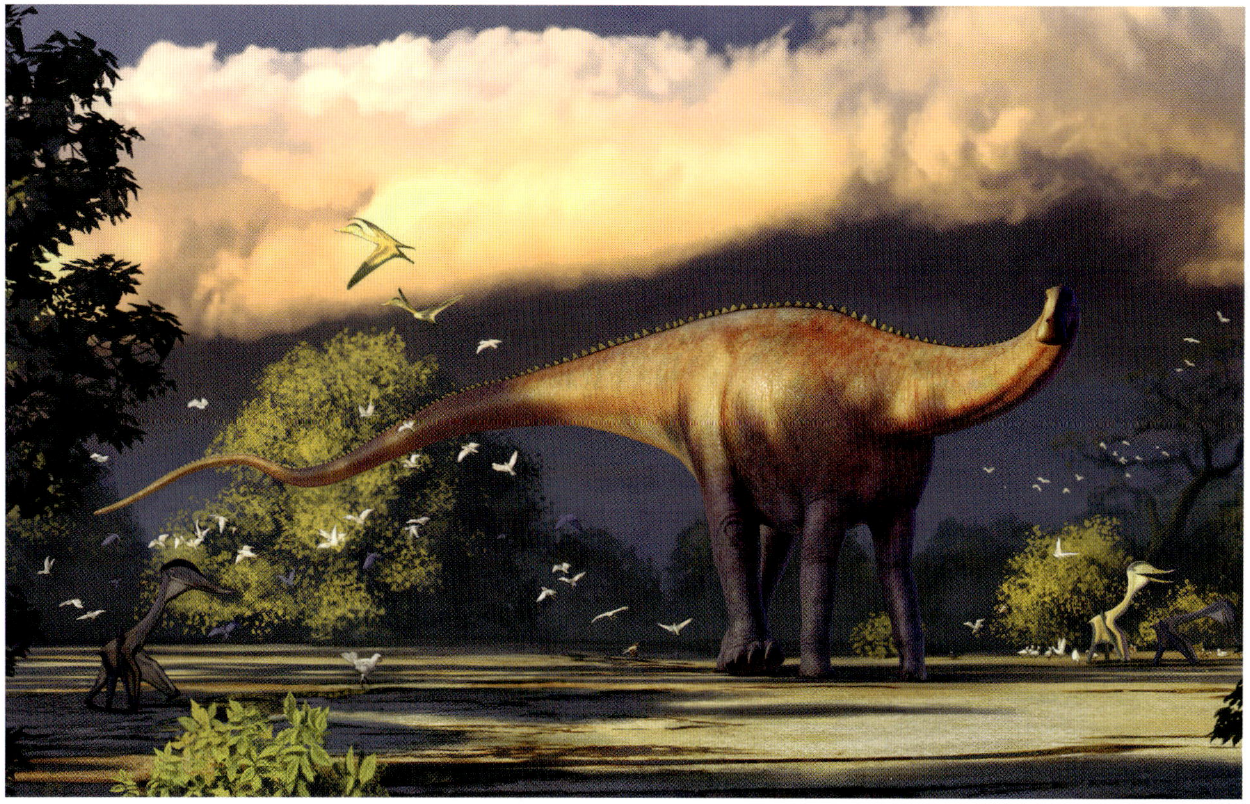

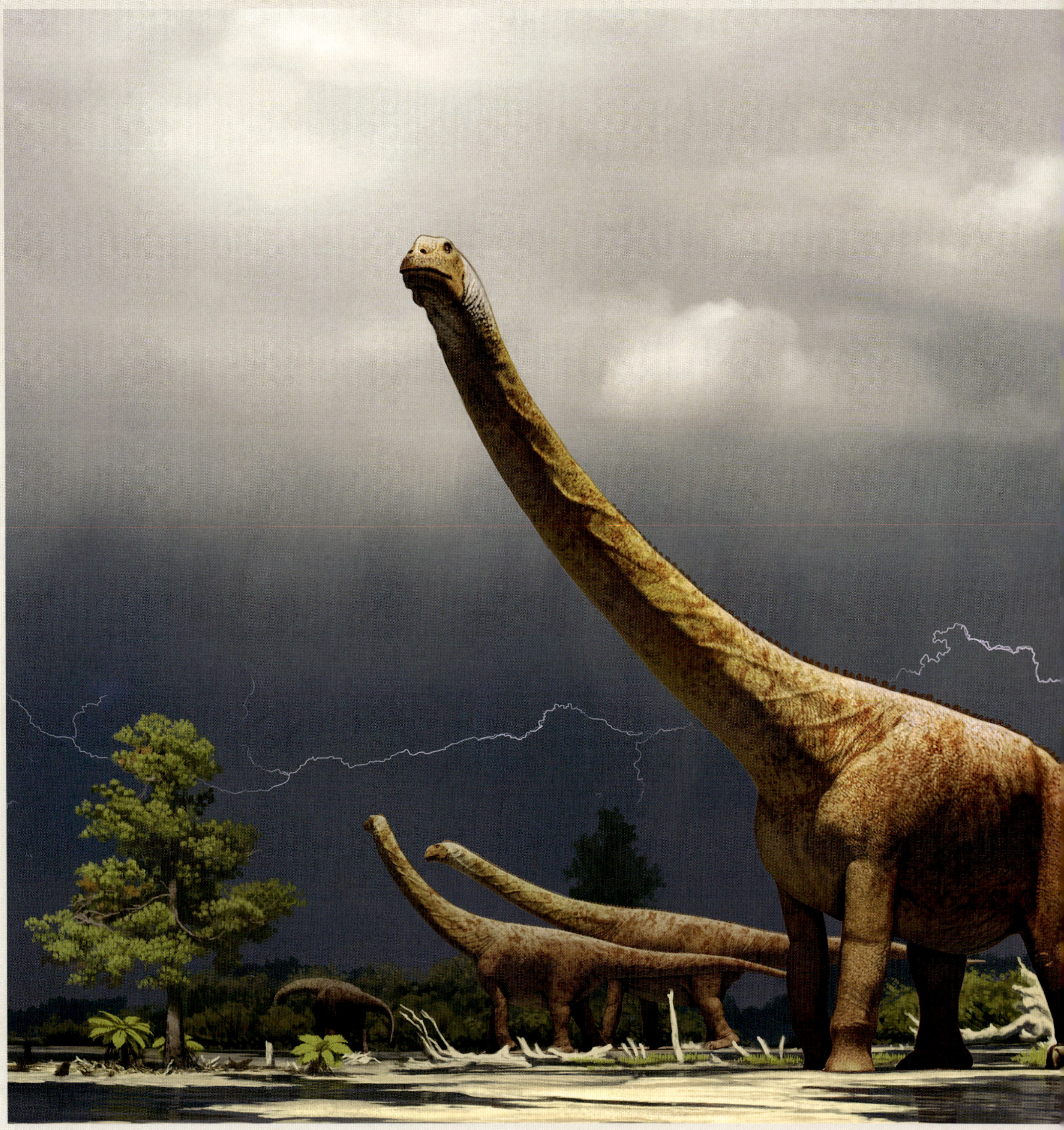

RUKWATITAN ▸

Several African nations have massive potential for dinosaur discoveries. Tanzania, in East Africa, is famous for its Jurassic dinosaurs, but its Cretaceous fossils remain poorly known. *Rukwatitan* is one of several Cretaceous sauropods recently reported from the country. It was a mid-sized titanosaur about 9m (29.5ft) long with slender forelimbs and long neck vertebrae.

• Digital | 2014

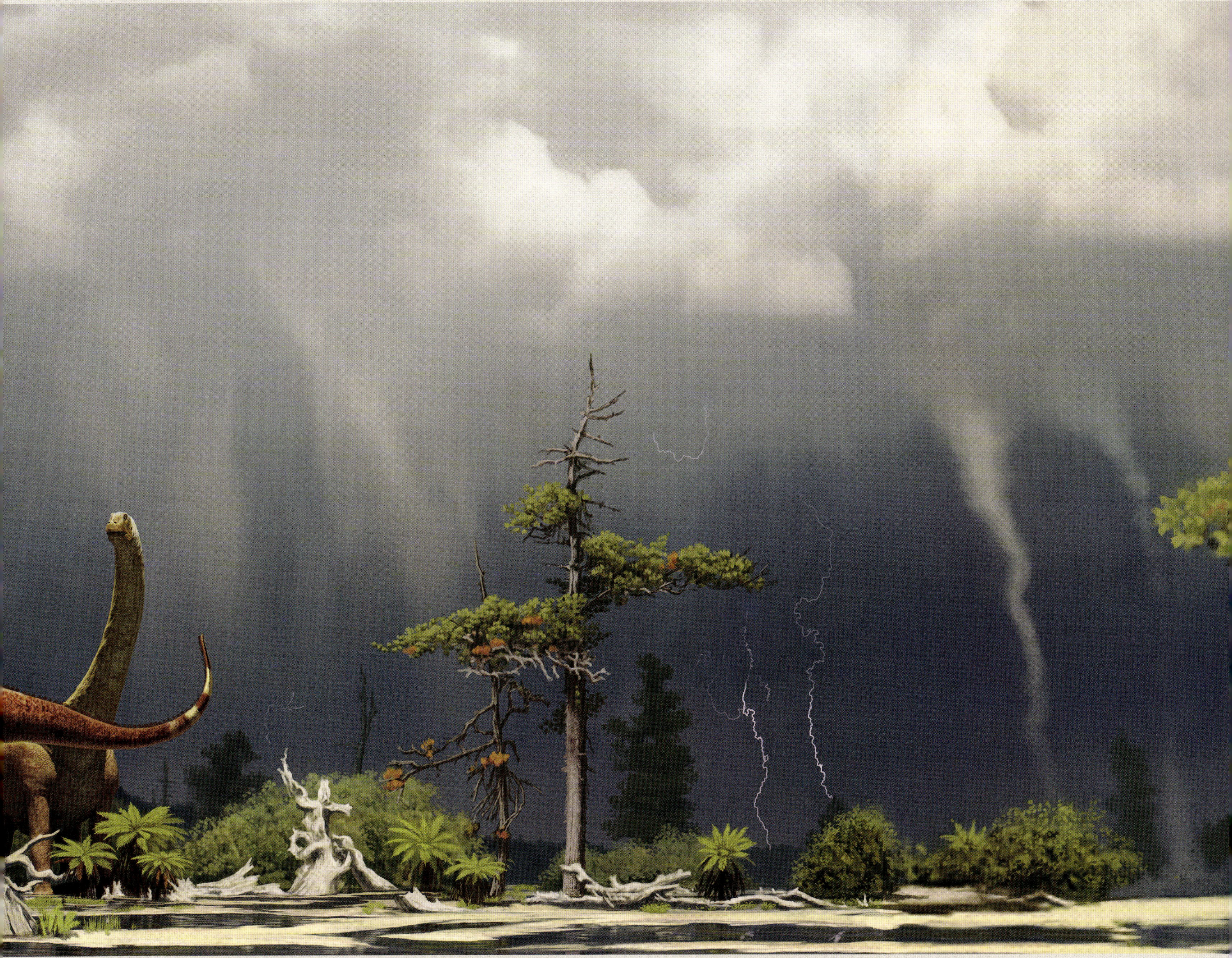

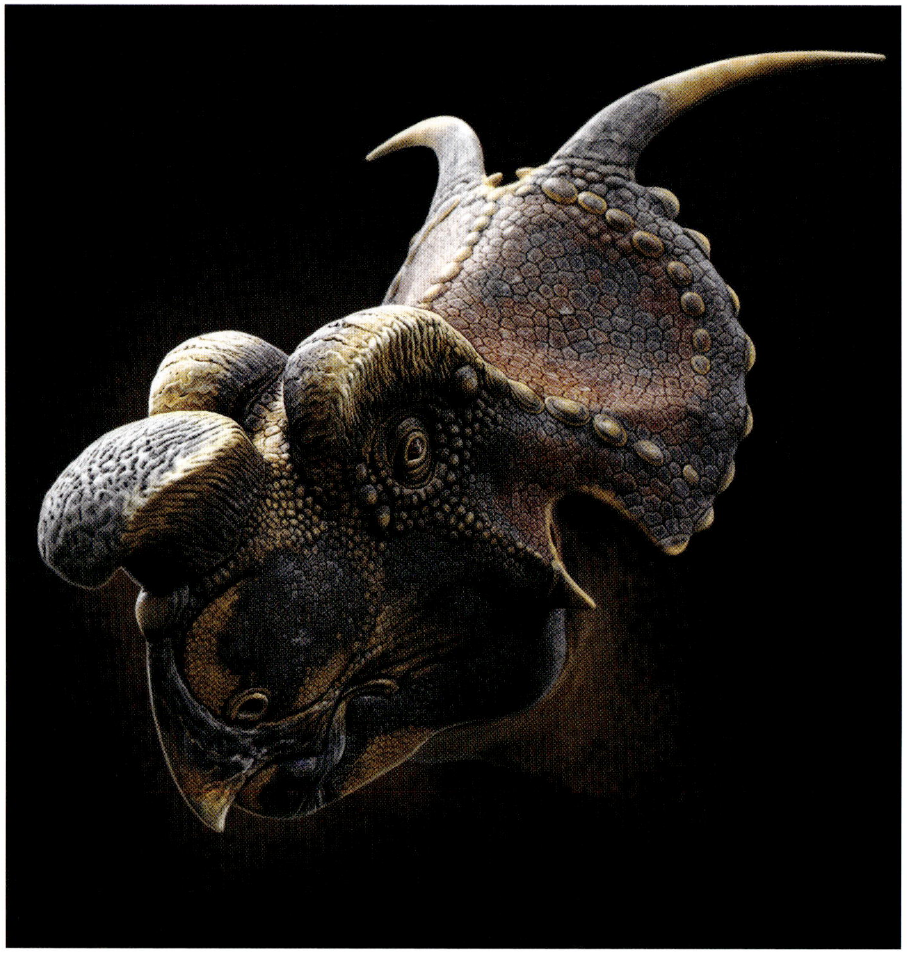

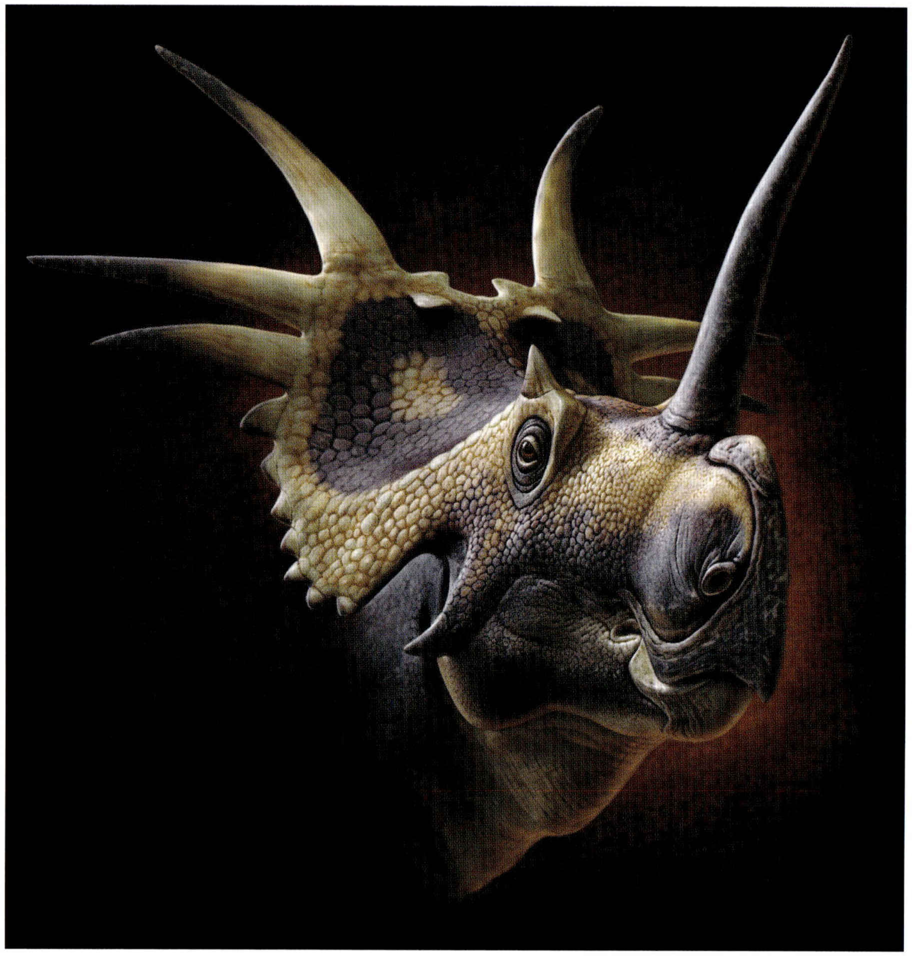

▲ *ACHELOUSAURUS*

Few artists can capture the anatomical intricacy of the ceratopsids – the group of large horned dinosaurs that includes *Triceratops* – as well as Andrey. This portrait shows *Achelousaurus* from Montana, a close relative of the famously spiky *Styracosaurus* and the more massive, thick-nosed pachyrhinosaurs.

• Digital | 2021

BISSEKTIPELTA ▶

Ankylosaurs – armoured dinosaurs – are notoriously difficult subjects for palaeoartists. Horns, hornlets and horn-covered lumps decorated the top and sides of the head. A large armour plate covered part of the side of the lower jaw, and semi-circular crests surrounded the nostrils. This image depicts *Bissektipelta* from the Late Cretaceous of Uzbekistan.

• Digital | 2021

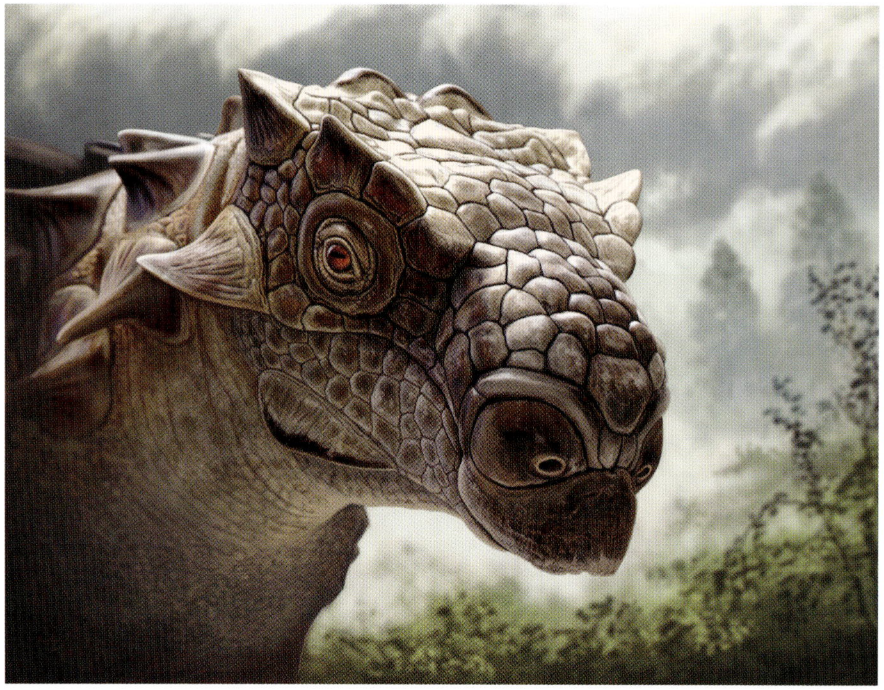

▲ *STYRACOSAURUS*

Styracosaurus albertensis – named in 1913 from Alberta, Canada – remains among the most iconic of horned dinosaurs. Many illustrations of this animal fail to capture the deep, narrow snout and configuration of spikes and projections around the edges of its frill. Andrey, of course, has depicted these details correctly.

• Digital | 2021

▼ ALASKAN DINOSAURS

The far north of Alaska was home to a diversity of dinosaurs during the latest part of the Cretaceous. The giant horned dinosaur *Pachyrhinosaurus* – noted for its massive, rock-like nasal boss – lived here alongside predatory dinosaurs of several sorts. These included saurornitholestines, distant cousins of *Velociraptor*. Seasonal darkness, low temperatures and the northern lights would have been typical of these environments.

• Digital | 2020

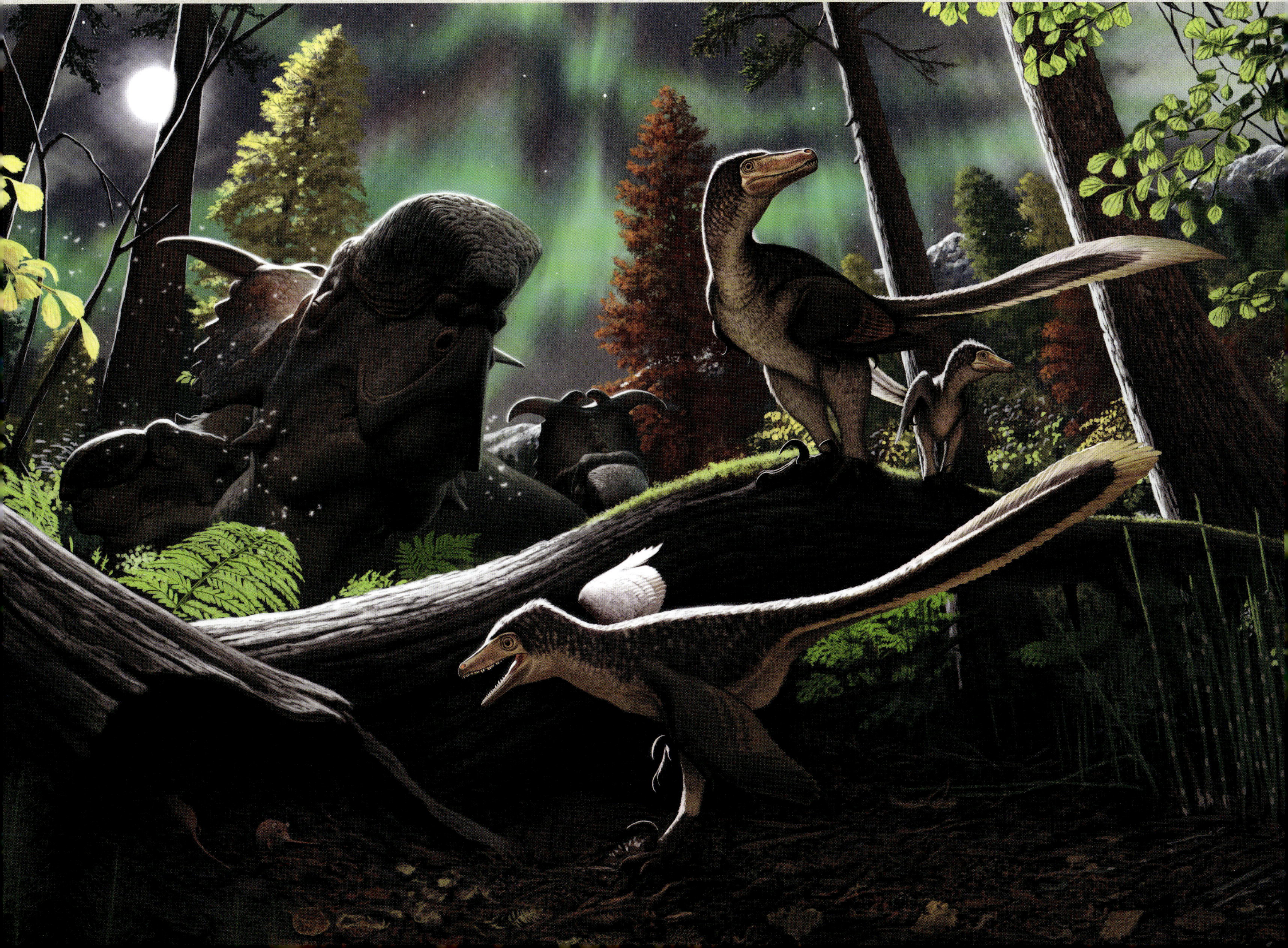

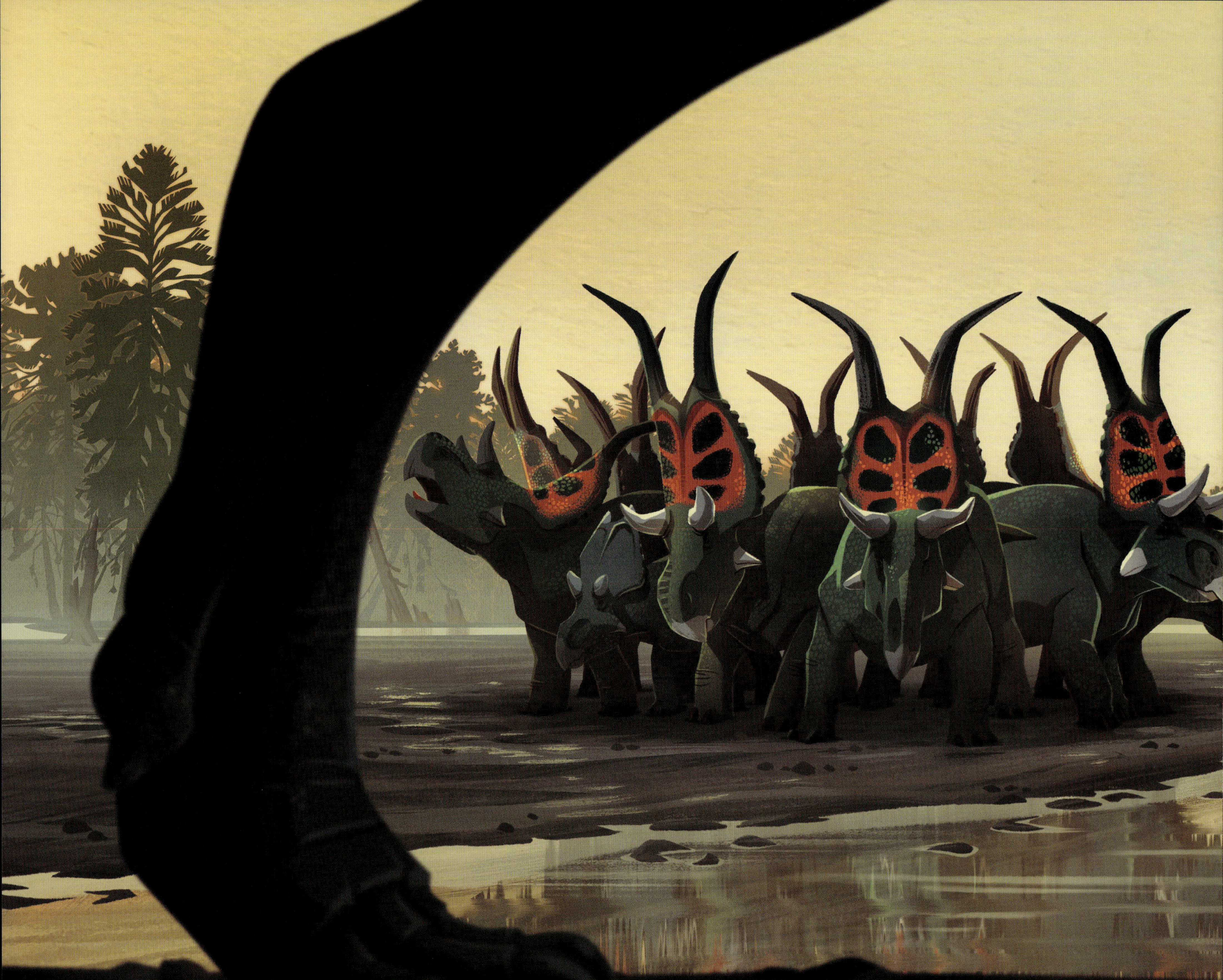

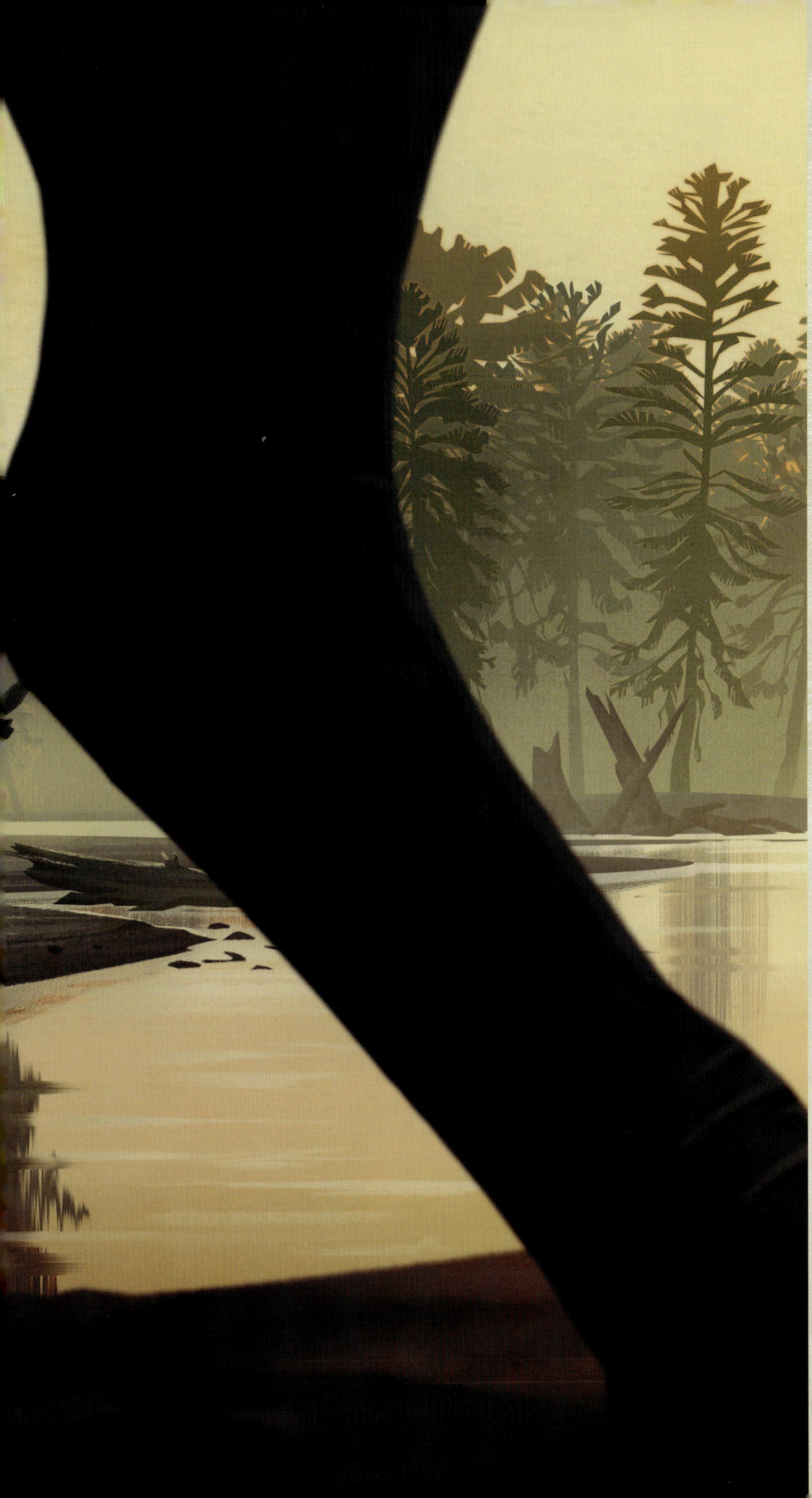

REBECCA DART

REBECCA IS AN ANIMATION art director who has worked in the animation industry in Vancouver, British Columbia, for over 27 years. She has contributed to such shows as *My Little Pony: Friendship is Magic* and the *Night at the Museum* animated special, amongst many others. She creates palaeoart in her spare time and her art has featured in exhibits at the Melbourne Museum and the Royal BC Museum.

Rebecca is especially interested in palaeobotany, an area she connects with her love of gardening and the keeping of carnivorous plants. The latter live with her in a tiny apartment, as do her husband and two tuxedo cats.

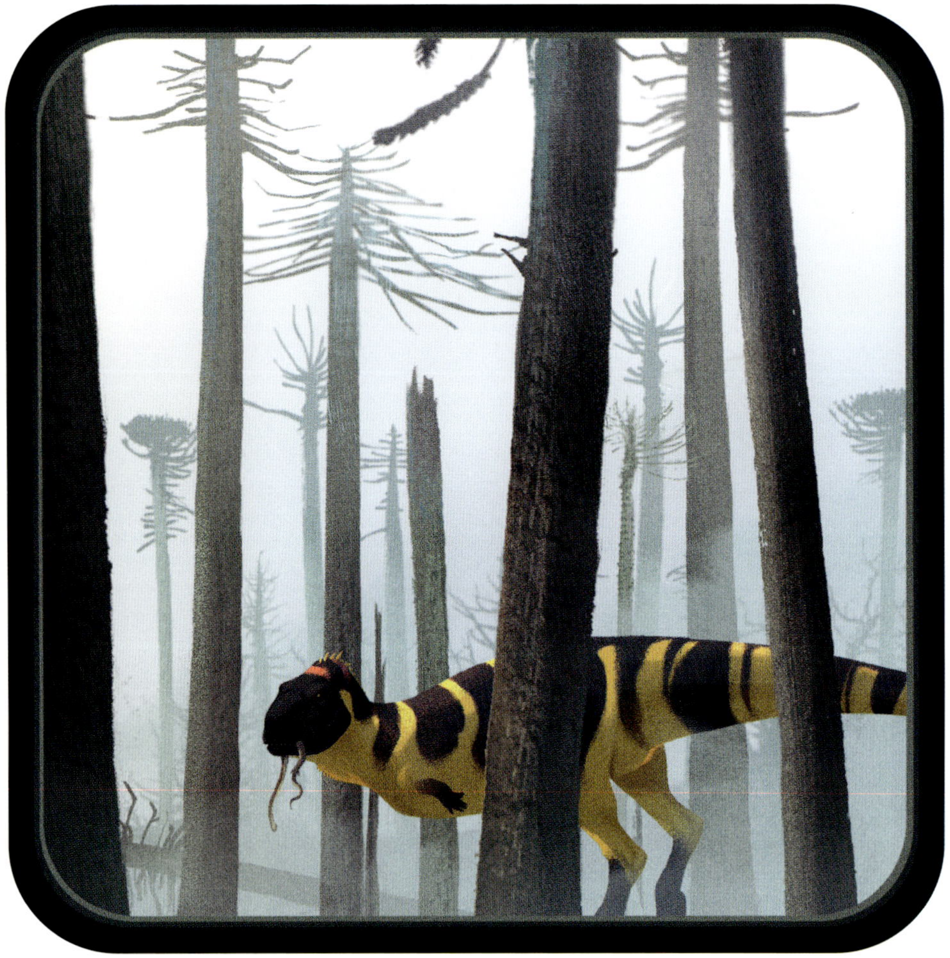

[Previous pages]

DIABLOCERATOPS

At the end of the 20th century, it became quite fashionable in palaeoart to show ceratopsians forming defensive circles in the manner of the modern muskox, often emulating the latter by having young within the circle. This notion is entirely speculative, but it looks cool. Here a herd of *Diabloceratops* face off against the tyrannosaurid *Lythronax* in Late Cretaceous Utah.

• Digital | 2019

◂ EKRIXINATOSAURUS

In Late Cretaceous Argentina, a storm surge has flooded a coastal forest, killing the trees and providing ideal hunting ground for the abelisaurid *Ekrixinatosaurus*. It has caught a *Najash*, an early snake still possessing hindlimbs. Known from the Candeleros Formation (therefore 100–97 million years old), *Ekrixinatosaurus* is one of the largest known abelisaurids, measuring up to 8m (26ft) in length. The rounded corners on many of Rebecca's pieces are intended to evoke the feeling of looking through an old View-Master Reel.

• Digital | 2022

T. REX & CERCIDIPHYLLUM ▸

Cercidiphyllum is a genus of trees that dates back to the Late Cretaceous. Called katsura or butterscotch trees, they emit a sweet smell during the autumn as the sugar in the leaves break down. Here, a young *Tyrannosaurus rex* catches a scent of caramel.

• Digital | 2021

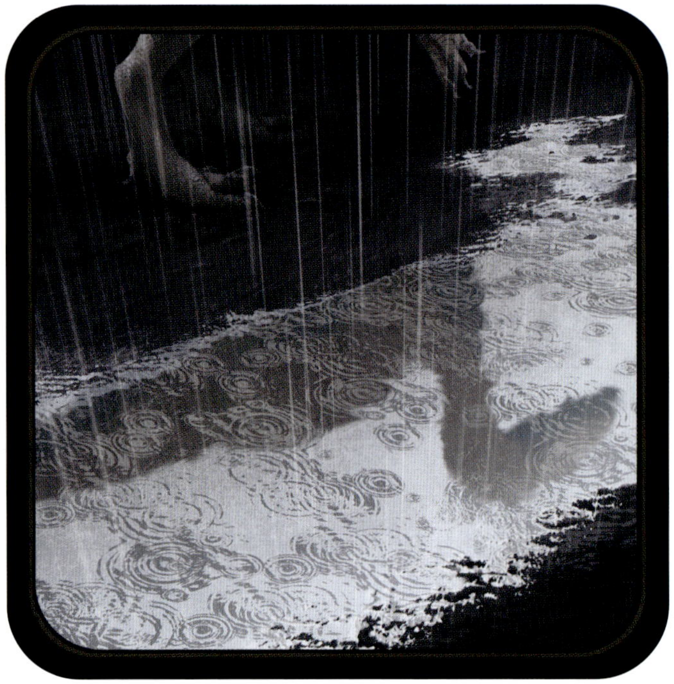

◂ REFLECTION

How do you illustrate a fossil animal known only from parts of its skeleton? Rebecca's solution is to show the animal via an opaque reflection. Shown here is *Bagualosaurus*, a Triassic dinosaur from the far south of Brazil and an early member of the great group that includes the long-necked sauropods and their kin.

• Digital | 2020

FABACEAE ▸

A member of the Fabaceae or legumes – which includes beans and peas – grows through the skull of the horned dinosaur *Coahuilaceratops*. A multituberculate mammal climbs in search of fruit. Fossils from Mexico, published in 2021, show that legumes were present in Late Cretaceous locations already famous for their dinosaurs.

• Digital | 2021

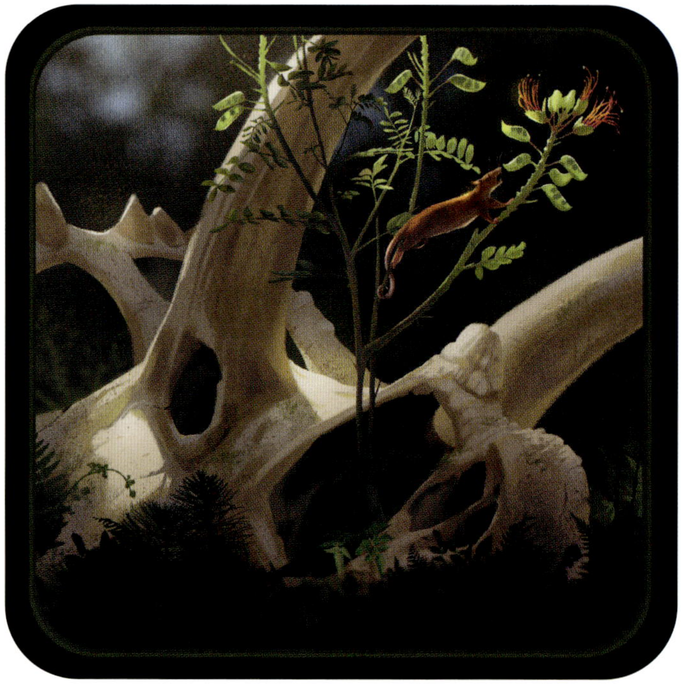

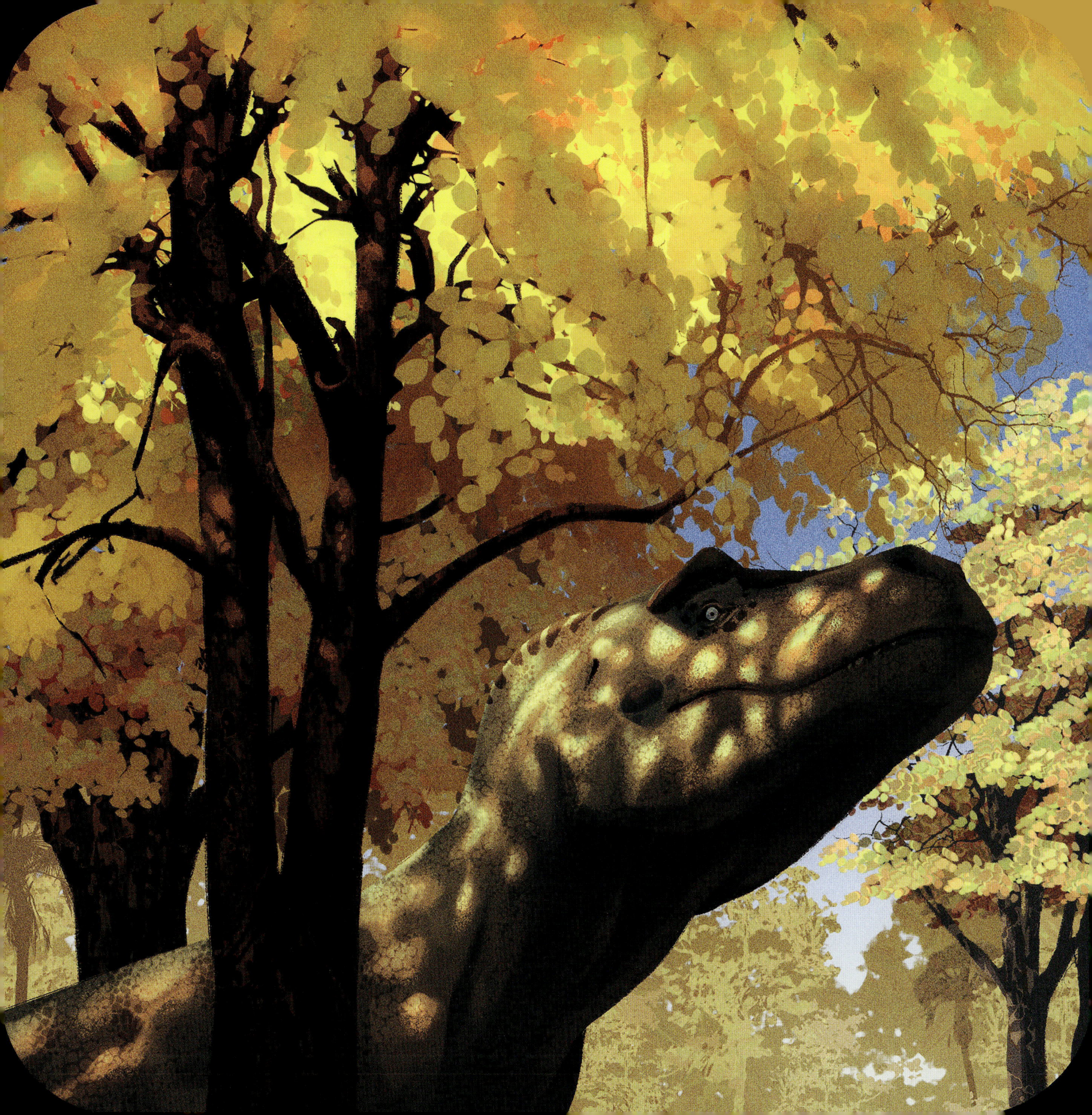

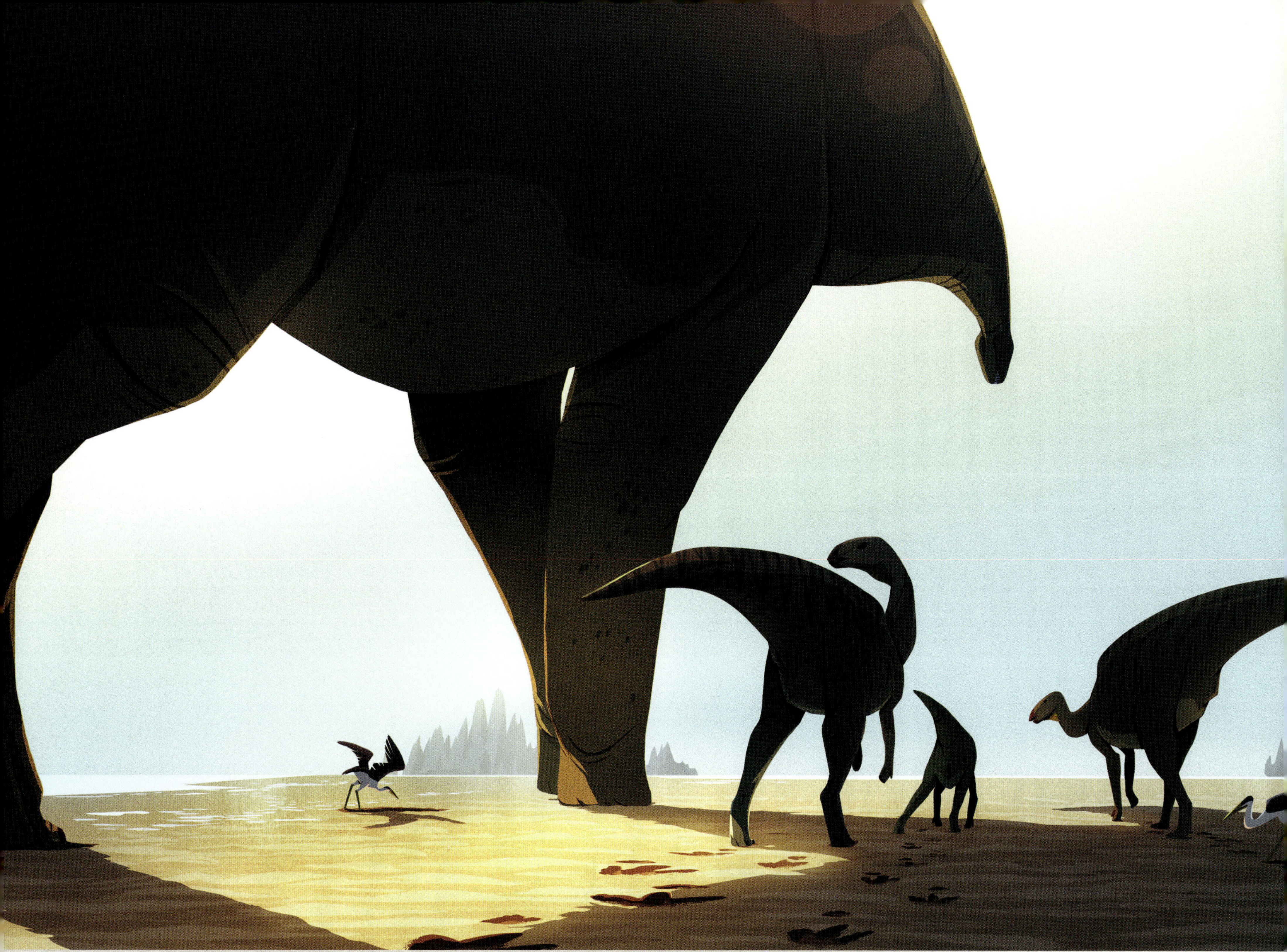

▲ TRACKS

During the Early Cretaceous, a huge *Cedarosaurus* and its escorting iguanodontians and wading birds leave their footprints in the mudflats of Utah's Cedar Mountain Formation. The gigantic size of the biggest sauropods invites all kinds of artistic experimentation involving perspective and composition.

• Digital | 2019

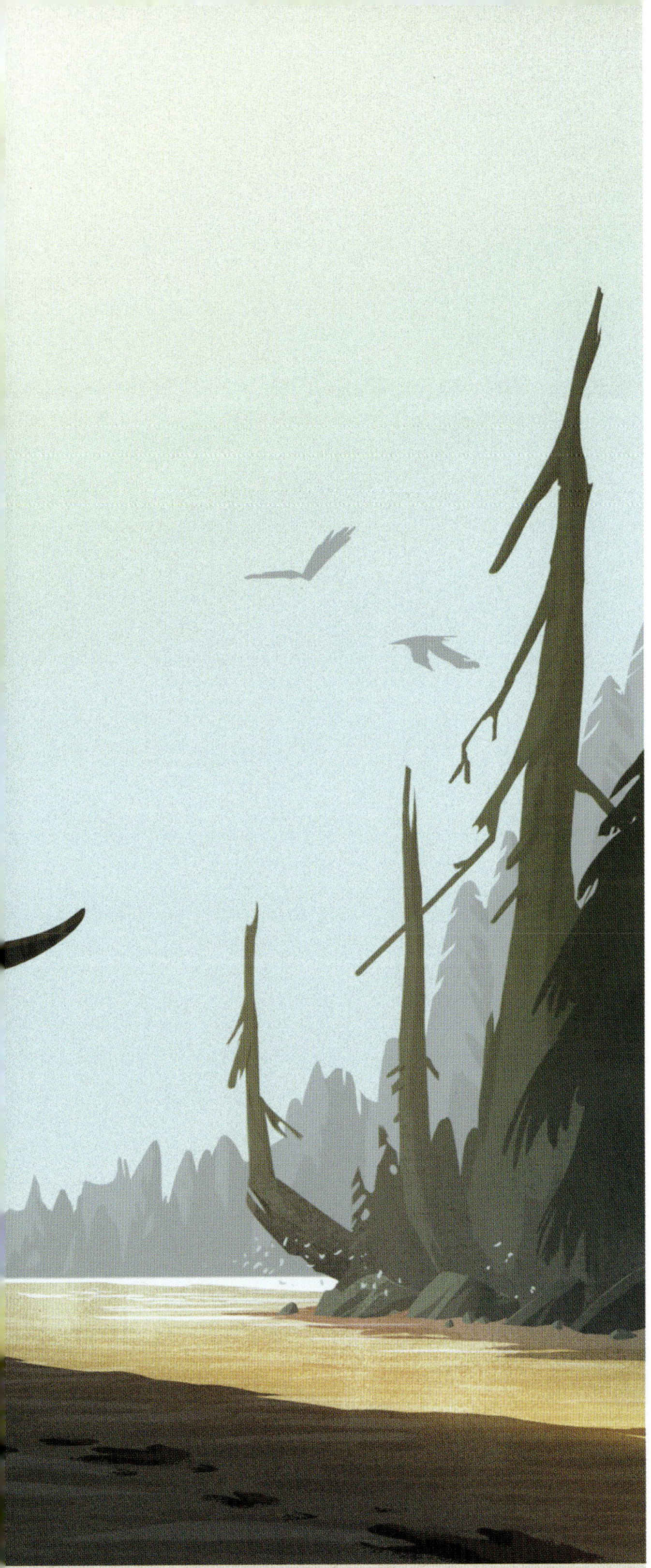

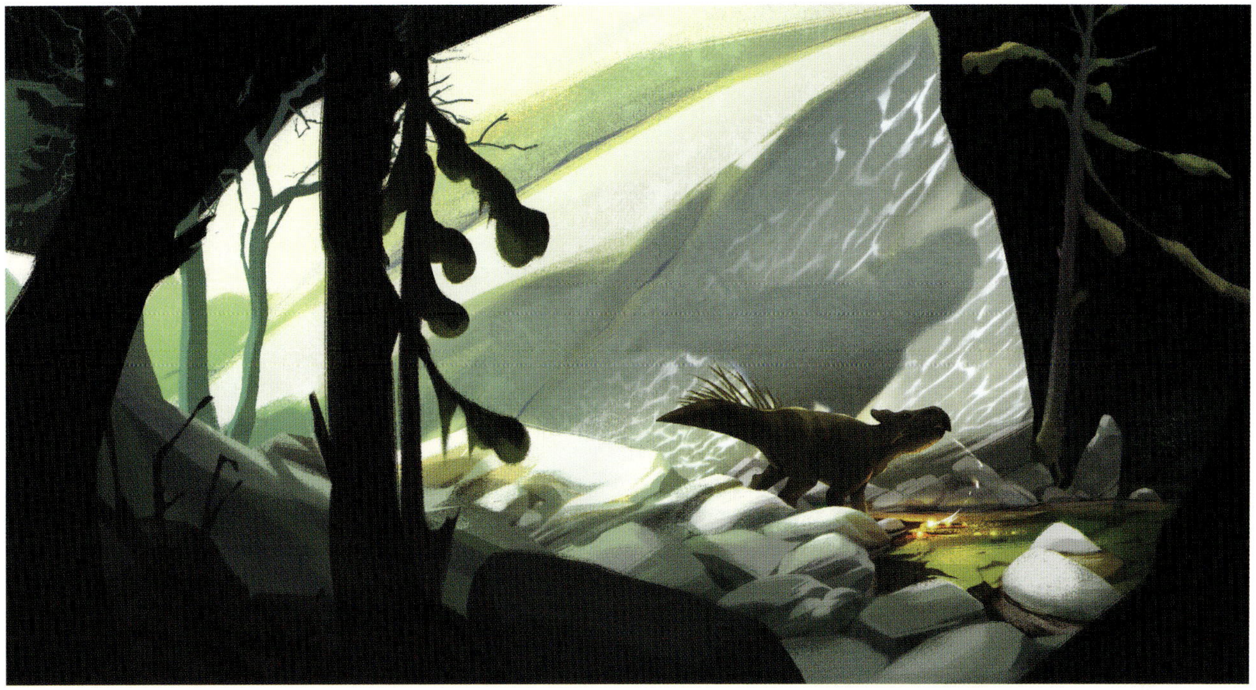

▲ *FERRISAURUS*

Ferrisaurus – a Late Cretaceous ceratopsian from British Columbia, Canada – approaches a secluded rocky pool. The recent discovery of tail quills on the ceratopsian *Psittacosaurus* raises the possibility that such structures were widespread in horned dinosaurs and their kin.

• Digital | 2020

▼ *RINCHENIA*

The Mongolian oviraptorosaur *Rinchenia* shelters in shade beneath a ginkgo tree. *Rinchenia* lived in an area that combined hot desert, lakes, rivers and abundant vegetation, and too little is known to say much about the dinosaur's habitat preferences.

• Digital | 2018

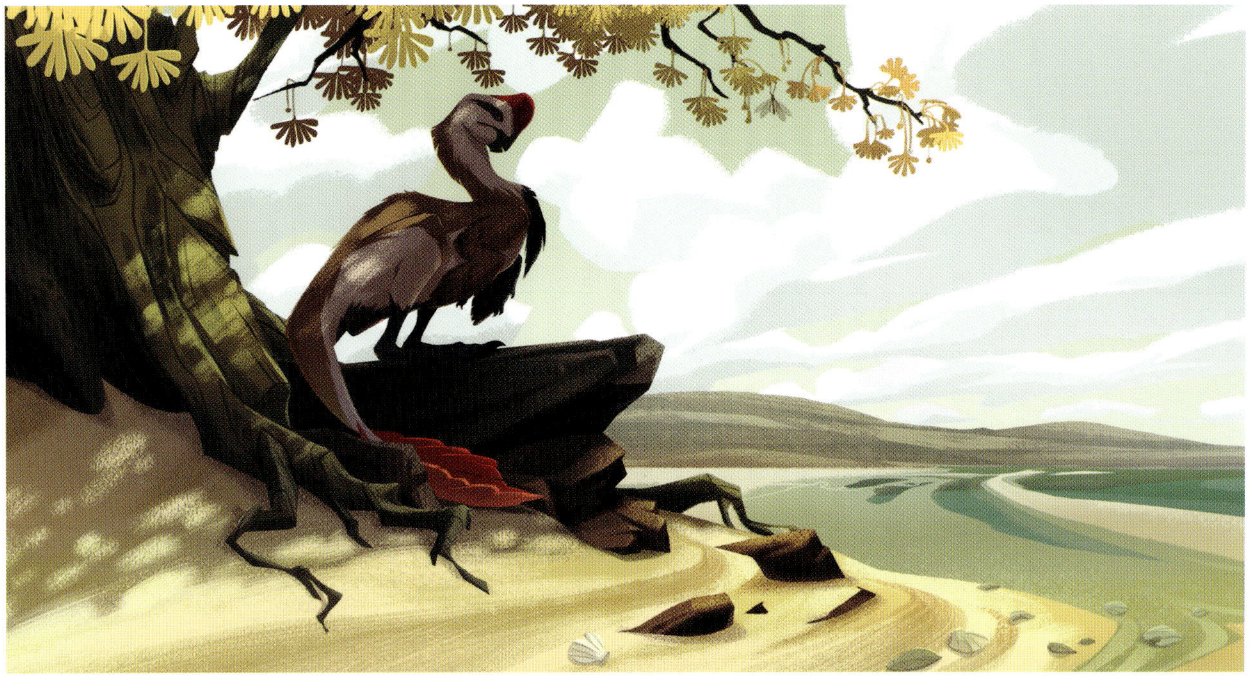

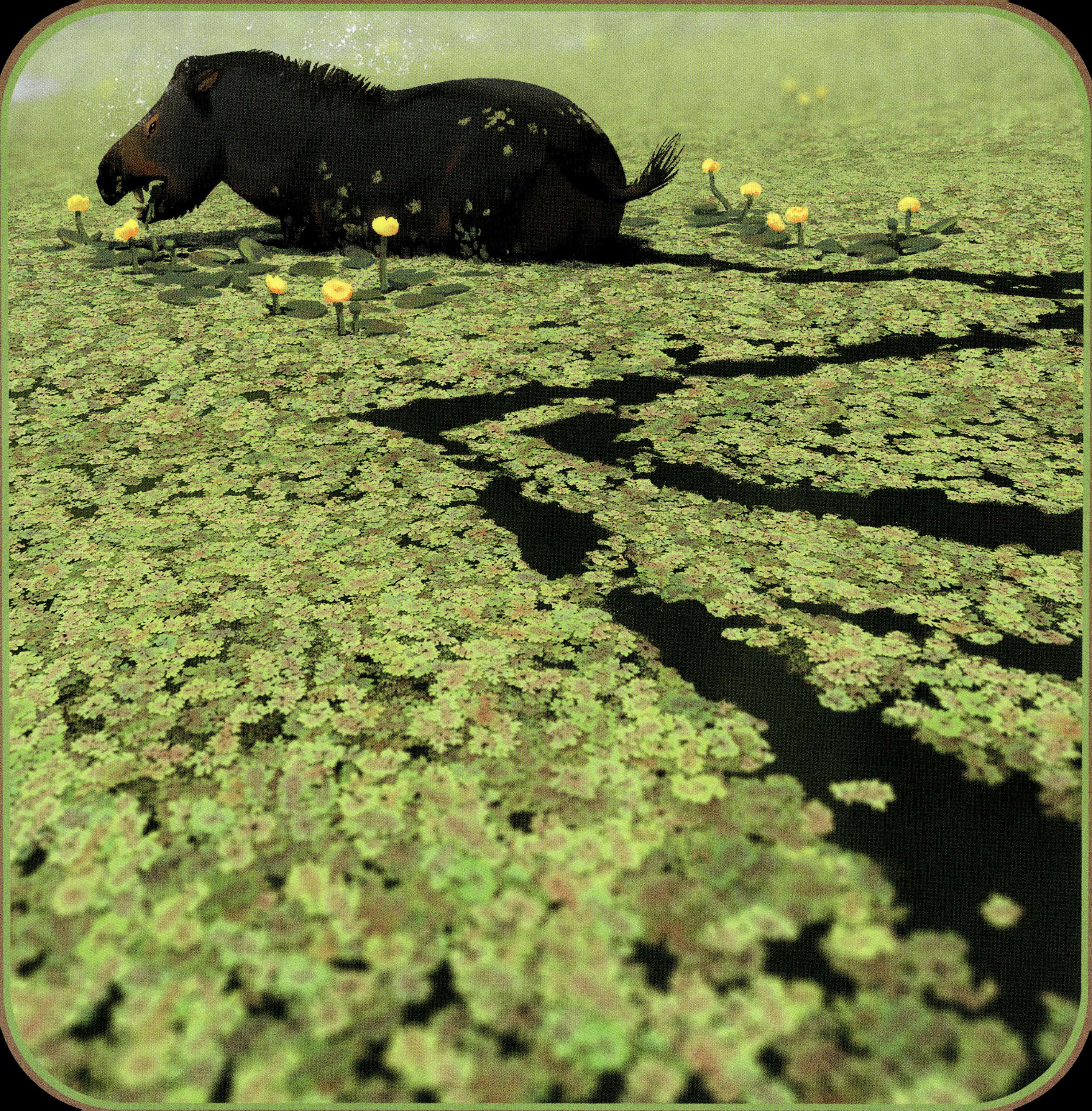

◀ *AZOLLA*

Coryphodon, a rhino-sized herbivore that lived in North America, Europe and Asia around 50 million years ago, wades through the aquatic plant *Azolla*. Fossil species of this fern are known from India, the USA, the UK and elsewhere.
• Digital | 2021

TITANICHTHYS ▶

A large *Titanichthys* cruises through schools of smaller fish during the Devonian, about 365 million years ago. Belonging to the placoderms, a class of early armoured fish, *Titanichthys* lacks cutting edges on its jaw plates and appears to have been a filter-feeder.
• Digital | 2021

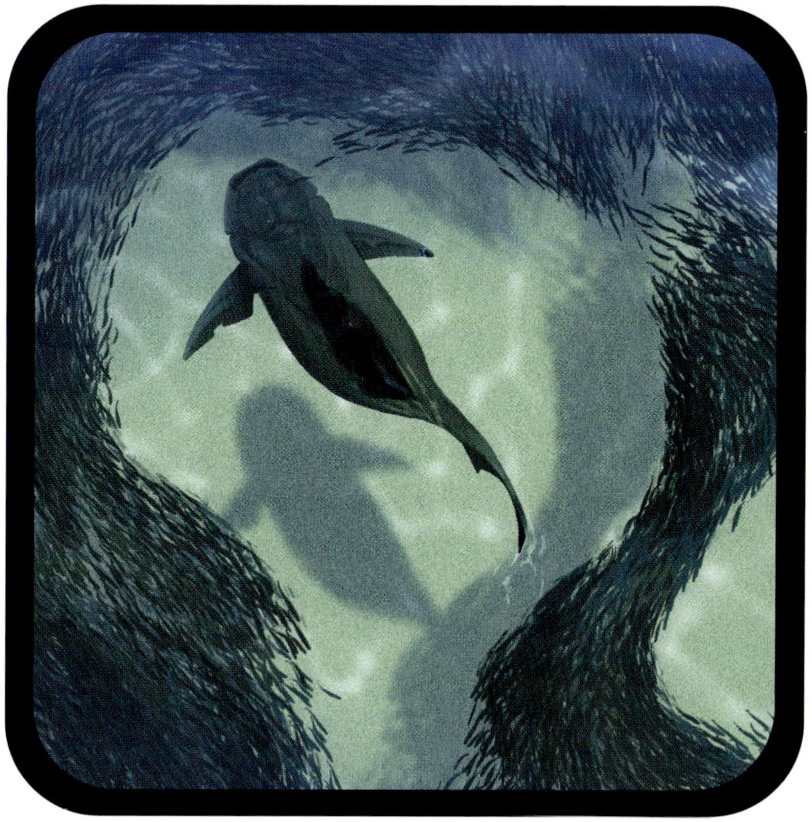

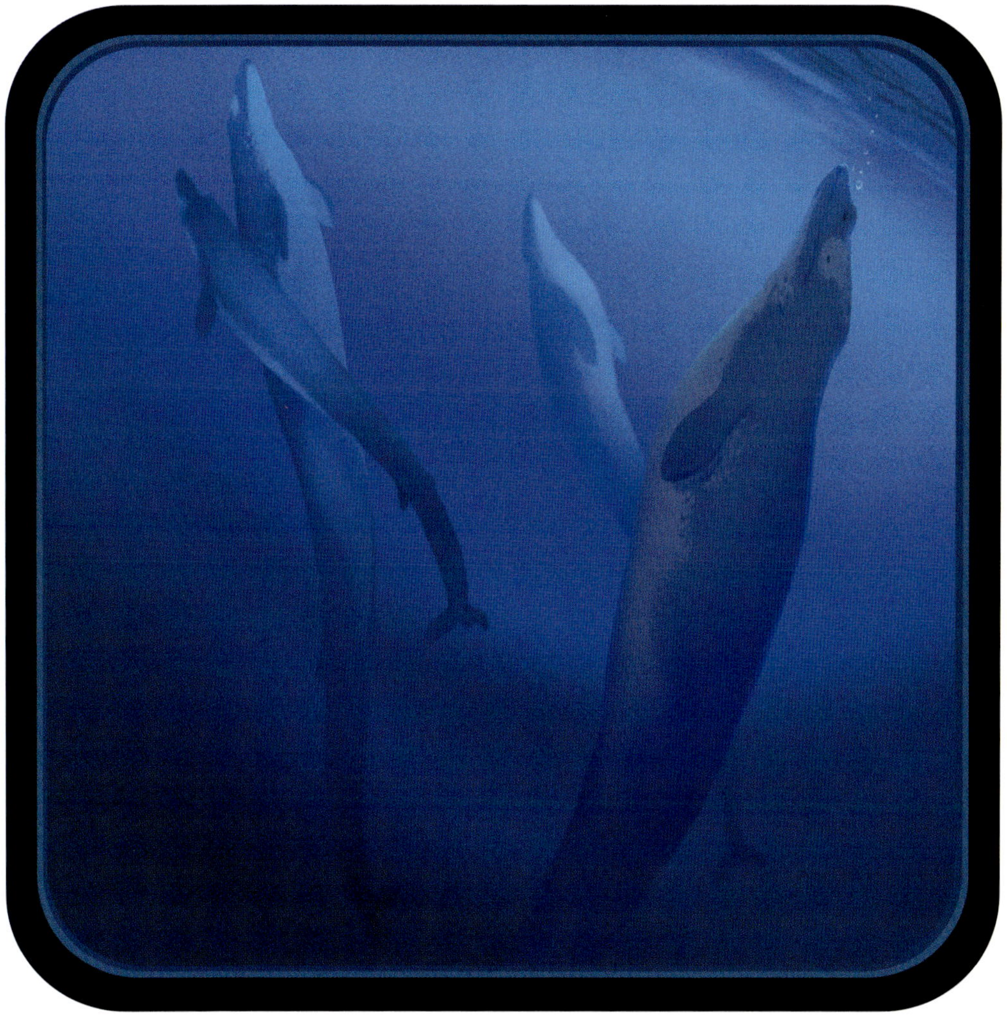

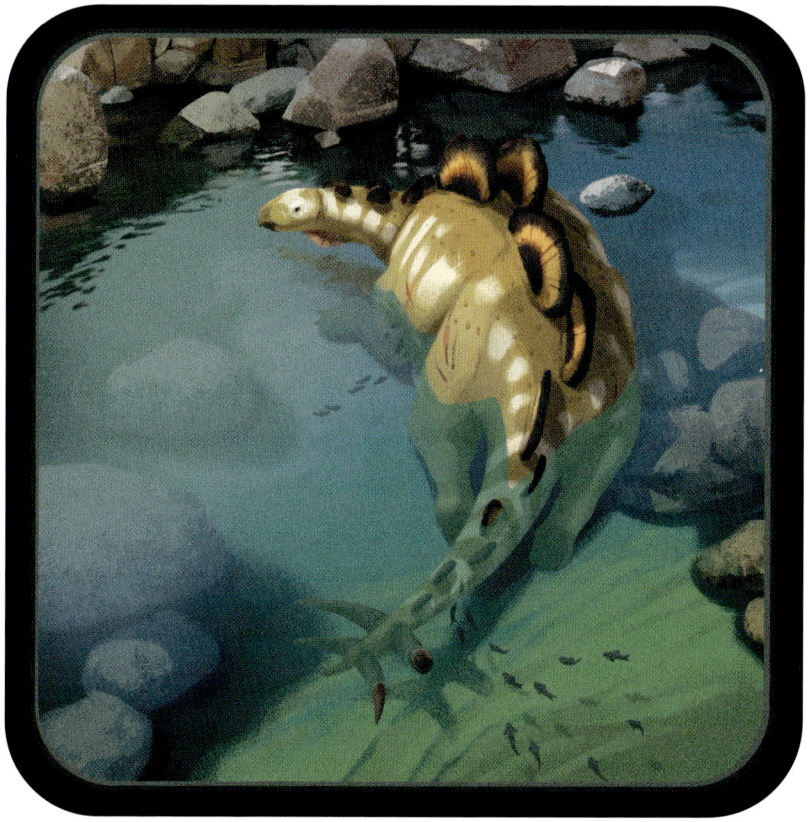

▲ *BASILOSAURUS* FAMILY

This image depicts a family group of *Basilosaurus*, a large, long-bodied whale known from the USA, Peru, northern Africa and southern Asia around 35 million years ago. Modern whales are highly social, so it's at least possible that archaic forms like *Basilosaurus* were too.
• Digital | 2022

HESPEROSAURUS ▶

Dinosaurs like *Hesperosaurus* were built for life on land, but individuals might well have visited coasts to rest, feed on marine plants, or, in this case, allow small fish to clean its wounds. *Hesperosaurus* is a close relative of *Stegosaurus*, noted for its lower, more rounded plates.
• Digital | 2022

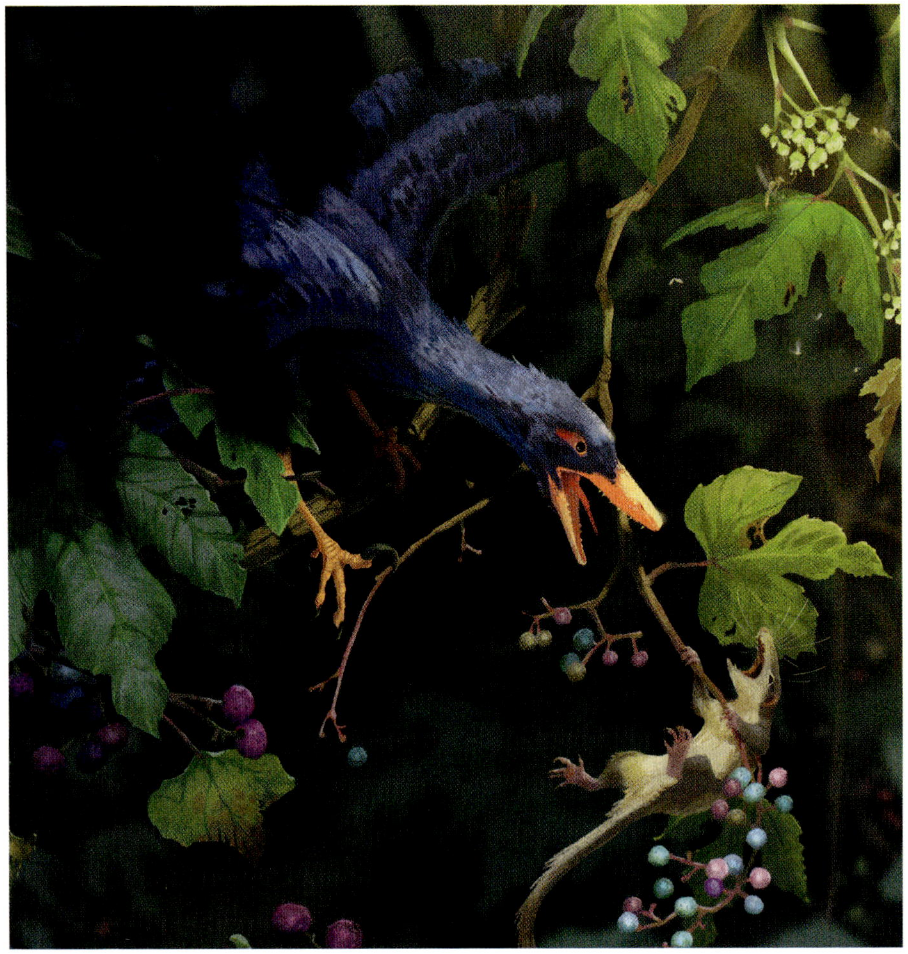

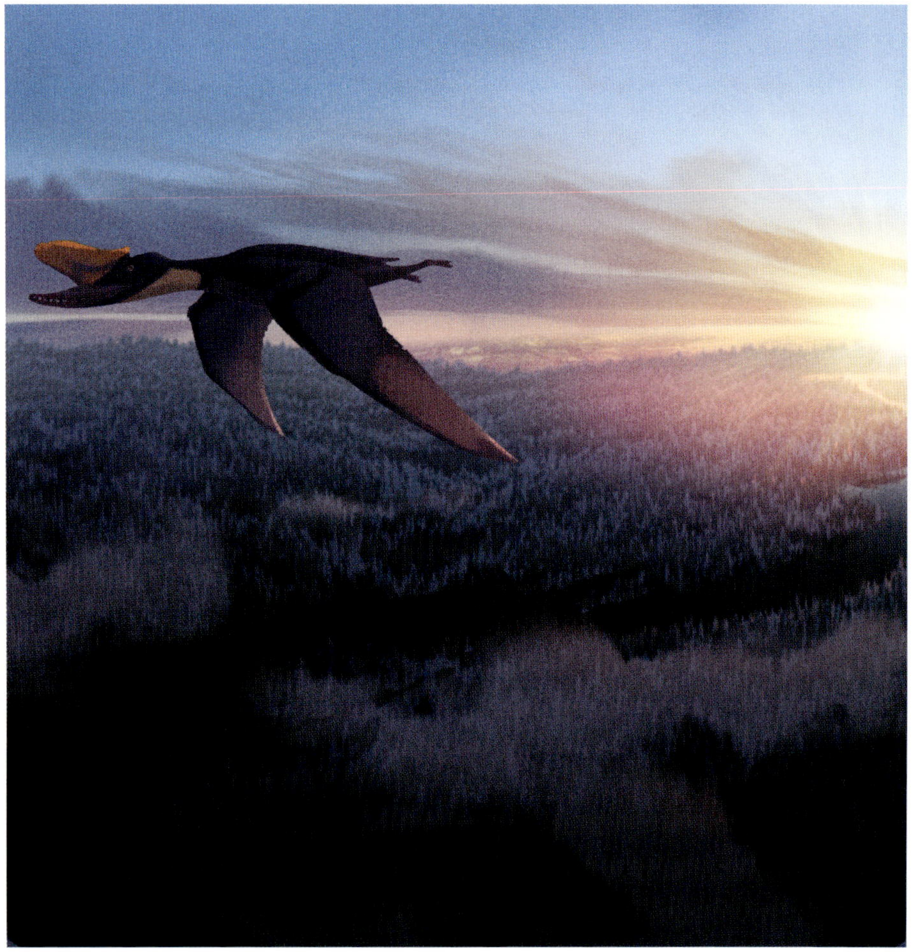

▲ *RAHONAVIS*

Vitaceae (grapevines and their kin) may have originated in India during the Cretaceous, when the subcontinent was an island, separate from continental Asia. Here *Rahonavis*, a crow-sized feathered dinosaur, pursues a mouse-sized climbing haramiyidan mammal through grapevines.
• Digital | 2021

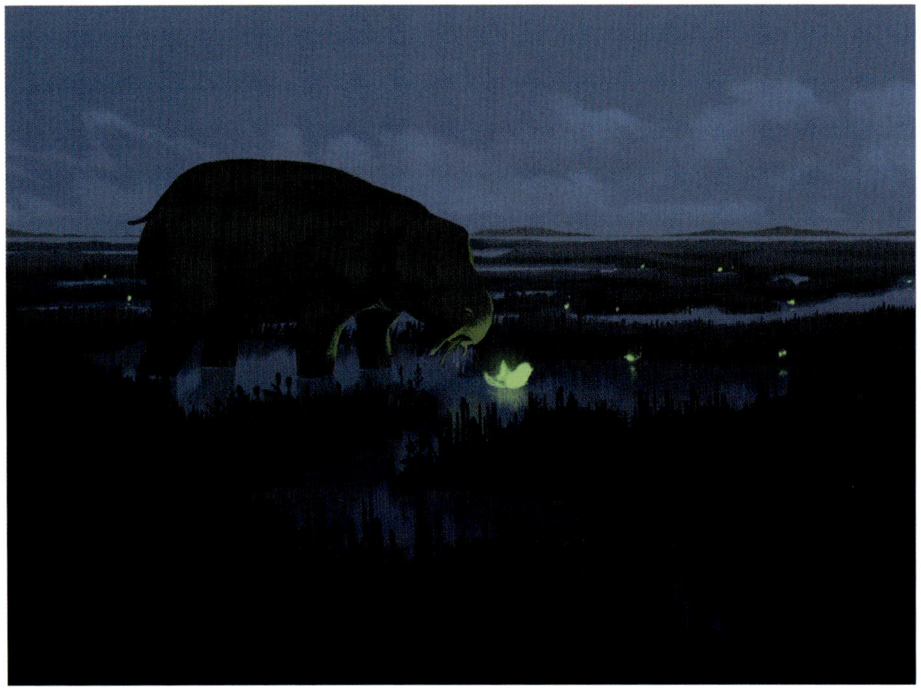

LISOWICIA & WILL-O'-THE-WISPS ▶

Dicynodonts were a diverse group of synapsids ('stem mammals') that occurred in many environments. Some were perhaps hippo-like waders, wallowing and feeding in water and wetlands. Here, the gigantic Polish *Lisowicia*, new to science as of 2019, stirs up swamp gases while feeding, creating *ignis fatuus* – or 'will-o'-the-wisps'.
• Digital | 2022

▲ *NORMANNOGNATHUS*

Studies of eye-size indicate that some pterosaurs had good vision in low light. Their warm-blooded physiology and fuzzy pelt also mean that they were able to remain active even in low temperatures. Some might have been nocturnal, an idea hinted at in this reconstruction of *Normannognathus* from the Late Jurassic of France.
• Digital | 2022

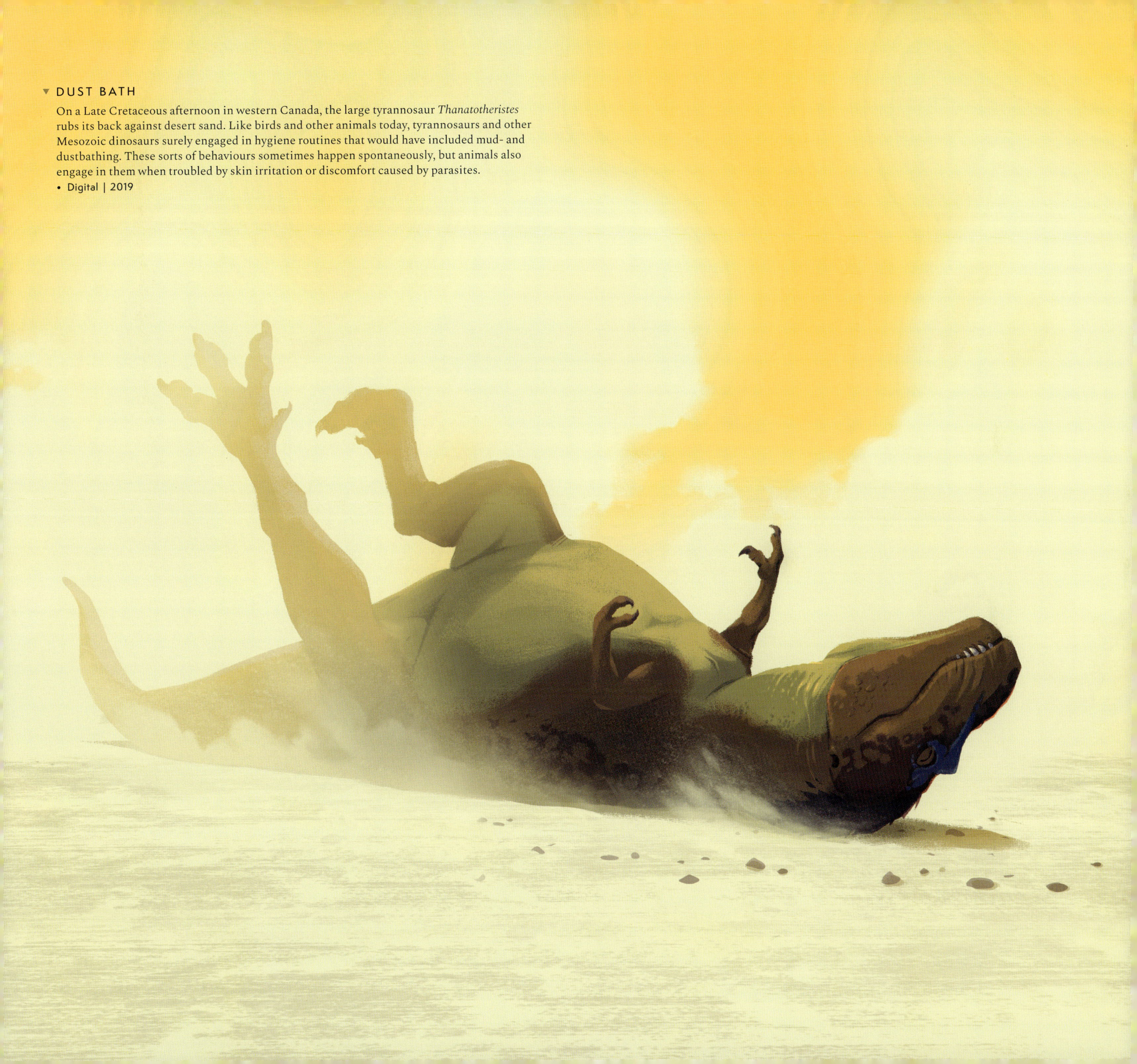

DUST BATH

On a Late Cretaceous afternoon in western Canada, the large tyrannosaur *Thanatotheristes* rubs its back against desert sand. Like birds and other animals today, tyrannosaurs and other Mesozoic dinosaurs surely engaged in hygiene routines that would have included mud- and dustbathing. These sorts of behaviours sometimes happen spontaneously, but animals also engage in them when troubled by skin irritation or discomfort caused by parasites.

• Digital | 2019

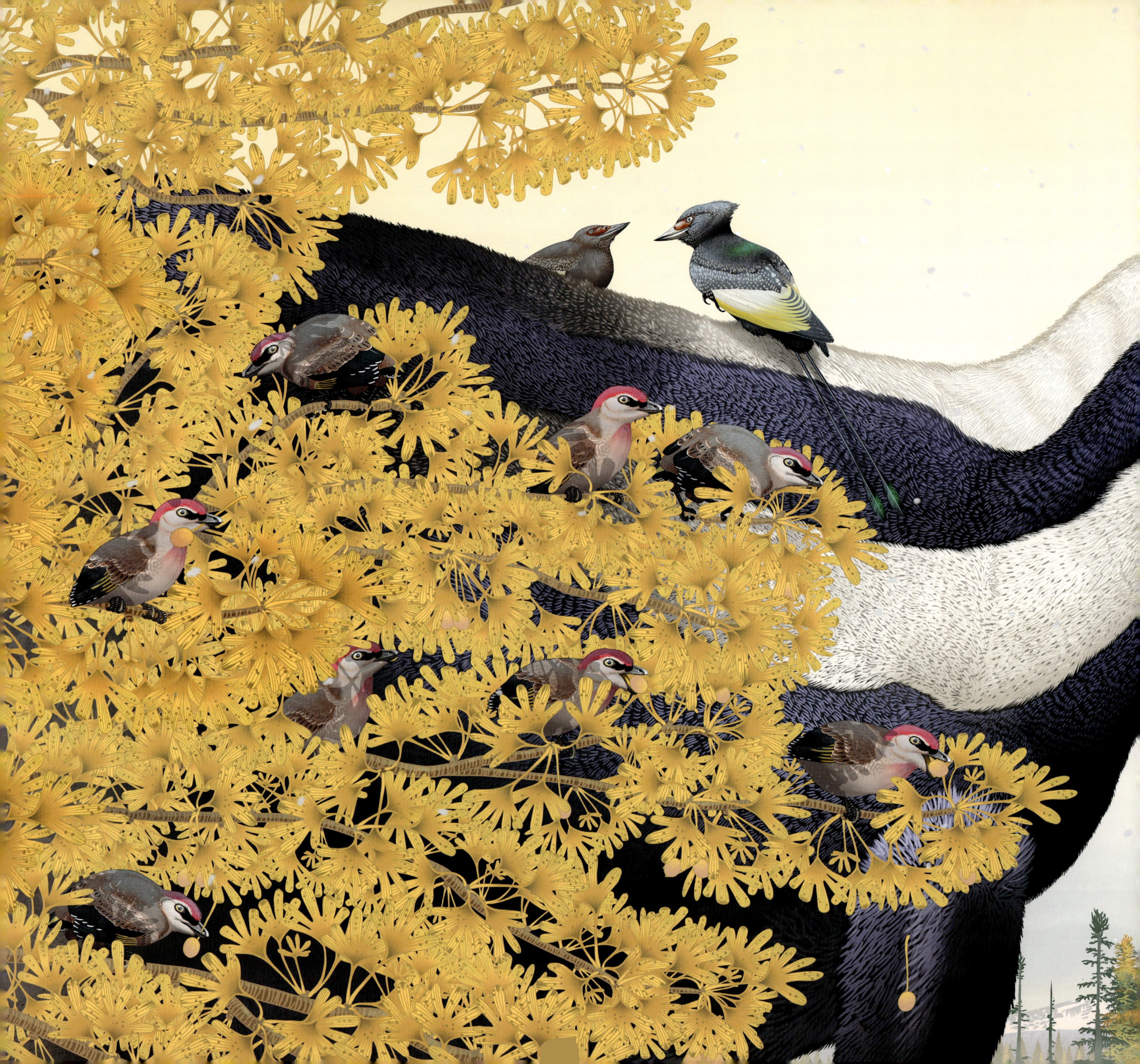

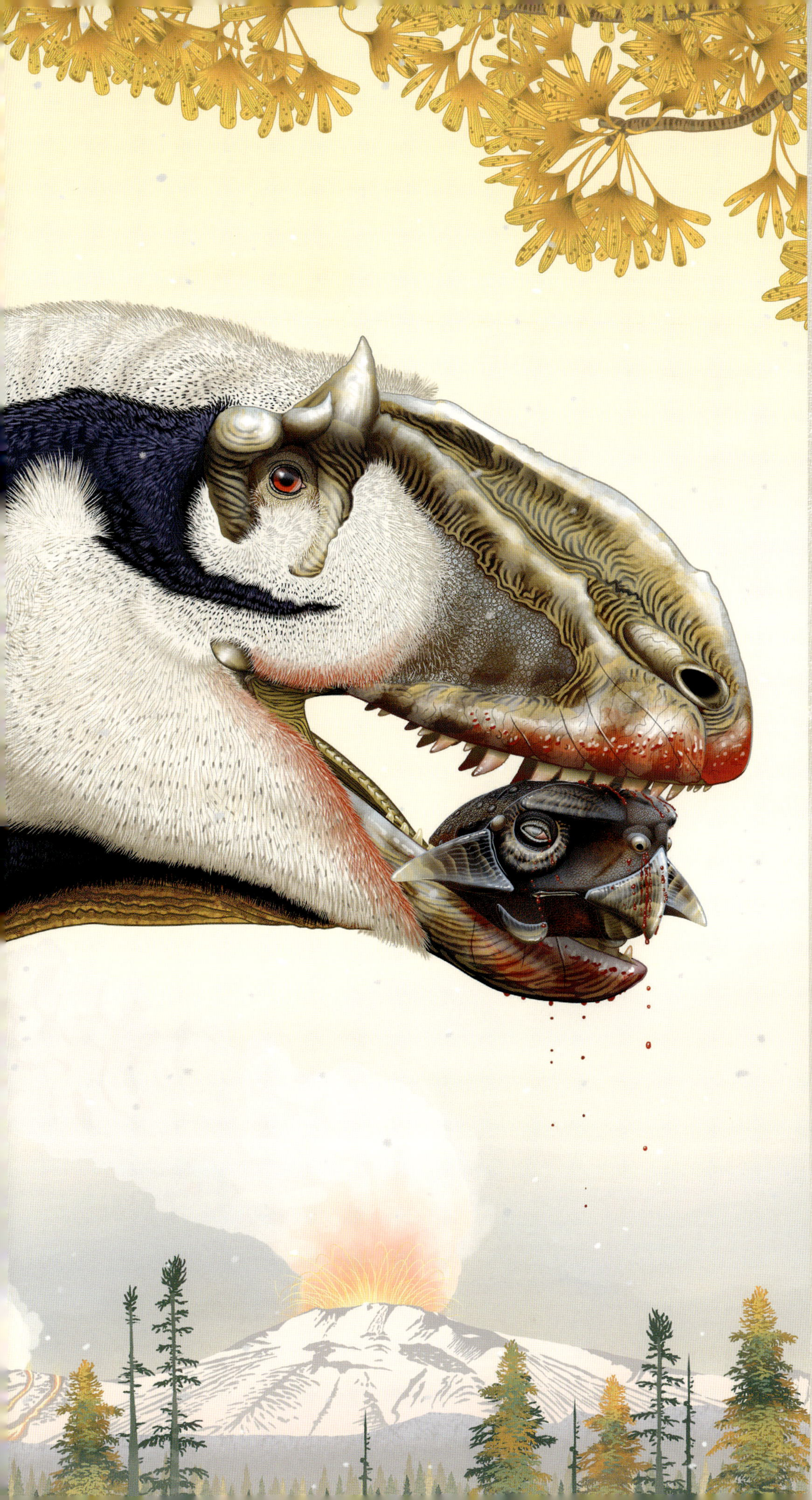

SIMONE ZOCCANTE

SIMONE IS AN ITALIAN natural history illustrator, palaeoartist and graphic designer. As a child, he developed a keen interest in both art and natural history. He started his career as a self-taught traditional artist but now mainly works in digital media. In the past ten years, he has been working in the visual arts and, from 2020 onwards, specialising in scientific and natural history illustrations. Over the past five years, Simone has also been involved in creating art for books and museums, as well as authoring books on illustrating animals and natural history subjects. He has also taught art classes and workshops.

Simone received awards in such prestigious palaeoart contests as the Premio Italiano di Paleoarte, organised by the Museum of Montevarchi in Italy, and the Lourinhã Paleoart Contest, hosted by the Museum of Lourinhã in Portugal.

Specialising in large format paintings, particularly digital murals for museums and exhibitions, Simone also works in smaller formats for illustrations in books and magazines. He also collaborates with Diorama Nature, a web project about natural history, as a writer and host in videos and podcasts.

[Previous pages]

DRAGON MOUNTAIN

The Early Cretaceous dinosaurs of China's Yixian Formation include a fascinating assemblage of feathered theropods large and small – birds of many sorts among them – in addition to various ornithischians, like *Psittacosaurus*. One of the latter has fallen prey to the large, fuzz-coated tyrannosaur *Yutyrannus*, on whose back is perched the bird *Confuciusornis*.

• Digital | 2024

MAGNOLIA-PAULIA ▸

In a flooded forest of Late Cretaceous Baja California, a group of the giant hadrosaur *Magnapaulia* have disturbed a young *Deinosuchus*. *Deinosuchus* is large – a big one could reach 7m (23ft) or more – but this individual is dwarfed by *Magnapaulia*, adults of which were 12m (39ft) long. Hey, it's not called 'Big Paul' for nothing. The plants are magnolias and members of the monkey puzzle group. This work was shown at an exhibition of the Italian Paleontological Society.

• Digital | 2021

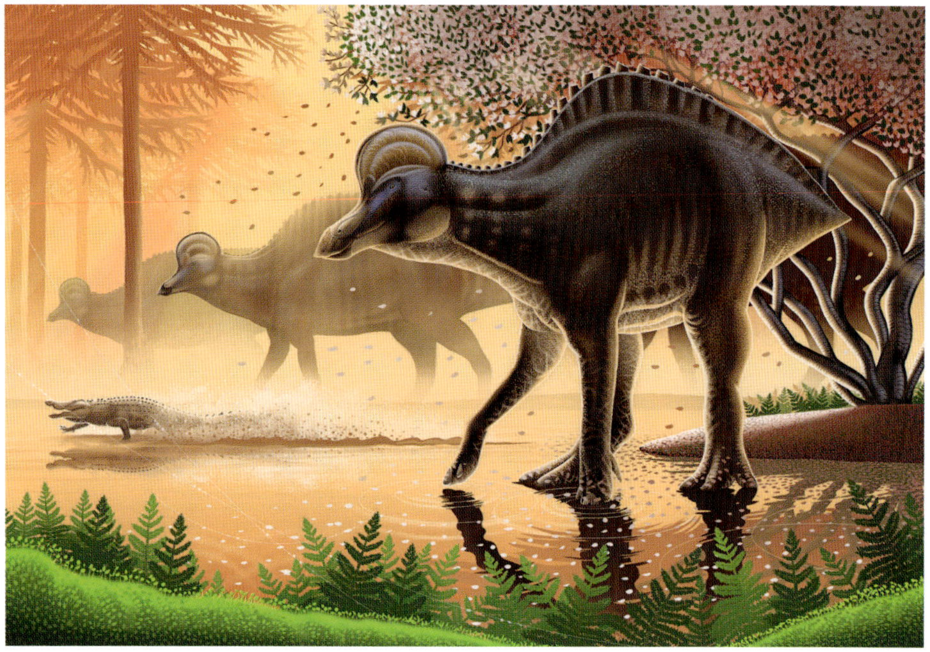

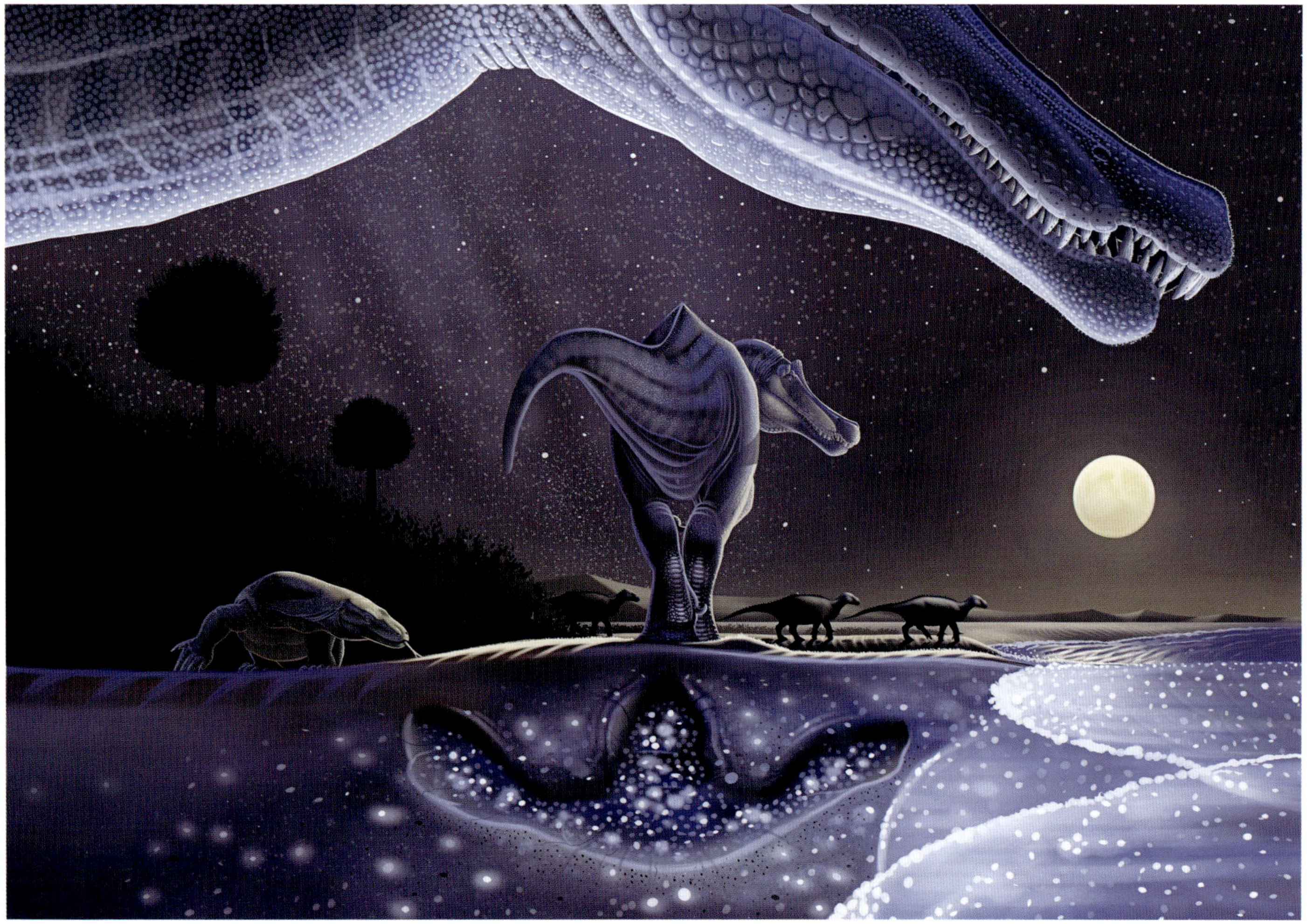

ANCIENT GADOUFAOUA ▸

This scene depicts animals of the Elrhaz Formation of Lower Cretaceous Niger, specifically that of Gadoufaoua in the country's centre. Gadoufaoua is famed for the spinosaurid *Suchomimus*, here shown consuming a large lungfish. Crocodyliforms like *Sarcosuchus* lived here too, as did the sail-backed ornithopod *Ouranosaurus*. Pterosaurs like *Anhanguera* occurred almost worldwide at this time.

• Digital | 2022

◂ IMMORTAL FOOTPRINT

During an Early Cretaceous night, a *Baryonyx* pair and the iguanodontian *Hypselospinus* walk an Iberian beach. Originally reported from England, both kinds of dinosaur are known from Spain as well. *Arcanosaurus* – a lizard related to modern monitors – is present too; the tree is the conifer *Frenelopsis*. This painting was selected for the XIII Castilla y León Paleoart Prize and was featured in the associated palaeoart exhibition at the Dinosaurs Museum of Sala de los Infantes in Burgos, Spain.

• Digital | 2021

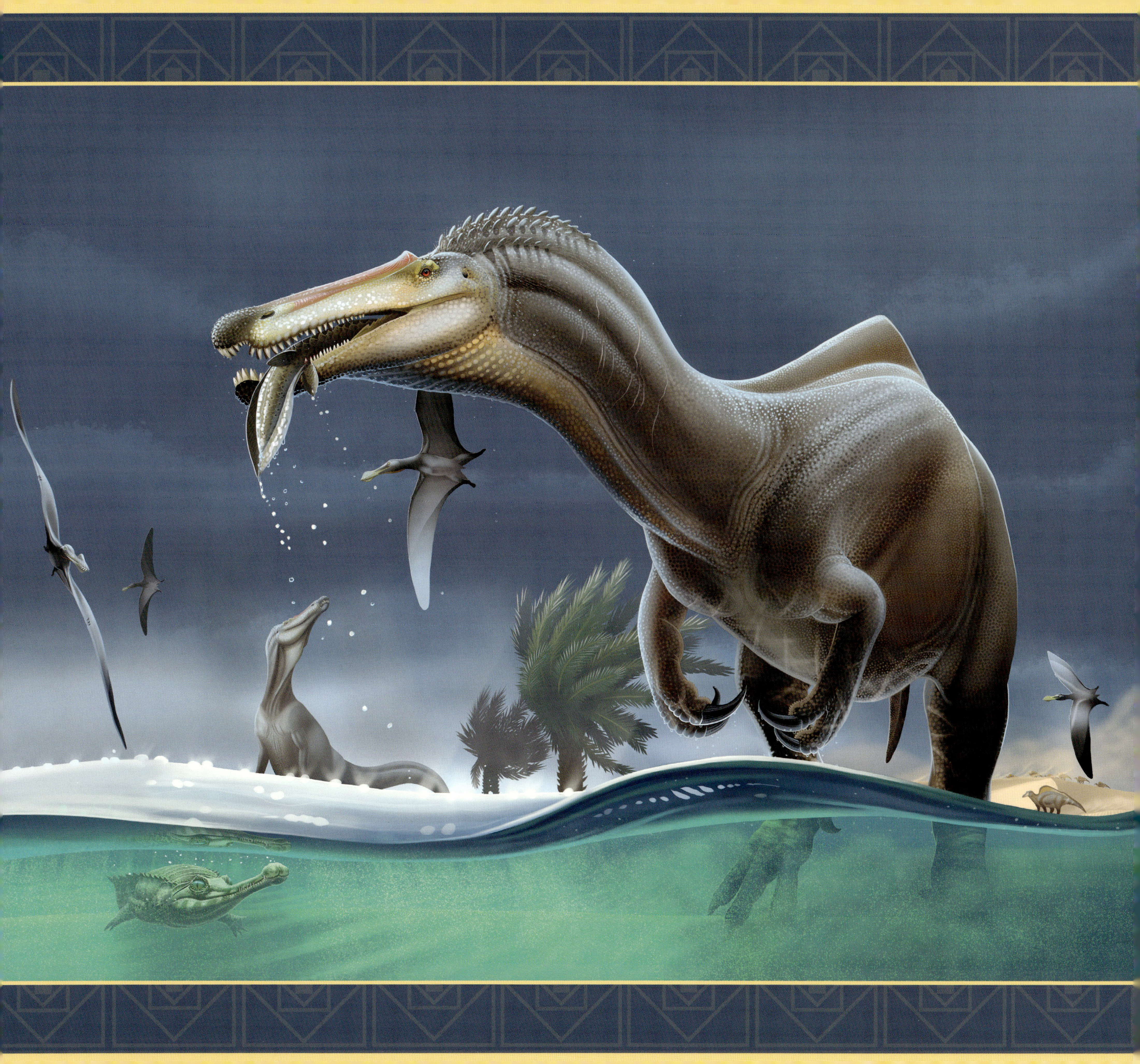

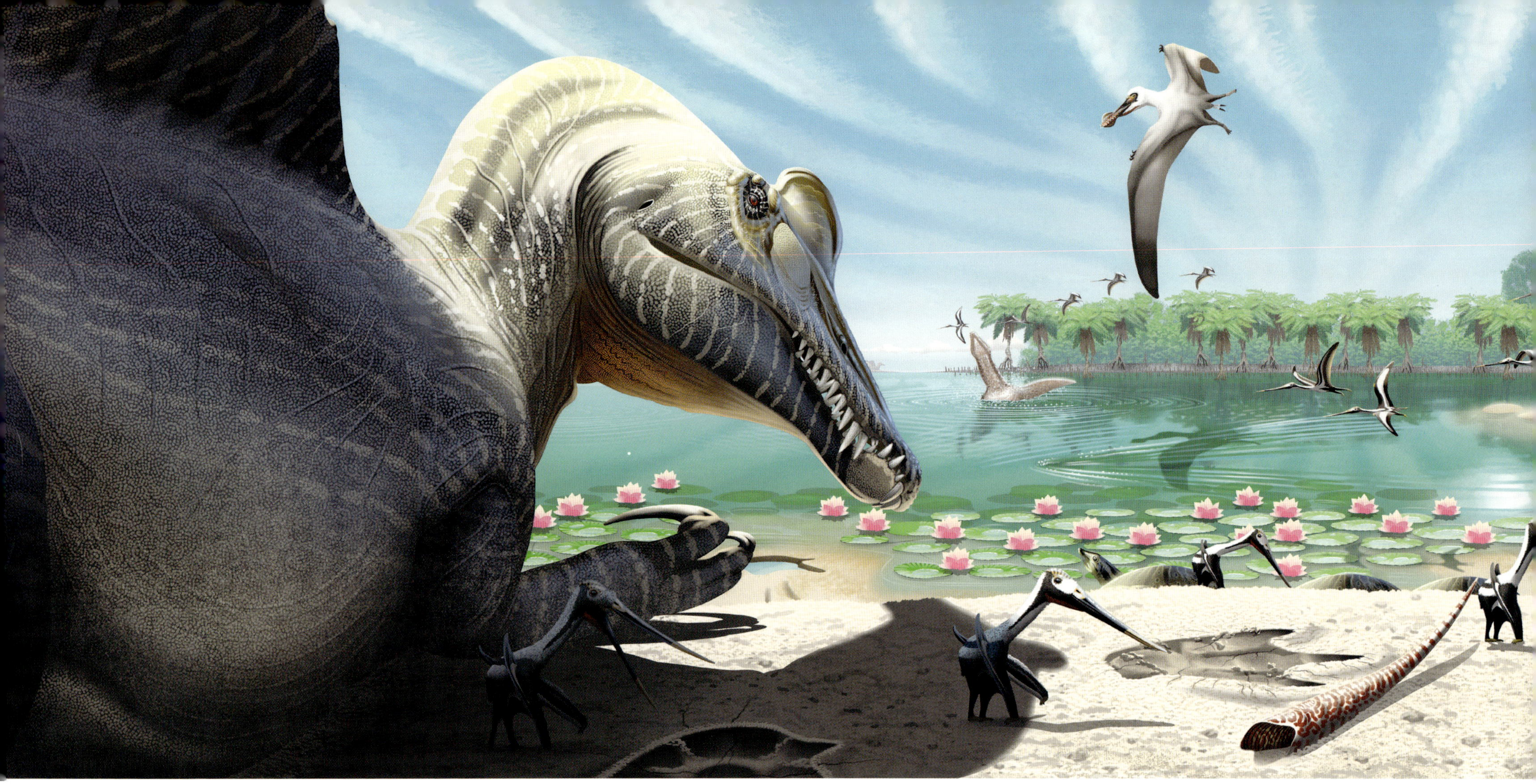

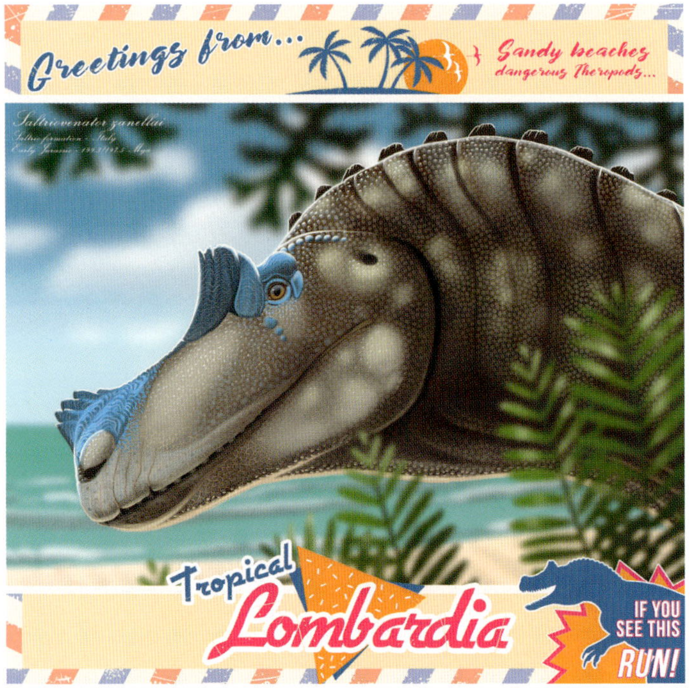

TROPICAL LOMBARDIA POSTCARD

Until just a few decades ago, Italy was supposedly a nation without (non-bird) dinosaurs. The Early Jurassic theropod *Saltriovenator*, found in 1996 and published in 2018, is one of several finds that have very much reversed this notion. Good skull remains are unknown for *Saltriovenator*, but an affinity with *Ceratosaurus* suggests that it possessed horns. When this animal was alive, the Saltrio region for which it was named was tropical and mostly covered by shallow sea. The skull image on the stamp is based on a reconstruction by palaeontologist Marco Auditore.

• Digital | 2022

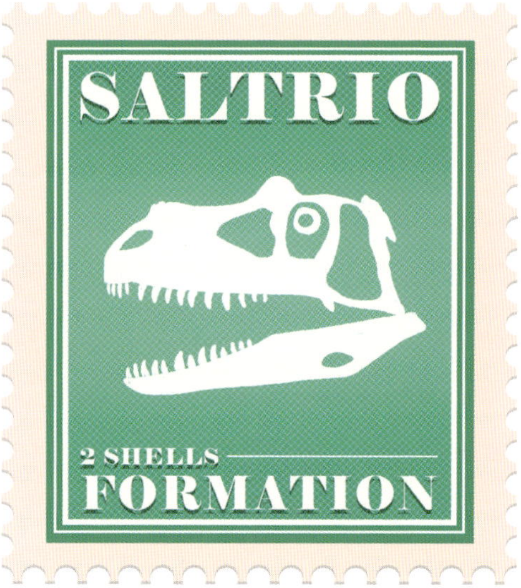

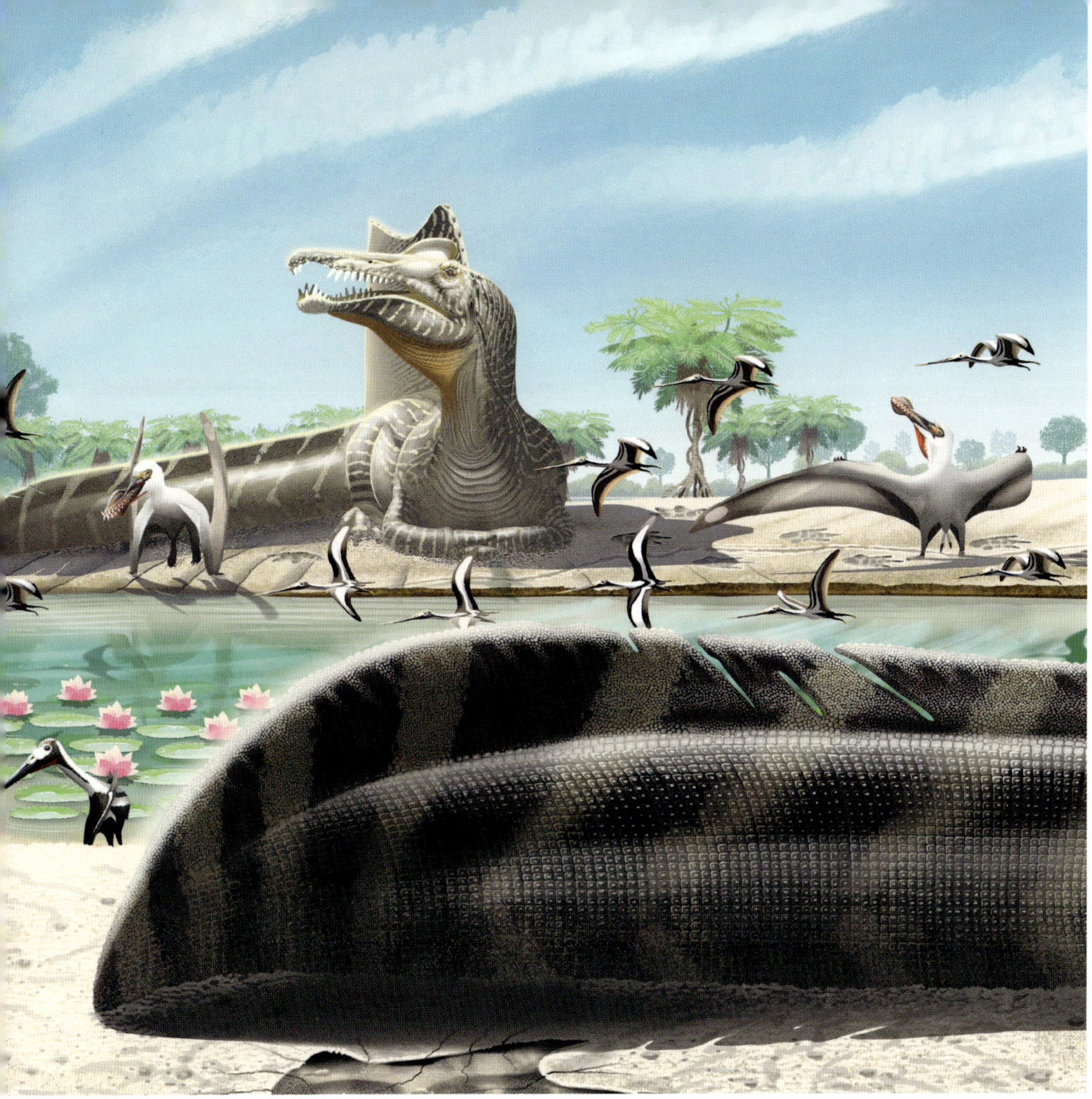

◄ THE FISHERS' BAY

A group of *Spinosaurus* relax on the shore of a Late Cretaceous Moroccan lagoon. Their footprints betray their amphibious specialisation. Turtles and pterosaurs – including the slender-jawed *Leptostomia* – are abundant. Leptocleidid plesiosaurs are here as well, one opportunistically lunging at a pterosaur. The sediments from this environment will form a geological unit known as the Kem Kem Group.

• Digital | 2024

▼ *LARIOSAURUS VALCERESII*

This reconstruction and portrait show *Lariosaurus valceresii*, a Middle Triassic nothosaur from Italy. Many *Lariosaurus* species are known from China and central Europe, though some may belong to other genera. *Lariosaurus* was less than 1m (3ft) long, had flipper-like limbs and interlocking, fang-like teeth. It's an early member of Sauropterygia, the great Mesozoic group that includes plesiosaurs.

• Digital | 2023

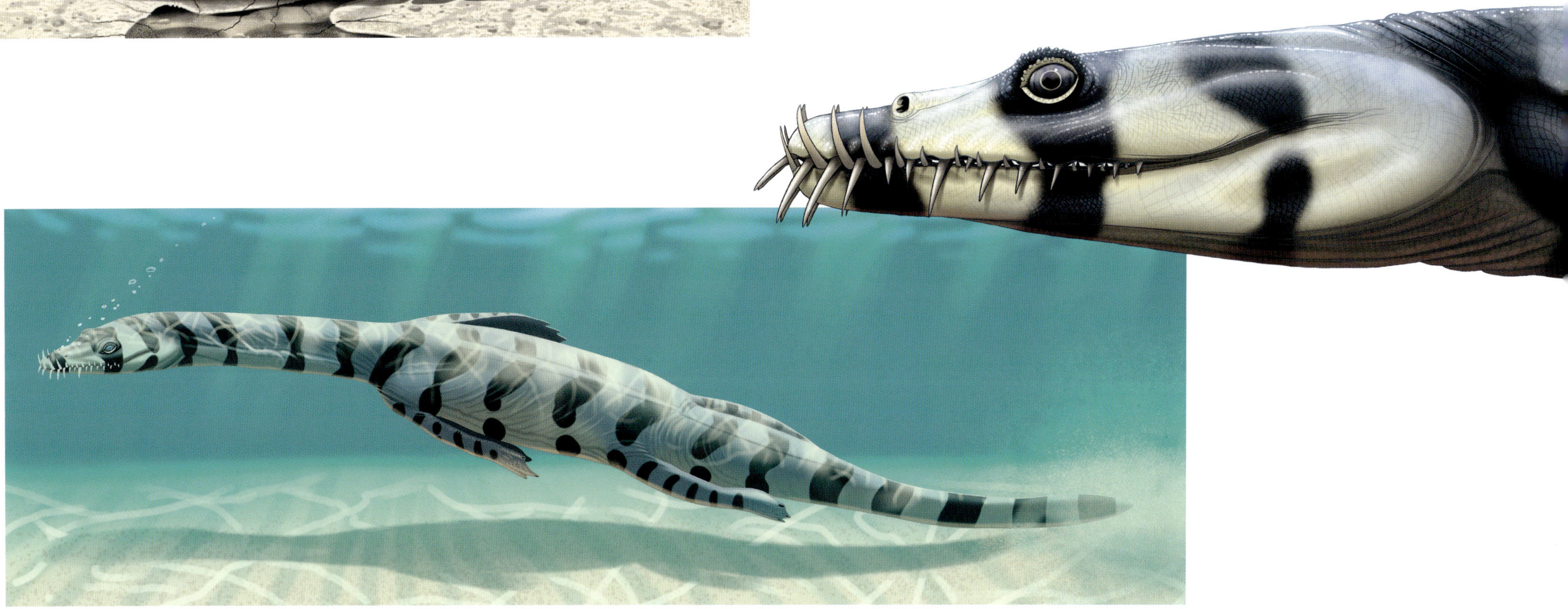

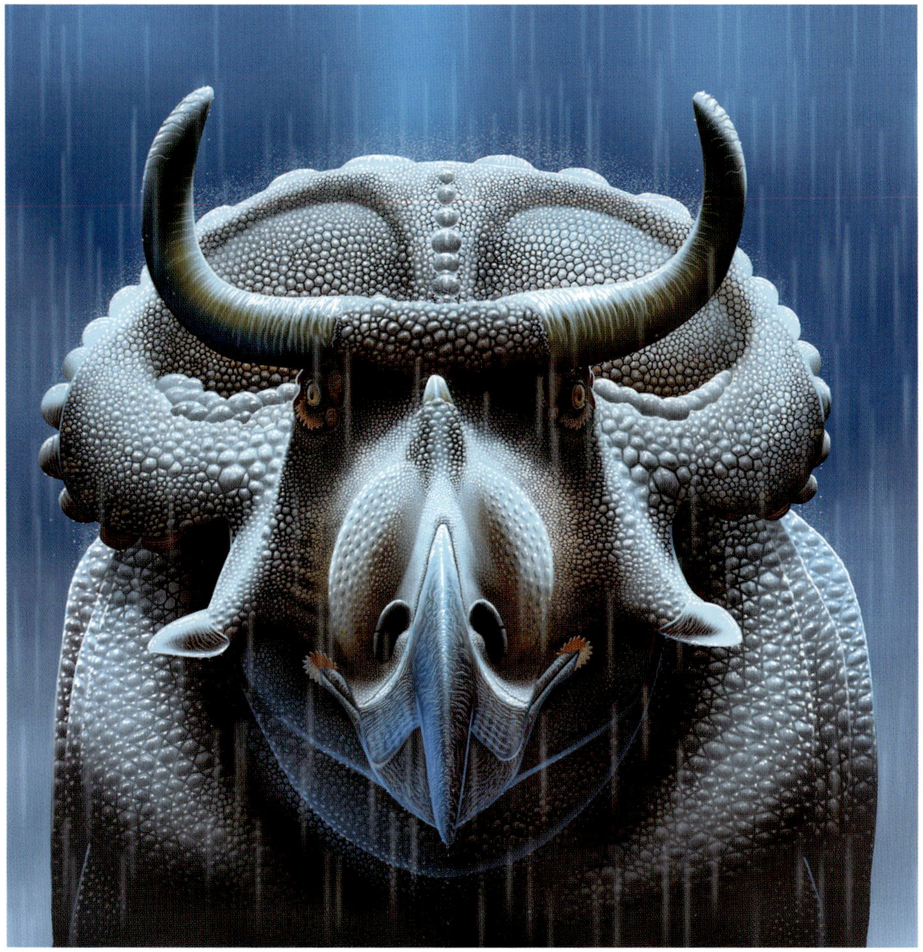

▲ BLUE BULL

The remarkable North American horned dinosaur *Nasutoceratops* from the Kaiparowits Formation of Utah, USA, was named in 2013. It revealed the presence of a hitherto unrecognised ceratopsian lineage, Nasutoceratopsini. Key traits include the deep, narrow, short snout, a massive, circular bony nostril and brow horns that curve outwards and forwards, recalling those of cattle more than those of other ceratopsians.
• Digital | 2023

MORRISON FORMATION ▶

In a landscape partly ruined – temporarily, at least – by a quick-burn forest fire, an adult *Stegosaurus* walks beside a sibling group of baby *Brachiosaurus*. The large theropod *Torvosaurus* has its eye on them. Despite the damage inflicted by the blaze, there are already signs of recovery. Plants include monkey puzzles and various conifers, cycads and ferns.
• Digital | 2024

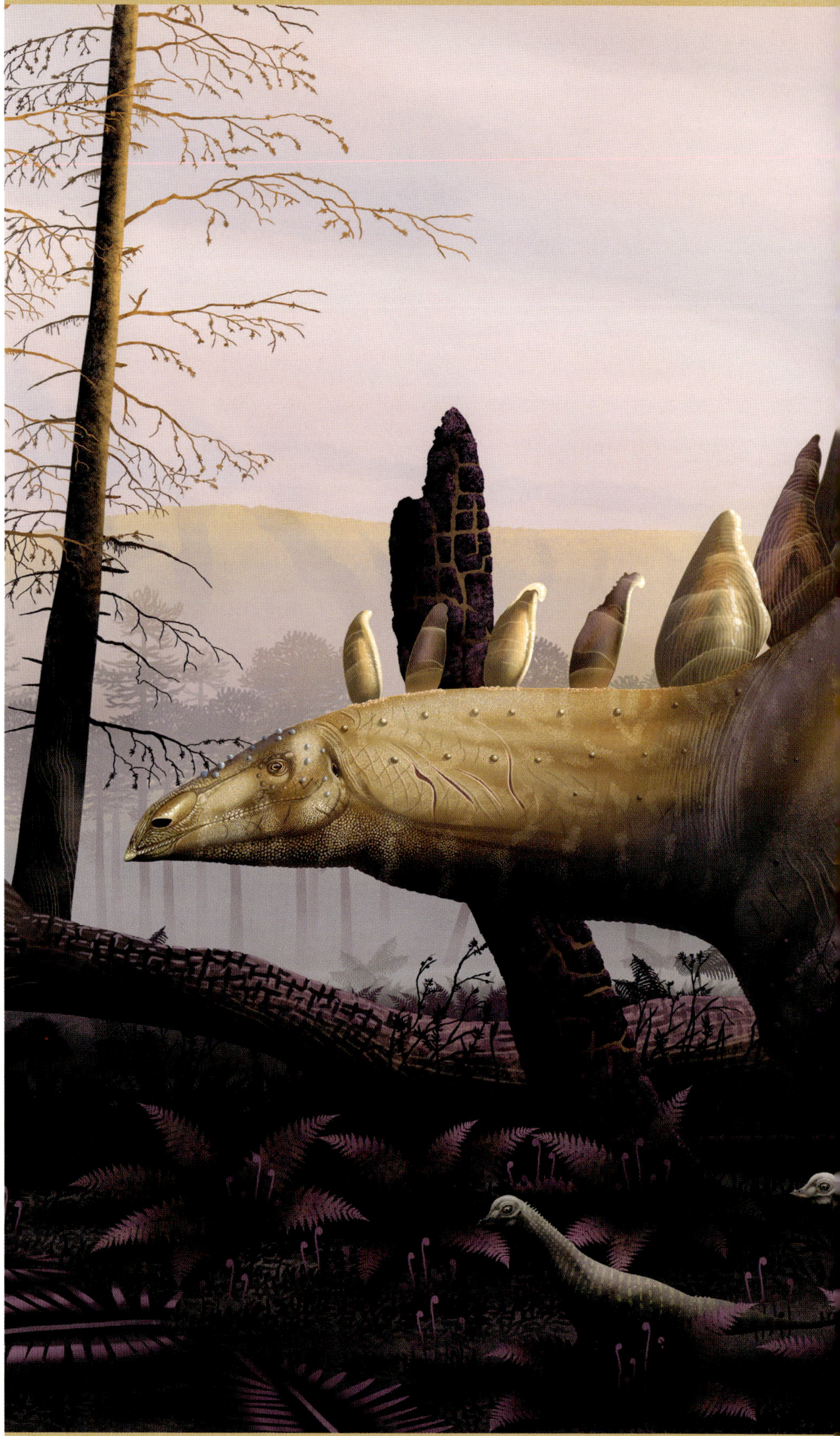

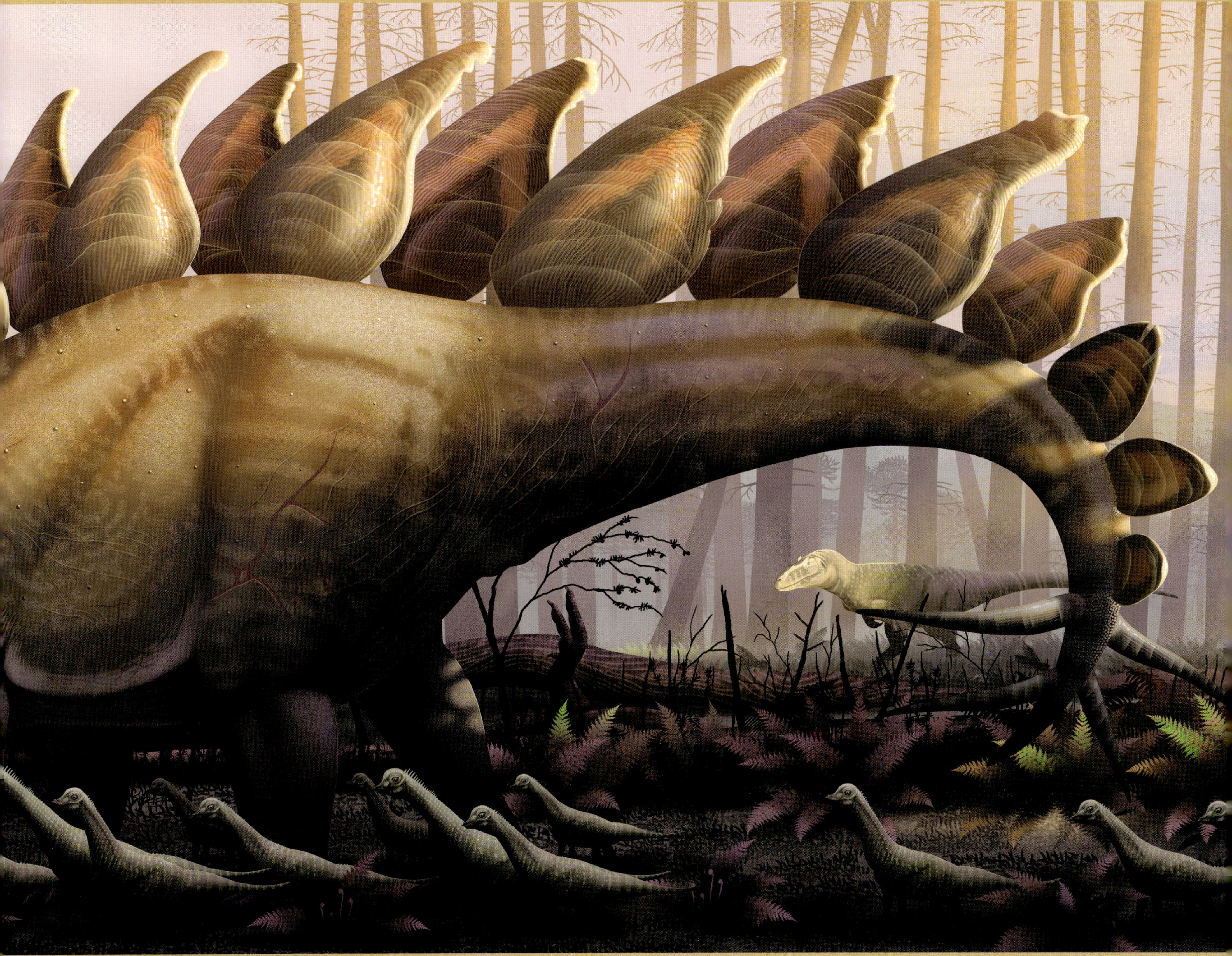

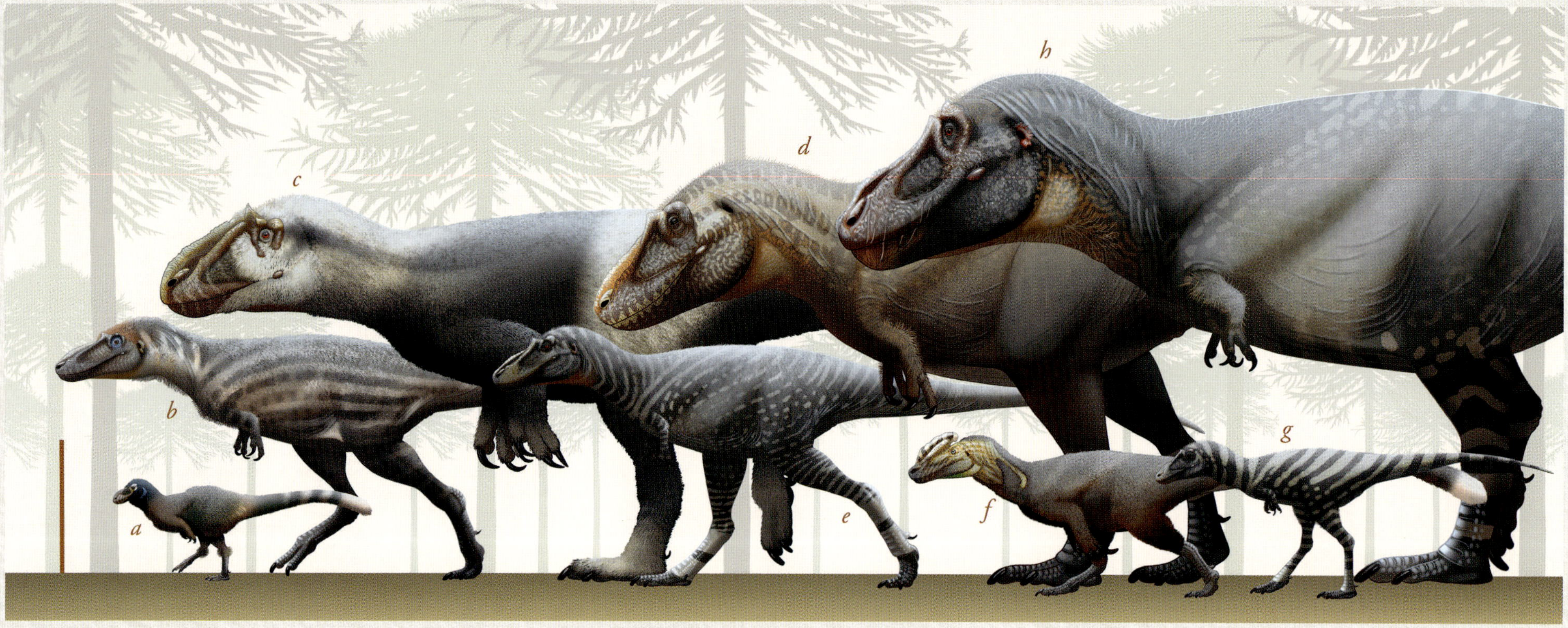

▲ THE TYRANT FAMILY

A tyrannosauroid motley crew, featuring diverse members of this theropod group large and small. These include: a. *Dilong paradoxus*. b. *Alioramus remotus*. c. *Yutyrannus huali*. d. *Daspletosaurus wilsoni*. e. *Tyrannosaurus rex*, juv. (BMRP 2002.4.1, Jane). f. *Guanlong wucaii*. g. *Tyrannosaurus rex*, young (MOR 6625, Chomper). h. *Tyrannosaurus rex*, adult.
• Digital | 2024

▼ TWO FACES OF DJADOCHTA

The Late Cretaceous dromaeosaurids *Halszkaraptor*, named in 2017, and *Velociraptor* (right) were both discovered in Mongolia's Djadochta Formation. Details pertaining to limb proportions, snout shape, neck anatomy and more show that *Halszkaraptor* was a diver or swimmer in lakes and pools. *Velociraptor* may have favoured more arid environments.
• Digital | 2023

ENCOUNTER UNDER THE RAIN ▶

The Canadian tyrannosaurid *Gorgosaurus libratus* is caught in a Late Cretaceous rainstorm. Integumentary details (meaning those concerning the animal's outermost surface) include horn-covered ridges along the snout, hornlets and bony lumps around the eyes and nostrils, and complex scales around the muzzle. The trees are the conifers *Metasequoia* (dawn redwood) and *Taxodium* (bald cypress), both of which survive today.
• Digital | 2023

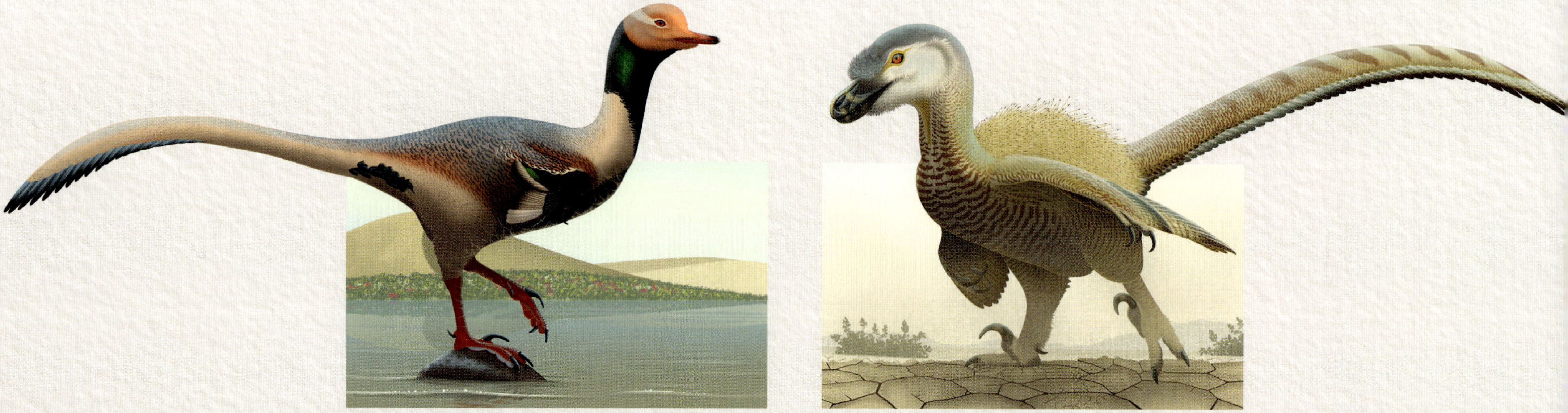

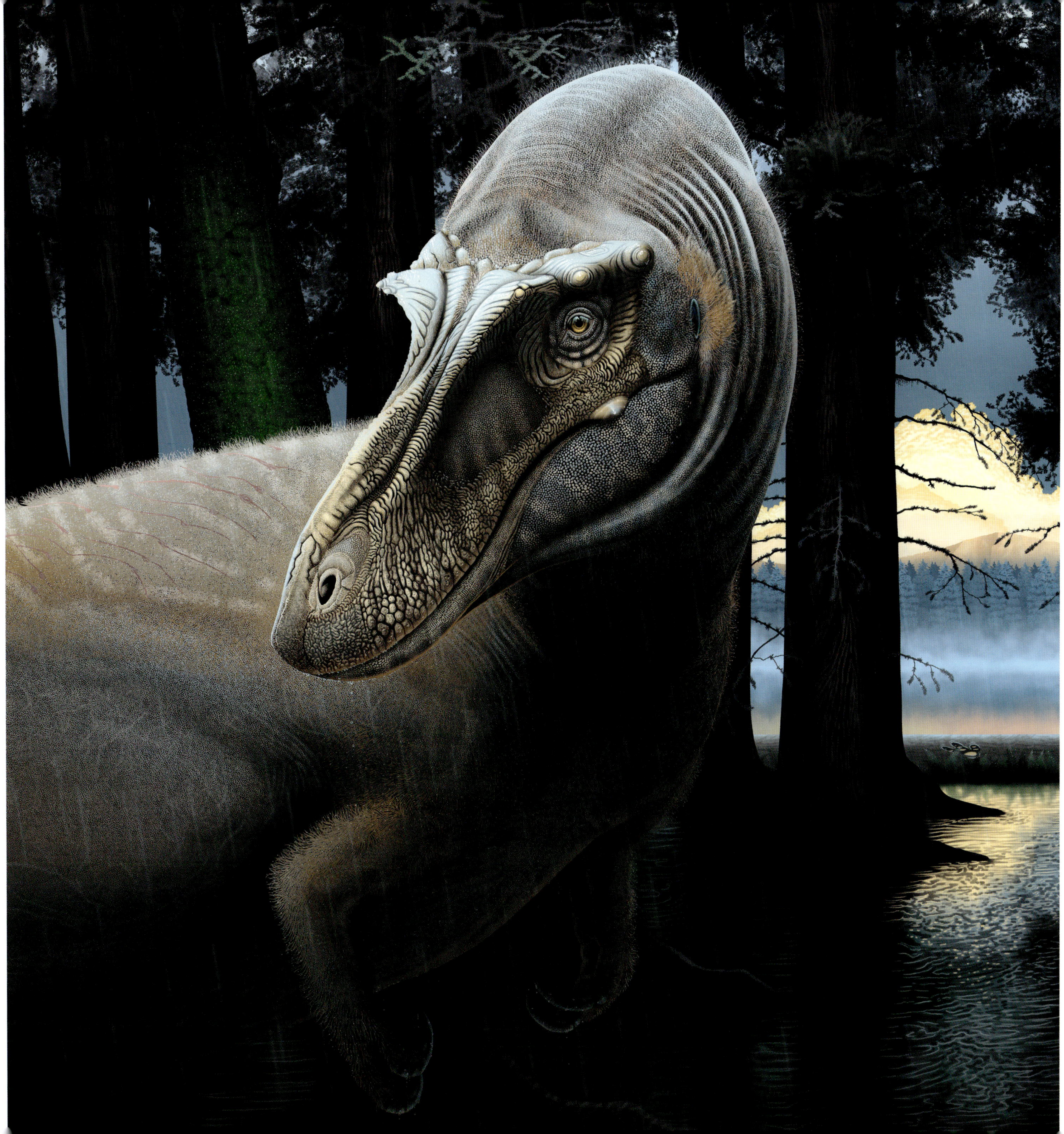

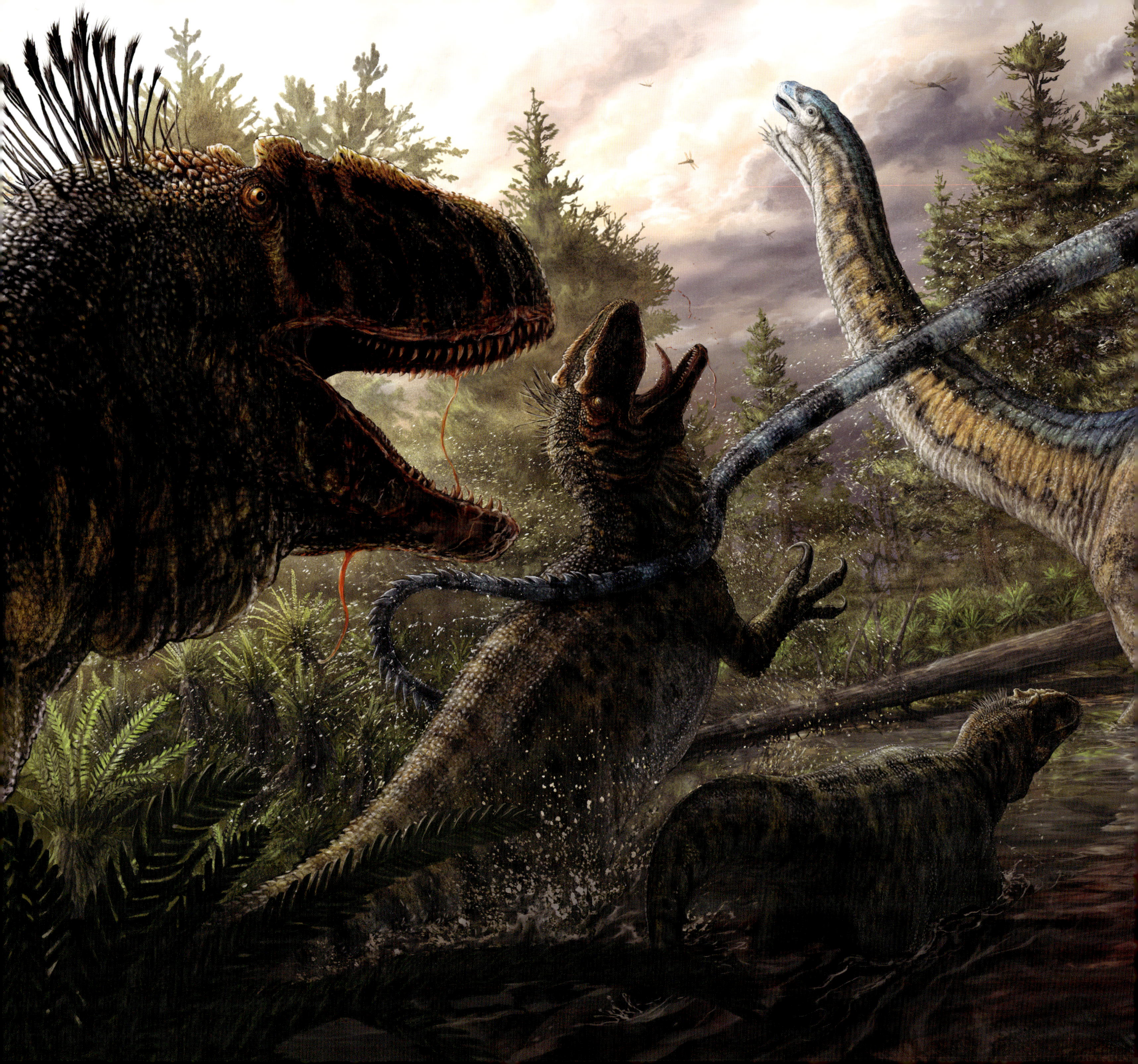

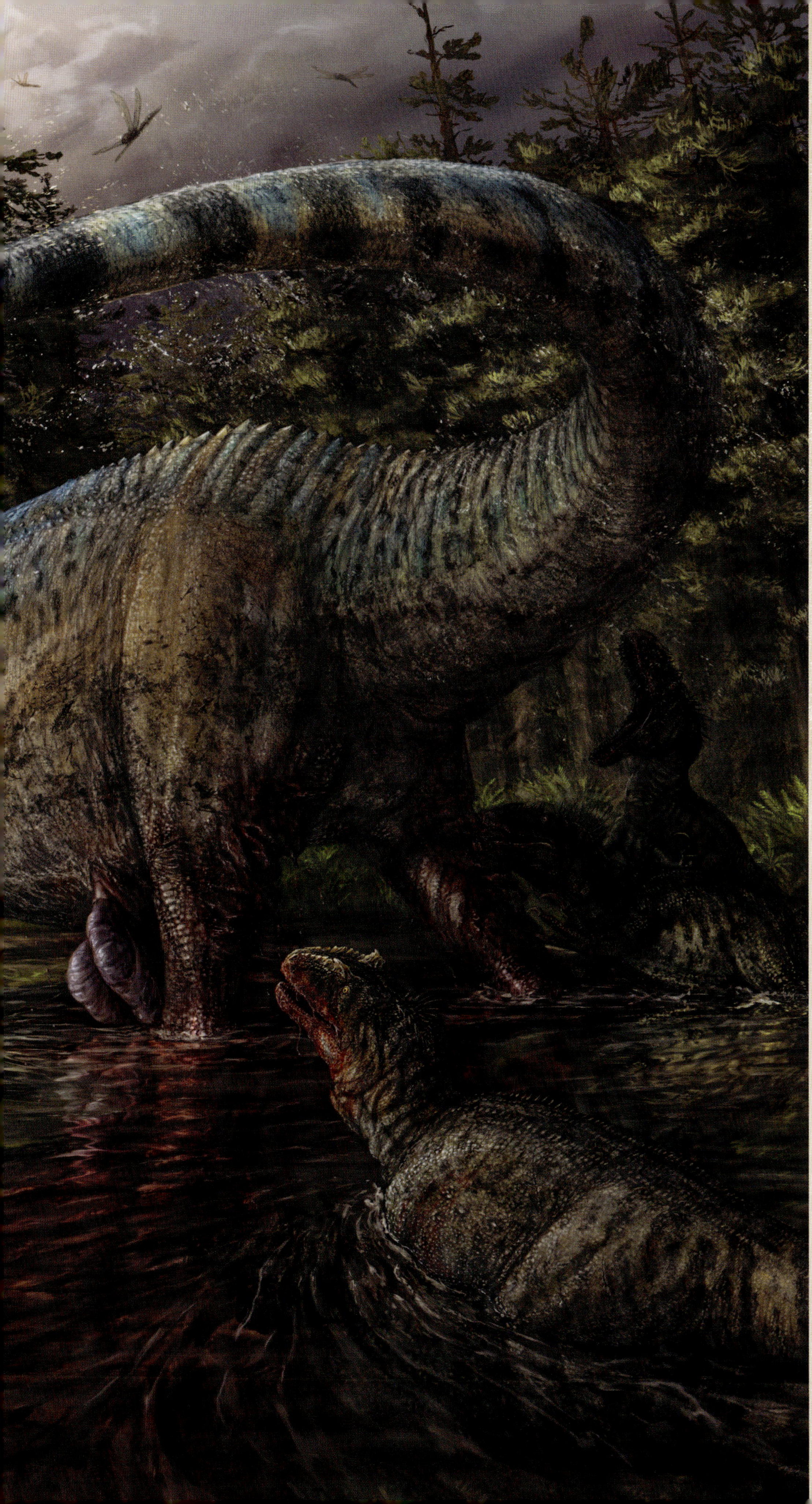

BRIAN ENGH

BRIAN ENGH IS A FILMMAKER and artist from California who has been working with palaeontologists since 2014, his aim being to incorporate new fossil discoveries and on-site palaeontological interpretation into art and storytelling. Brian's production company, Living Relic Productions, specialises in working directly with researchers to help them tell their stories through murals, illustrations, sculptures and animation for museum exhibits, interpretive sites, documentaries and publications.

Brian's lifelong interest in strange creatures and wild ecosystems has led him to participate in wildlife field surveys, as well as palaeontological surveys and excavations. He has created numerous educational videos documenting and explaining his palaeoart process, and his work in the Late Jurassic Morrison Formation has led to the development of an independent documentary film project called *Jurassic Reimagined*. Several of the pieces featured in *Mesozoic Art II* are stills from animated scenes that will be featured in the documentary.

Brian strives to create palaeoart that will engage the public in the hope that palaeontological perspectives will open a door to a deeper understanding and appreciation for living ecosystems and our role within them.

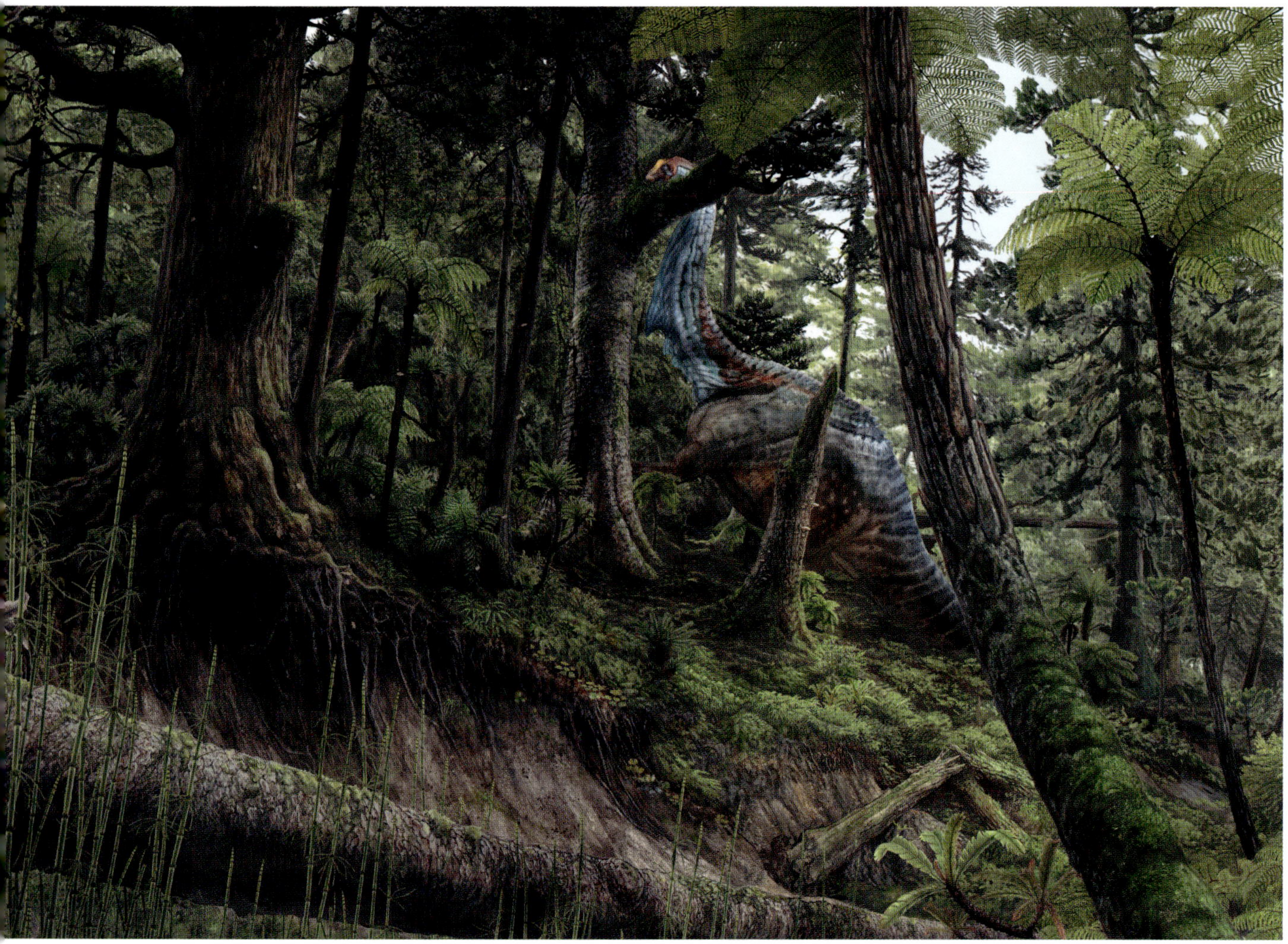

◂ MORRISON MURAL

This segment is from a giant, 7.2m (23.6ft)-wide, animated mural at the Natural History Museum of the University of Zurich. It is based specifically on the fossil assemblage from northern Wyoming's Howe-Stephens Quarry , famous for its sauropod dinosaurs. Brian completed an extraordinary amount of research on the plants featured in this scene, all of which can be identified right down to species level.

• Acrylic on paper and linen, animated digital composite | 2024

SQUID BAITBALL ▸

In the Western Interior Seaway of the Late Cretaceous, swarms of breeding bioluminescent squid attract numerous predators, including the swordfish-like *Protosphyraena*, the filter-feeding *Bonnerichthys* (centre left), long-necked elasmosaurids, the mosasaur *Dolichorhynchops* and (top) *Pentanogmius*, a ray-finned fish. This piece was part of a travelling exhibition called 'Savage Ancient Seas'.

Used courtesy of Triebold Paleontolgy Inc.

• Graphite on paper, digital colours | 2019

▾ HELL CREEK UNENLAGIINE

In a Late Cretaceous swamp now preserved as part of the Hell Creek Formation of North Dakota, a family of an as-yet-undescribed unenlagiine species (unenlagiines are a subfamily within the dromaeosaurids) hide at the water's edge. The other animals and plants shown are from the same quarry as the unenlagiine, or from nearby. They include the big-leaved angiosperm *Gunnera*, a plant that still occurs in wetlands today. The animals include *T. rex*, the dromaeosaurid *Acheroraptor* and – under the water – a gar, paddlefish, bowfin and baenid turtle.

• Graphite on paper, digital colours | 2024

[Previous pages]

ALLOSAURS VS APATOSAURINE

Evidence shows that *Allosaurus* may well have attacked apatosaurines in interactions like the one shown here, depicting both disembowelment and use of the sauropod tail in self-defence. This scene, commissioned by the Museum of Western Colorado, shows fossils of the Mygatt-Moore Quarry, part of the Upper Jurassic Morrison Formation.

• Graphite on paper, digital colours | 2017

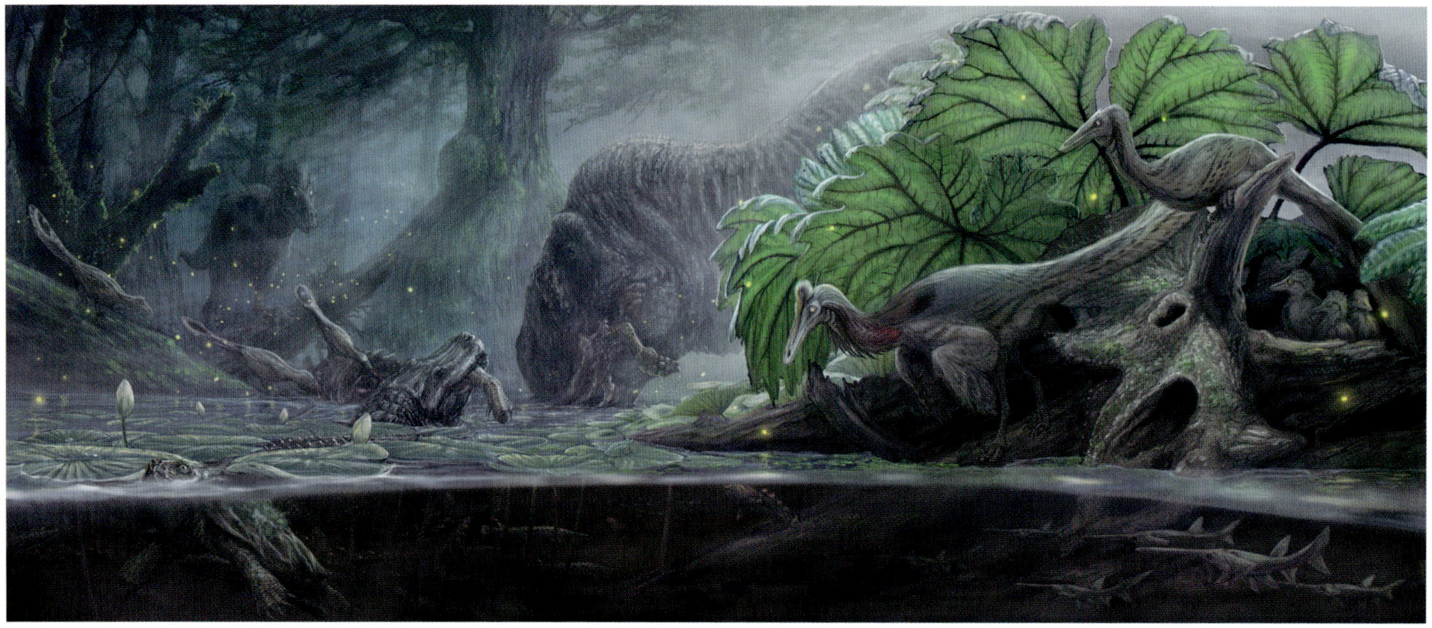

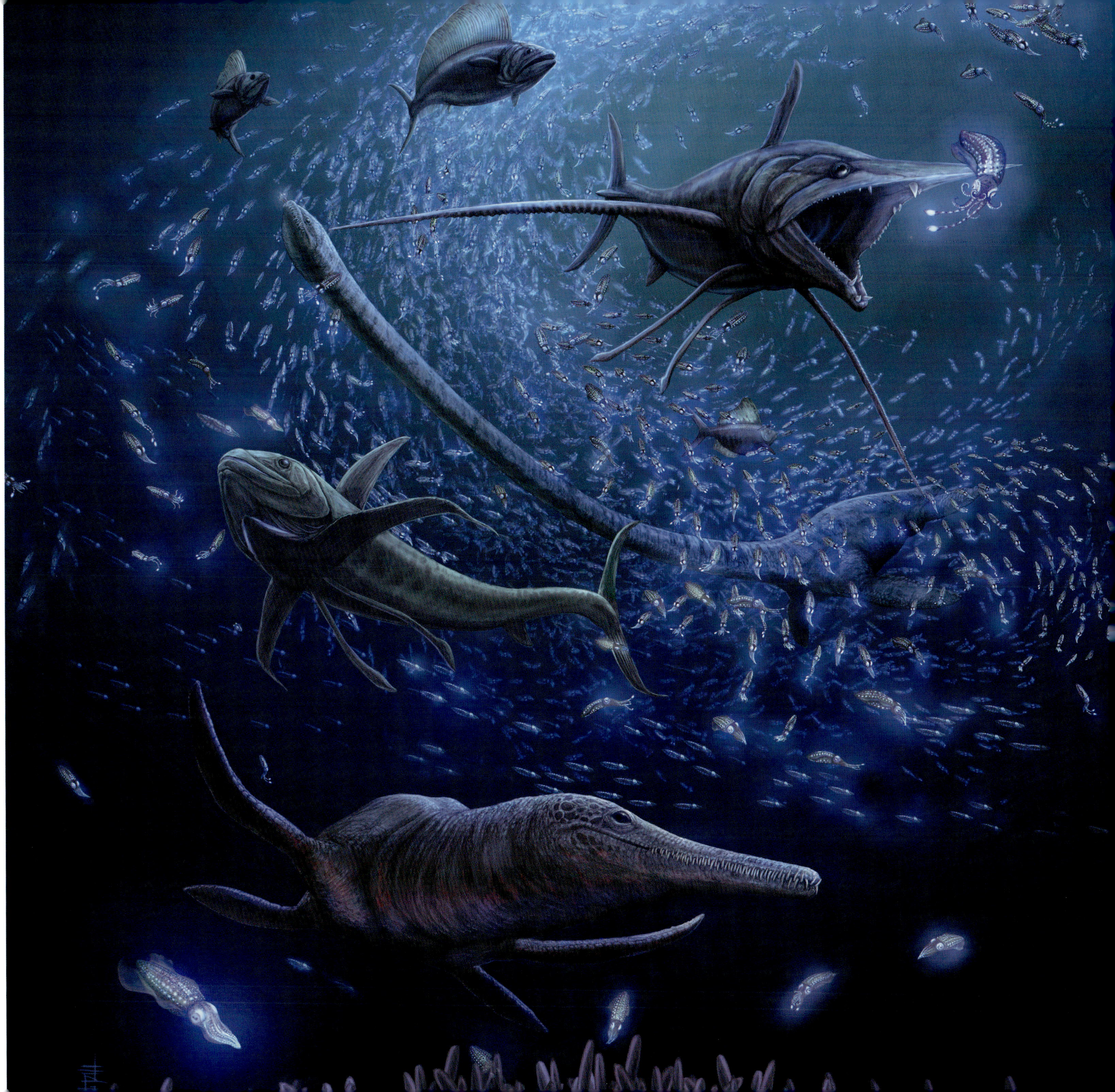

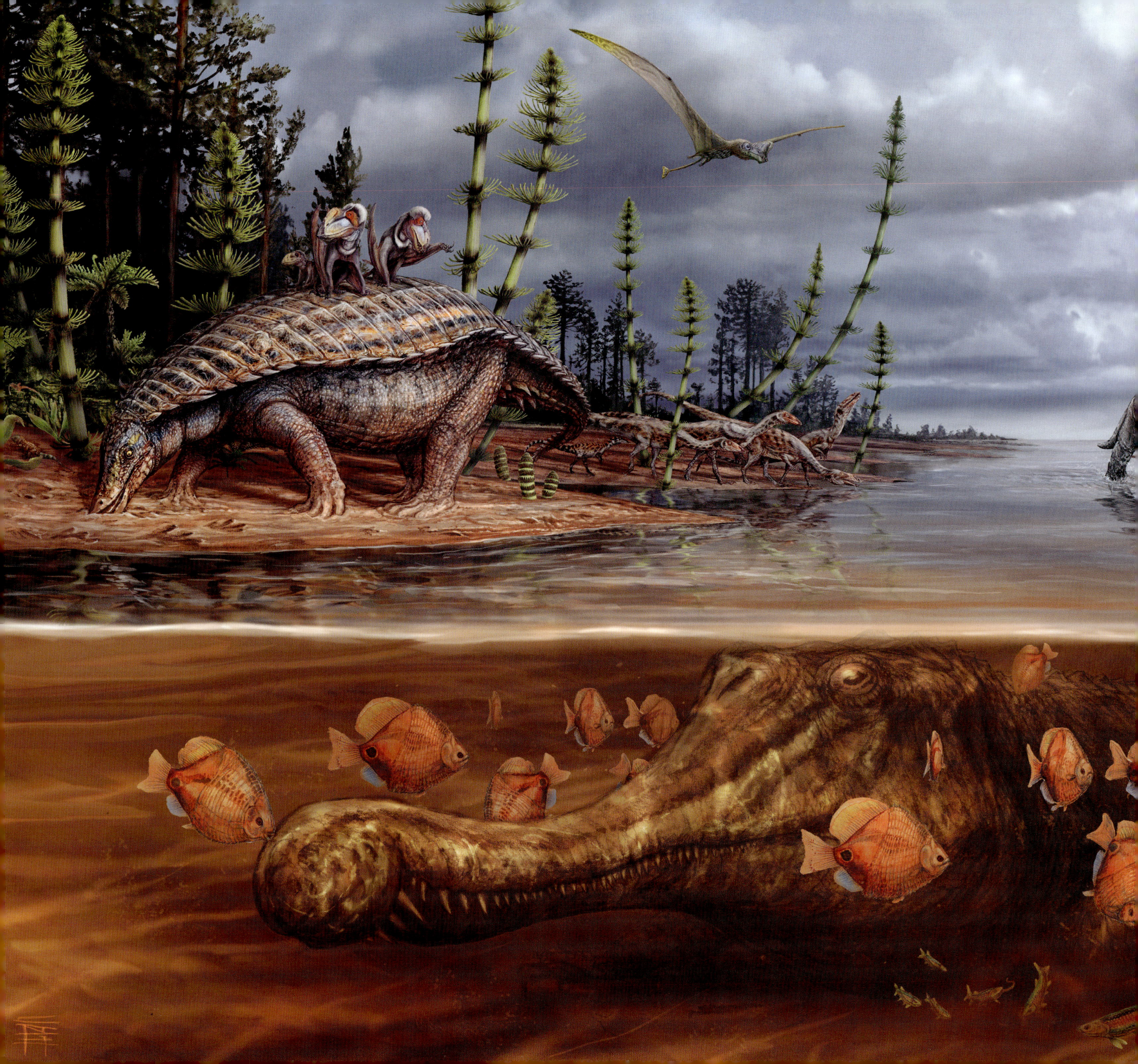

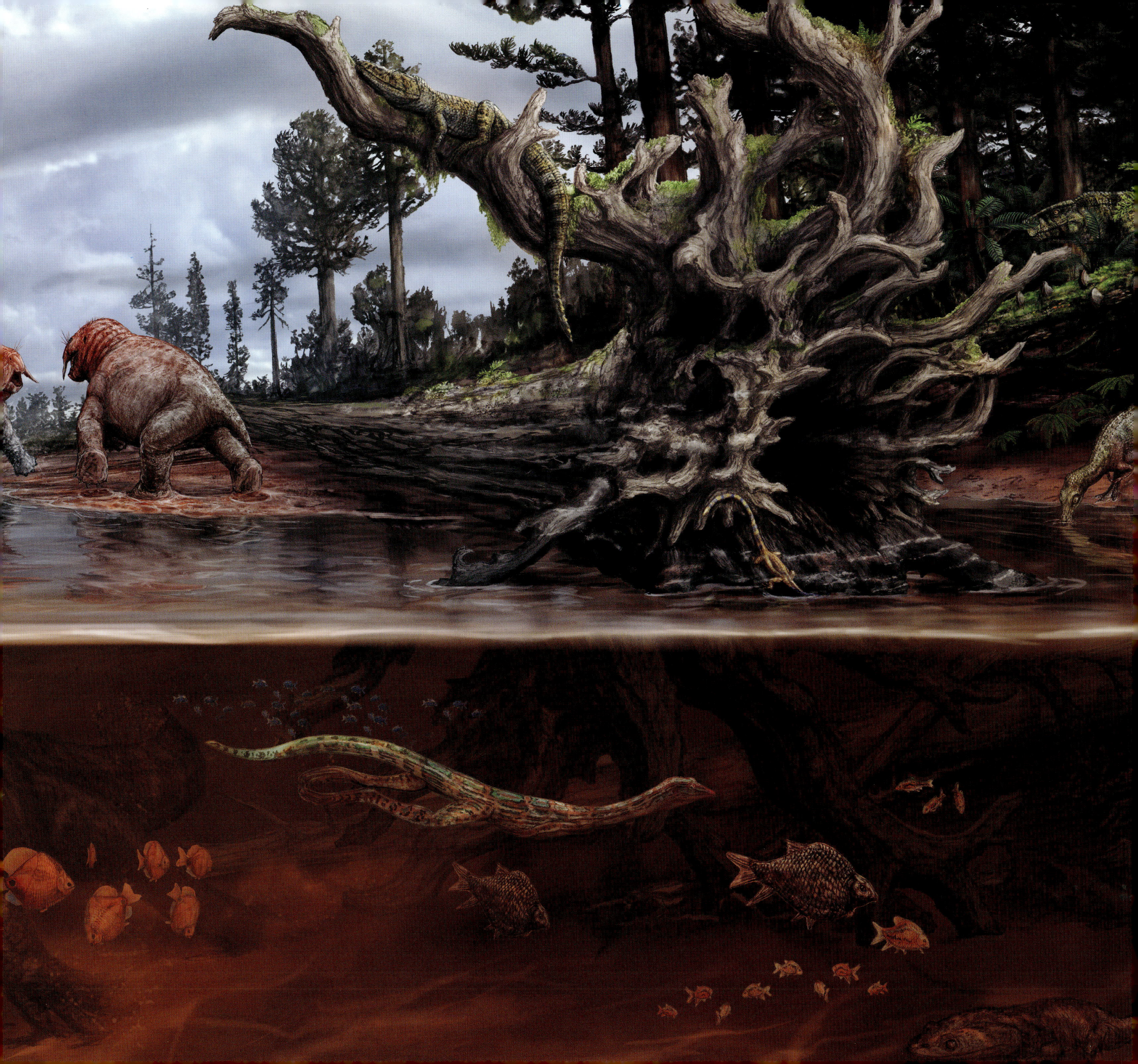

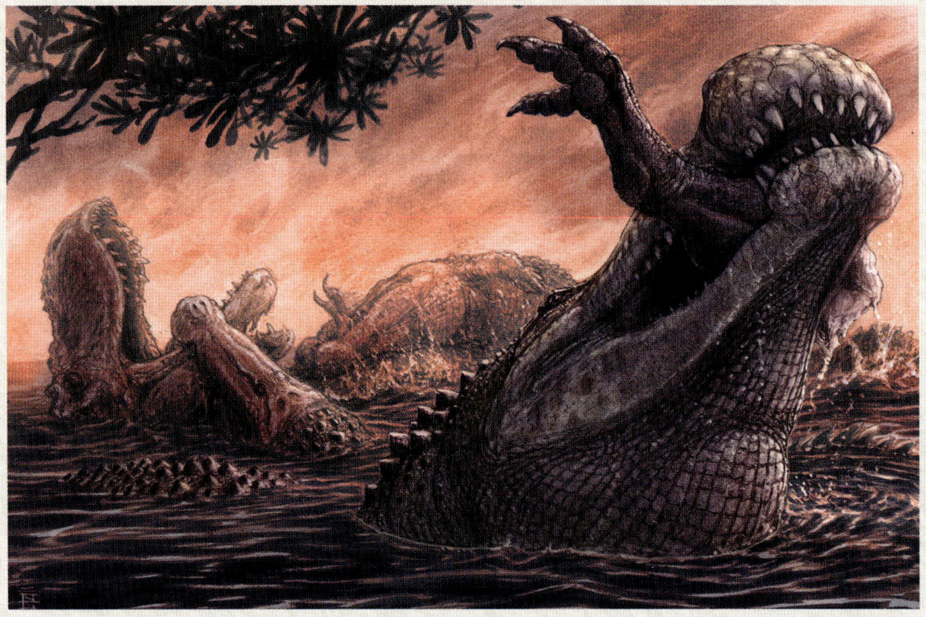

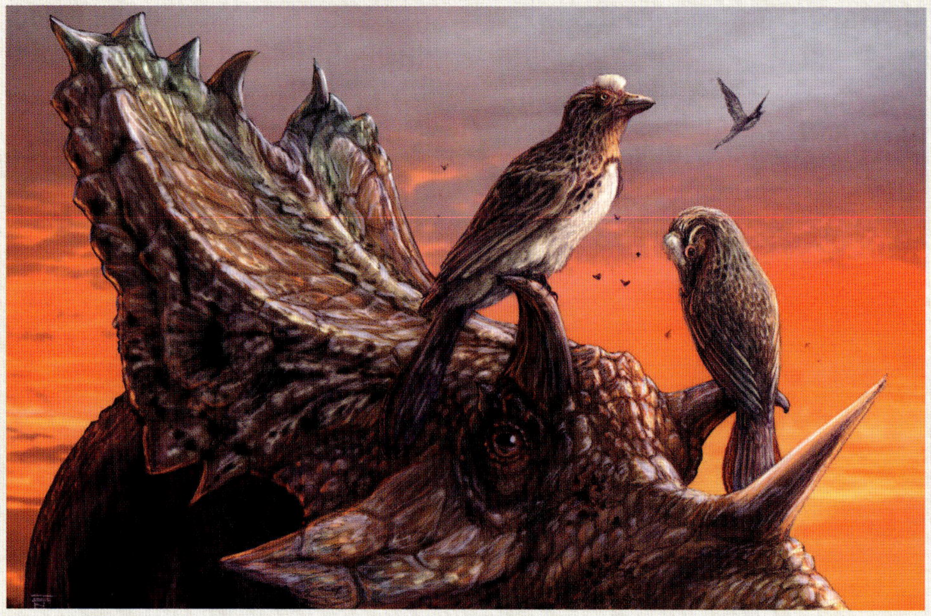

[Previous pages]

TRIASSIC RIVER SCENE

A Triassic riverside scene, showing animals of the Chinle Formation of Utah, 207–210 million years ago. On the left are an armoured aetosaur, pterosaurs and coelophysoids. In the water is a crocodile-like phytosaur, ray-finned fishes, long-necked tanysaurs and a temnospondyl amphibian. Reptiles on the right include a *Hesperosuchus*-like croc, a predatory postosuchid and a toothless, *Effigia*-like shuvosaurid.
• Graphite on paper, digital colours | 2018–2023

▲ *DEINOSUCHUS*

The giant crocodylian *Deinosuchus* lived on both sides of the Western Interior Seaway – the seaway that split North America in two – during the Late Cretaceous. Its size, broad snout and stout teeth indicate that it was probably well capable of grabbing and dismembering large dinosaurs like the tyrannosaur *Dynamoterror* shown here. This image was commissioned by the Western Science Center.
• Graphite, ink wash, digital colours | 2023

▲ *MIRARCE*

Mirarce, named in 2018, belongs to a group of Cretaceous birds termed avisaurids. With a wingspan of just over 1m (3ft), it was similar in size to a large hawk. Like hawks, avisaurids had powerful legs, and feet and probably hunted small animals. Like many birds today, avisaurids may have taken advantage of big animals – like this ceratopsian, *Utahceratops* – as mobile perches and hunting platforms.
• Graphite on paper, digital colours | 2023

SEITAAD ▶

Commissioned by Zion National Park, Utah, this shows a mother *Seitaad* and her youngsters. *Seitaad* is an Early Jurassic sauropodomorph, an archaic cousin of sauropods. Finds made in South Africa and Argentina show that the tiny juveniles of these dinosaurs differed in proportions from adults. In addition, the discovery of adults and juveniles together suggests that these dinosaurs cared for, and perhaps even fed, their babies.
• Acrylic on paper | 2022

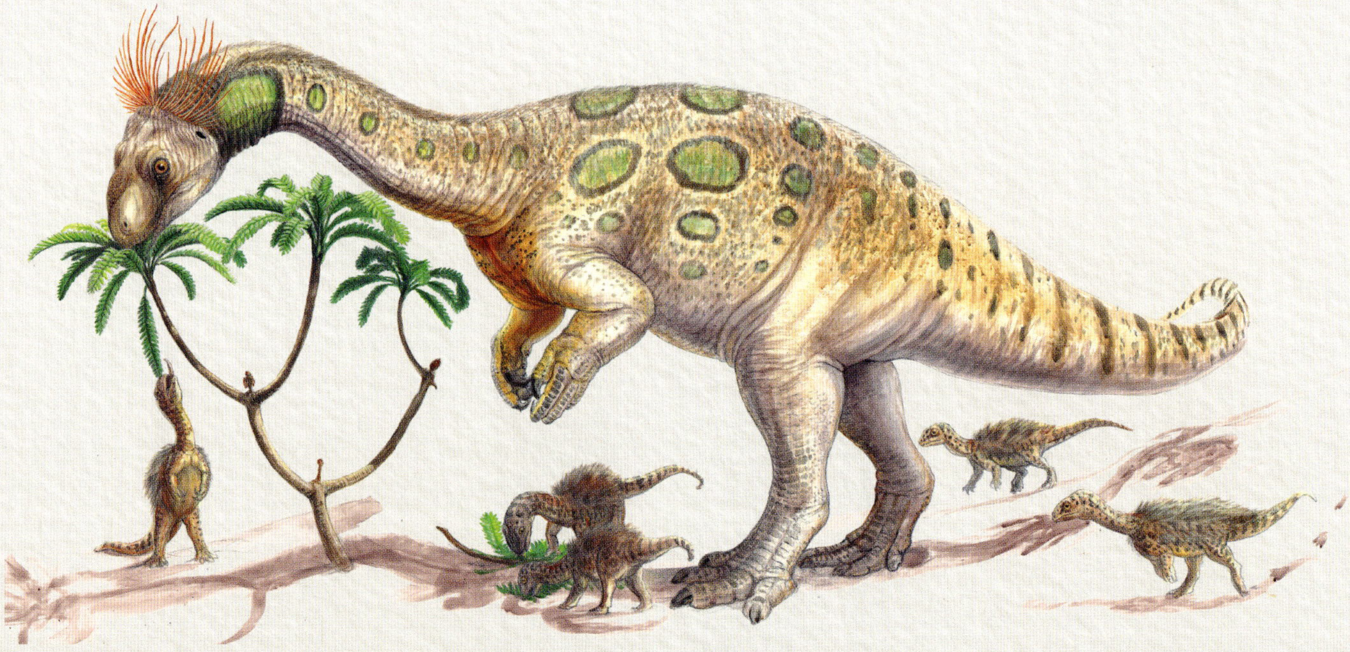

▼ TRITYLODONTID STRONGHOLD

This scene reconstructs possible events at 'Tritylodontid Cove', an Upper Triassic site at Lake Powell, Utah, discovered in 2023. It preserves two groups of tritylodontids, one in the Kayenta Formation and one in the overlying Navajo Sandstone, and it appears that both formed after the animals died in flooded burrows. Tritylodontids are mammal-like synapsids noted for their superficially rodent-like teeth. Both aggregations feature theropod and tritylodontid footprints, suggesting that these predators took advantage of the unlucky synapsids.

• Graphite on paper, digital colours | 2023

SARAHSAURUS ▶

A profile illustration of *Sarahsaurus*, an Early Jurassic sauropodomorph from Arizona, USA, named in 2011. This reconstruction highlights several aspects of sauropodomorph anatomy that remain the topic of discussion and speculation. These include the form of the nasal covering and shape of the nostrils, the presence and distribution of beak-like tissue around the mouth, and the overall skin texture.

• Graphite on paper | 2018

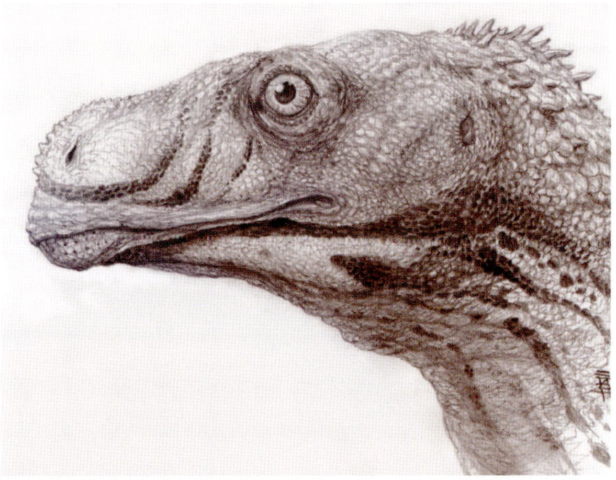

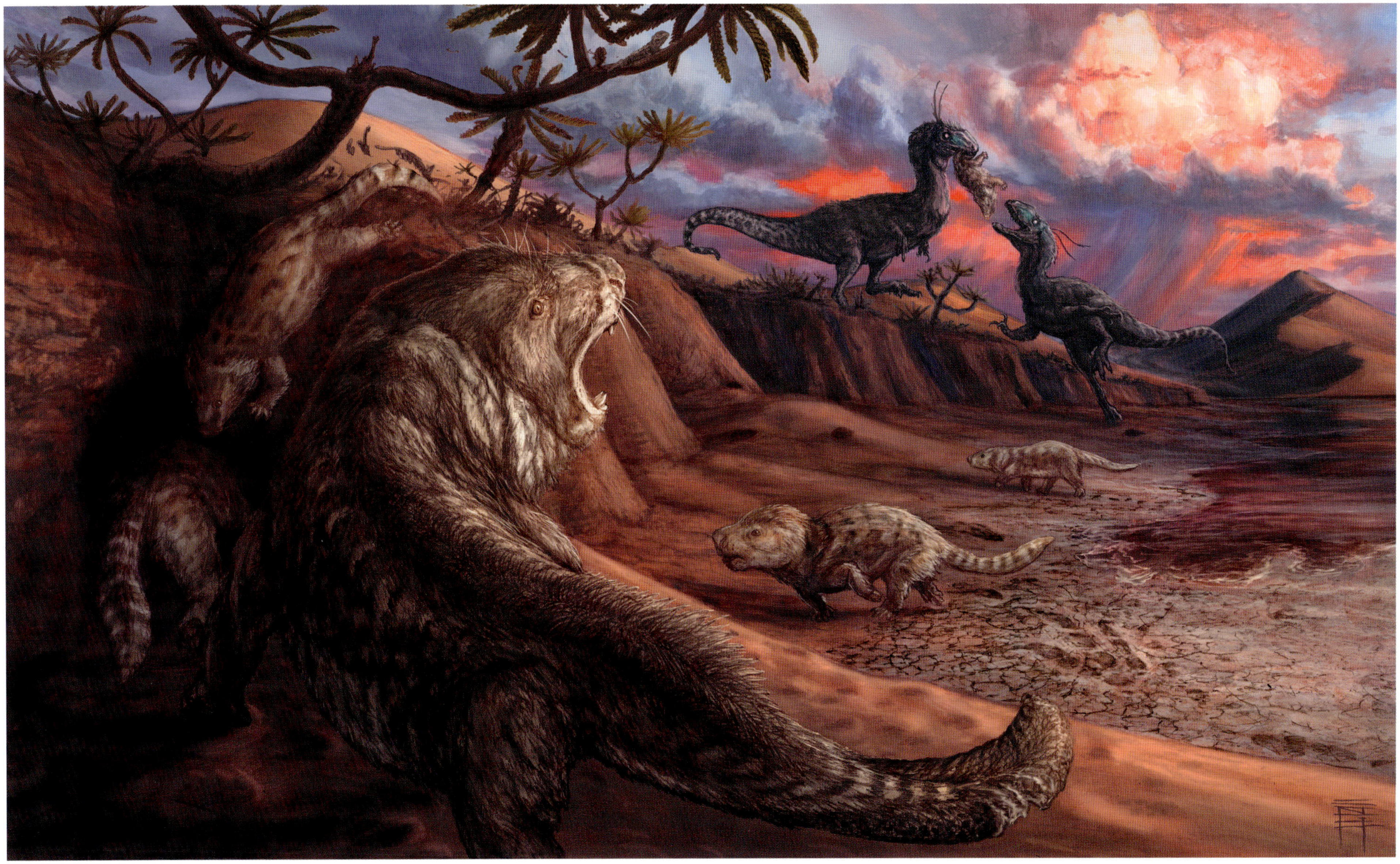

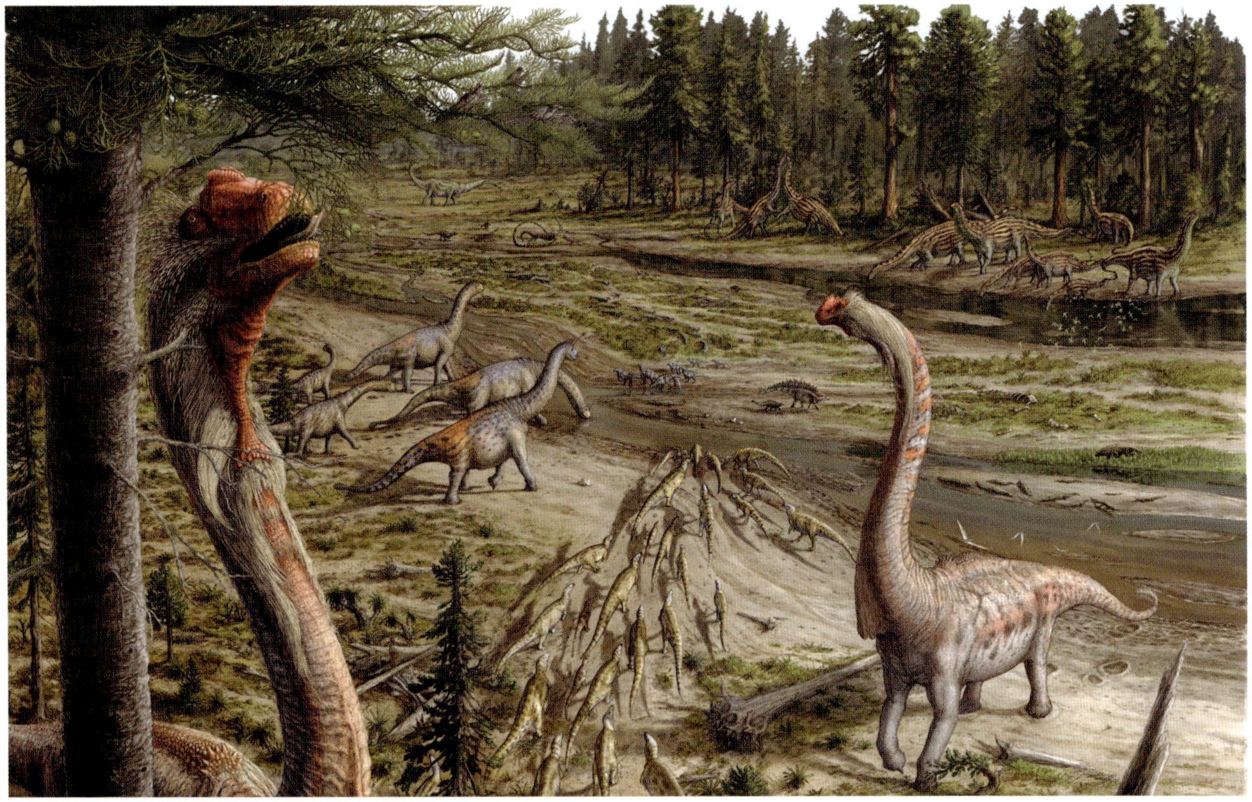

▲ MORRISON DROUGHT

Drought conditions in what is today Southwest Utah in the USA has caused these Morrison Formation dinosaurs to gather in a dry river valley. The scene might look crowded, but animals of this many species will gather together under such adverse conditions. To fully realise the lighting and perspective required for this scene, Brian arranged scale models on a modern lake shore.

• Graphite on paper, digital colours | 2016

MORRISON MORNING LIGHT ▶

This is another part of the Natural History Museum of the University of Zurich mural shown on p. 38. Again, the large dinosaur – *Stegosaurus* – here is dwarfed by the gigantic trees known from the same environment. These include representatives of still-living families such as the Araucariaceae, Podocarpaceae and Ginkgoales.

• Acrylic on paper and linen, animated digital composite | 2024

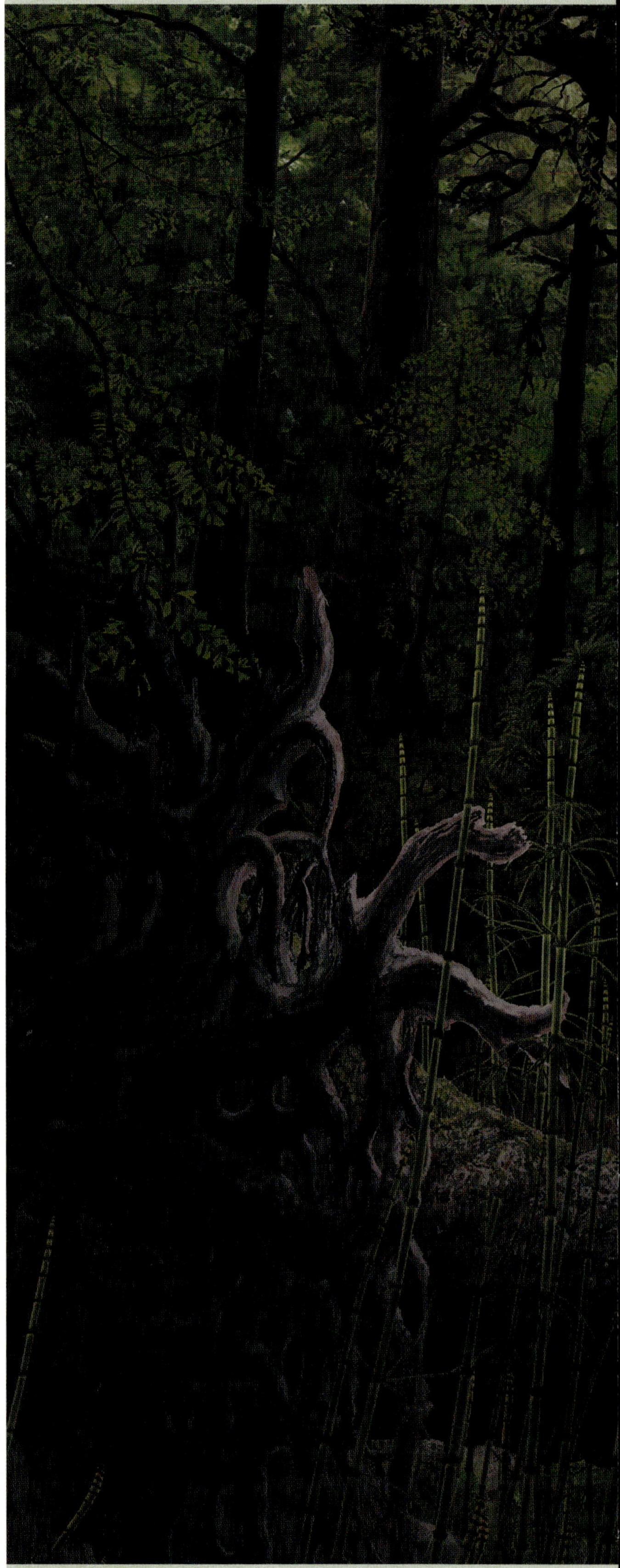

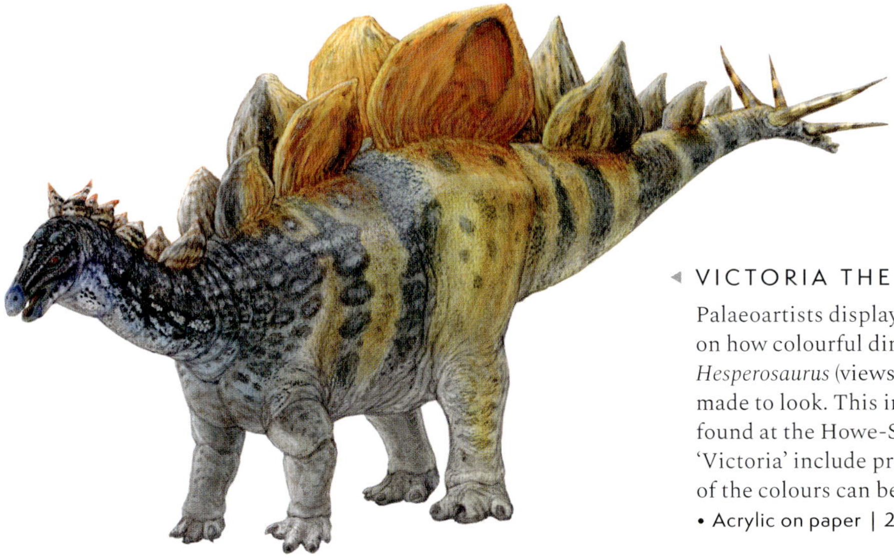

◀ VICTORIA THE STEGOSAUR

Palaeoartists display marked variation in their views on how colourful dinosaurs, like this *Stegosaurus* or *Hesperosaurus* (views differ on its taxonomy), should be made to look. This image depicts 'Victoria', a stegosaur found at the Howe-Stephens Quarry. The remains of 'Victoria' include preserved skin but sadly no hints of the colours can be discerned.

• Acrylic on paper | 2024

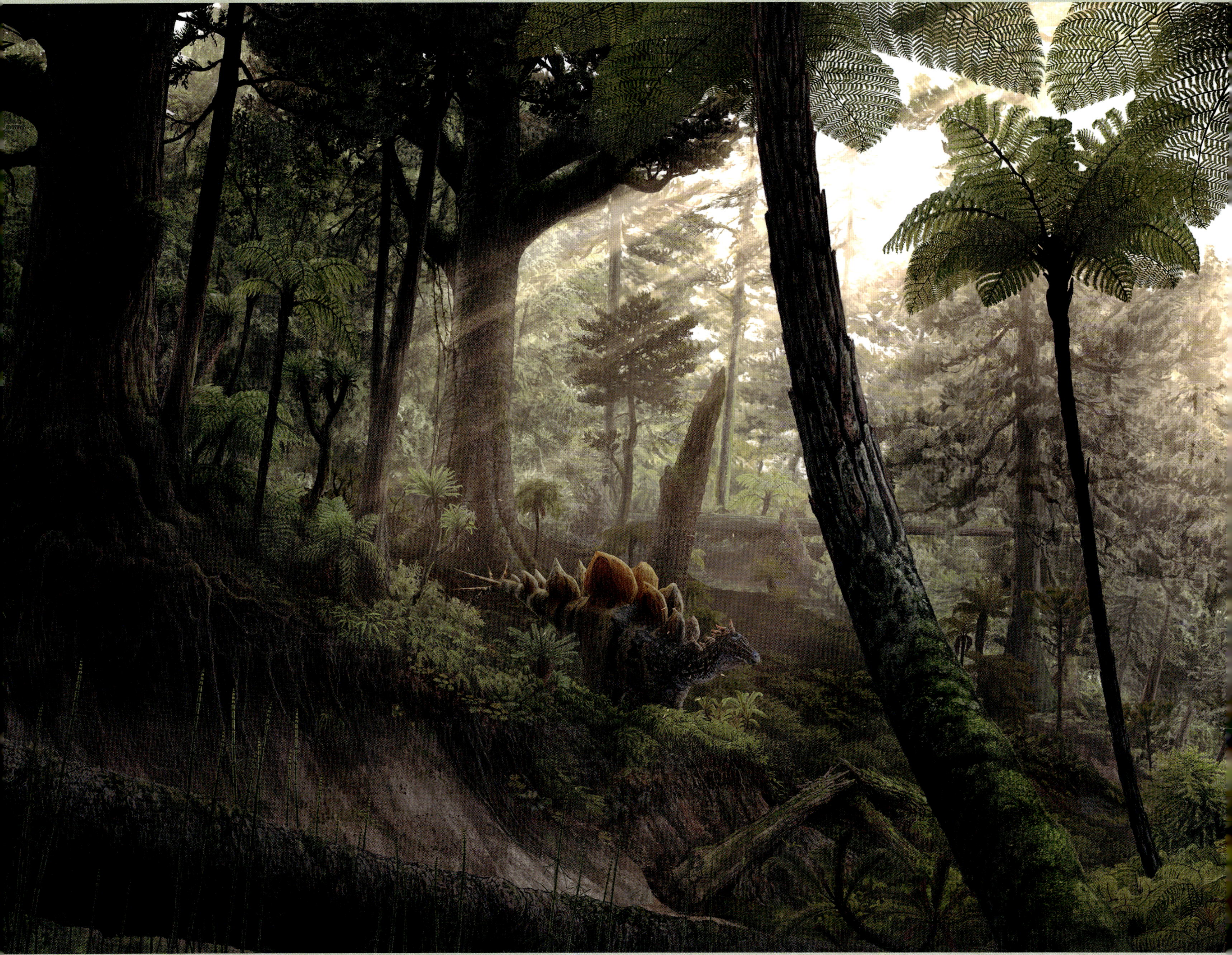

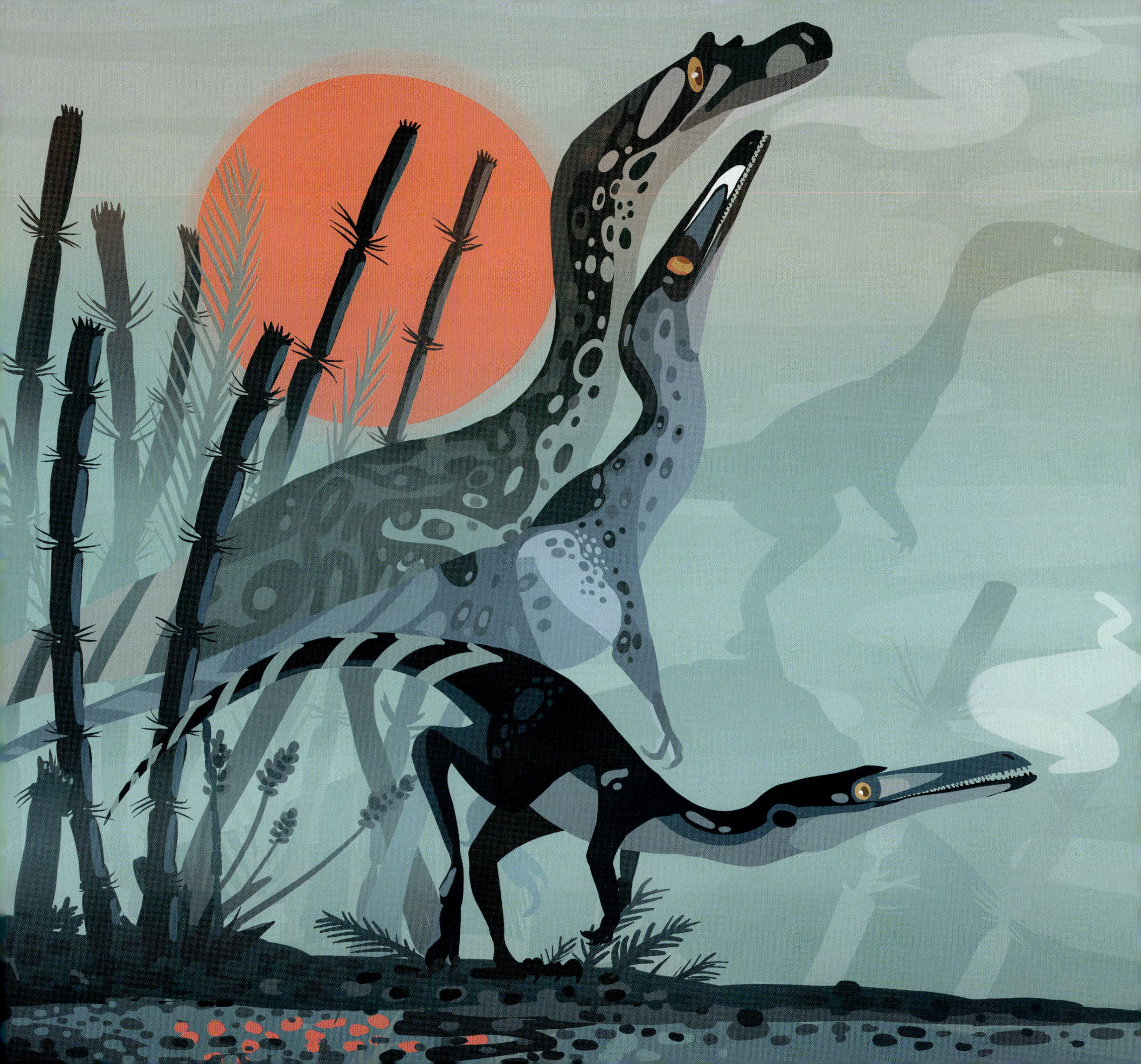

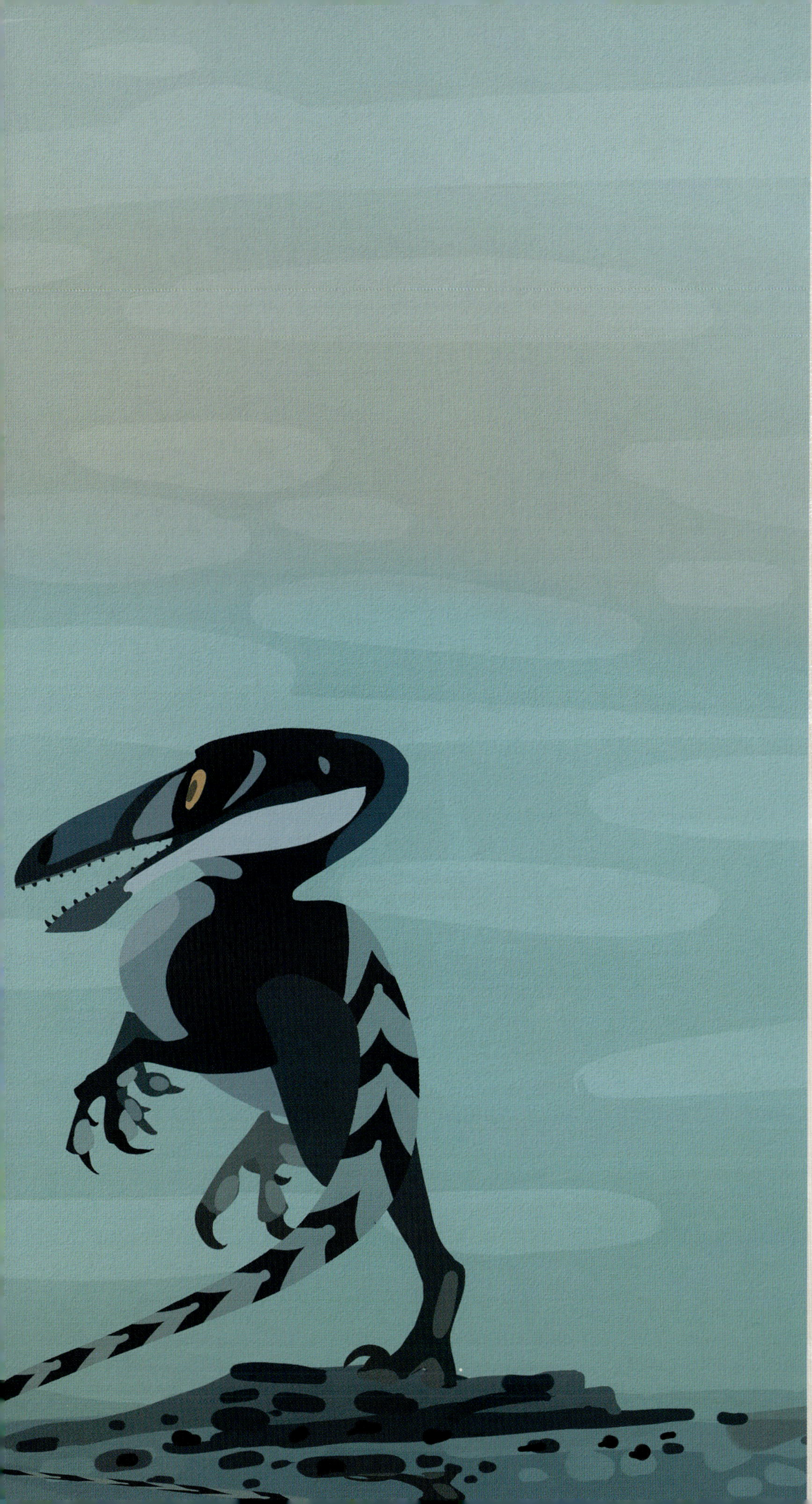

NATALIA JAGIELSKA

JURASSIC PARK III, while not revered by the wider palaeontological community, was the gateway drug to palaeontology for Natalia Jagielska, a Polish-born palaeontologist and artist.

Surprisingly, Natalia's childhood was more geared towards an interest in big cats, her tween years spent creating fan art of *Lion King* characters of questionable quality. Her career and artistic trajectory changed after completing a degree in Geology at the University of Manchester and being awarded a doctorate in palaeontology from the University of Edinburgh for her study of *Dearc sgiathanach*, the first named Scottish pterosaur. Natalia held the position of collections curator at the Lyme Regis Museum, the birthplace of Mary Anning, and, arguably, the field of palaeontology itself.

In addition to her pursuit of academia, Natalia has illustrated children's books and journal articles. Her works have been featured in museum exhibits and art galleries worldwide, from the Polish Geological Institute in Warsaw to the *Rediscovering Dinosaurs* exhibition at the Tianjin Science and Technology Museum, China and the *Nihonga to Kyōryū* exhibition in Tokyo, Japan.

Since 2025, she has lived in China, researching pterosaur flight membranes at the Chinese University of Hong Kong.

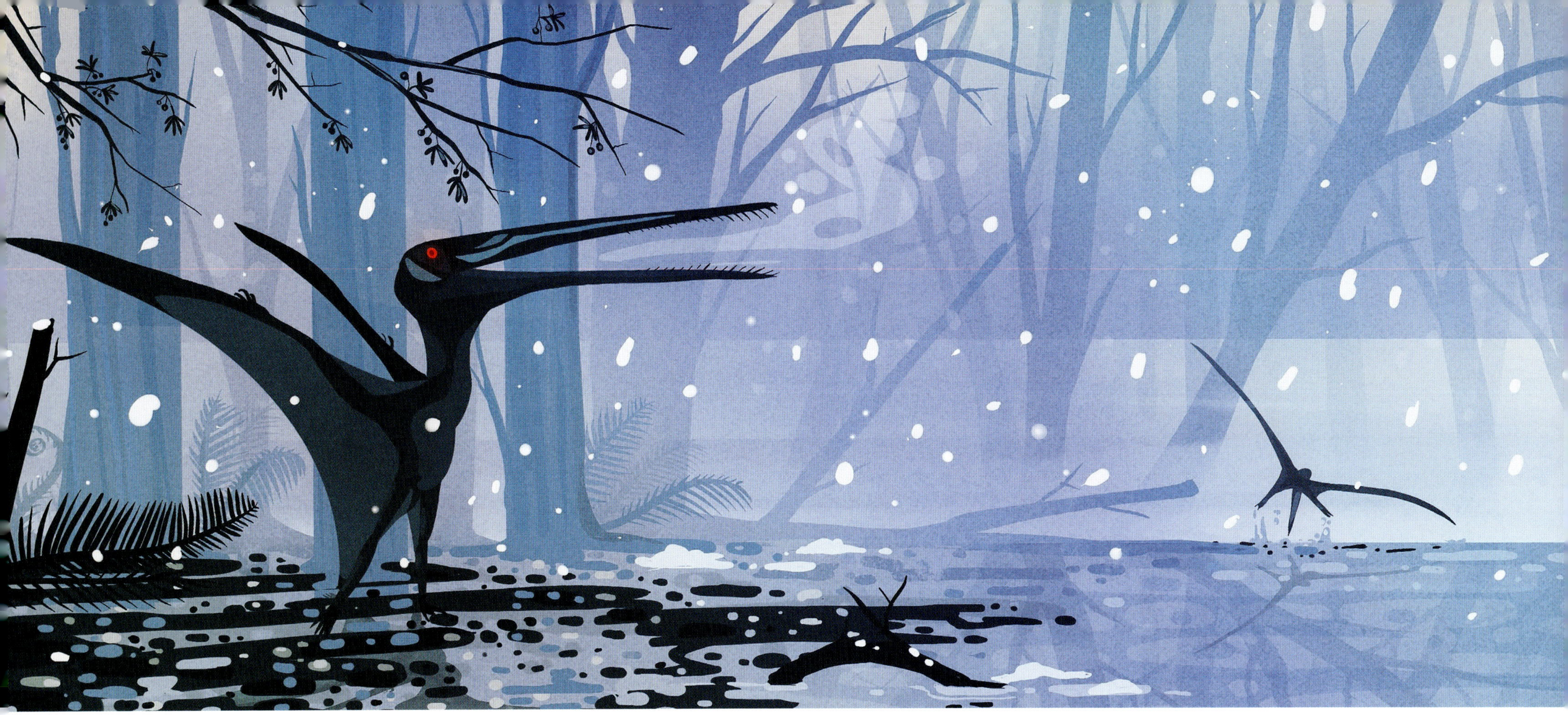

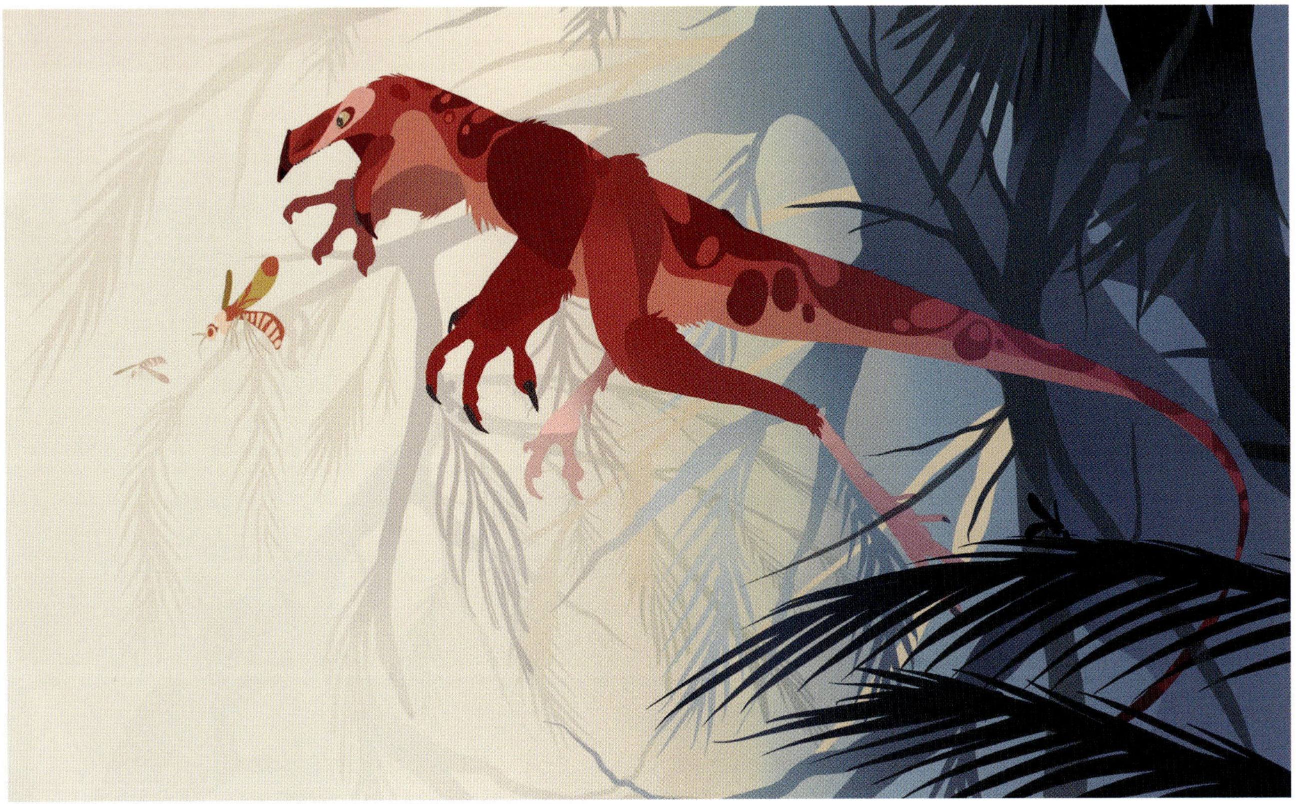

[Previous pages]

GHOST RANCH

The discovery of hundreds of specimens of the Early Jurassic *Coelophysis* at Ghost Ranch, New Mexico, invites many questions. *Why* so many of these animals were gathered together remains the main one. Artists have depicted these aggregations many times. This novel image shows the animals in a damp, misty setting, the shades of the cool morning contrasting with the red of the rising sun.
• Digital | 2023

◂ LEAP OF FAITH

Pterosaur origins are famously enigmatic, but recent studies show that the lagerpetids – a Triassic archosaur group known from South America, the USA and Madagascar – are probably their closest cousins. *Venetoraptor* from Brazil, named in 2023, is an unusual lagerpetid with a toothless, hooked upper jaw and strongly curved hand claws. It might have used these claws when climbing in pursuit of treetop prey.
• Digital | 2023

◀ CRETACEOUS FOREST

A Chinese pterosaur of the Early Cretaceous – *Gladocephaloideus* – exhales and stretches its wings before taking flight. Geochemical evidence shows that the average temperature of the region during this time was similar to that of modern-day England, so snow was certainly possible during the cool winter. Animals like pterosaurs must have dealt with cold weather.

• Digital | 2024

THEN THEY FELL SILENT ▶

During a night at the very end of the Cretaceous, a *Dakotaraptor* watches an incandescent fireball tear through the atmosphere. This evocative image is inspired by the composition and colours used in the wartime artwork of Japanese ukiyo-e artist Kobayashi Kiyochika. His work is famed for its red-blue contrasts and use of atmospheric lights and reflections.

• Digital | 2021

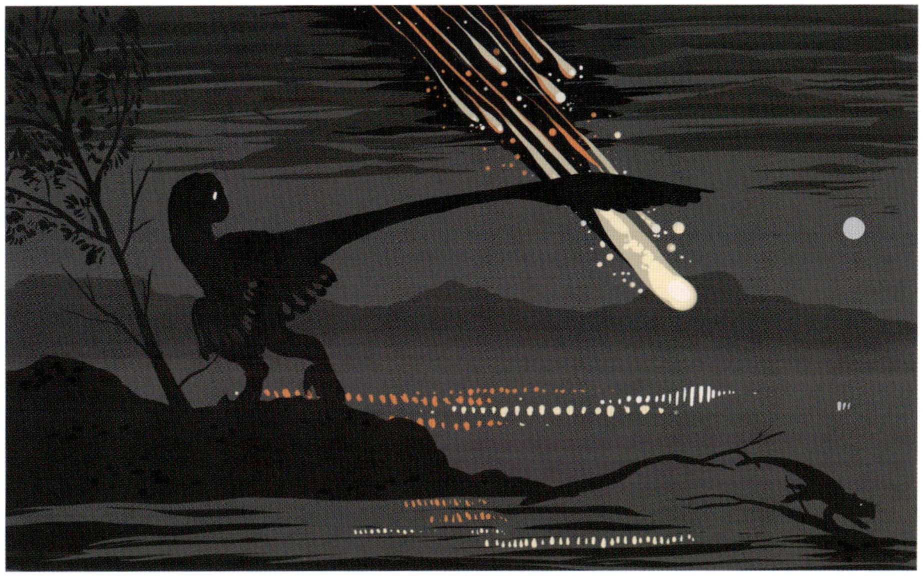

▼ APOCALYPSE 66

It's depressing and sobering to imagine what those animals alive during the great cataclysm at the very end of the Cretaceous saw and experienced. Animals living in those American forests affected by the heat released from the end-Cretaceous impact would have been unable to escape a world ablaze. This azhdarchid might be luckier than most, if it can escape the flames in time.

• Digital | 2021

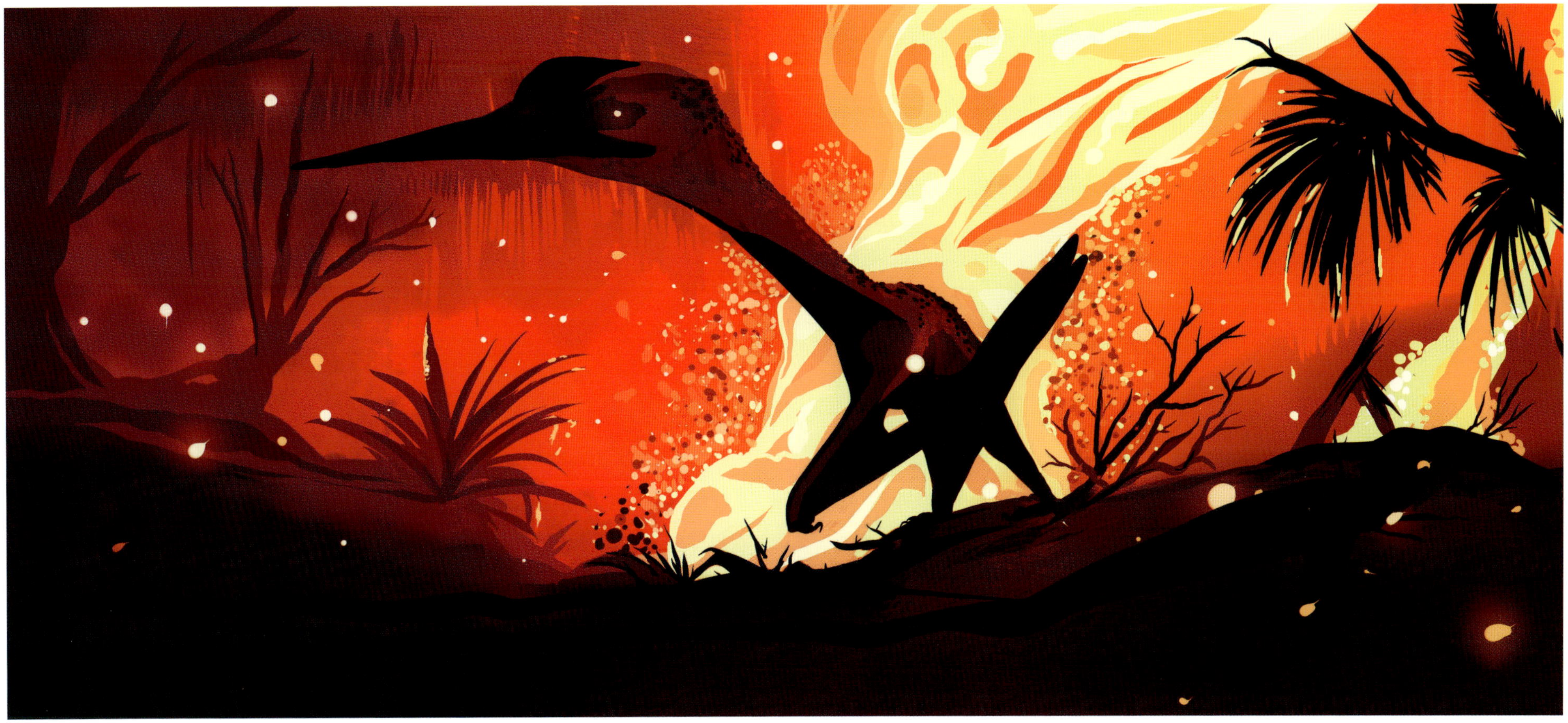

▼ *PTERODACTYLUS*

Pterodactylus, scientifically named in 1809, is the first pterosaur known to science and it might be assumed that there's little new to learn about it. Not so: recent studies have resulted in much discussion on its appearance (it had head crests and facial ornaments, but of what sort exactly?) and on how many species the known fossil specimens represent.
• Digital | 2021

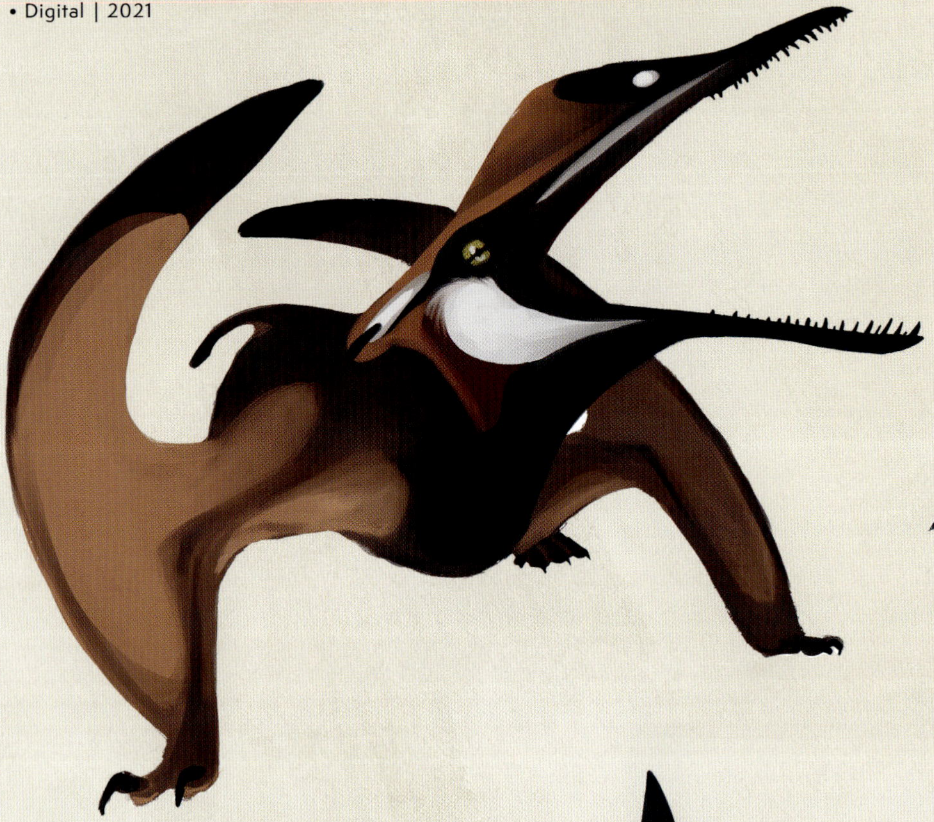

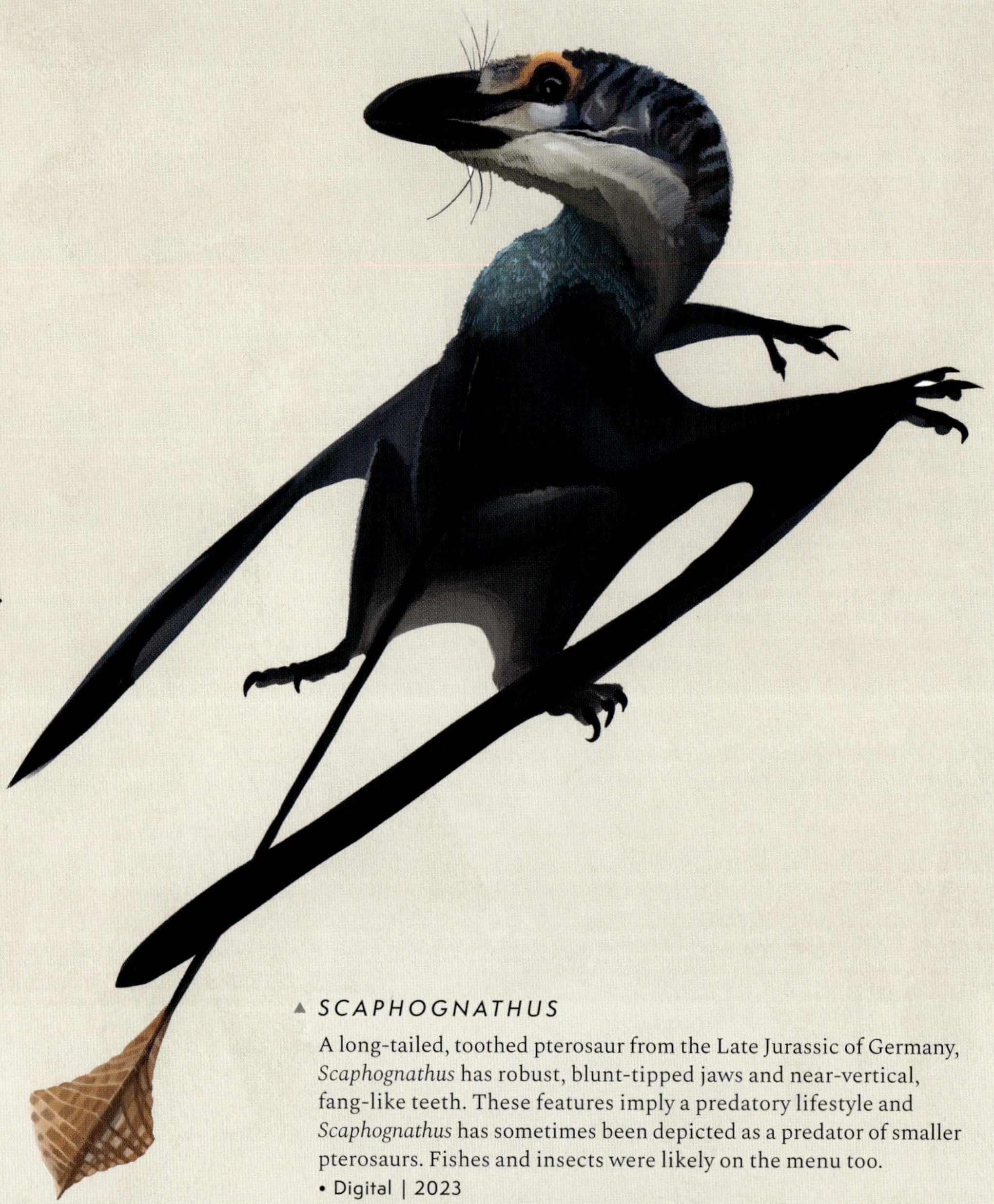

▲ *SCAPHOGNATHUS*

A long-tailed, toothed pterosaur from the Late Jurassic of Germany, *Scaphognathus* has robust, blunt-tipped jaws and near-vertical, fang-like teeth. These features imply a predatory lifestyle and *Scaphognathus* has sometimes been depicted as a predator of smaller pterosaurs. Fishes and insects were likely on the menu too.
• Digital | 2023

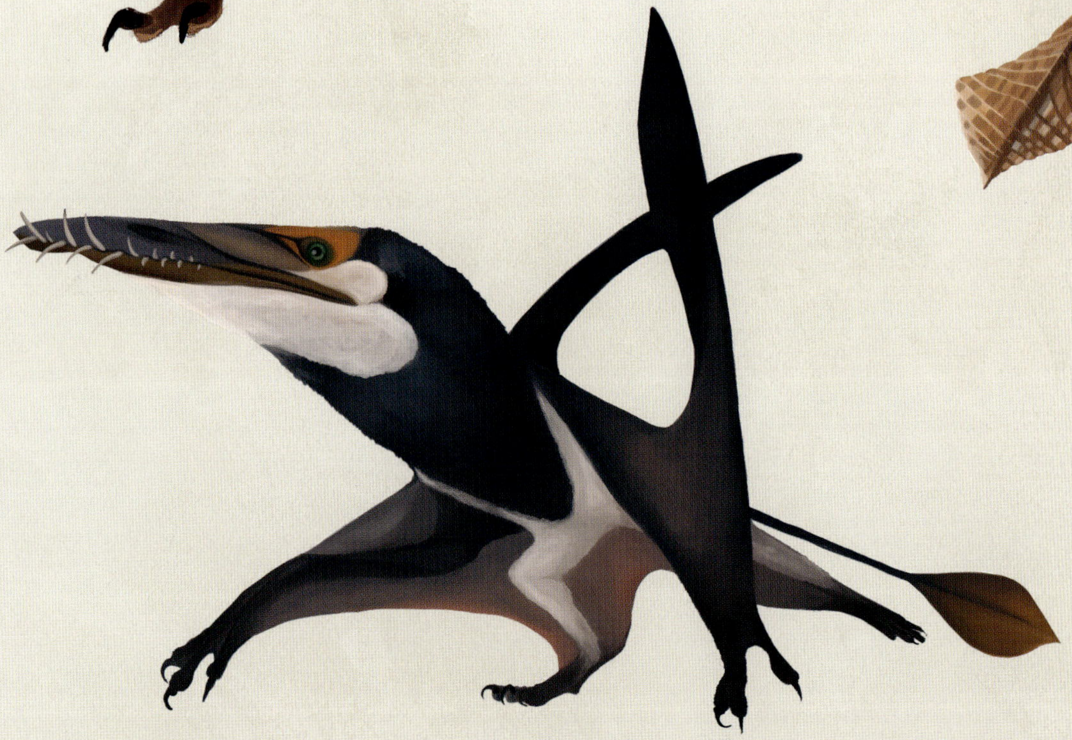

◀ *DEARC*

In 2023, Natalia and colleagues named *Dearc*, an archaic pterosaur from the Early Jurassic of Skye, Scotland. Study of this significant animals formed the basis of Natalia's PhD thesis. The pose depicted here is consistent with arguments that pterosaurs like this were capable climbers and walkers and could even have foraged on foot.
• Digital | 2022

MICRORAPTOR ▶

A dynamic pose, great composition and striking background are combined in this children's book illustration. The black, apparently iridescent plumage possessed by *Microraptor* is consistent with a mischievous, crow-like attitude that we're unlikely to ever see confirmed by the fossil record. However, stomach contents imply an opportunistic, adaptable lifestyle.
• Digital | 2023

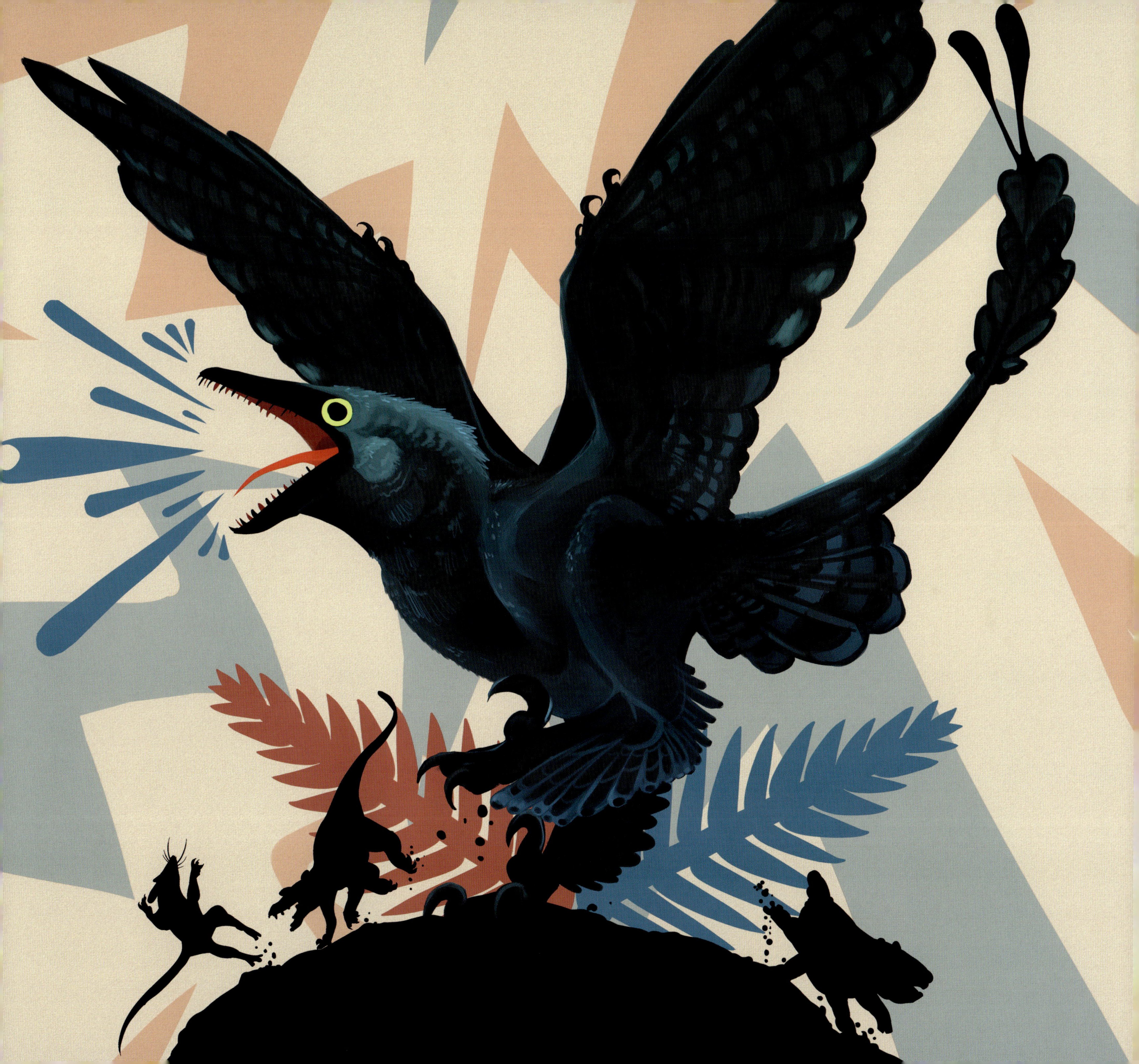

RHAMPHORHYNCHUS ▸

Natalia was one of a group of researchers that studied the tail vane of *Rhamphorhynchus*, the well-known Late Jurassic pterosaur from the famous Solnhofen limestones of Germany (possible remains are also known from England, Spain and Tanzania). Usually shown as diamond-shaped, the tail vane varied in shape, likely reflecting different species and life stages. Studies show a lattice of fibres kept the vane stiff, keeping it stable in flight.
• Digital | 2024

▾ *HAOPTERUS* SUNSET

Known from crushed remains found in China's Early Cretaceous Yixian Formation, *Haopterus* seems to have been a fisheater, as suggested by its sharp, backward-curved teeth. The narrow shape of its hindfeet indicate it probably moved on all fours, bat-style, as shown here.
• Digital | 2024

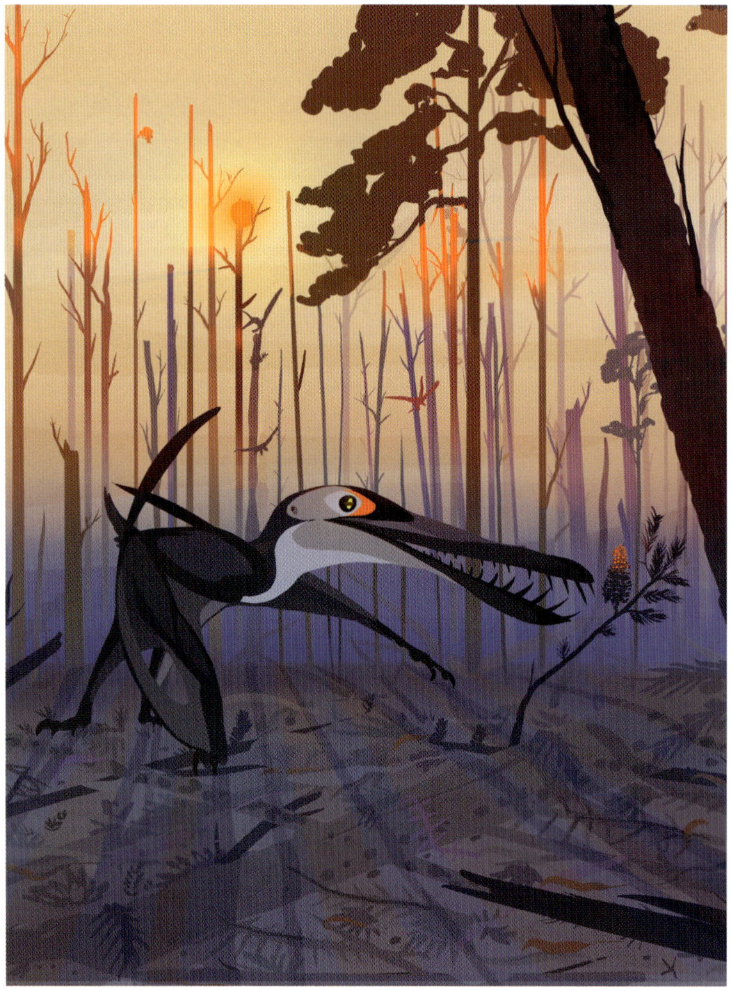

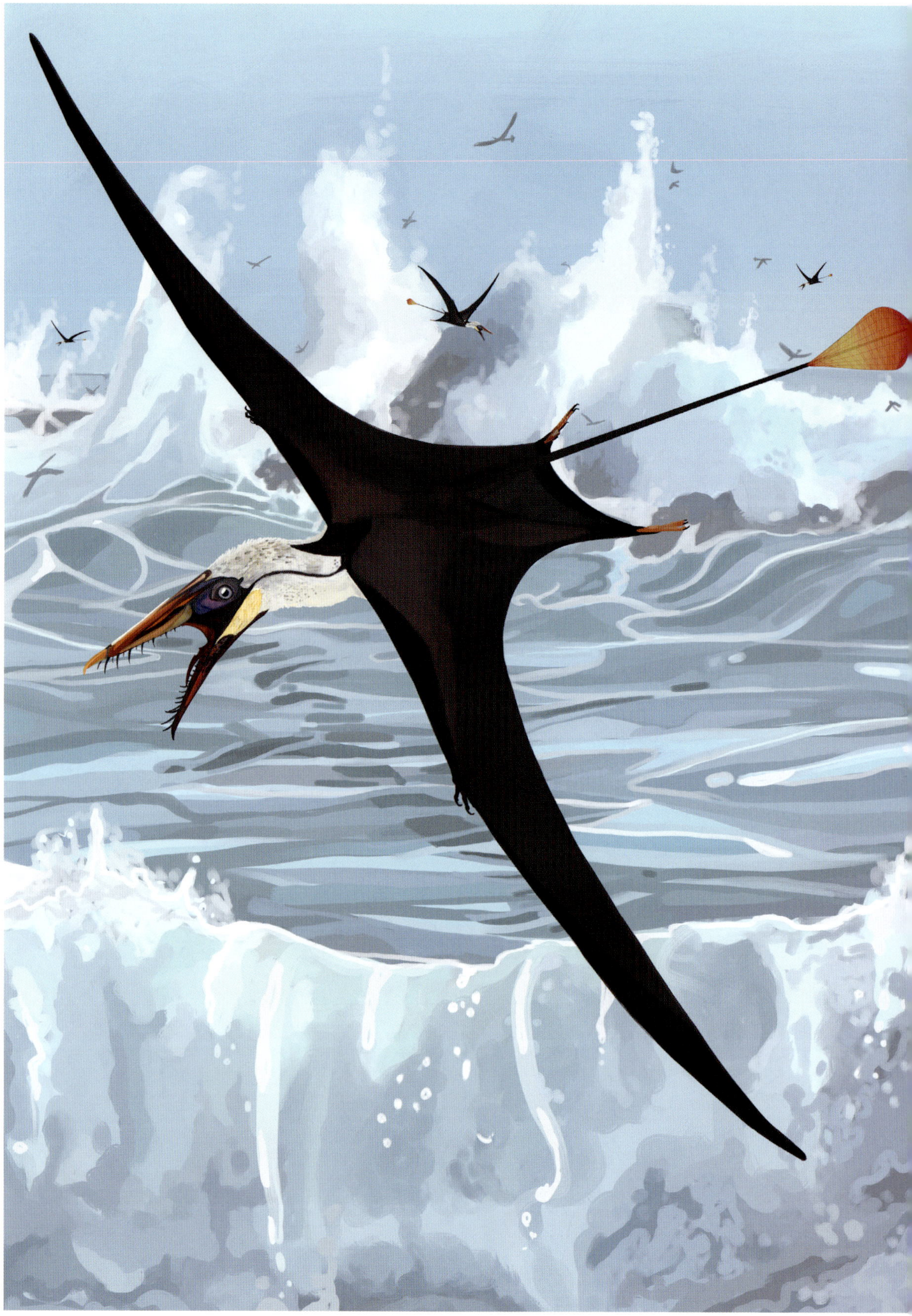

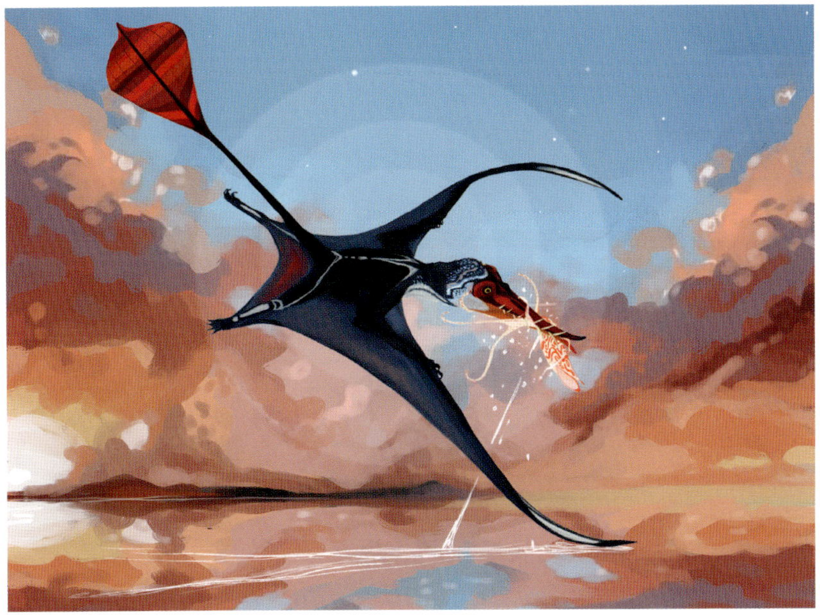

▲ *RHAMPHORHYNCHUS* & SEAFOOD

Rhamphorhynchus is known from dozens of species, including several dating back to Victorian times. The wingspan ranged from 1.8m (5.9ft) in large adults to just 290mm (11.4in) in small juveniles, termed 'flaplings'. *Rhamphorhynchus* likely preyed on fish and cephalopods, its eyes suggesting that hunting might have happened at night.

• Digital | 2024

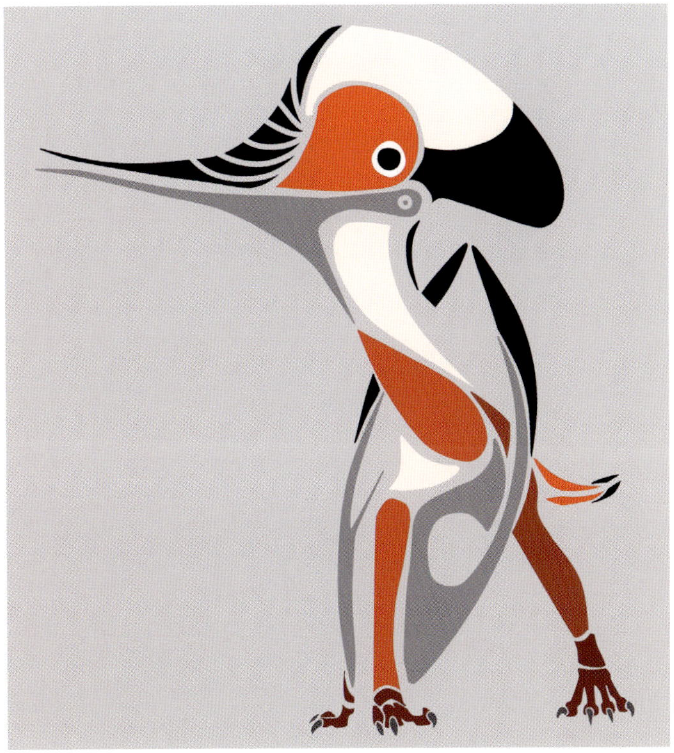

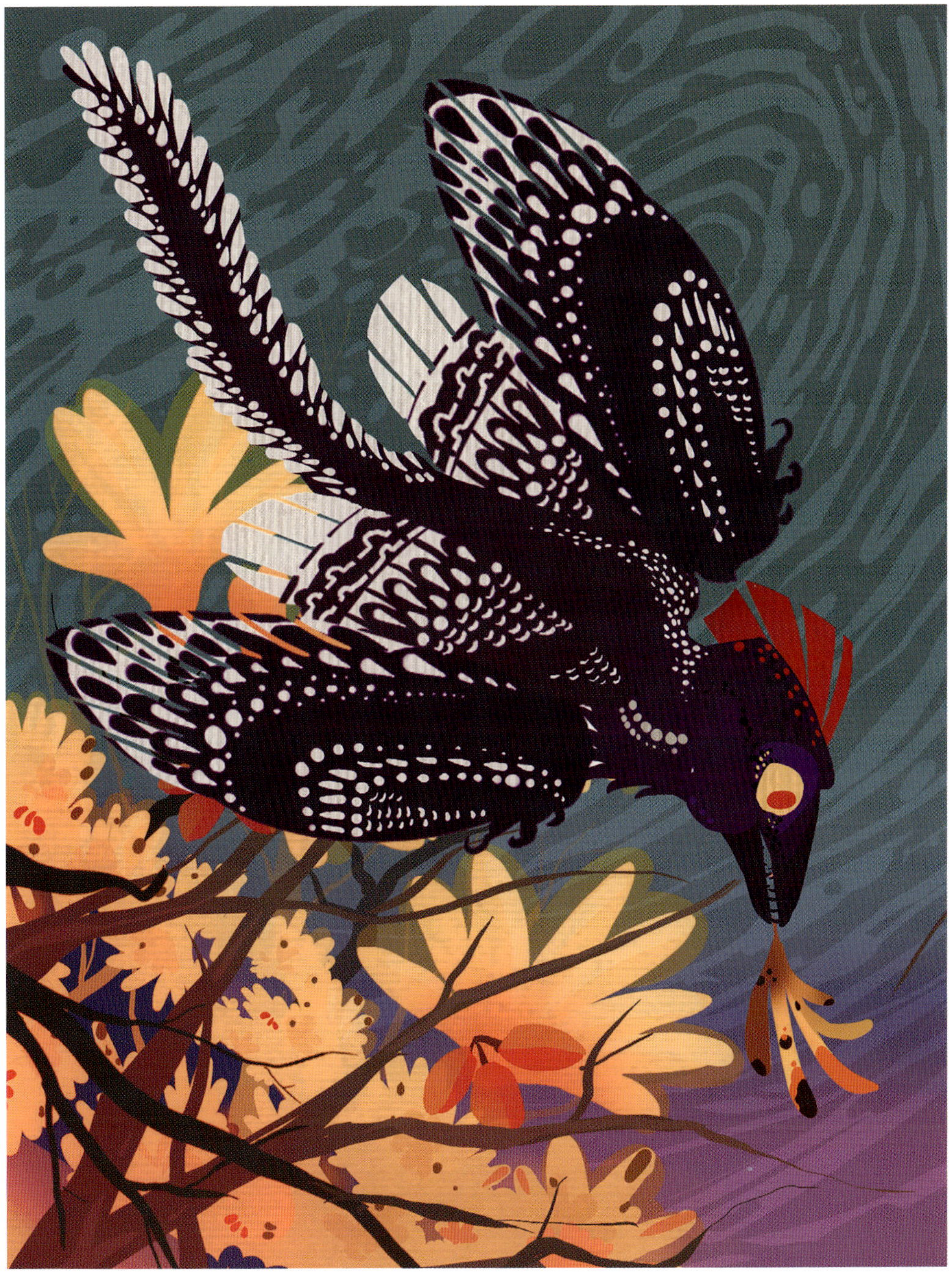

◀ *MEILIFEILONG*

A new Chinese pterosaur named in 2023, *Meilifeilong* ('beautiful dragon') is a slender-jawed, toothless animal with a 2m (6ft) wingspan. It's among the best understood of its group – the chaoyangopterids – and resembles the others in possessing a strongly convex upper part of its skull that sweeps over the eyes and terminates with a backward-pointing spike.

• Digital | 2024

▲ THE LEAF COLLECTOR

Many extinct dinosaurs built nests on the ground, in trees and in burrows, and some species *must* have incorporated leaves and other kinds of vegetation in their nests. The dark and white plumage and red feathery head crest show that this image portrays the Jurassic Chinese maniraptoran *Anchiornis*.

• Digital | 2023

▼ *MEGALOSAURUS*

The archetypal non-bird theropod, Middle Jurassic England's *Megalosaurus* is relatively under-illustrated, despite its fame. Scientifically named in 1824, it has gradually become better known but remains enigmatic. Argument remains over how similar it was to related forms like *Torvosaurus* of the USA and Portugal. The swirling, patchwork look here emphasises that art depicting dinosaurs need not always lean into uber-realism.
• Digital | 2023

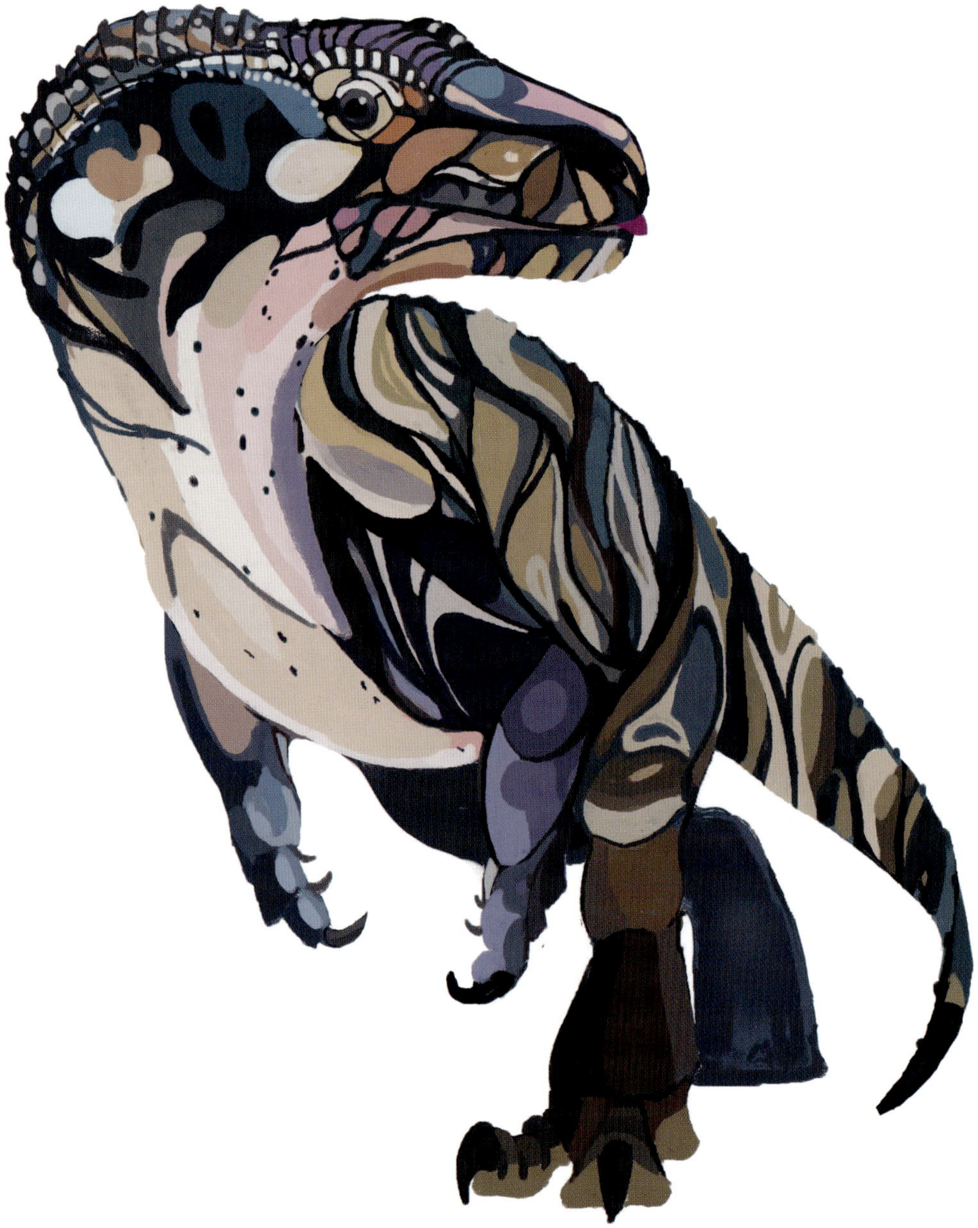

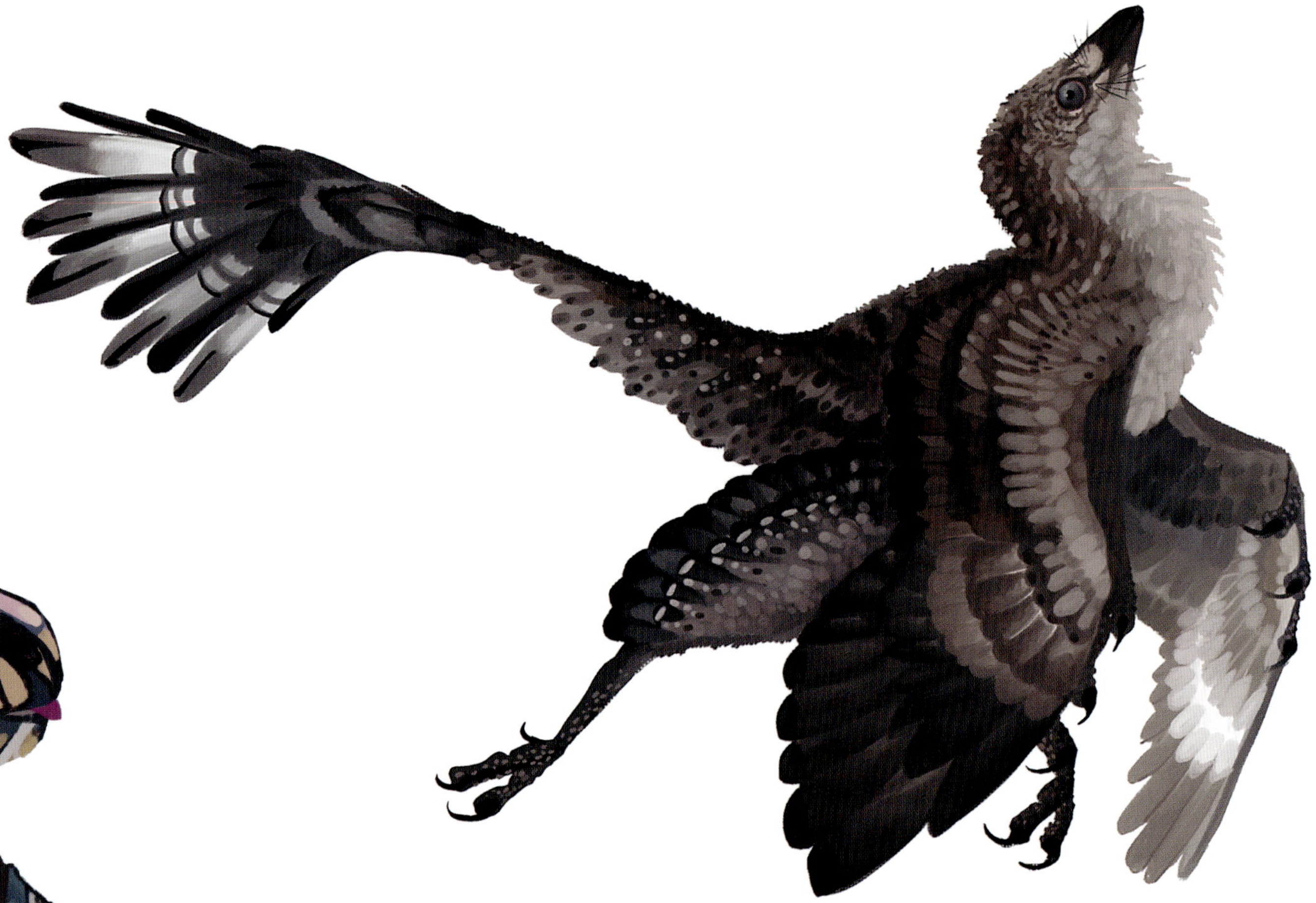

▲ GREY MANIRAPTORAN

This image depicts an as-yet-undescribed Chinese maniraptoran that's probably a relative of *Anchiornis*. Several Jurassic maniraptorans are known to be part of, or allied to, the bird lineage. Some well-preserved specimens allow the accurate reconstruction of their plumage, even down to their colours and patterning. It's worth remembering that colours can vary between genders and juveniles, even between same-aged individuals.
• Digital | 2022

DEARC ►

Dearc has proved to be the gift that just keeps on giving. Originally published in 2022 as a member of the same group as *Rhamphorhynchus* (and thus not much relevant to the origin of the short-tailed pterosaur group termed Pterodactyloidea), a 2024 study found it to be one of several animals that bridges the gap between *Rhamphorhynchus* and pterodactyloids.
• Digital | 2022

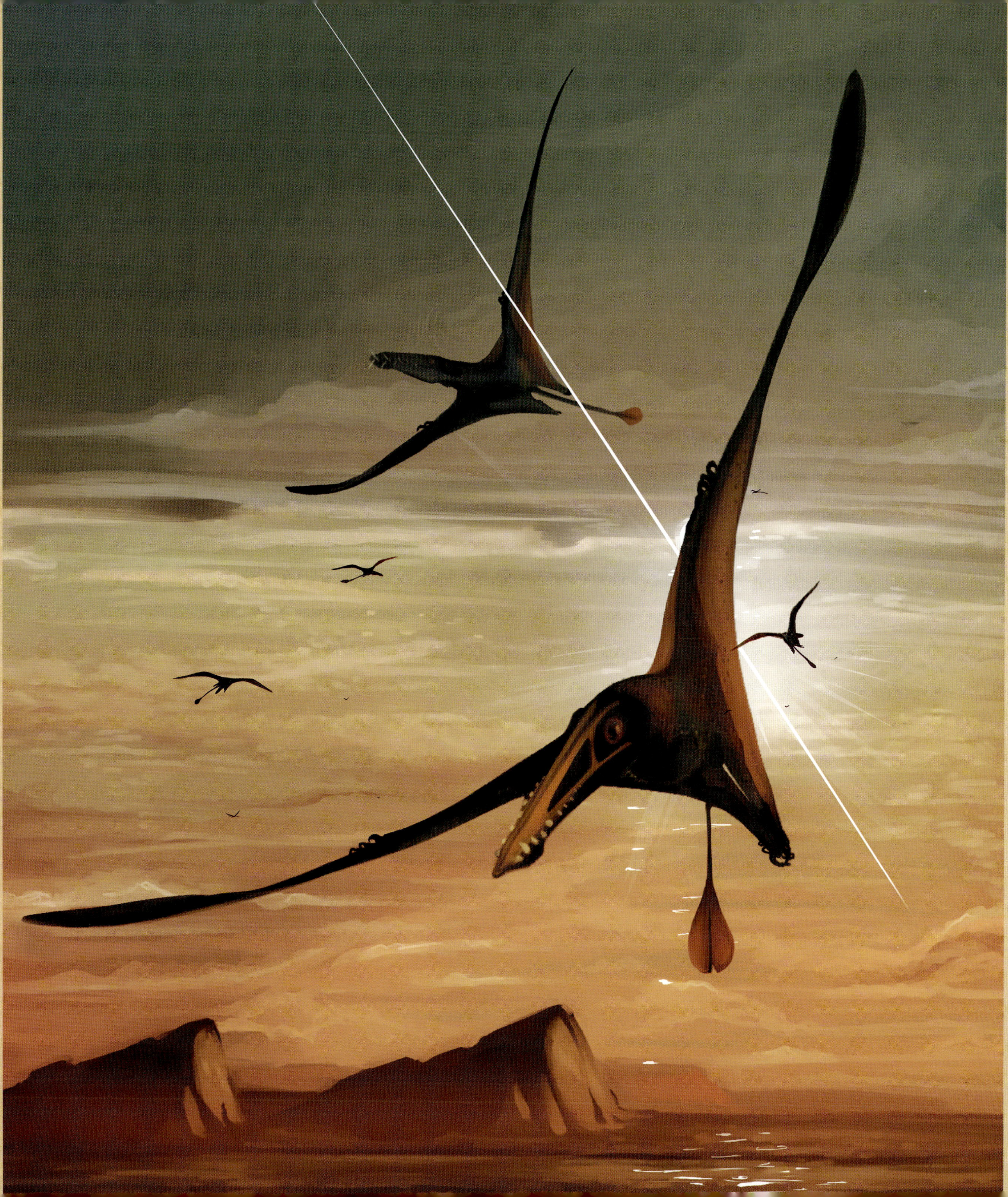

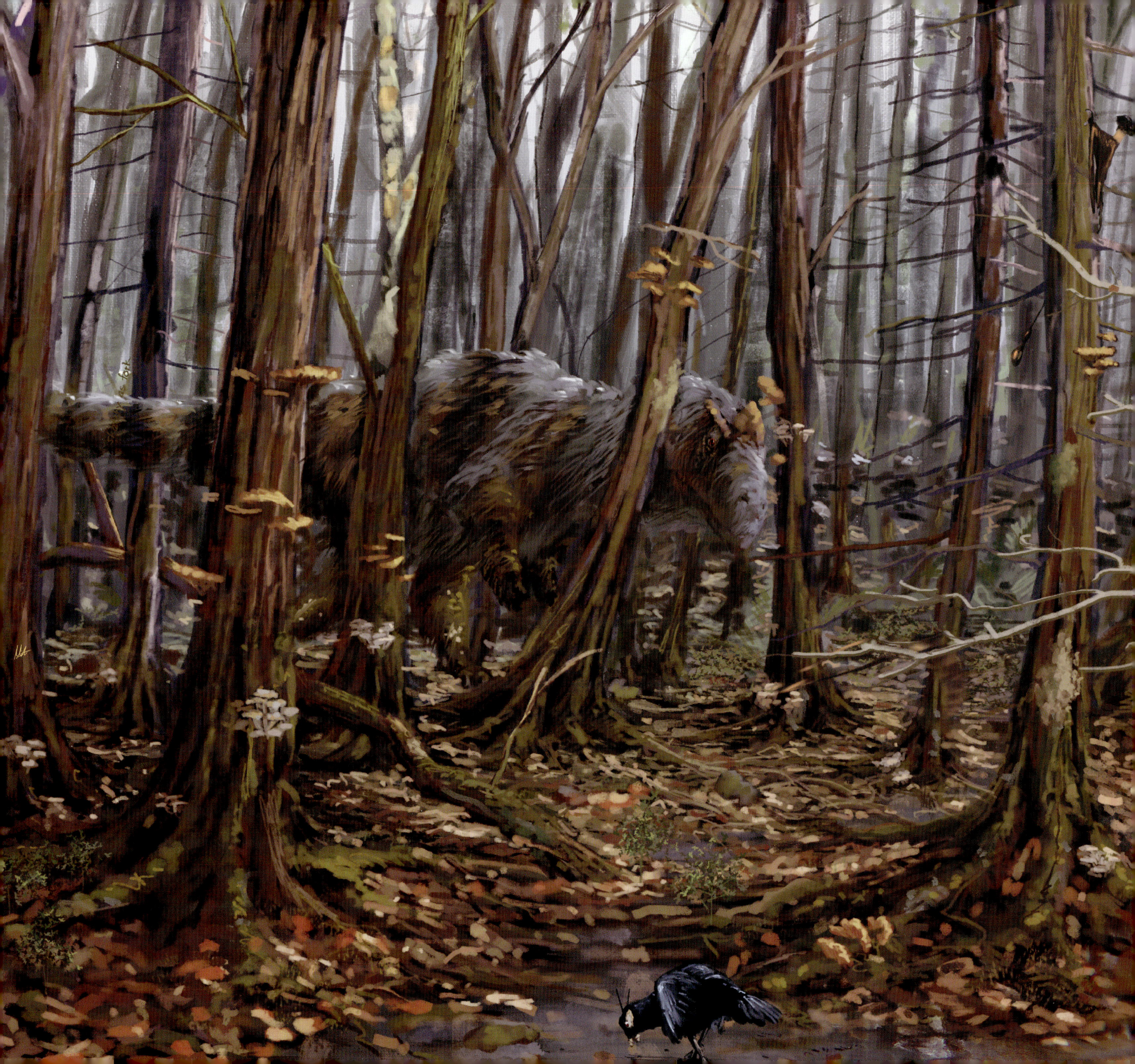

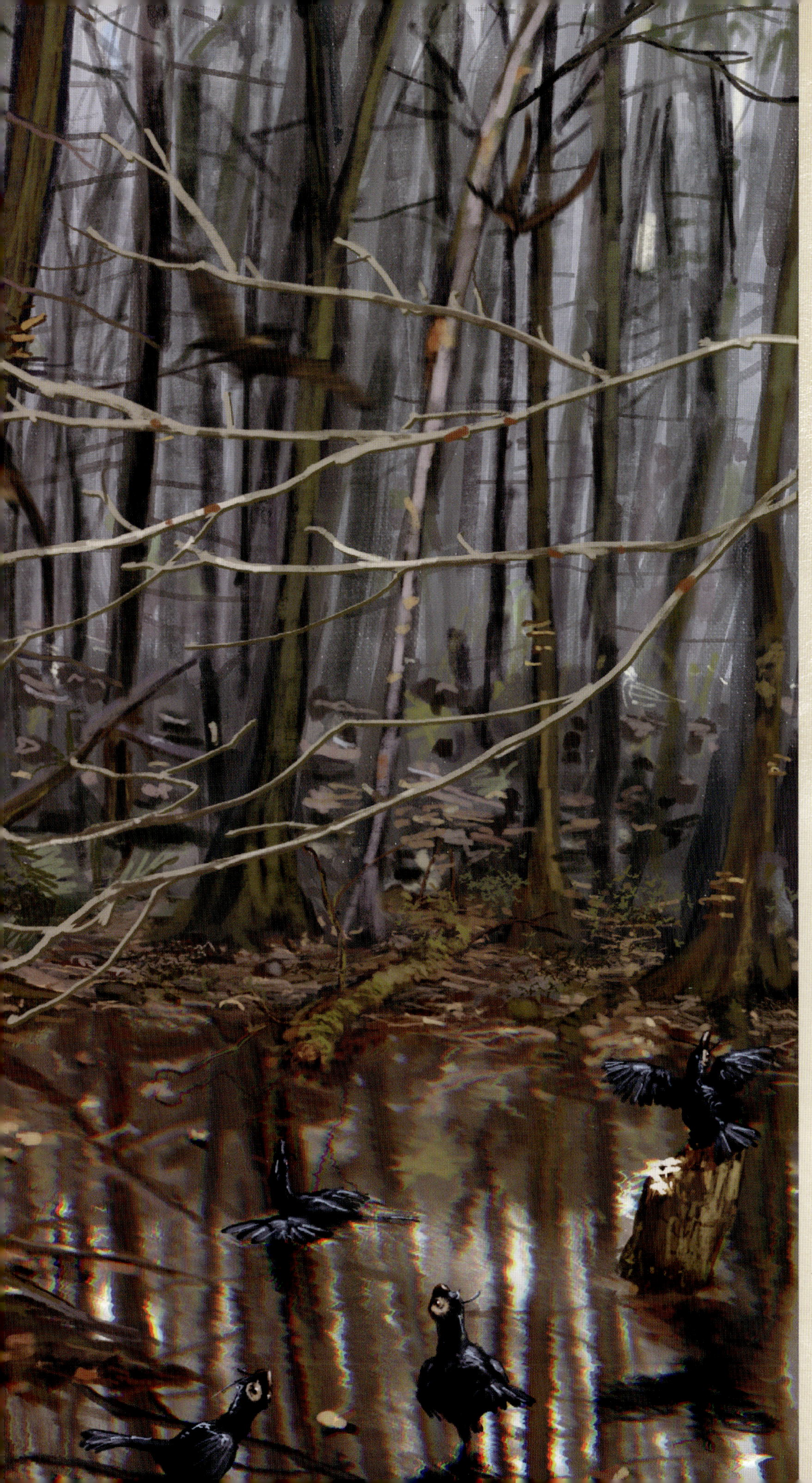

MIN TAYZA

MIN TAYZA IS A 21-YEAR-OLD Burmese artist who has been drawing since as far back as he can remember. His passion for palaeontology has driven an interest in living animals as well, and he is currently studying zoology at the University of Sheffield.

Min Tayza sees his work as a blend between art and science, using prehistoric forms to create unique compositions. His art is also inspired by real stories about living animals (see, for instance, the mourning *Elasmosaurus* on pp.60–61), which he feels can connect contemporary viewers with the distant past. In doing so, he hopes to give glimpses of the rich, full lives prehistoric animals no doubt led.

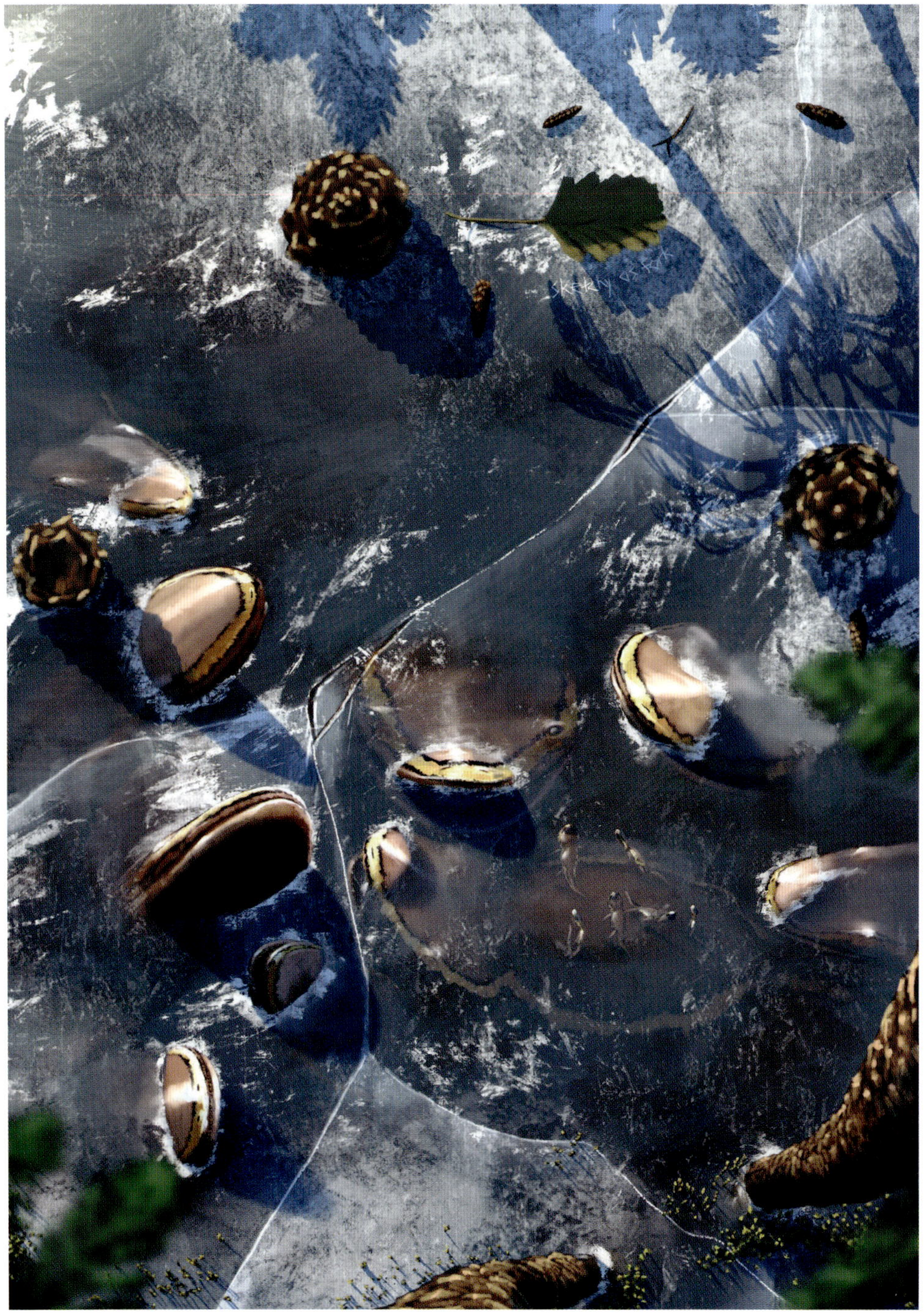

◄ BRUMATING TEMNOSPONDYLS

For 200 million years, between the Carboniferous and the Cretaceous, a dynasty of amphibians, the temnospondyls, occurred in wetlands worldwide, as well as in certain dry and marine places. While fossil evidence shows that certain of these animals underwent seasonal torpor in burrows, did those living in seasonally cold places undergo torpor (technically termed brumation) while frozen in ice? Modern alligators, turtles and amphibians do, so it's plausible.

• Digital | 2023

NEST DISPUTE ►

On a Late Jurassic beach, two *Pterodaustro* – a slender-jawed, Argentinian pterosaur noted for its hundreds of needle-like teeth – squabble over access to a possible nesting site. Fossils from China show that pterosaurs of this sort nested colonially and buried their leathery-shelled eggs in sediment. The different colour schemes of the animals imply different sexes or species.

• Digital | 2022

▼ WATERFALL SAUROPODS

The giant size and relatively long legs of sauropods suggest they were capable of wading through lakes, deltas and even raging rivers. But wandering close to a waterfall can be dangerous even for these animals, especially for the youngsters among them. Several modern cases record hippos, elephants and other animals doing exactly this sort of thing – and dying.

• Digital | 2022

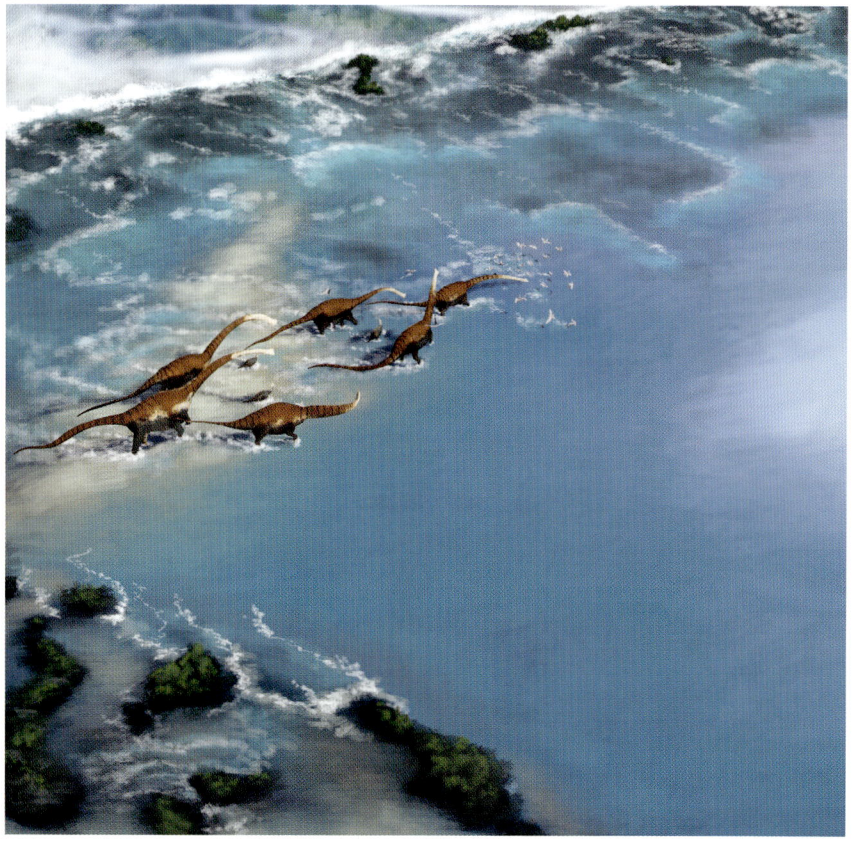

[Previous pages]

SPRING OPPORTUNITY

In a springtime Chinese deciduous forest of the Early Cretaceous, the large tyrannosauroid *Yutyrannus* prepares to ambush a group of the turkey-sized oviraptorosaur *Caudipteryx*. The idea that Mesozoic dinosaurs – especially in the Cretaceous – inhabited cool, damp forests of a sort familiar to us in the Northern Hemisphere has become more accepted in recent years.

• Digital | 2023

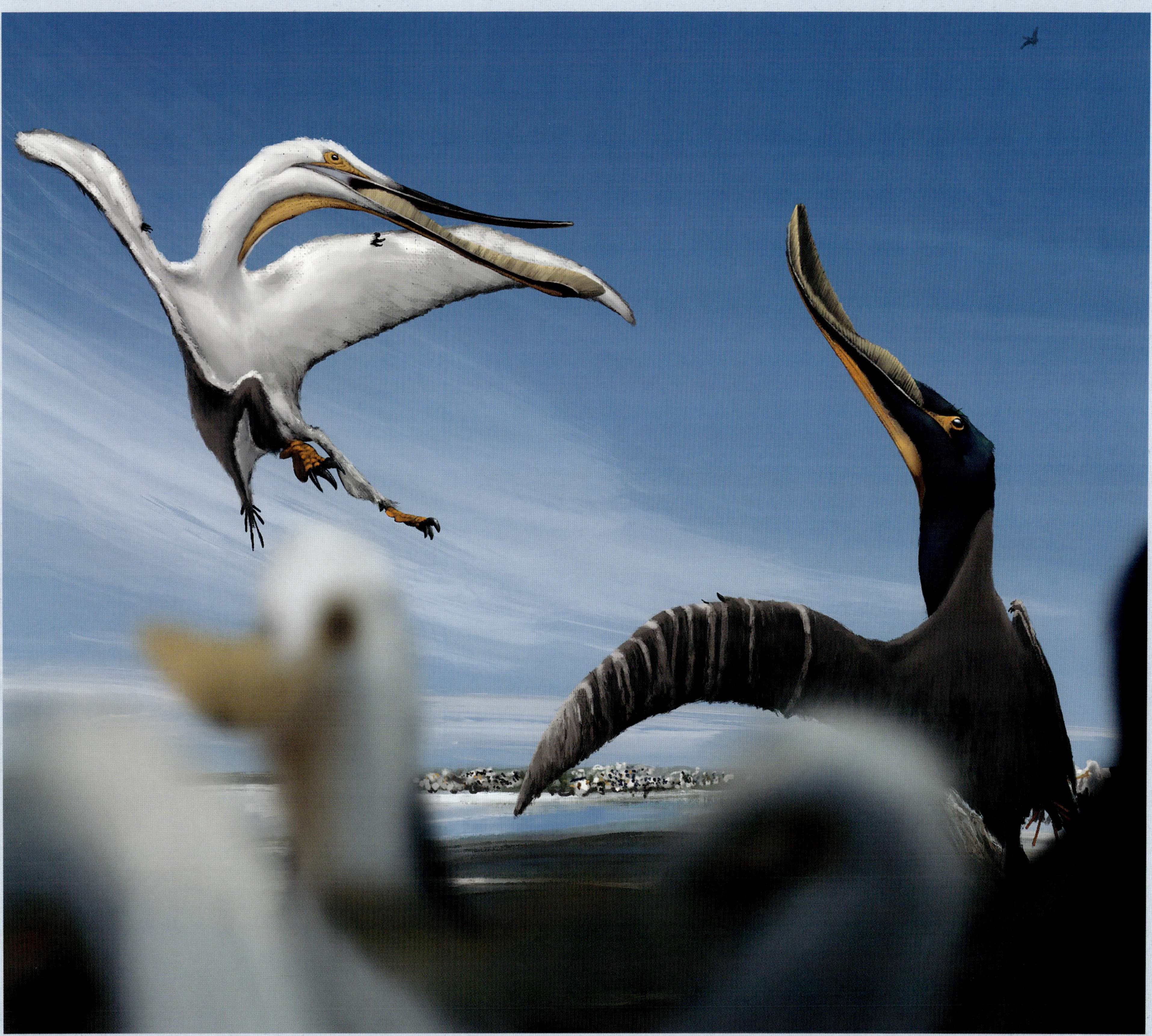

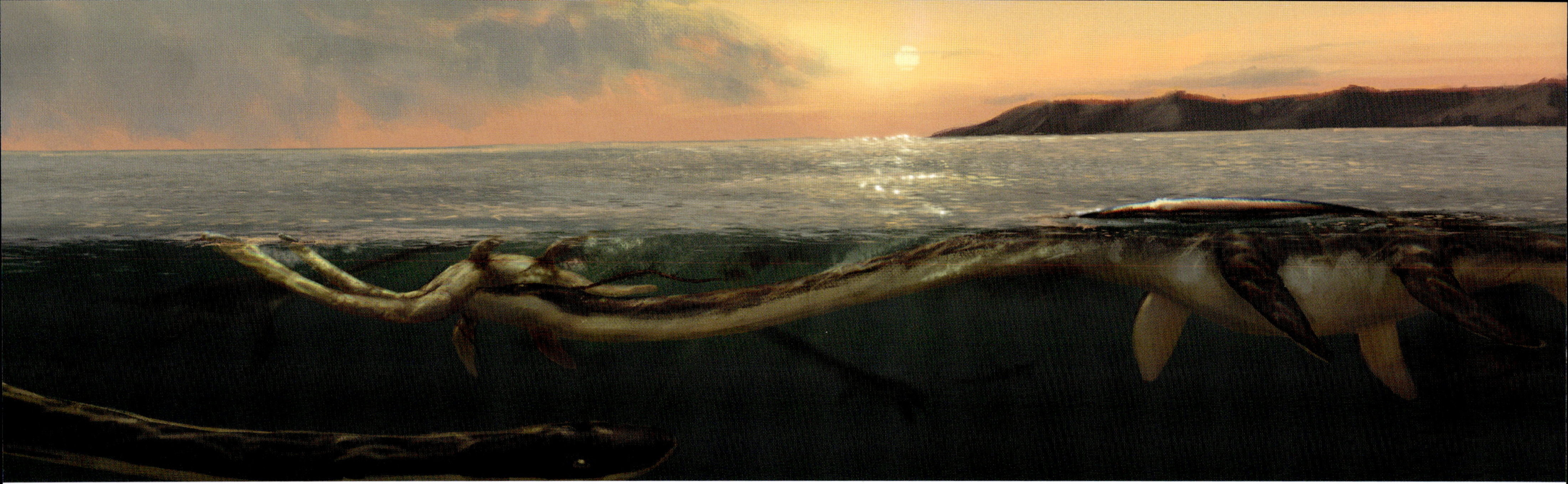

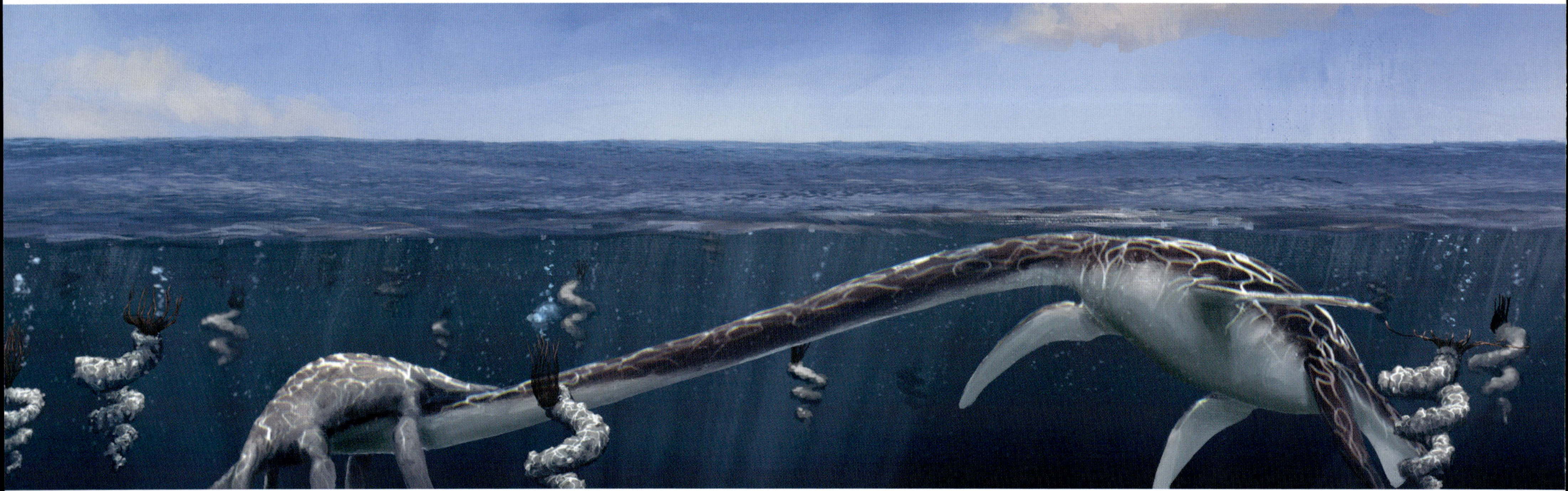

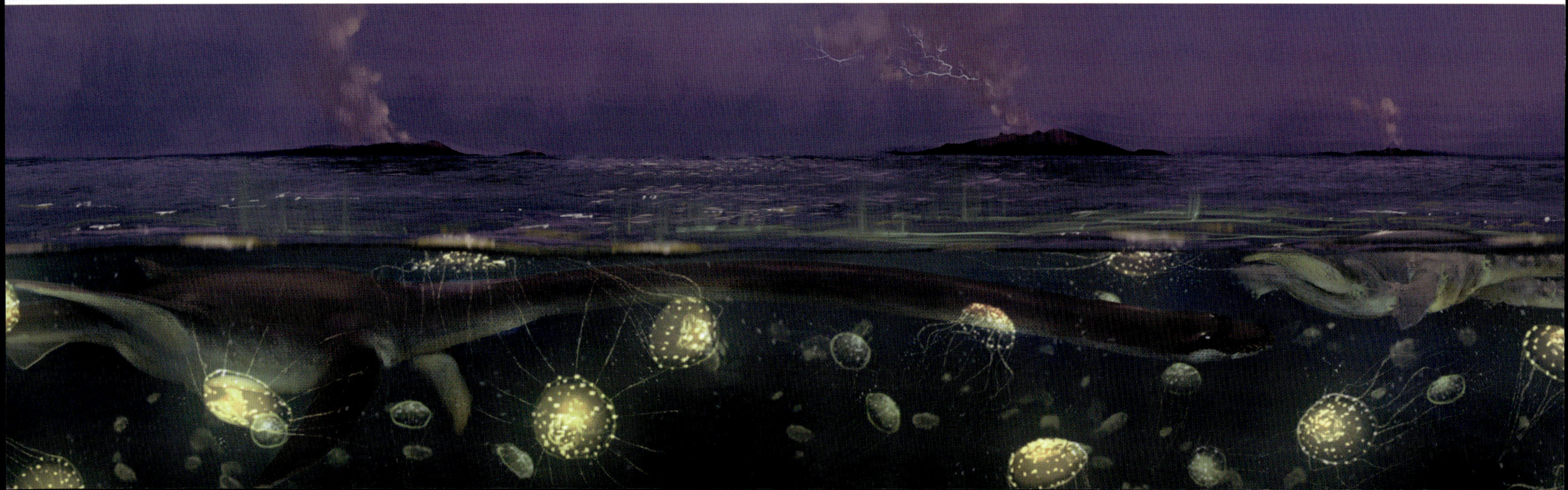

NYCTOSAURS ▸

Many of the seas associated with oceanic pterosaurs – like these nyctosaurs – were not as stormy or seasonally variable as the high latitude seas of today. However, animals like nyctosaurs lived almost worldwide for tens of millions of years and *must* have experienced storms on occasion. Like the seabirds of today, they were likely able to survive such events. However, storms today sometimes kill even the most supremely adapted marine species.

• Digital | 2022

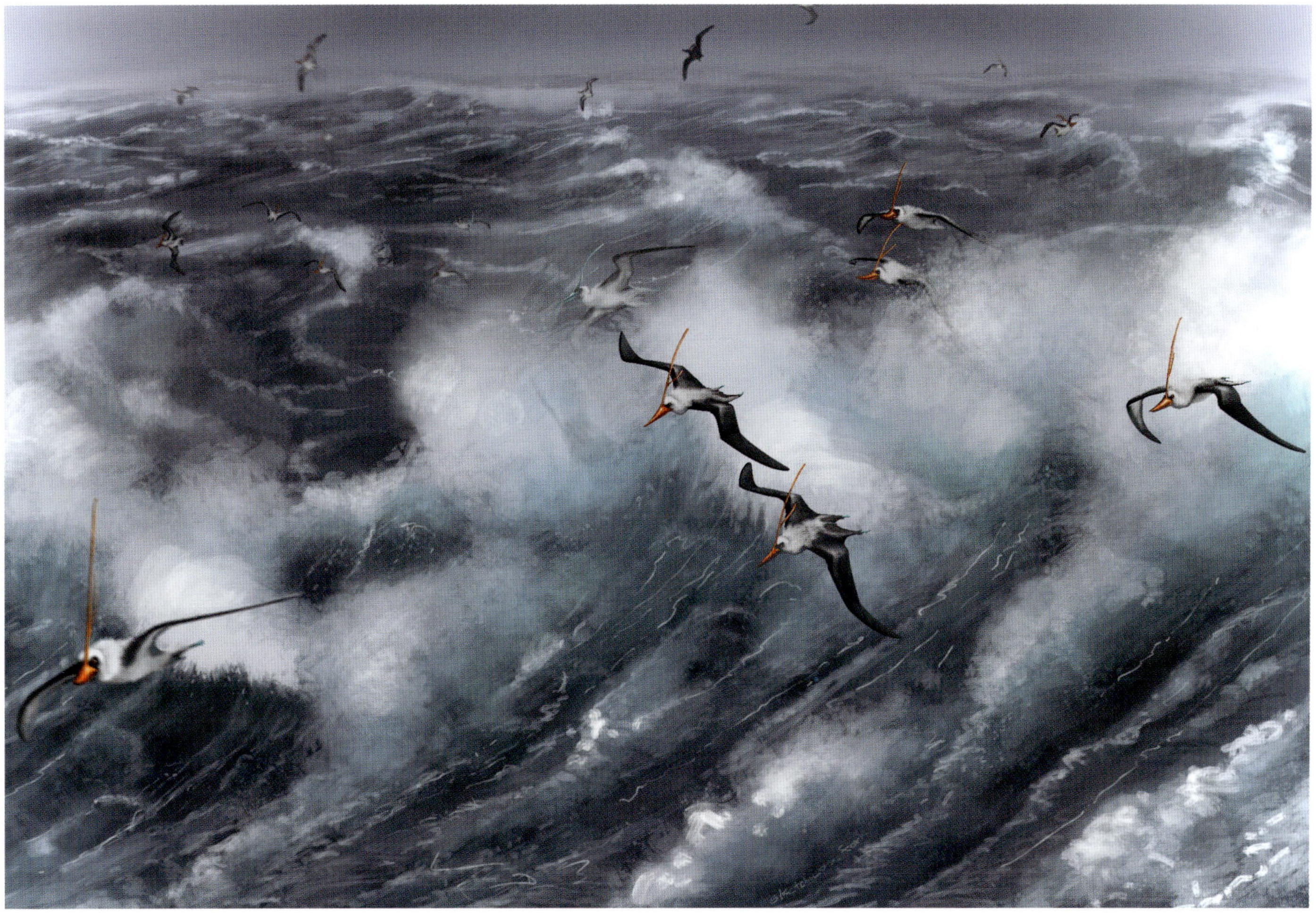

◂ MOURNING *ELASMOSAURUS* ▾

Some experts think plesiosaurs like this Late Cretaceous *Elasmosaurus* were somewhat whale-like in behaviour and perhaps tightly bonded to their newborns. In these images, an aberrant baby – born with the two-headed condition seen, rarely, in modern animals – has died, and the mother has remained with it. They were inspired by Tahlequah, a British Columbian killer whale, who, in 2018, stayed with her newborn for almost three weeks after its death. In the final, fourth image (at right), the mother finally abandons the body.

• Digital | 2023

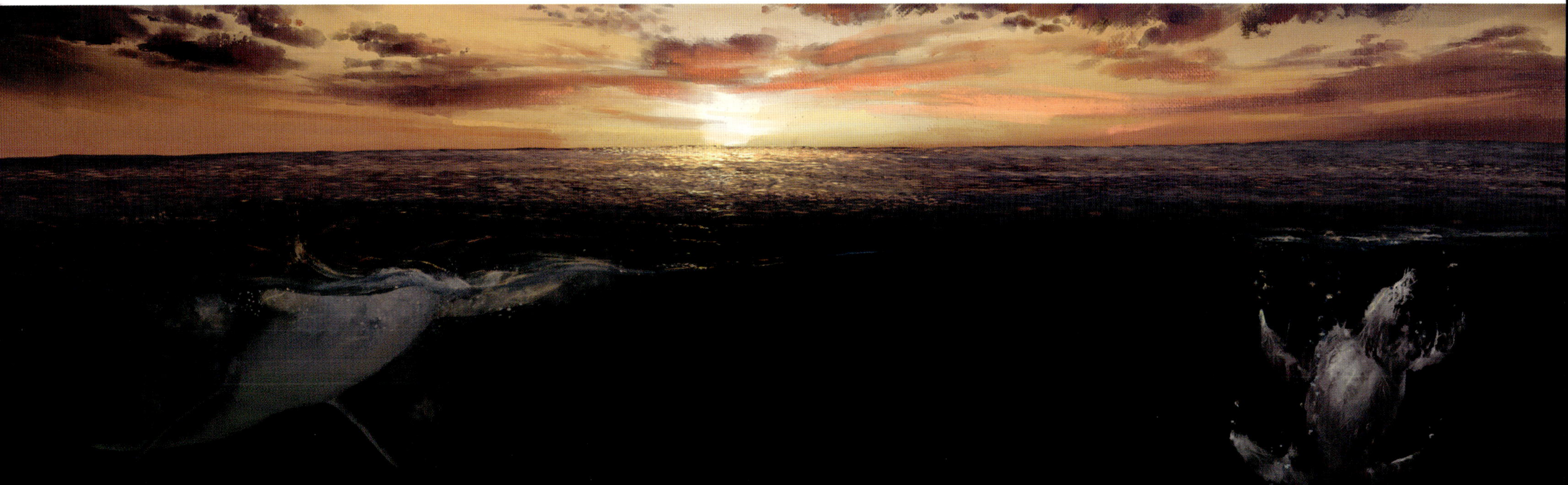

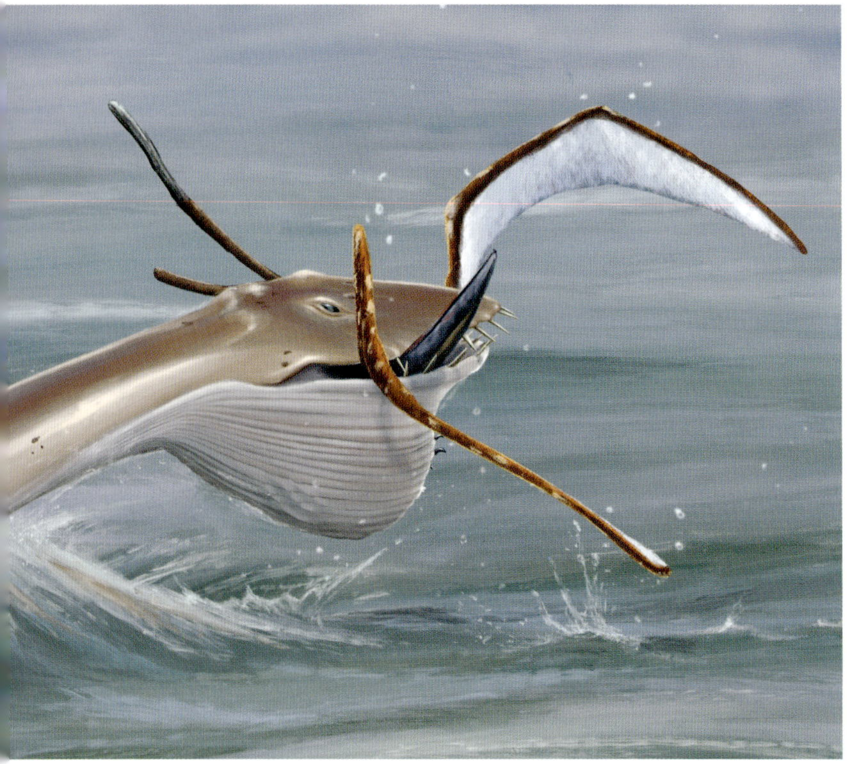

◂ *MORENOSAURUS* KILL

Morenosaurus is a little-known elasmosaurid plesiosaur from the Late Cretaceous of California, USA. Its skull was about 40cm (1.3ft) long and equipped with interlocking, fang-like teeth. The animal would surely haved have been capable of grabbing and consuming relatively large prey – even pterosaurs like this nyctosaur, whose bony crest and wings might make for an awkward eating experience.
• Digital | 2022

▾ DEAD LIGHTS

In cloudy, twilight conditions of the Late Cretaceous north, the azhdarchid pterosaur *Cryodrakon* inspects a dinosaur skull trapped in a frozen lake. *Cryodrakon* was named in 2019 for fossils from Canada, and numerous remains from the Upper Cretaceous Dinosaur Park Formation suggest that it was quite common in the relevant habitats. Gas escaping from beneath the ice glows eerily.
• Digital | 2022

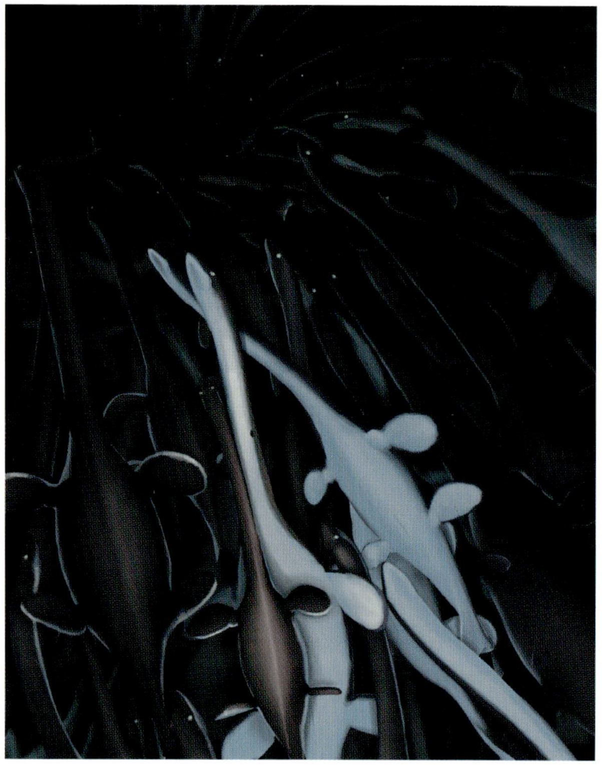

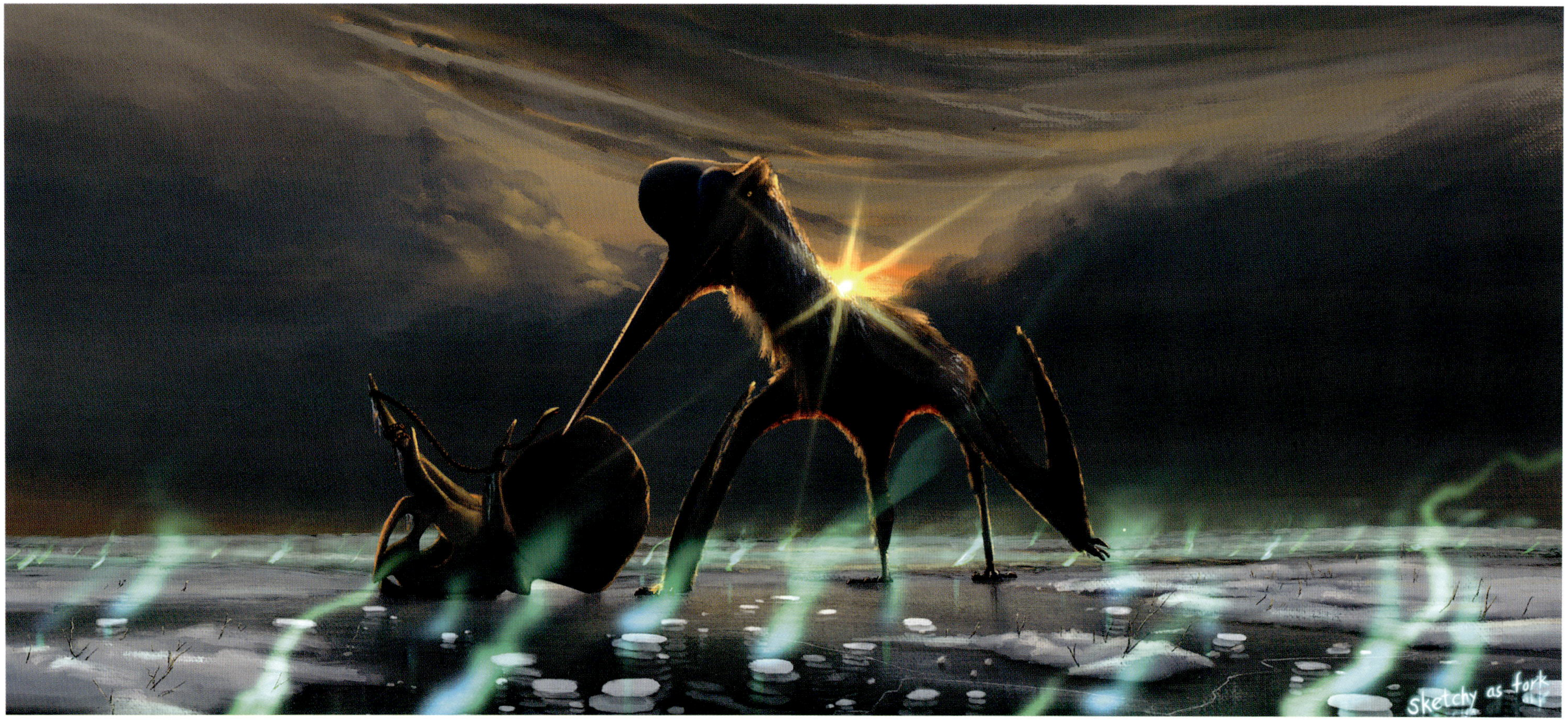

◂ RIVER OF SOULS

Animals of all kinds sometimes form massive aggregations, sometimes because of emergency or disaster, but also because it's a normal part of their regular lives. Here, the Late Cretaceous plesiosaur *Styxosaurus* – named for the river of Greek mythology – form a river of bodies. Social behaviour is inferred for these animals, who might have lived and hunted in groups.
• Digital | 2022

▾ CLEANING STATION

Ichthyosaurs – the famed 'fish lizards' of Mesozoic seas – evolved giant size early in their history. This Late Triassic giant is a *Shonisaurus*, here rolling on its side so that individuals of the pterosaur *Carniadactylus* can pluck parasites from its skin. Extant giant fish – specifically ocean sunfish – deliberately solicit the interest of albatrosses for parasite removal. The ichthyosaur's nasal exhalation is catching the light.
• Digital | 2023

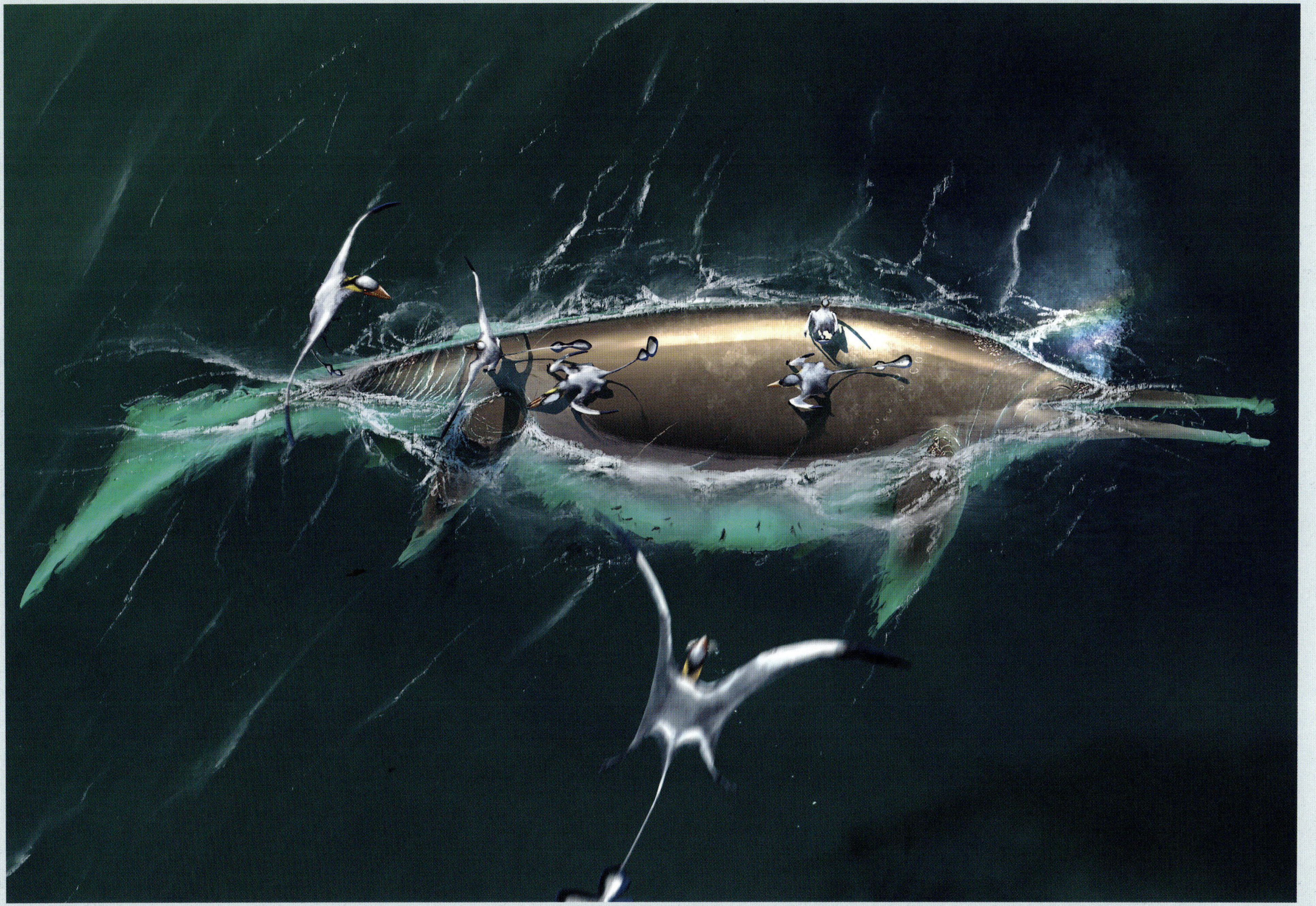

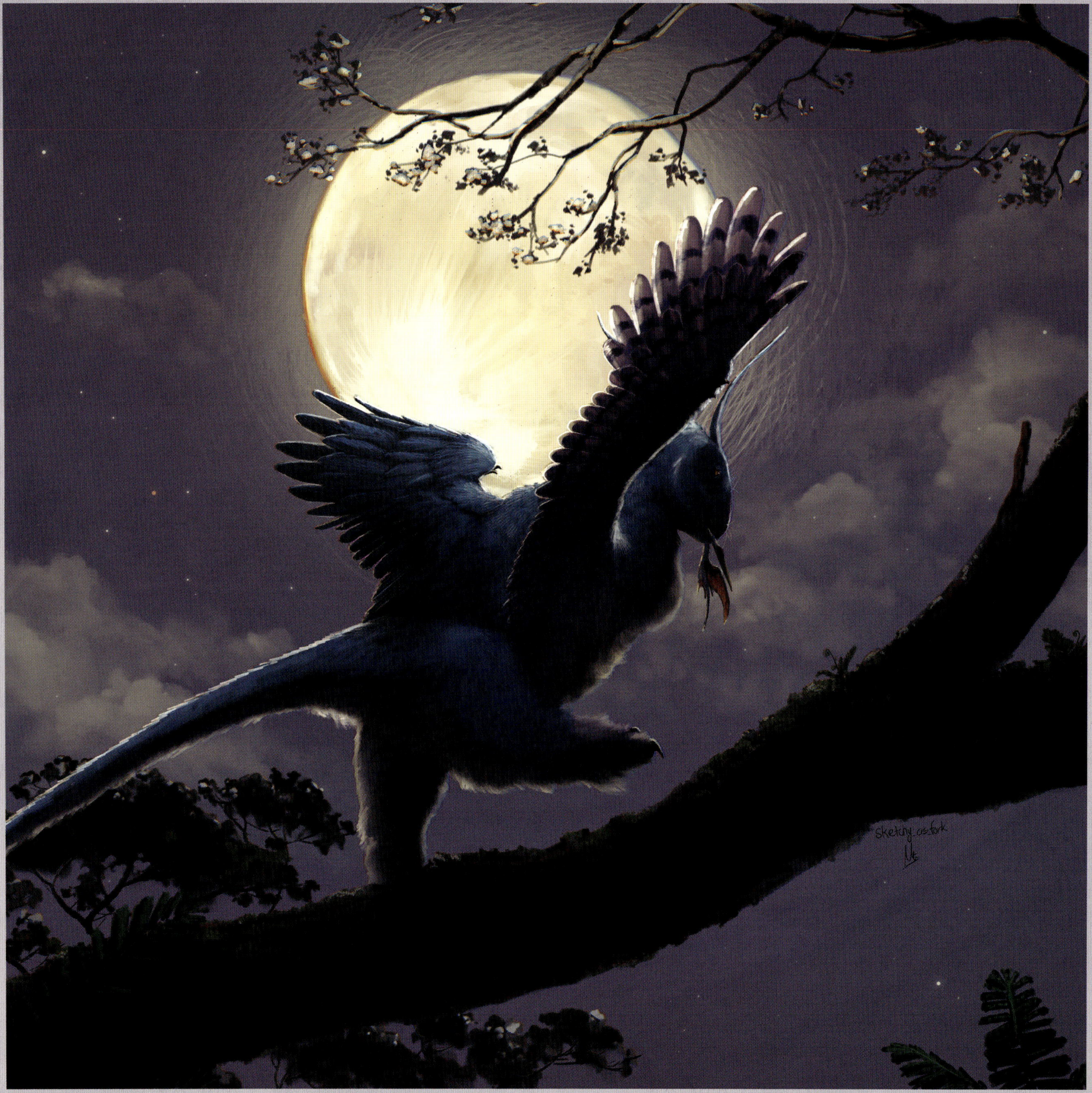
sketchy_as_fork

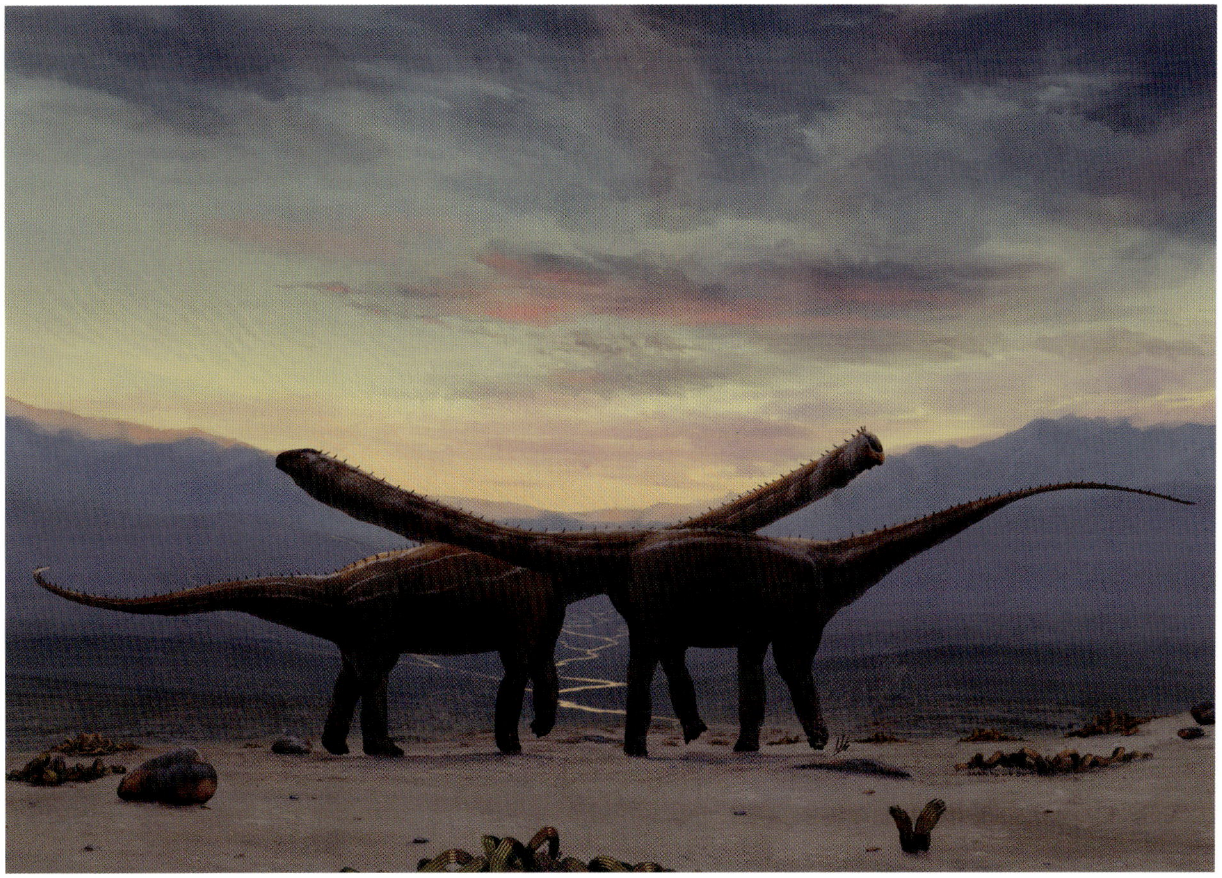

▲ *ATLASAURUS*

A pair of sub-adult *Atlasaurus* exchange low infrasound greetings. This Middle Jurassic mid-sized sauropod is named after the Atlas Mountains in Morocco, where its remains were found. The long necks (and associated trachea), extensive network of air sacs and huge lungs suggest that sauropods could generate deep, low sounds and even infrasound, as elephants do today.

• Digital | 2023

◄ *BALAUR*

The Late Cretaceous Romanian maniraptoran *Balaur* remains fascinating and enigmatic. In the absence of its skull, we wonder whether it was a *Velociraptor*-like predator or a very different, omnivorous or even herbivorous creature. Its unusual feet suggest that it was good at clambering and perching. As reconstructed here, *Balaur* has a vaguely owl-like character, with fluffy feet, a mostly concealed snout and an implied nocturnal habit.

• Digital | 2023

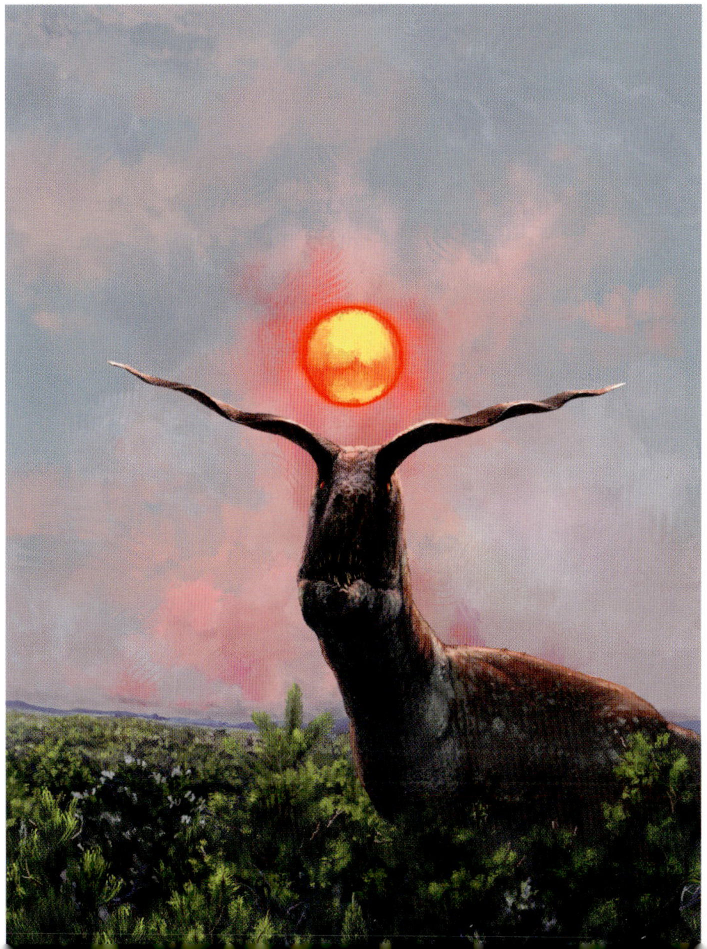

▼ AZHDARCHID

This ground-eye view of an azhdarchid pterosaur bestriding a Late Cretaceous forest epitomises just how colossal these animals were. The long legs and long, superficially stork-like jaws of some species indicate that their hunting strategies resembled those of modern storks and ground hornbills. Some, like the Romanian *Hatzegopteryx*, appear to have filled the role of apex predator in the places where they lived.

• Digital | 2023

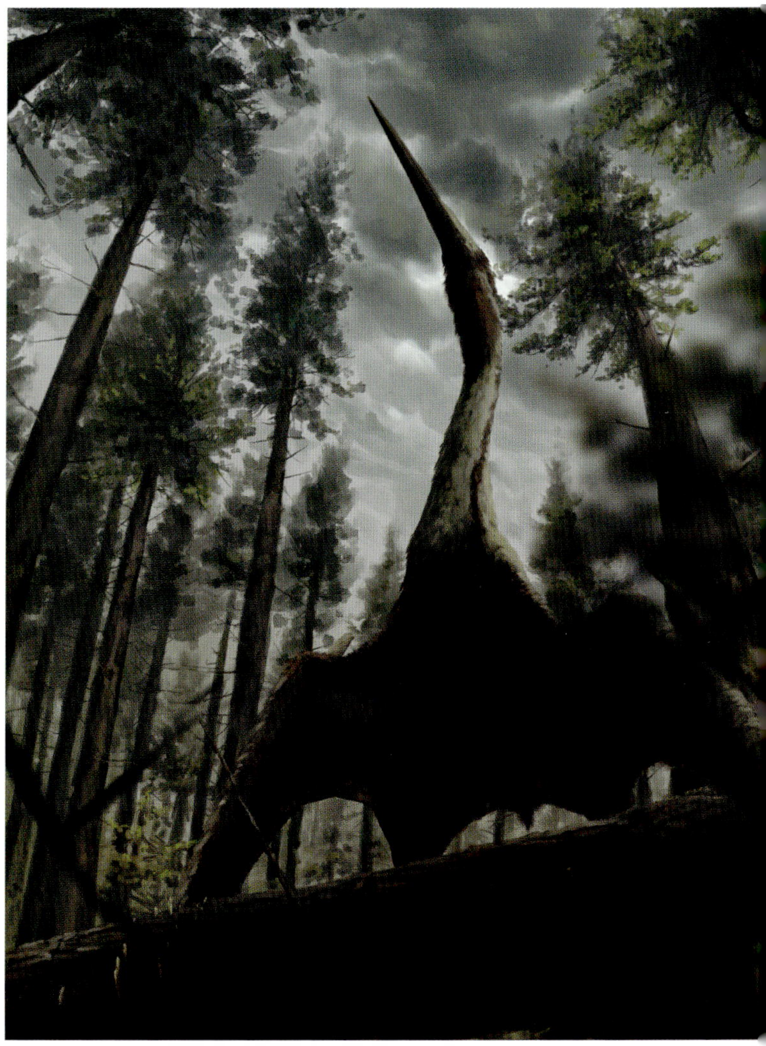

◄ EMISSARIES OF THE LOW SUN

The Late Cretaceous abelisaurid Carnotaurus is known from just a single, albeit well-preserved, specimen. Its horns are enigmatic, and we remain unsure how long and complex they were in life. The possibility that one sex had longer, more complex horns than the other exists too. We hope that additional specimens will provide insight on these issues.

• Digital | 2024

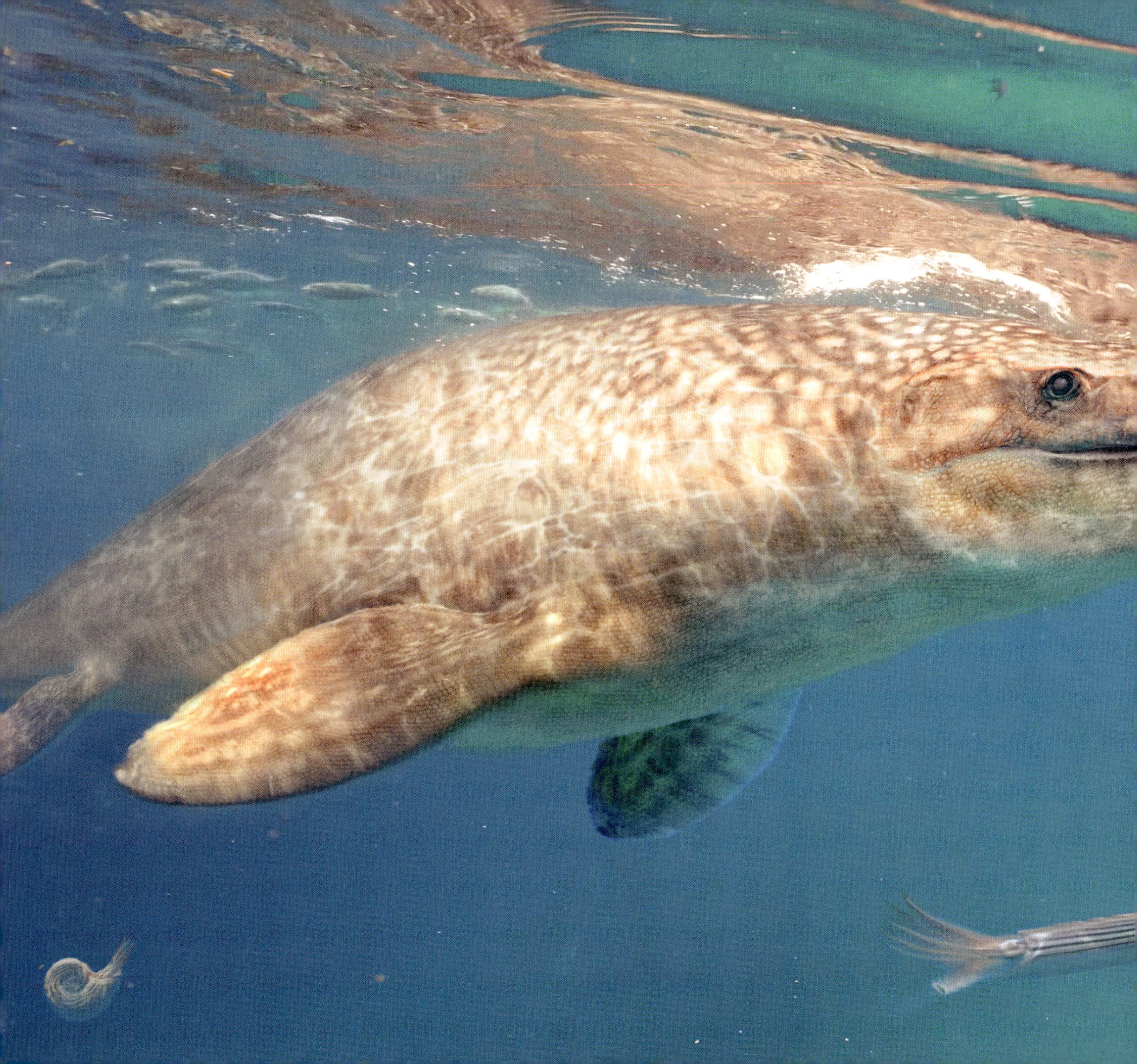

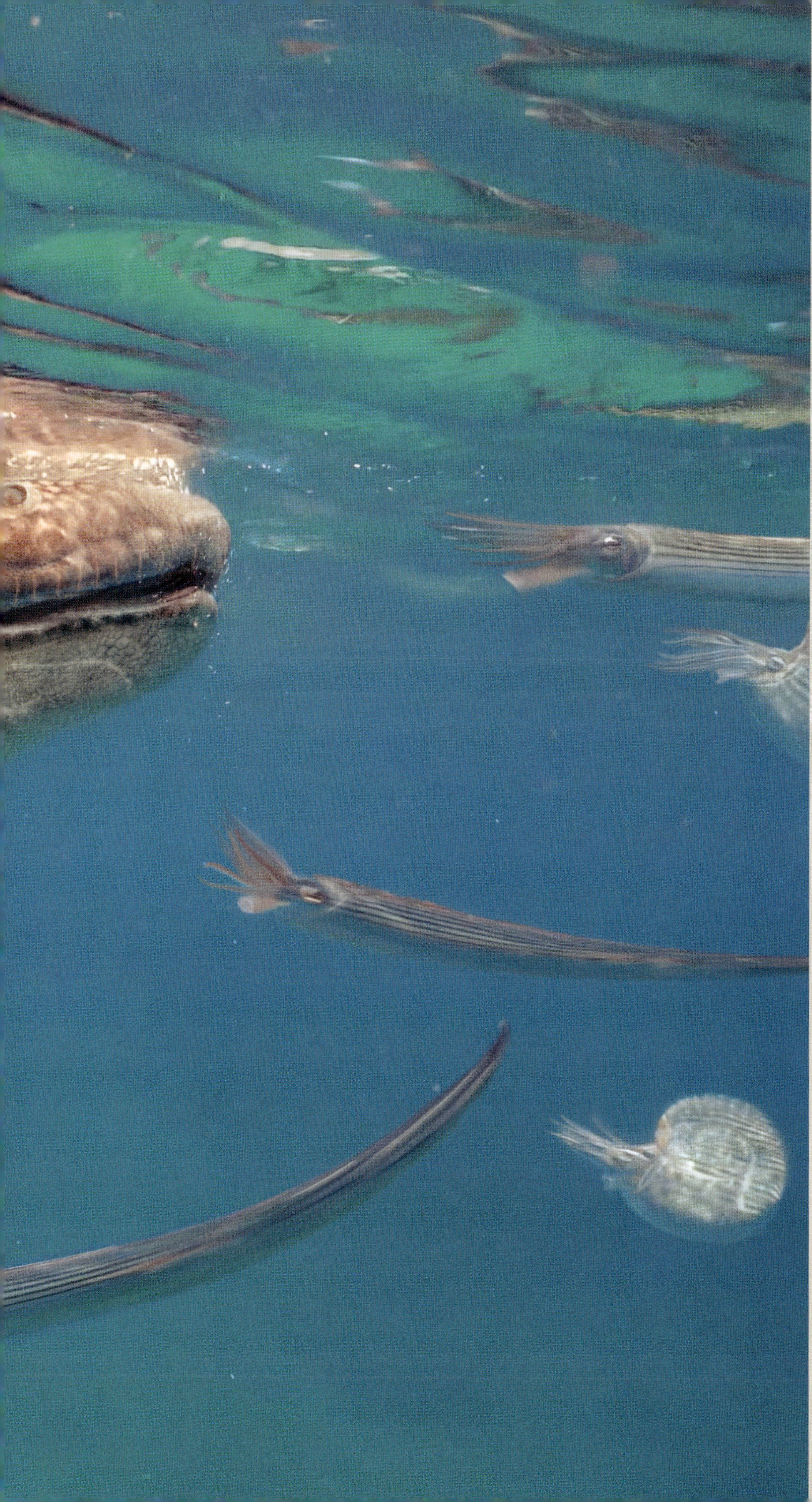

BILL UNZEN

EVER SINCE HE WAS A CHILD, Bill has been fascinated by animals of all kinds, living and extinct. This innate and ever-expanding interest in the natural world, coupled with a penchant for drawing, led him to palaeoart, which he has produced sporadically ever since. A self-taught artist, he works in various traditional and digital media (often mixing the two) and strives to produce a spectrum of art that ranges from heavily stylised to photo-realistic. This interest in nature has also led Bill to work conducting Breeding Bird Atlas surveys, owl banding, and Colonial Waterbird surveys.

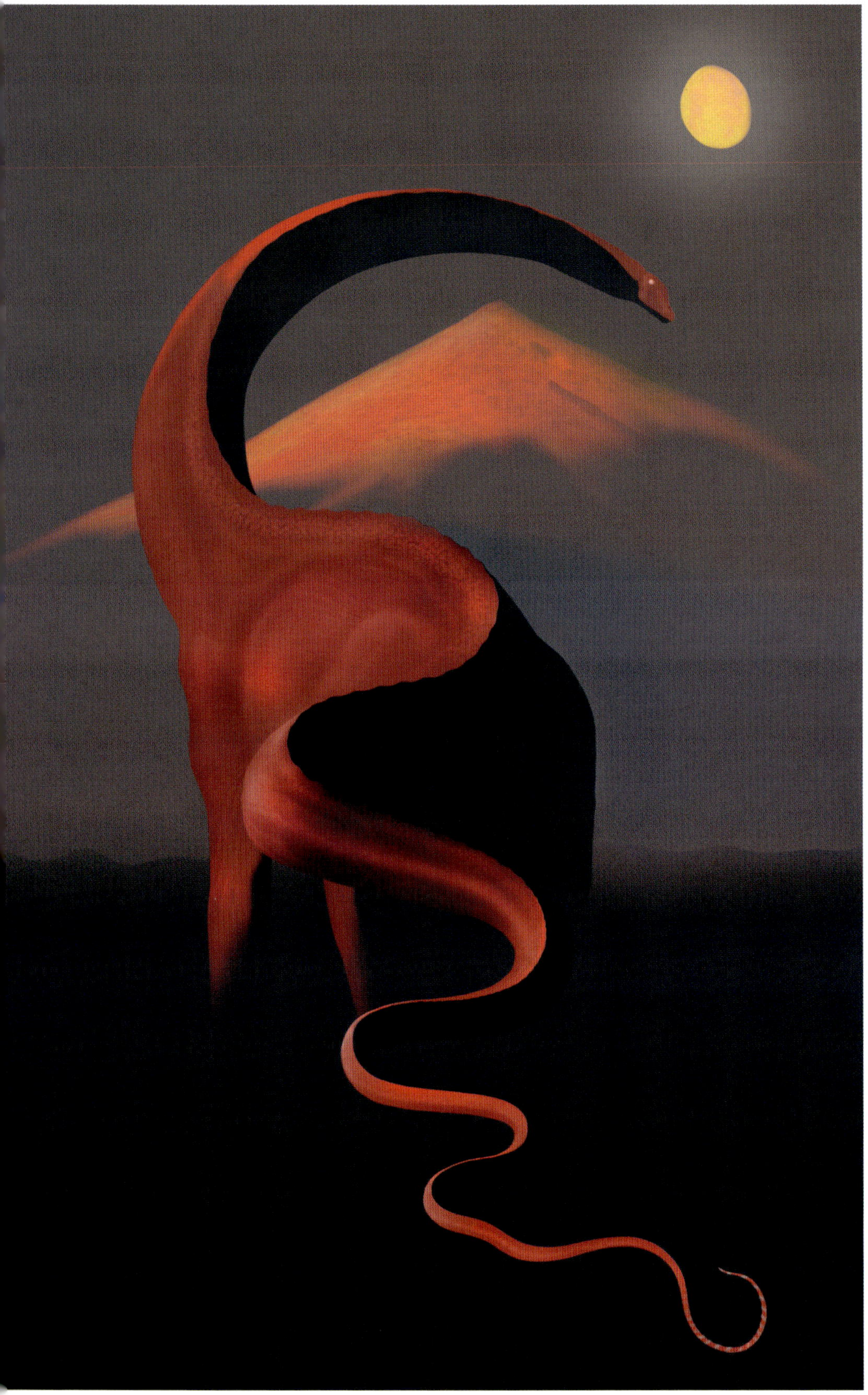

[Previous pages]

GLOBIDENS

The mosasaur *Globidens* is known from the USA, Brazil, northern and southern Africa, and the Middle East. *Globidens* is named for its rounded teeth, and artists often try and depict them in their art. However, if mosasaurs had the lips and extensive gum tissue typical of lizards – as evidence suggests – they would have had a far 'fleshier', more streamlined look.

• Pencil and digital | 2023

HATZEGOPTERYX & PALUDITITAN ►

The great azhdarchid pterosaur *Hatzegopteryx* observes a wading *Paludititan*, one of several small and mid-sized titanosaurian sauropods that shared its habitat in Late Cretaceous Romania. *Paludititan* reached 6m (20ft) in length, suggesting a height at the head of around 3m (10ft) – about the same as that of *Hatzegopteryx*. *Paludititan* hatchlings would almost certainly have been potential food items to this giant pterosaur.

• Digital | 2023

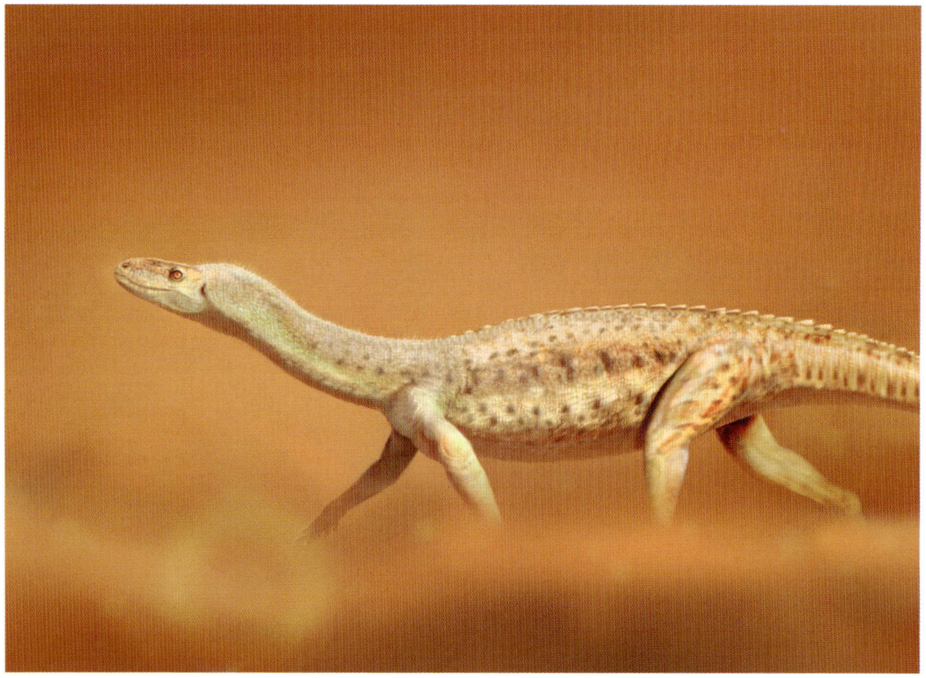

◄ SUPERSAURUS

The sheer size of mega-sauropods almost defies belief, and the impression conveyed here is that they were virtually landscape features of their world. This shows the superbly named *Supersaurus*, originally named from the USA's Morrison Formation but more recently identified in the Upper Jurassic of Portugal as well. *Supersaurus* is a diplodocoid, and part of the group Flagellicaudata, literally meaning 'whip-tails'.

• Digital | 2022

▲ MAMBACHITON

Among the most surprising recent discoveries from the Triassic is that early members of the bird lineage – ancestors to pterosaurs and dinosaurs – were quadrupedal with armour plates running along the backbone. *Mambachiton* from Madagascar is one of them. Little is known of what these animals were like when alive – perhaps they were 'archosaurian monitor lizards' as shown here.

• Digital | 2023

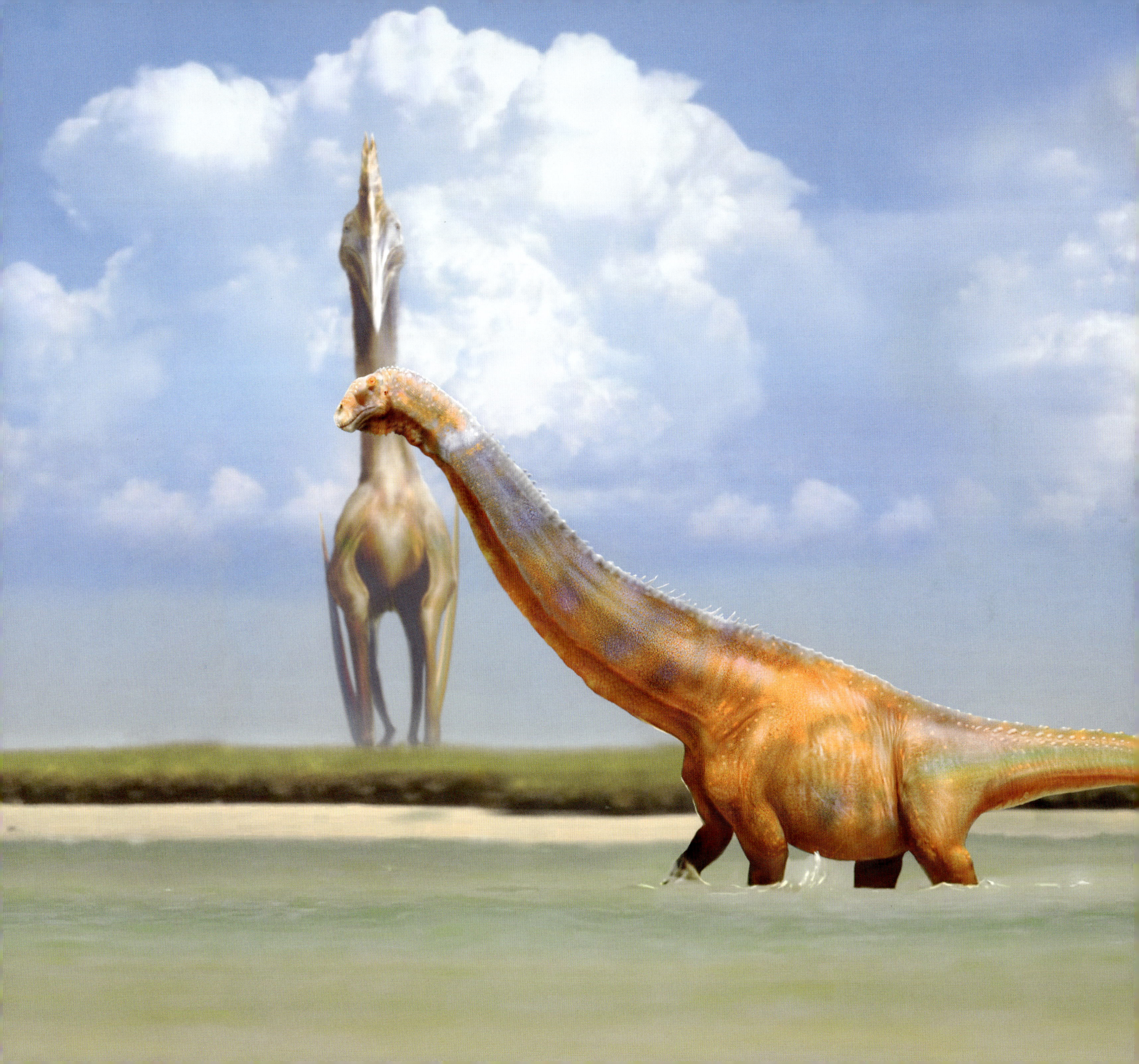

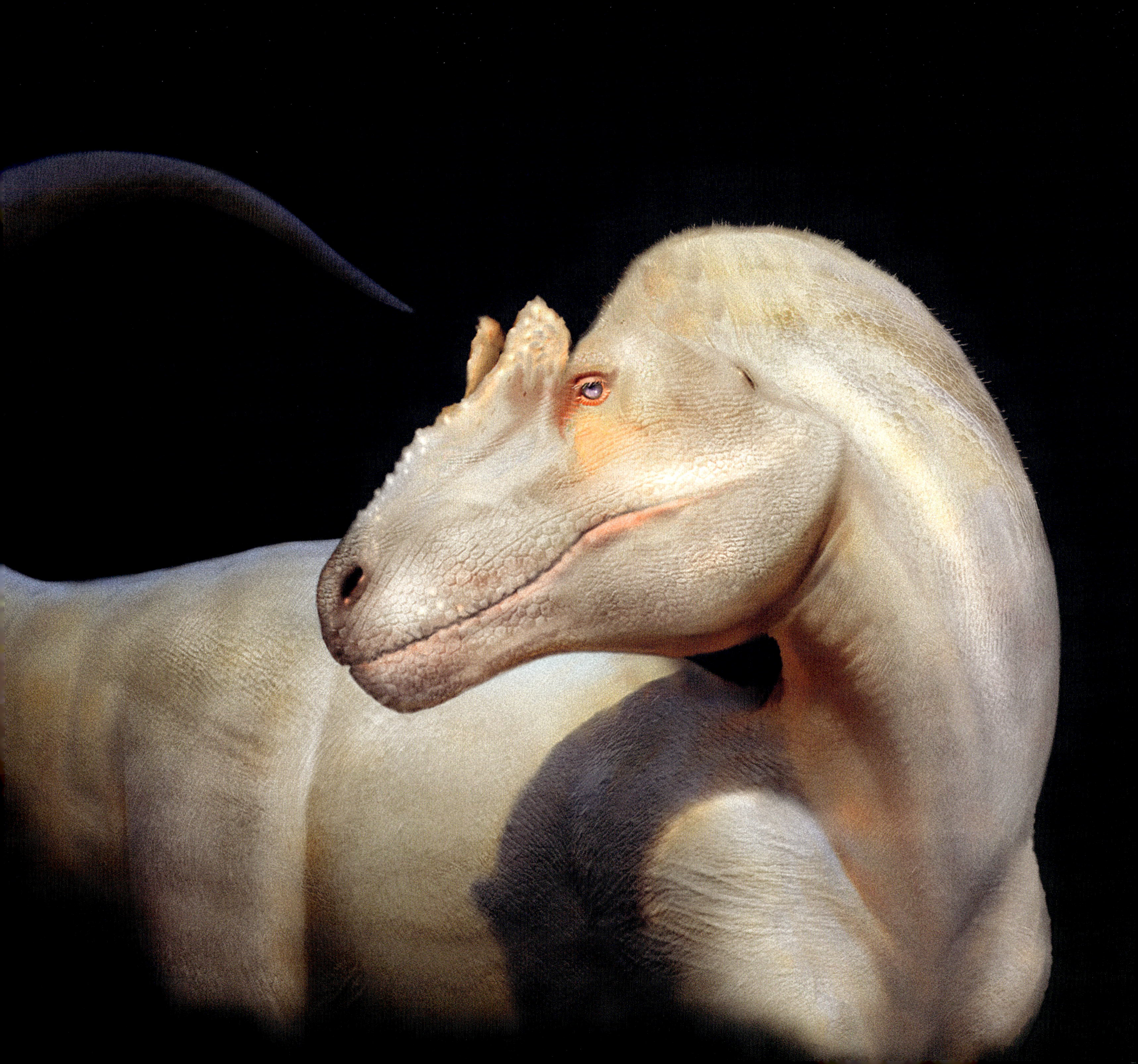

ALBINO ALLOSAUR

Albinism occurs across all animal groups today and we can *guarantee* that it was present in the animals of the past too, dinosaurs included. A big Jurassic theropod like *Allosaurus* was surely an awesome, arguably beautiful animal in life. Now imagine an albino one! Of course, we don't know whether such animals would really survive to adulthood.

• Pencil and digital | 2023

ZHUCHENGTYRANNUS

China's *Zhuchengtyrannus*, named in 2011, was a giant tyrannosaurid, close to *Tarbosaurus* and *Tyrannosaurus* and differing from them in minor details of skull structure. Like its relatives, it almost certainly had prominent lumps and hornlets above and around its eyes, as well as a thick, horny crest running along the snout.

• Digital | 2023

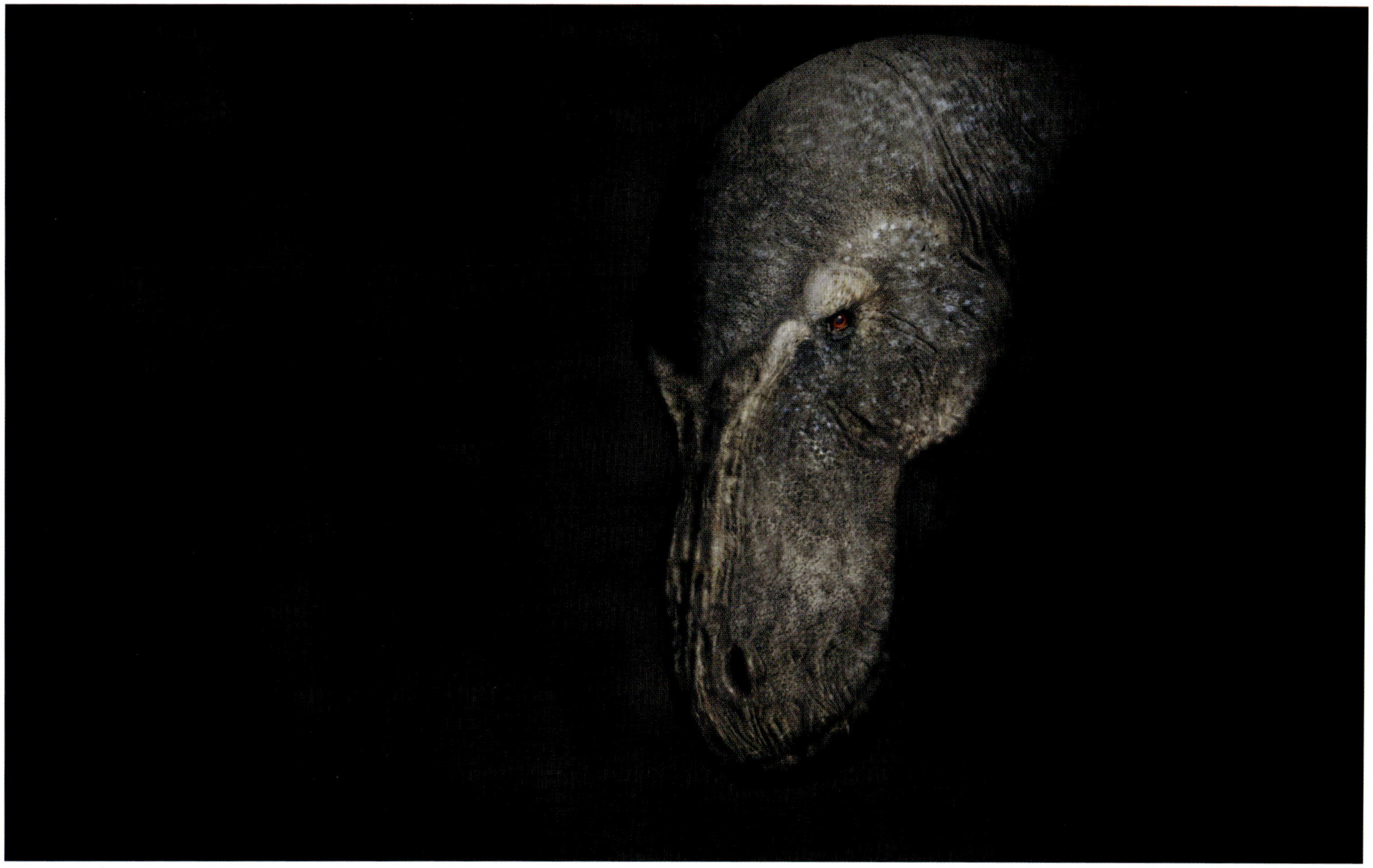

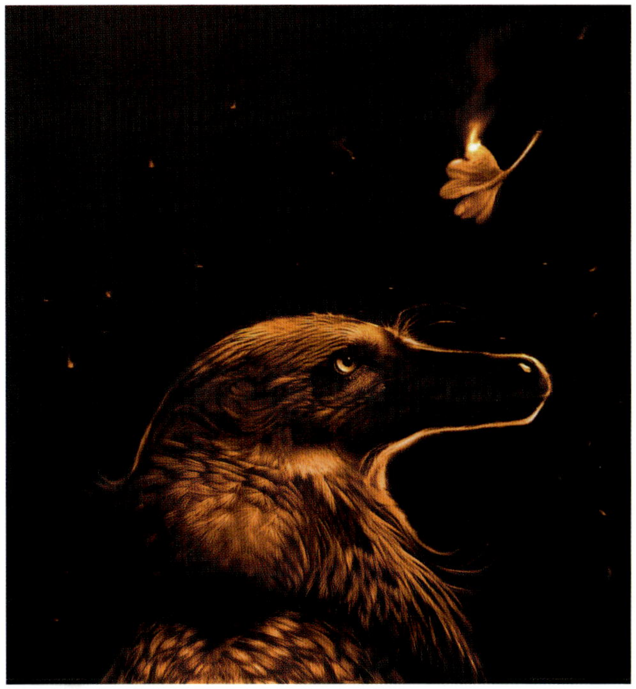

KURU KULLA

The dromaeosaurid *Kuru kulla*, a Late Cretaceous, Mongolian relative of *Velociraptor*, was named in 2021. It appears to have longer hindlimbs and a more robust skull than related species. Here, a burning ginkgo leaf and the dinosaur's plumage are illuminated by fiery surroundings.

• Pencil and digital | 2023

TUPANDACTYLUS & MAYFLIES

The sail-crested tapejarid pterosaur *Tupandactylus* catches the sunset light as it soars through a swarm of *Astraeoptera*, a Cretaceous mayfly contemporary with this pterosaur. These insects were superficially similar to the mayflies of today and their abundance as fossils indicates that they too formed great but short-lived mating swarms.

• Digital | 2023

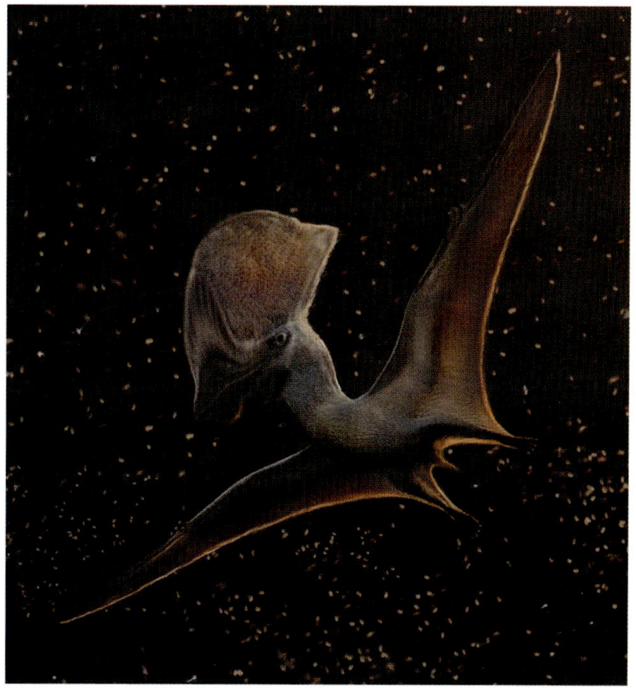

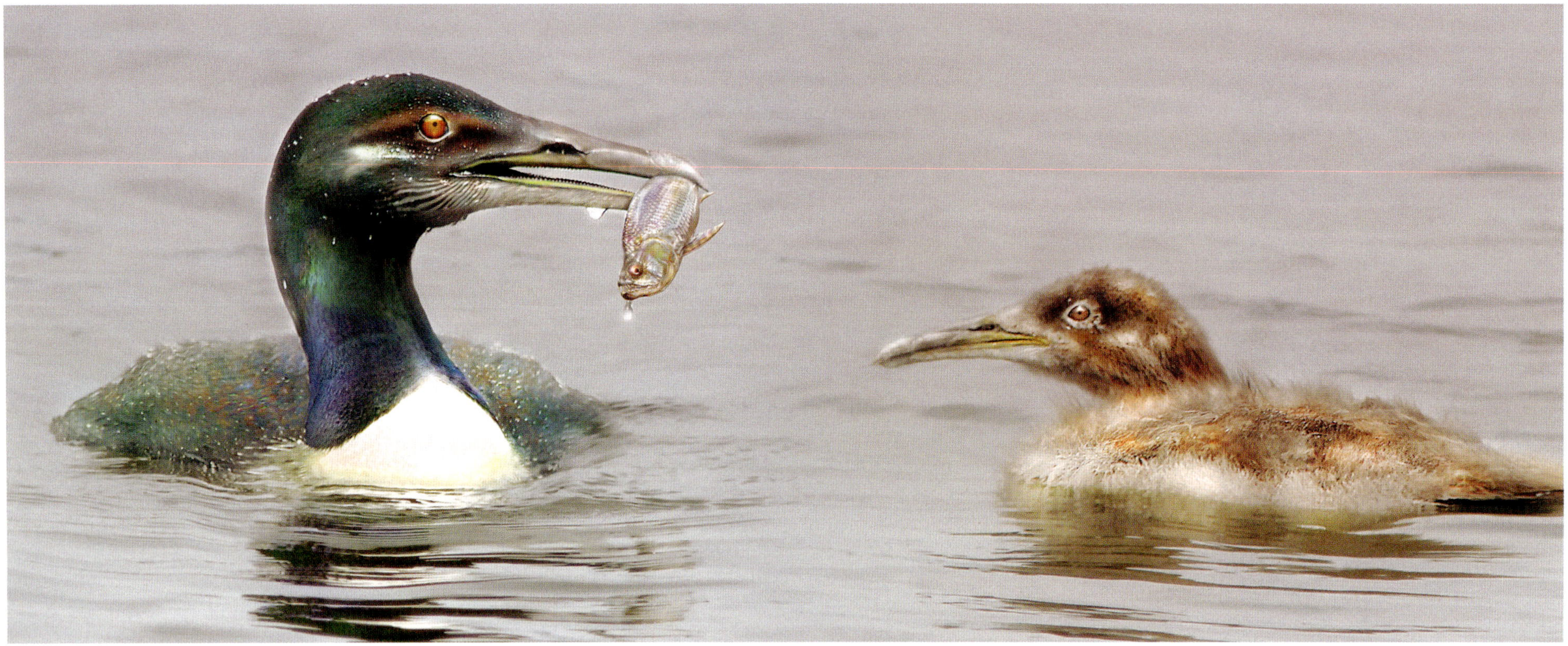

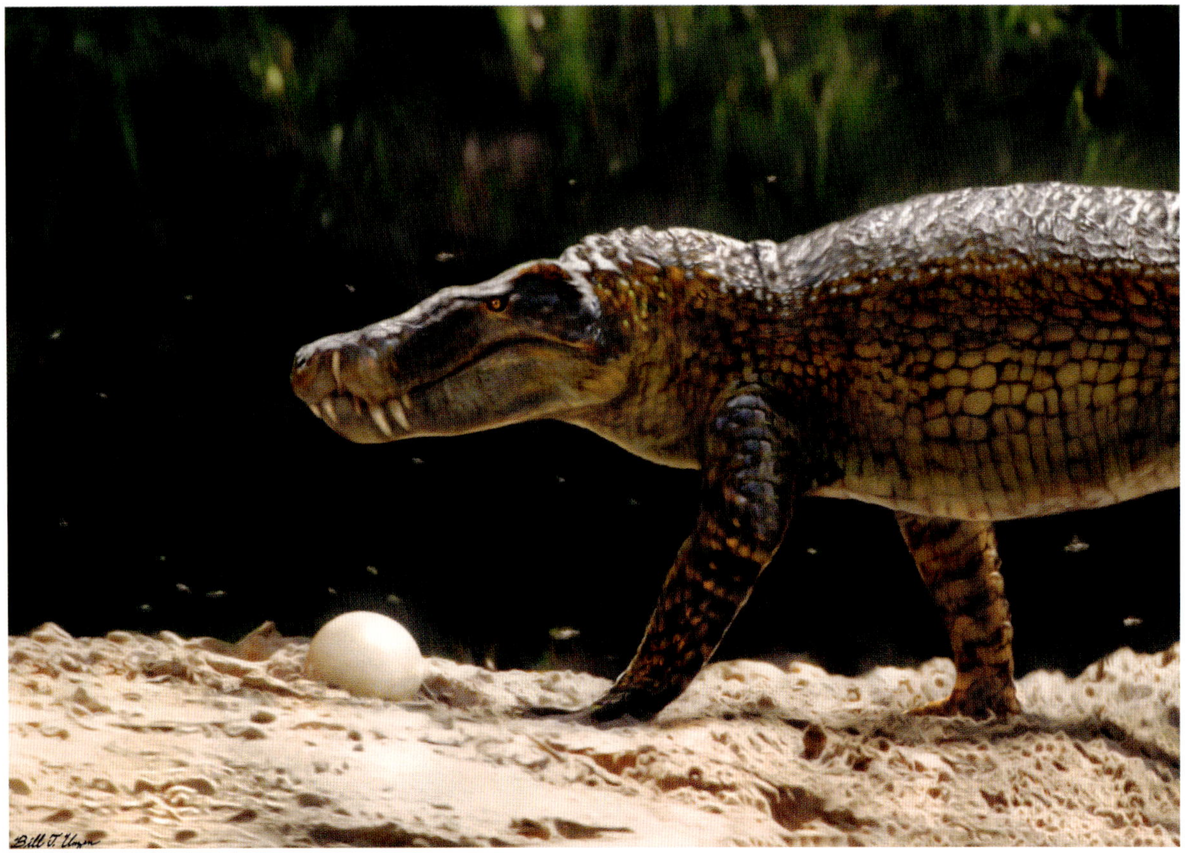

▲ *HESPERORNIS* VS *XIPHACTINUS*

We remain unsure which Cretaceous birds cared for their young. This is the big, flightless, vaguely loon-like seabird *Hesperornis*. The food item here is a juvenile *Xiphactinus*, a ray-finned fish that reached 6m (19.5ft) in length. An awesome predator sometimes depicted as hunting *Hesperornis*, *Xiphactinus*, like all giants, started small, and this image turns the tables.

• Pencil and digital | 2022

◄ *BAURUSUCHUS*

Baurusuchus of Late Cretaceous Brazil is among the best known of the sebecosuchians, a group of predatory terrestrial crocodyliforms equipped with narrow, deep faces and massive teeth. A partial *Baurusuchus* skeleton described in 2010 confirmed long, slender limbs and other features consistent with a terrestrial, active and swift-moving way of life.

• Digital | 2022

DRIZZLY DAY *DILOPHOSAURUS* ►

The large Early Jurassic theropod *Dilophosaurus* (it was 7m or 23ft long) must sometimes have been out in the open when it rained. Was their skin adapted for this? Did they have furrows in their scales, or even 'drip lines' on their plumage? Or was rain infrequent enough in its ecosystem that they lacked such features?

• Digital | 2023

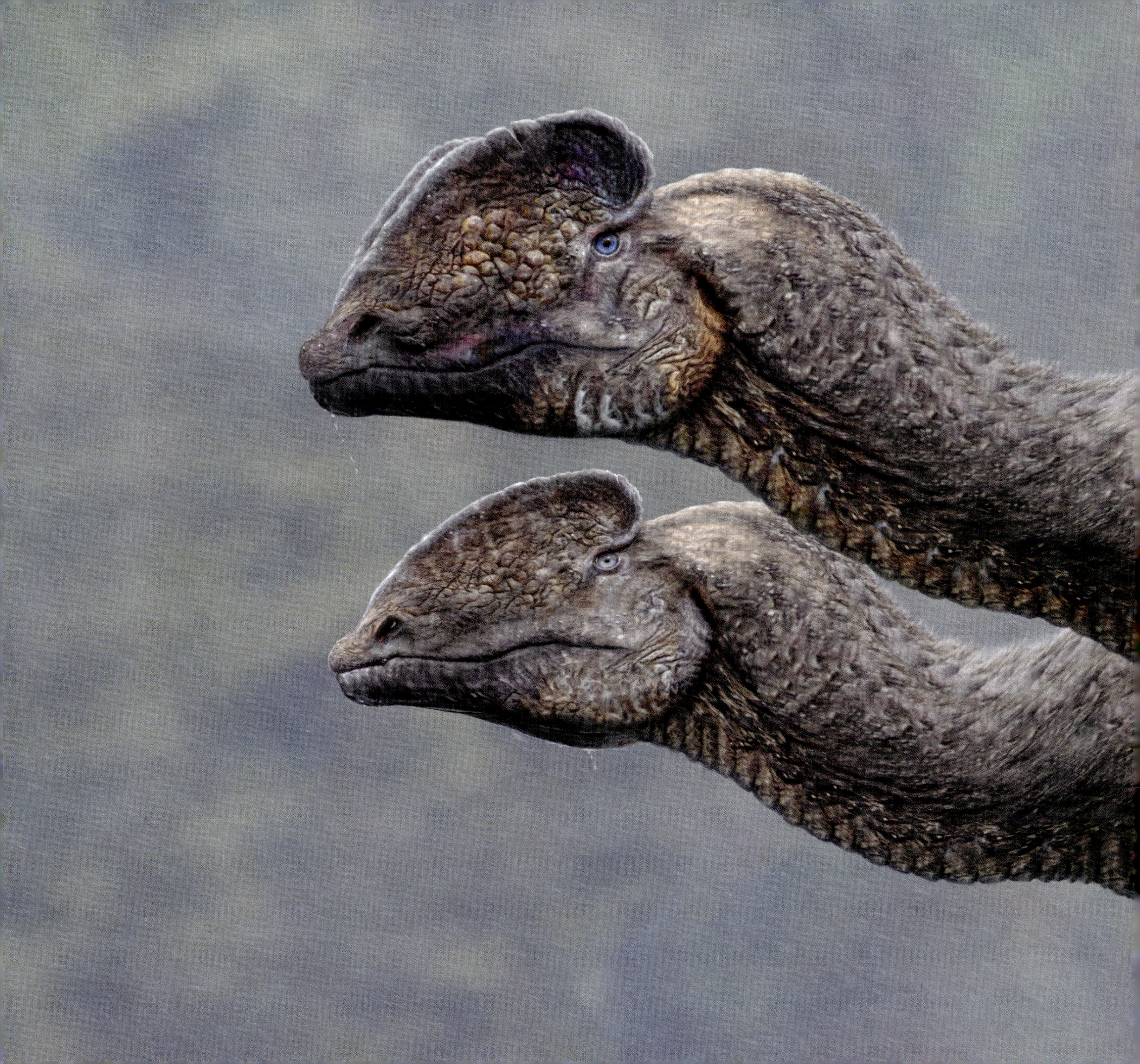

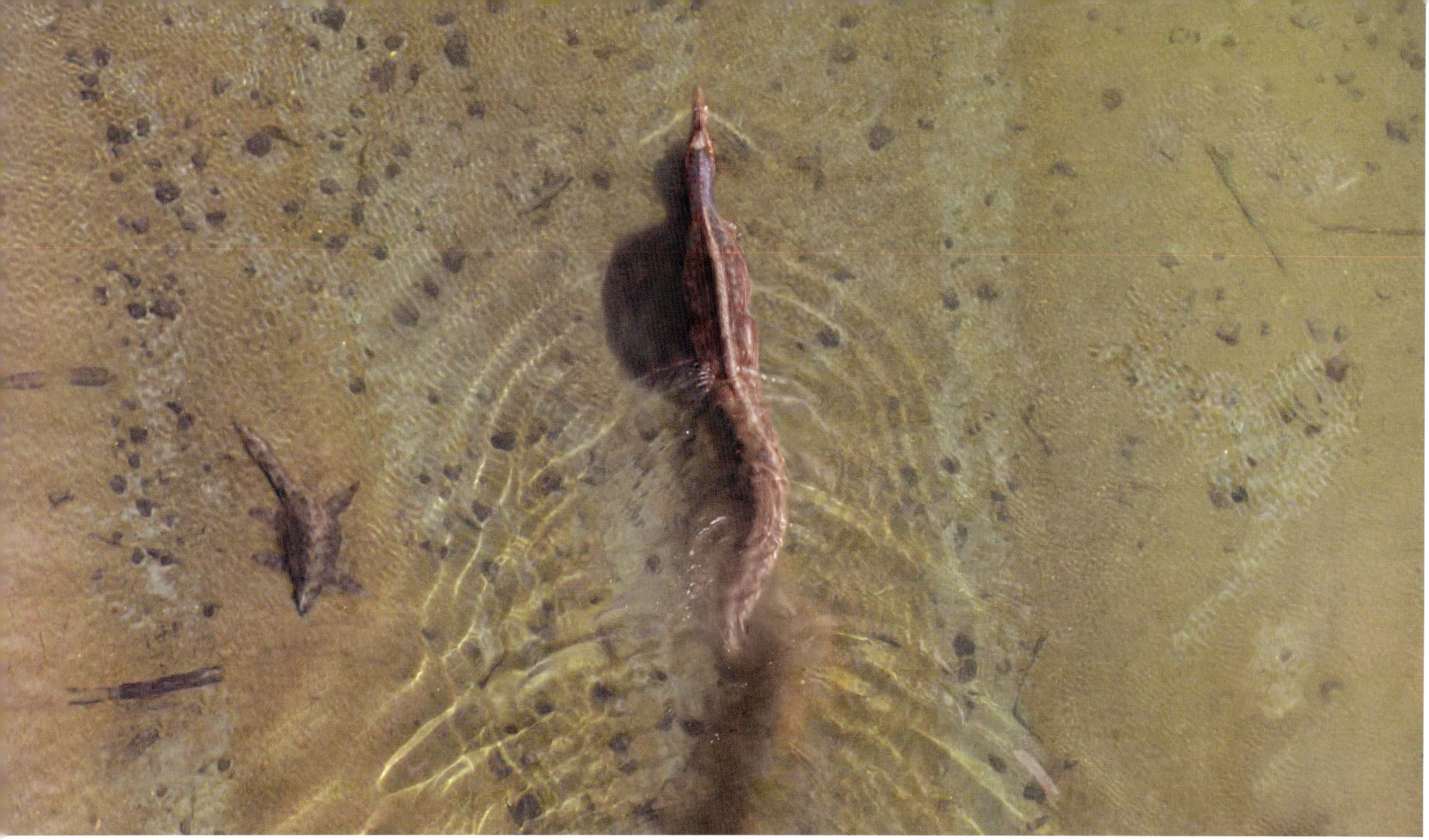

▲ SWIMMING *SPINOSAURUS*

Whatever we make of the lifestyle and ecology of *Spinosaurus*, it surely swam on occasion. Inspired by views of crocodylians photographed from drones, this image suggests what *Spinosaurus* might have looked like from above. Nearby, a leptocleidid plesiosaur keeps its distance. Leptocleidids were small plesiosaurs often associated with lagoons, deltas and rivers.

• Digital | 2022

ALLOSAURUS ►

The more detailed our understanding of *Allosaurus* has become, the more appealing it has become as a subject in art. Its facial crests and horn-lined ridges are striking. In addition, the combat and predation that these great predators took part in must have left them bloodied, injured and torn at times.

• Digital | 2023

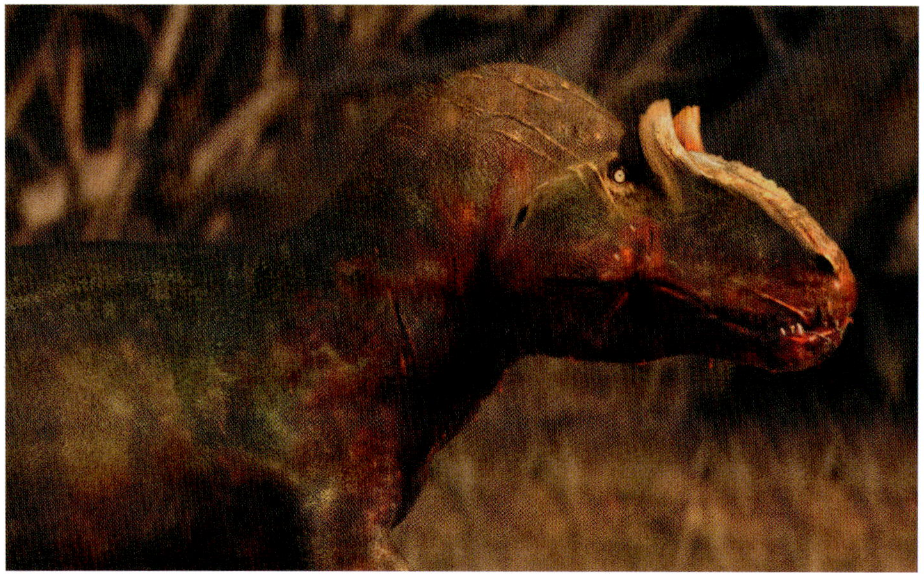

▼ *MIRAGAIA* MIRAGE

If viewing a dinosaur across a hot, sunlit plain, we should expect heat haze, shimmer, and mirages. That concept is captured to great effect in this very innovative illustration. The dinosaur is *Miragaia*, a famously spiky stegosaur originally described from the Upper Jurassic. Some North American fossils might also represent this dinosaur.

• Digital | 2023

MERAXES FOOT ▶

Like modern birds, the feet of some – though not all – extinct theropods were lined with large, overlapping scales. This foot is that of *Meraxes*, a gigantic Argentinian allosauroid named in 2022 after one of the dragons from George R. R. Martin's novel series *A Song of Ice and Fire*. A special feature of this dinosaur is the presence of an unusually large, strongly curved claw on the second toe. Perhaps *Meraxes* was in the habit of pinning prey down during capture and restraint.

• Pencil and digital | 2023

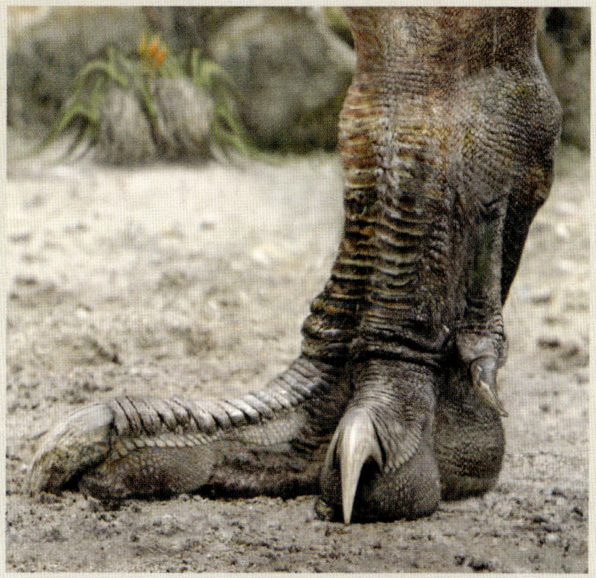

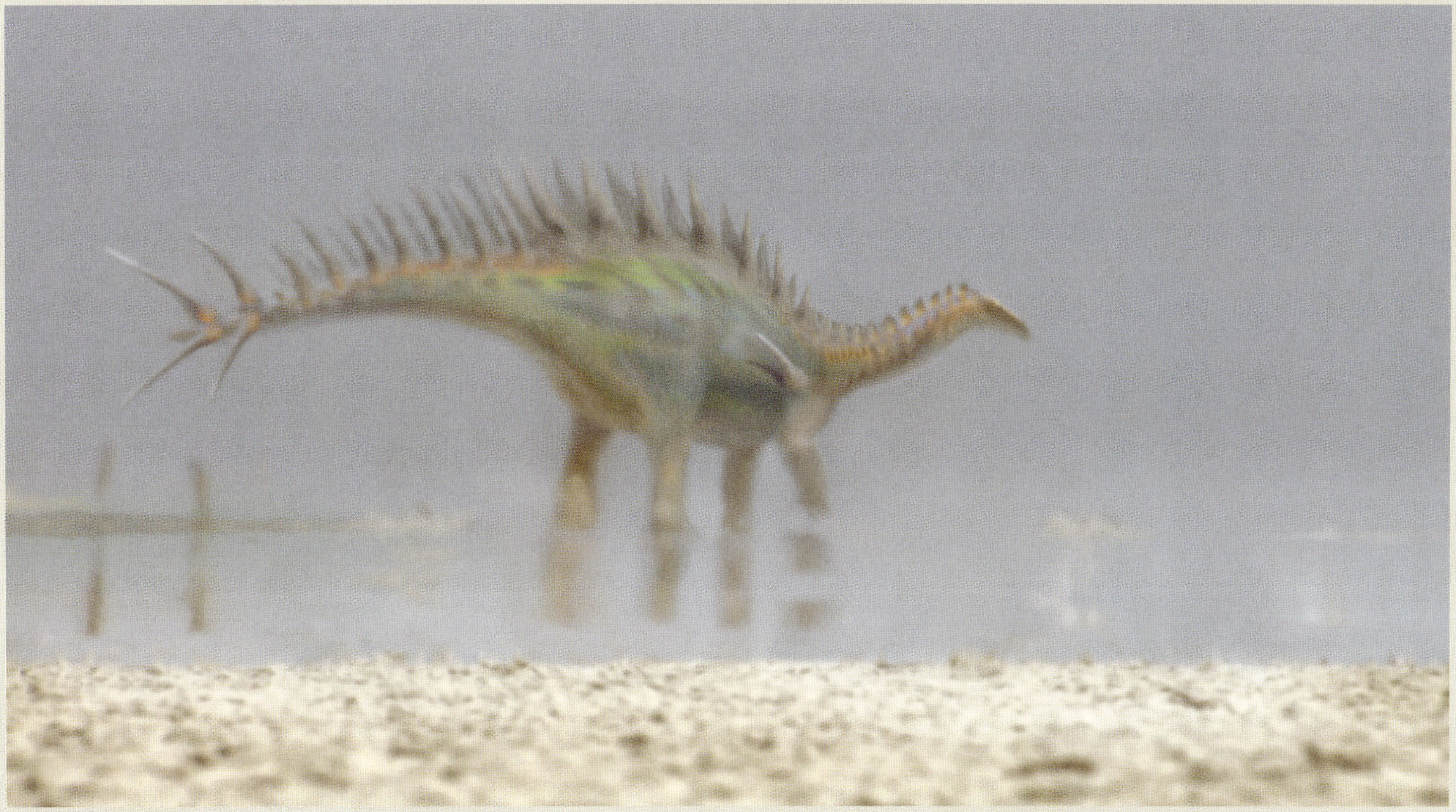

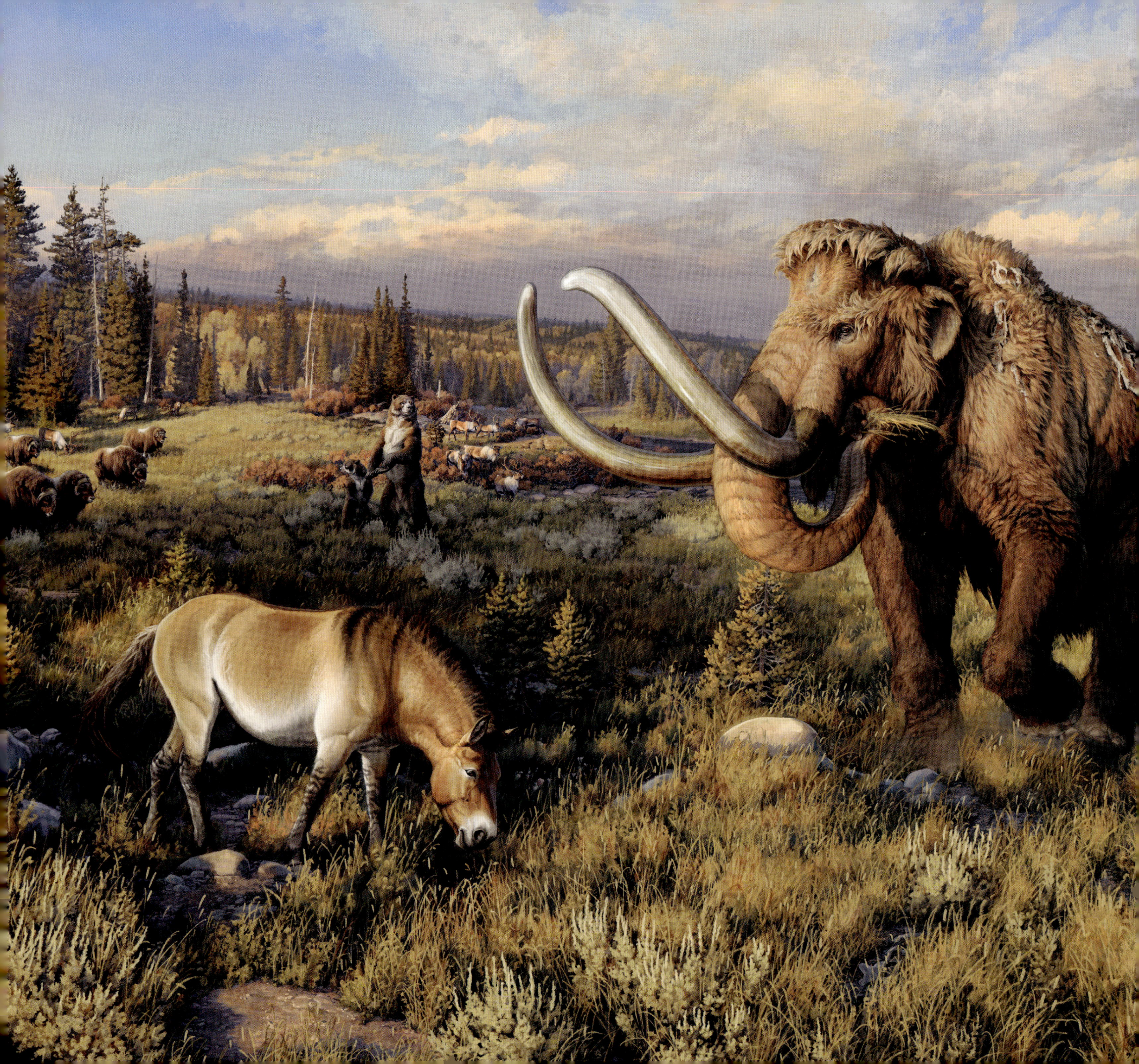

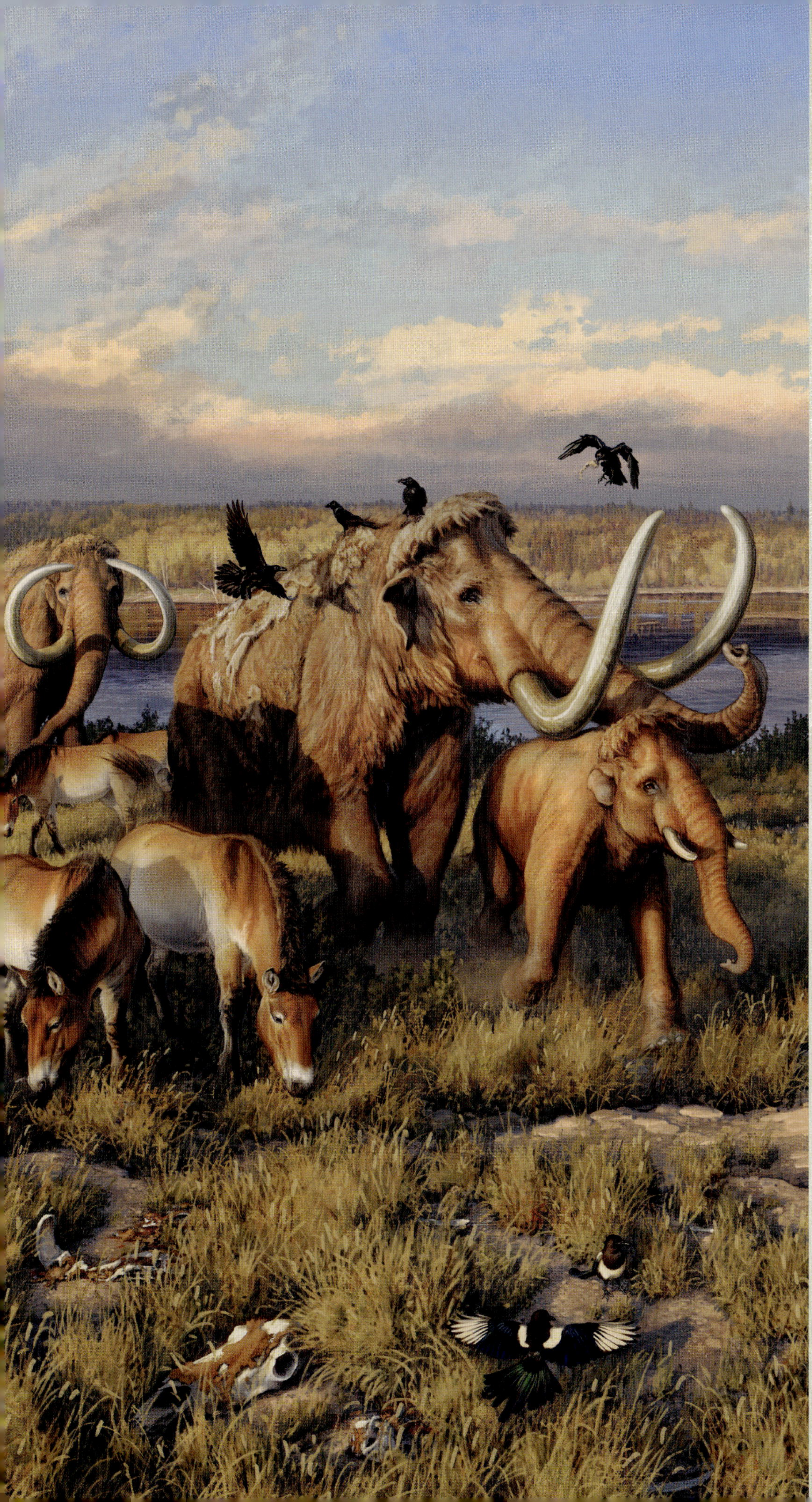

BETH ZAIKEN

BETH ZAIKEN IS A MINNESOTA-BASED professional artist specialising in the interpretation and communication of natural history and science for museums and educational venues, academic research and publication. She is best known as a mural painter and palaeoartist, particularly of Quaternary megafauna. Beth graduated from the Rhode Island School of Design with a B.F.A. in Illustration. Since 2008, she has worked as an artist-fabricator and lead muralist for Blue Rhino Studio (a prominent museum fabrication studio in Minneapolis, Minnesota), where she has helped to design and create a broad range of custom 2D and 3D art features for museums, zoos and cultural centres around the world. Beth has been awarded the Lanzendorf-National Geographic PaleoArt Prize for 2D Art (presented by the Society of Vertebrate Paleontology) twice: in 2020, and again in 2023. Her permanent exhibit artwork can be seen at such institutions as the Royal Alberta Museum, Edelman Fossil Park & Museum, Chicago's Field Museum, the Royal Saskatchewan Museum, the Ancient Ozarks Natural History Museum, San Diego Natural History Museum, Ice Age Fossils State Park, the New York State Museum and the Minnesota Zoo.

In addition to her monumental painterly murals, she also illustrates for small publications and academic projects, creating science communication tools for research as well as multiple pop-science books. The latter include 2020 *Mega Rex: A* Tyrannosaurus *Named Scotty* by Dr. W. Scott Persons IV and the 2022 title *Otherlands* by Thomas Halliday. Recently, her work has graced the covers of major publications such as *Scientific American* and the international scientific journal *Nature*.

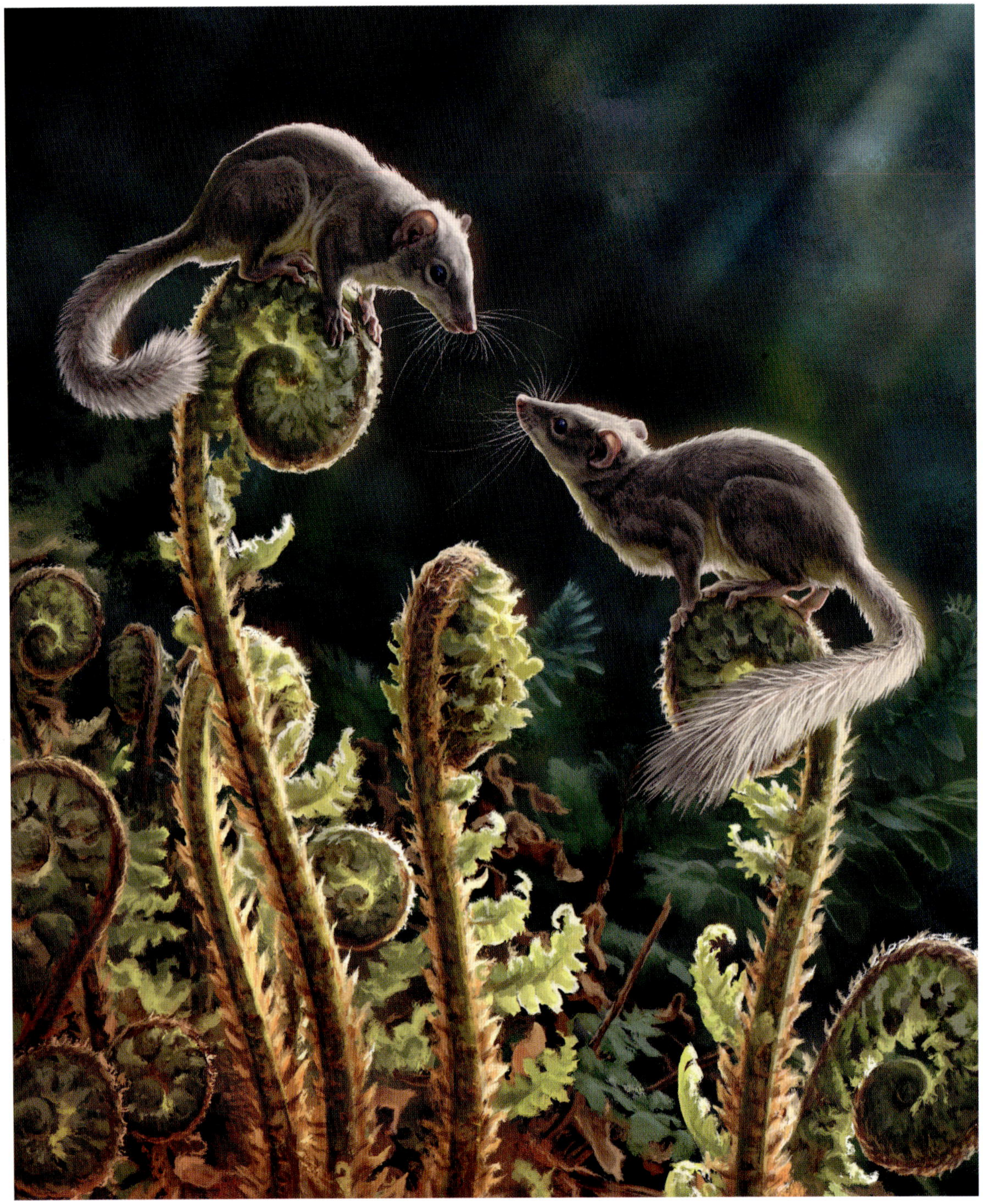

[Previous pages]

ICE AGE EDMONTON

Just part of a huge mural 30m (60ft) long and 3.9m (13ft) tall, this section depicts woolly mammoth (note the moulting fur, no doubt popular nesting material for the ravens and magpies shown), Ice Age horses, and, in the background, both musk ox and the short-faced bear *Arctodus*. This amazing piece won the Lanzendorf-National Geographic PaleoArt Prize for 2D Art in 2020. Used courtesy of the Royal Alberta Museum.

• Digital | 2019

PTERANODON COLONY ▸

Few pterosaurs have been as intensively studied as *Pteranodon*, a denizen of the Western Interior Seaway that covered a great swathe of North America during the Late Cretaceous. *Pteranodon* males and females differed in size and crest shape, and we infer that these sea-going giants nested on islands, coasts and shores, as shown here. Used courtesy of Edelman Fossil Park & Museum.

• Digital | 2022

◂ *CIMOLESTES* ON FERNS

Cimolestes is an archaic placental mammal known from both the Late Cretaceous and post-Cretaceous rocks. The affinities of *Cimolestes* remain uncertain but it is assumed that these animals superficially resembled such modern omnivorous mammals as tree-shrews. This artwork was used on the cover of the June 2022 edition of *Scientific American*.

• Digital | 2022

▾ HELL CREEK HUNT

Fossil evidence shows that *T. rex* and *Triceratops* fought, as revealed by bite marks on both the brow horns and frills of *Triceratops* specimens, some of which subsequently healed. That the Trike survived indicates that an adult *Triceratops* was a tough opponent for even the biggest *T. rex*. This was the September 2024 cover for *Scientific American* magazine.

• Digital | 2024

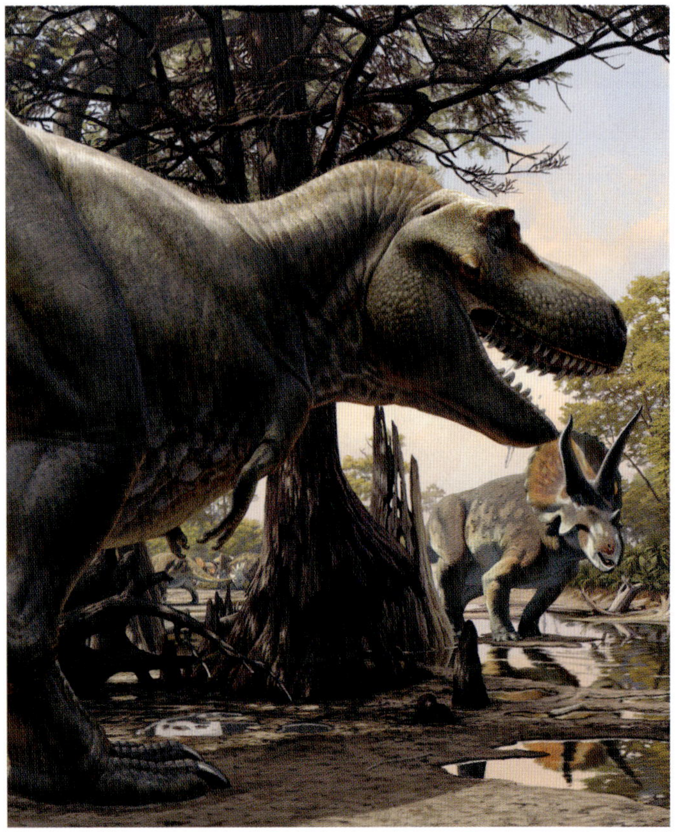

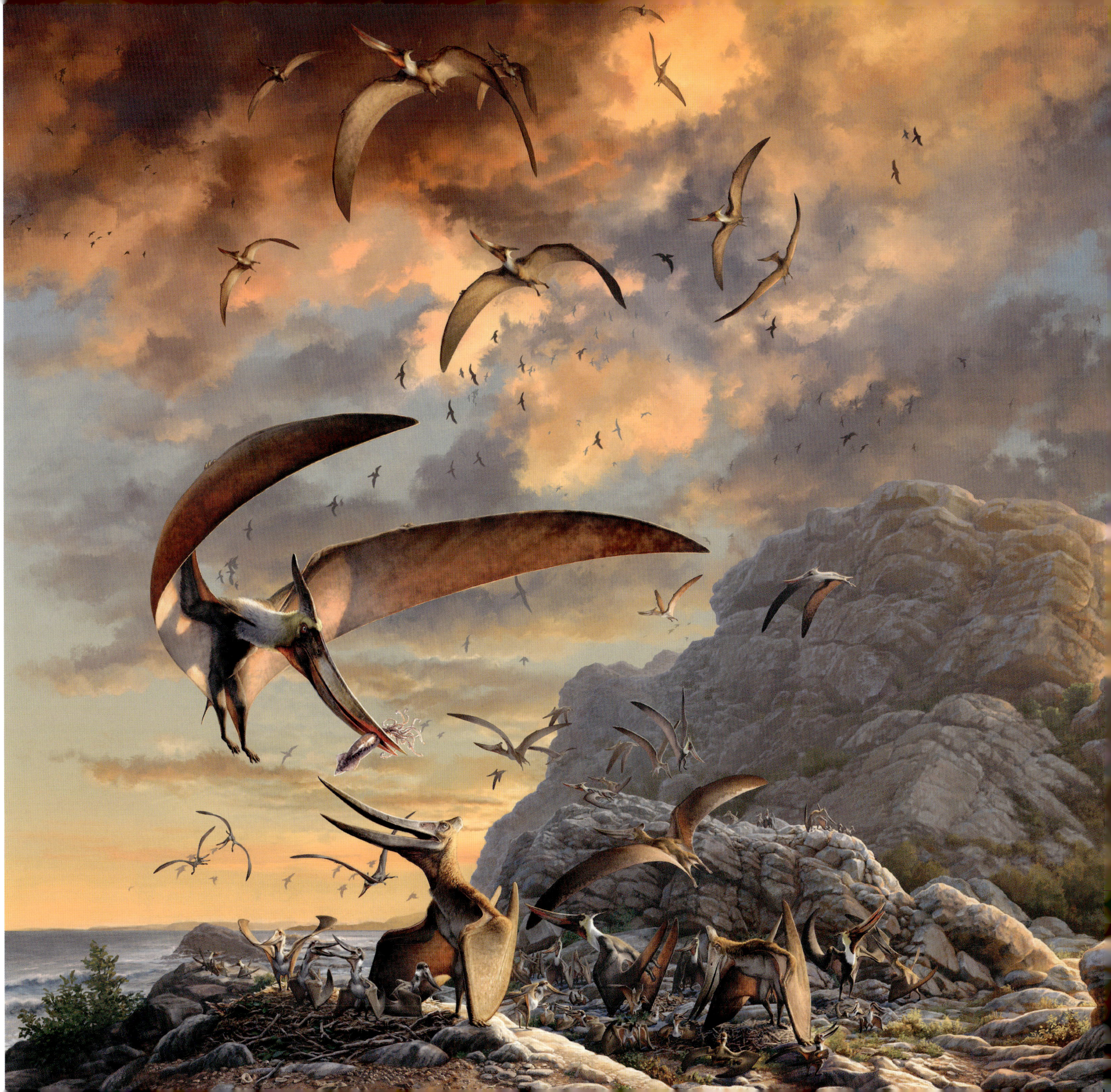

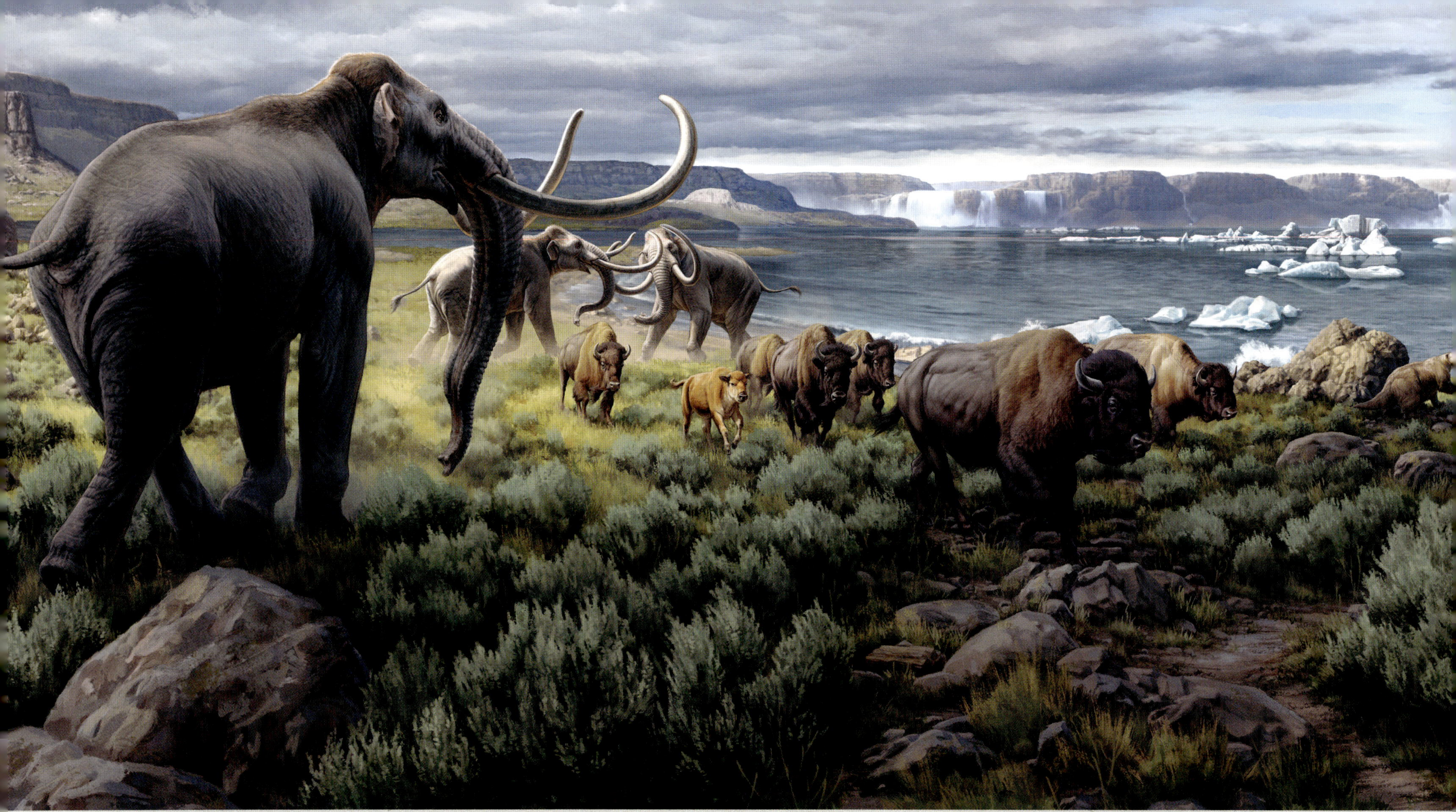

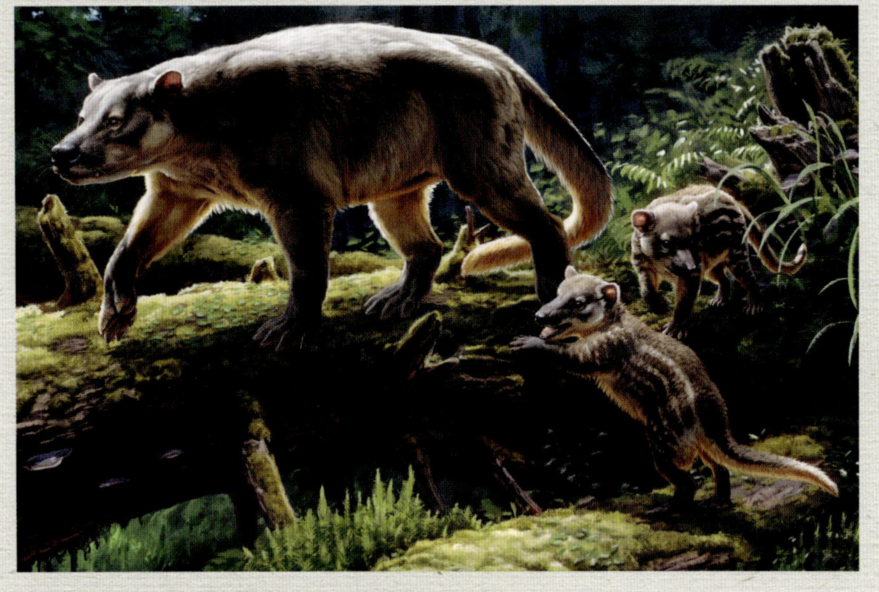

◄ *ECTOCONUS*

Somewhere in the southern part of the USA about 65 million years ago, a mother *Ectoconus* leads her stripey youngsters across a fallen log. *Ectoconus* is a member of an enigmatic mammal group called the periptychids. Though obscure and poorly known outside of specialist groups, *Ectoconus* is known from good remains, and illustrations of its skeleton and life appearance are a mainstay of books and articles on fossil mammals.

• Digital | 2022

▲ THE GREAT FLOOD

Another of Beth's incredible murals – and, again, just part of it – illustrates cataclysmic flooding in what is now the Channelled Scablands region of Washington state, USA. This mini-Grand Canyon was carved by the waters of Glacial Lake Missoula as they were released by the intermittent melting of a glacial ice dam. These repeated flooding events had a devastating impact on the landscape and its wildlife, among which were the Columbian mammoths, *Bison antiquus* and ground sloth *Megalonyx* depicted here.

Used courtesy of the Yakima Valley Museum.

• Digital | 2019

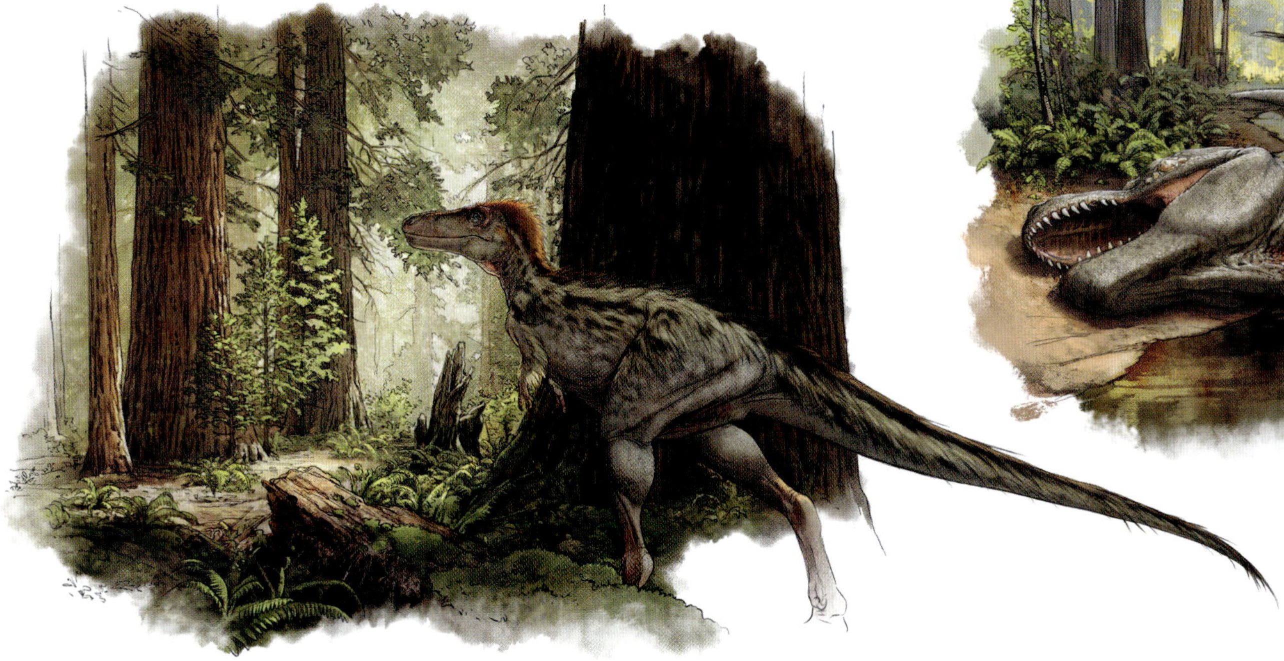

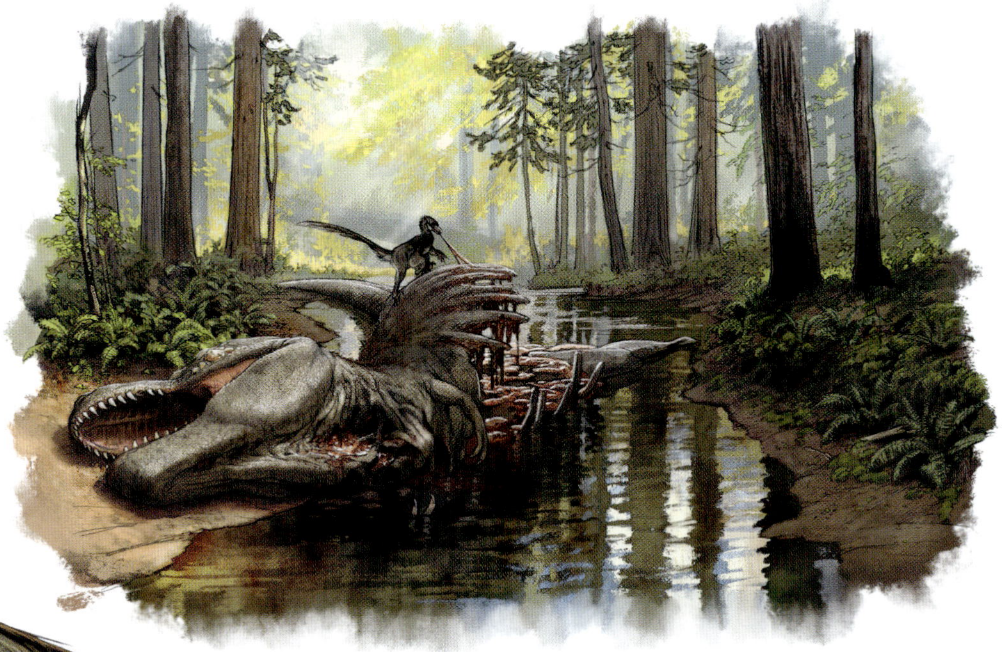

▲ SCOTTY'S LIFE CYCLE

Fossils show that *T. rex* spent its early life as a lightweight, slender-limbed predator of small and mid-sized animals. Studies of bone growth reveal that *T. rex* lived to about 30 years of age, though death due to disease, injury and accident likely claimed some individuals long before this time. The images here are featured in Scotty's home exhibit at the Royal Saskatchewan Museum as well as in the book *Mega Rex: A* Tyrannosaurus *Named Scotty*, by Dr W. Scott Persons IV.

• Digital | 2019

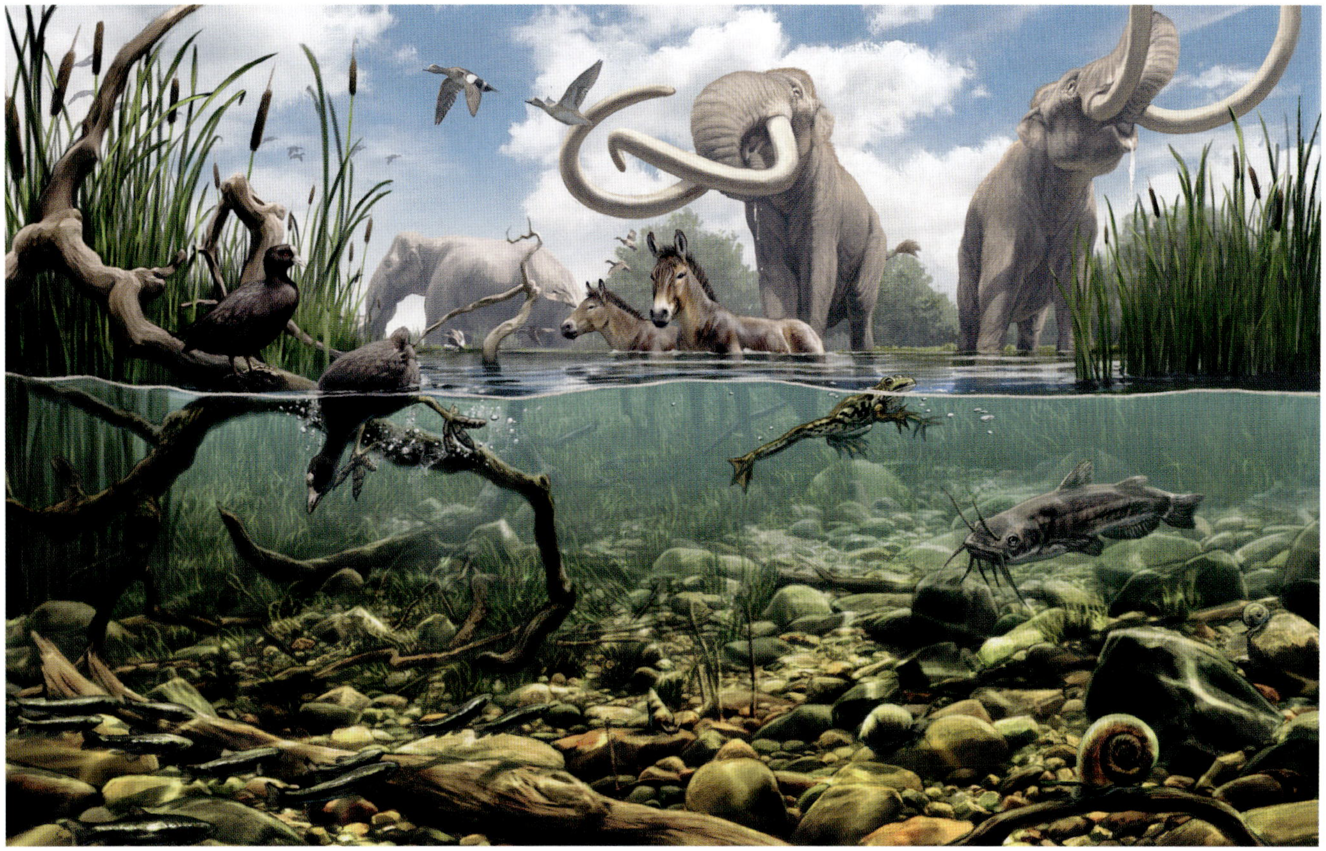

BENEATH THE SURFACE ▶

This remarkable 'frog's eye view' depicts animal life known from Ice Age Fossils State Park, Las Vegas. Big mammals like Columbian mammoths and *Equus scotti* – a large horse resembling the modern Przewalski's horse – visited the location, while freshwater animals like coots, leopard frogs and channel catfish lived here year-round.

Used courtesy of Ice Age Fossils State Park.

• Digital | 2020

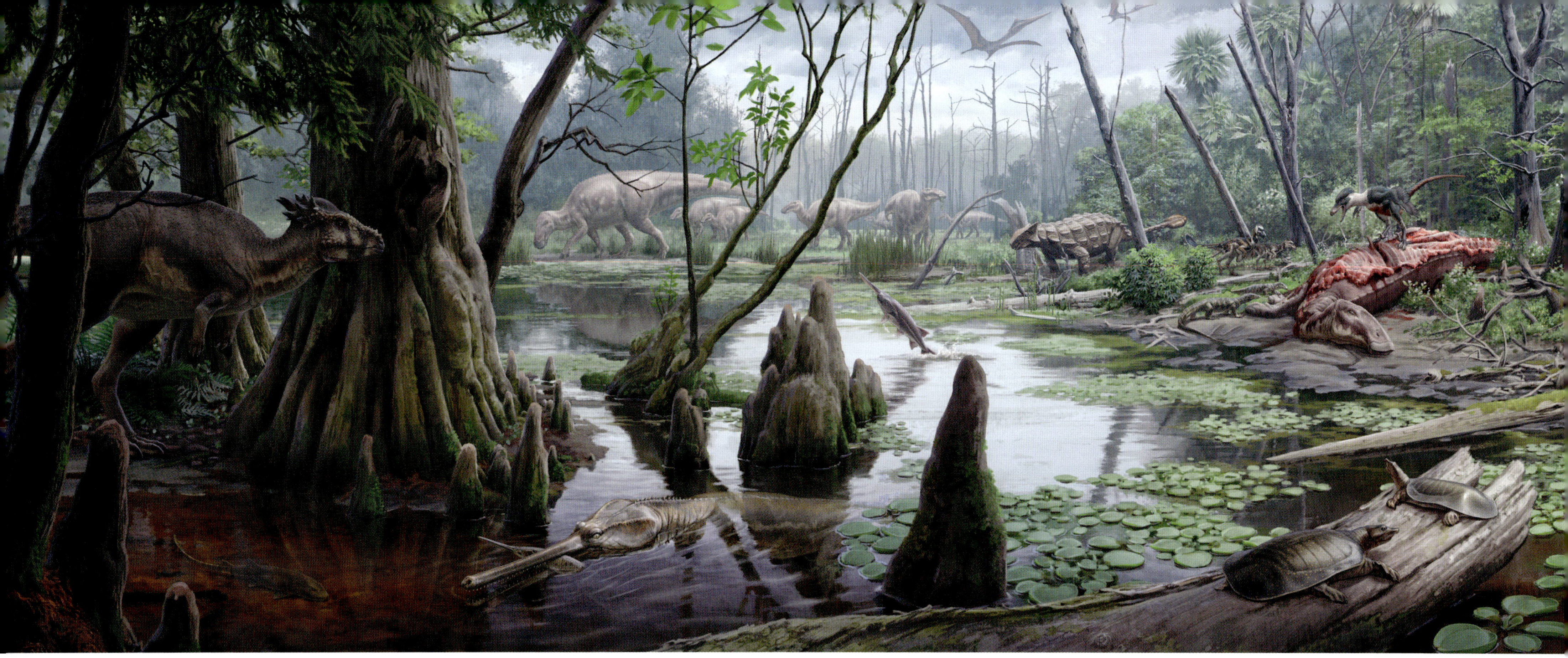

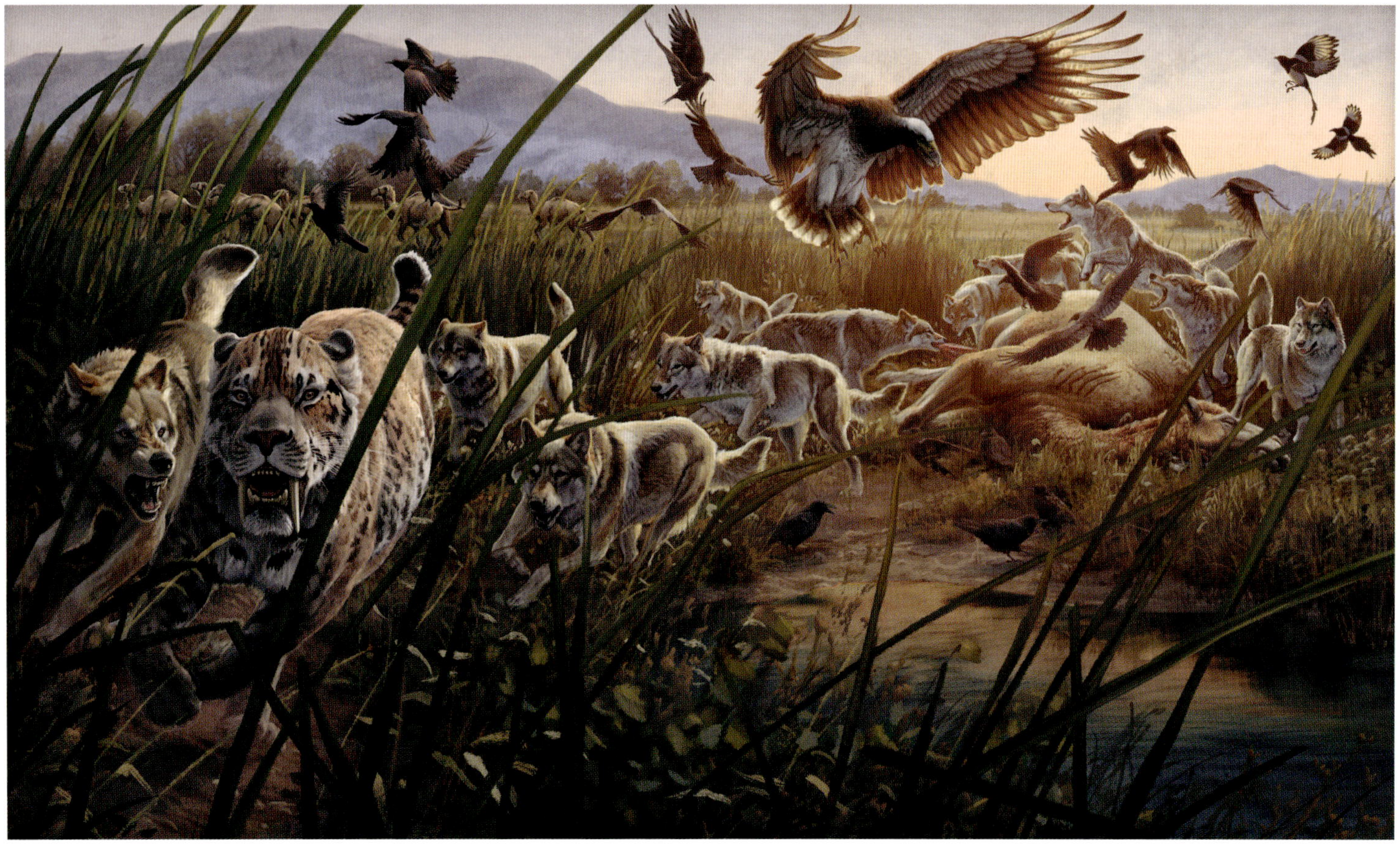

◂ DIRE WOLVES VS *SMILODON*

The iconic American sabretooth *Smilodon* loses its kill – a camel – to a park of dire wolves, a large member of the dog family, known technically as *Aenocyon dirus*. Once thought to have a close affinity with the living wolf (*Canis lupus*), a 2021 study showed otherwise, the dire wolf lineage evolving in isolation in the Americas. The dire wolf was once thought to have a close affinity with the living wolf (*Canis lupus*), but a 2021 study showed the lineage evolving in isolation in the Americas. Interactions like that shown were surely not unique to North America, since big sabretooths and large members of the dog family co-occurred in many places.

Used courtesy of Ice Age Fossils State Park.

• Digital | 2020

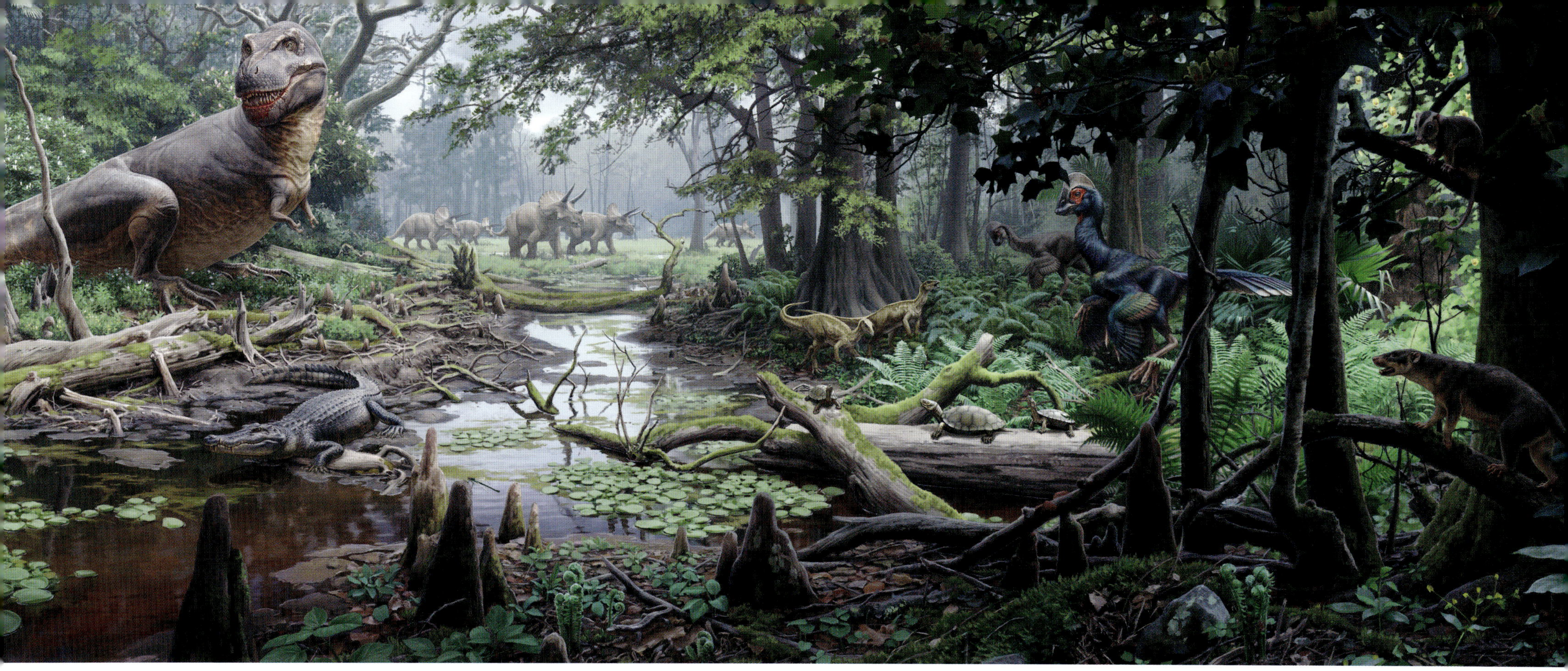

▲ HELL CREEK

The Late Cretaceous Hell Creek fauna of the western USA includes the world's most famous animal assemblage of its time. *T. rex*, *Ankylosaurus*, *Edmontosaurus* and *Triceratops* were part of this, but so were crocodylians, turtles, small mammals and fishes like gar. Fossil plants and studies of the sediments reveal a swampy, heavily forested landscape. This mural was designed for the exhibition *Sue: The* T. rex *Experience* at Chicago's Field Museum.

• Digital | 2020

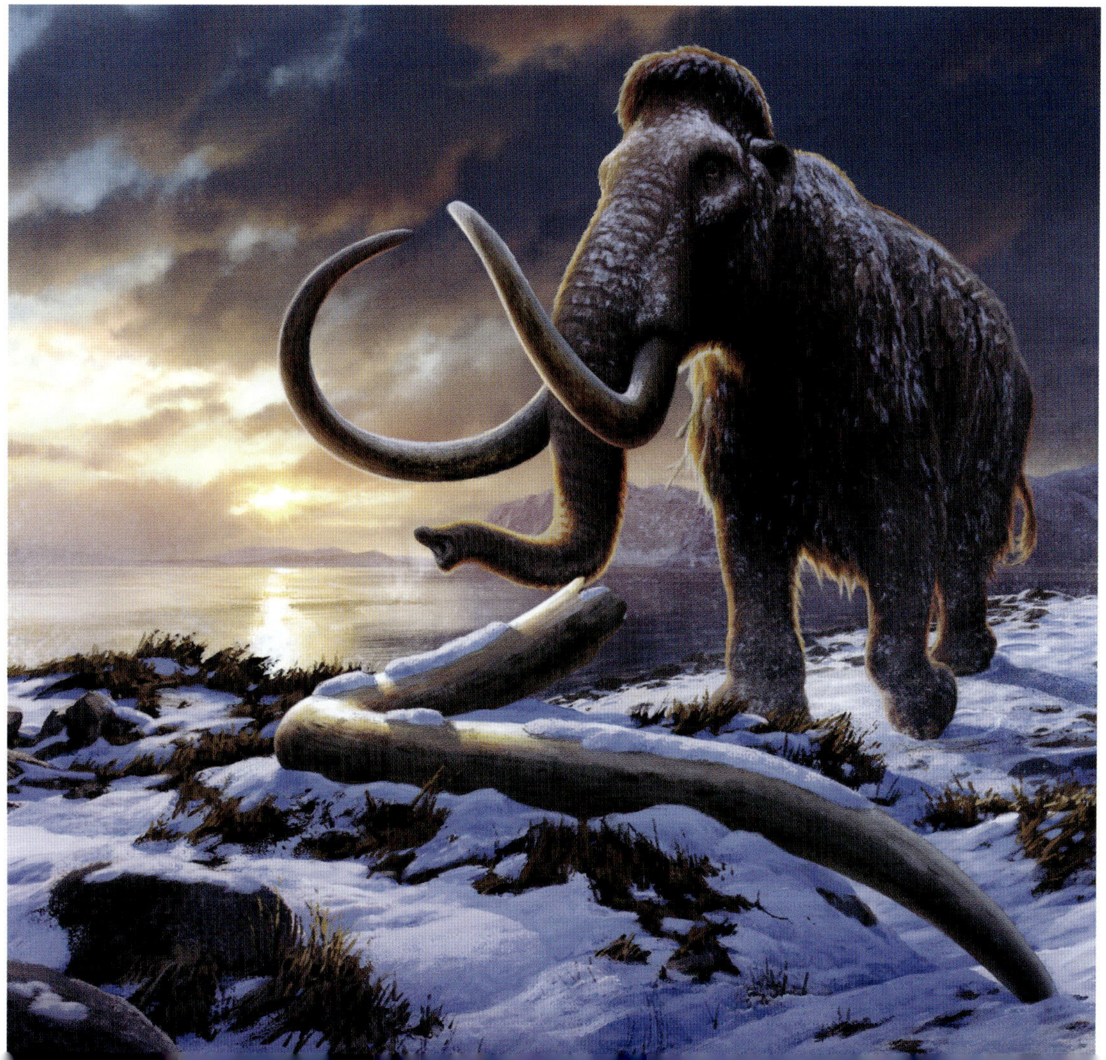

◀ THE WRANGEL MAMMOTH

Located in the Arctic Ocean to the north of Russia's eastern tip, Wrangel Island was home to the very last woolly mammoths. Arguments continue regarding their final extinction, about 4,000 years ago. Studies have suggested genetic meltdown – an accumulation of harmful mutations in an organism's DNA – of the isolated population, but other researchers favour hunting by recently arrived humans as the cause. Beth's art appeared with a large genomic analysis by the Centre for Palaeogenetics in the scientific journal *Cell*.

• Digital | 2024

MOSASAUR DIVERSITY ▶

Mosasaurs – the great sea lizards of the Cretaceous – include at least five distinct groups, each with their own anatomical specializations and long evolutionary histories. That fact is emphasised here as the gigantic, long-snouted *Tylosaurus* prepares to dispatch and consume a specimen of the very differently proportioned and far smaller *Plioplatecarpus*.

• Digital | 2021

▼ ORDOVICIAN SEABED

This Ordovician scene – showing the marine life of Minnesota, USA, about 470 million years ago – shows a seabed decorated with ancient corals, sea lilies (relatives of urchins and sea-stars) and ancient molluscs. Swimming overhead are orthoceratoids, a group of shelled, squid-like predators, the biggest of which could reach 5m (16.4ft) or more in total length.

Used courtesy of Quarry Hill Nature Center.

• Digital | 2018

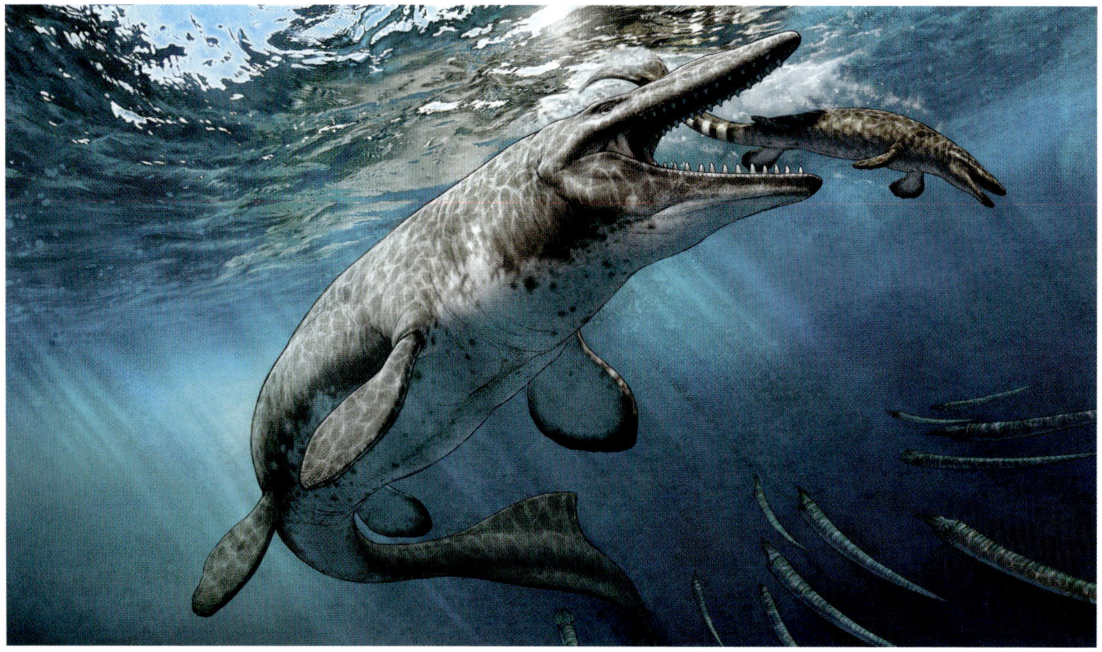

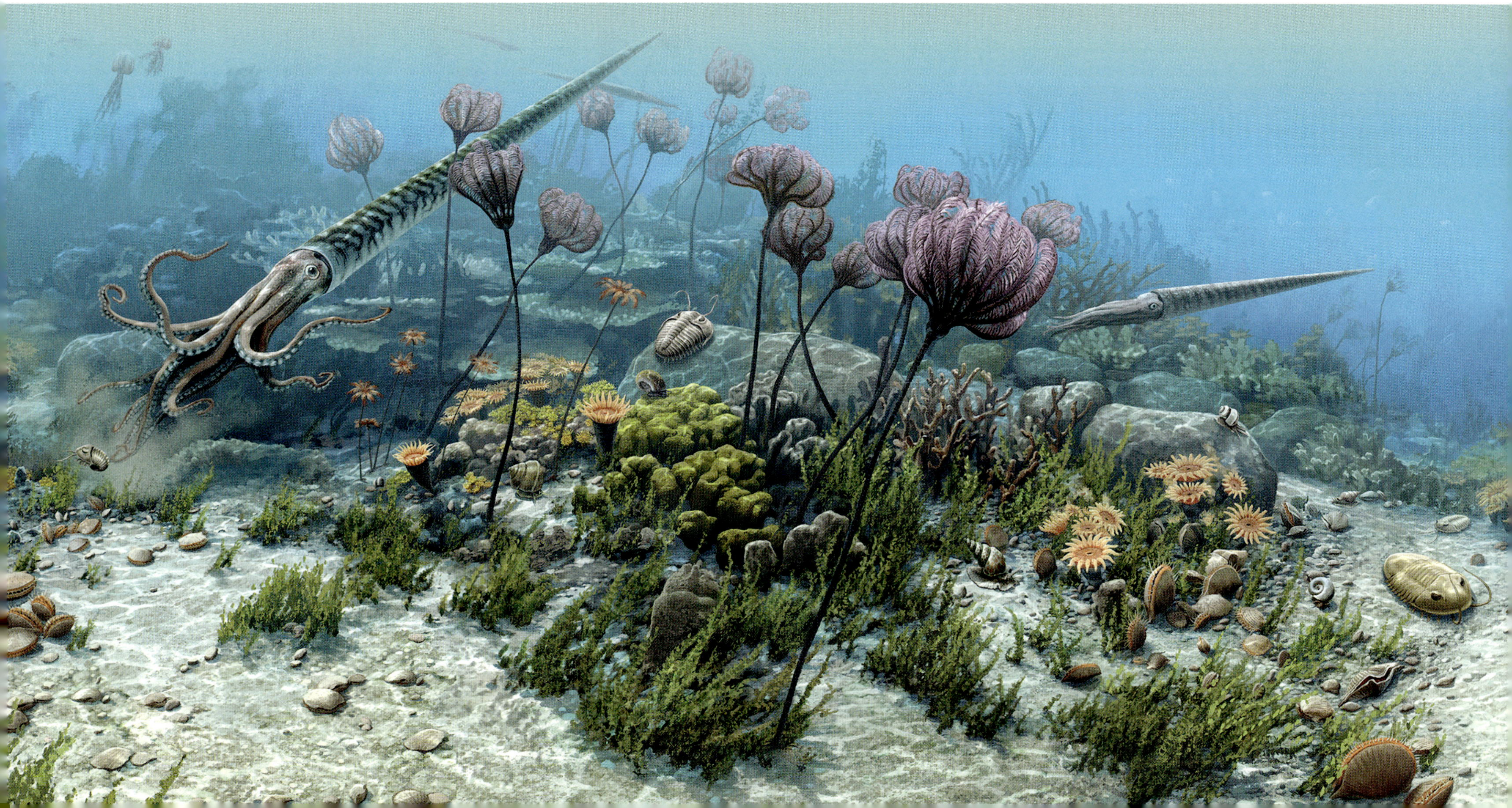

TERMINONATATOR ▸

For all our uncertainty over how they used their absurdly long necks, elasmosaurid plesiosaurs were a successful, long-lived group. *Terminonatator* is from the Late Cretaceous of Saskatchewan, Canada, and was among the last members of its group to inhabit the Western Interior Seaway. This is reflected in its name, which means 'last swimmer'. Stomach contents show that elasmosaurids consumed large predatory fish like the ray-finned *Enchodus*, shown here.

• Digital | 2021

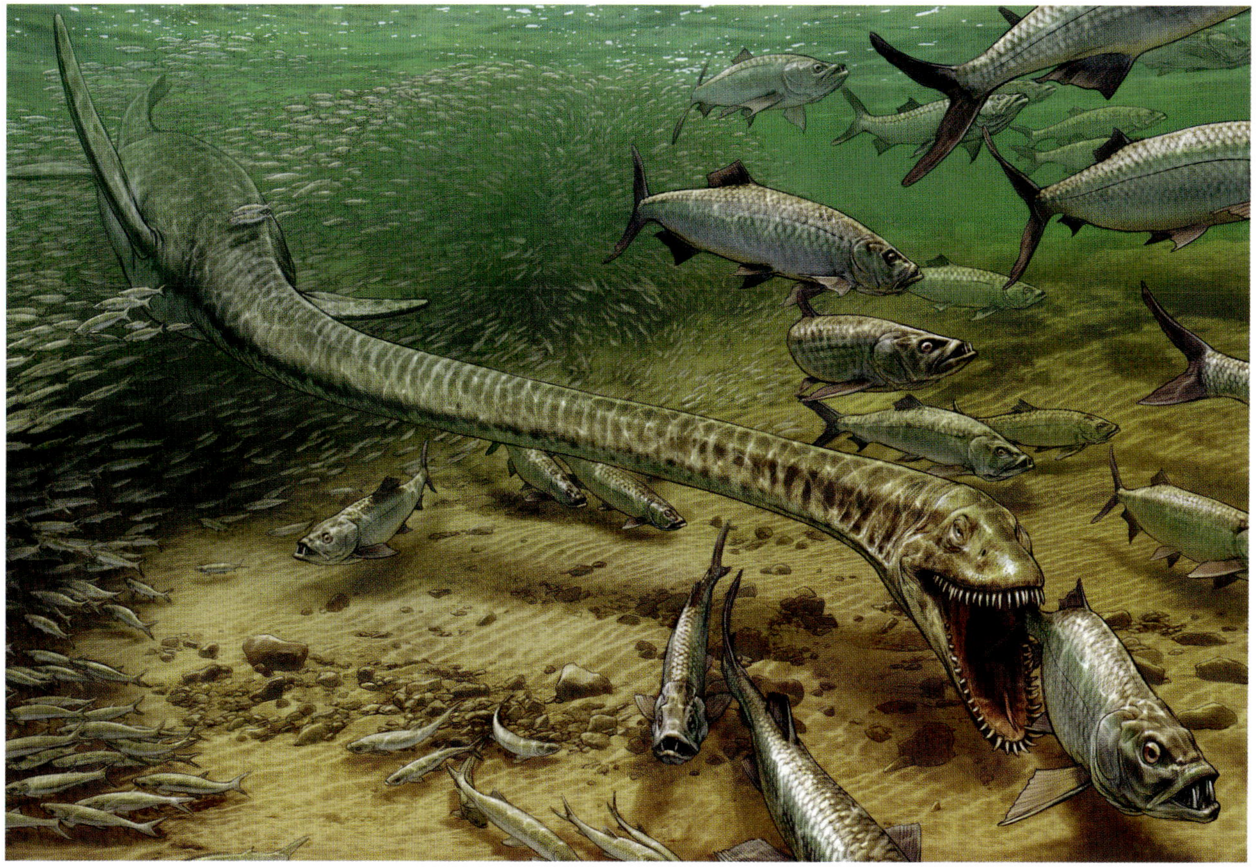

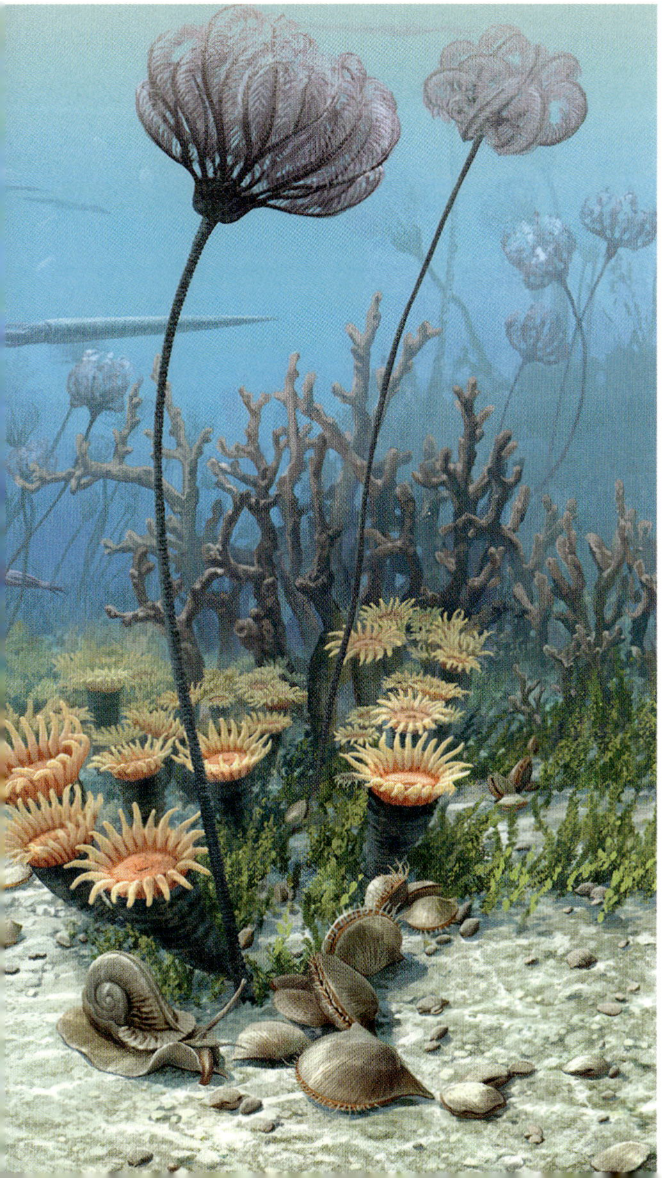

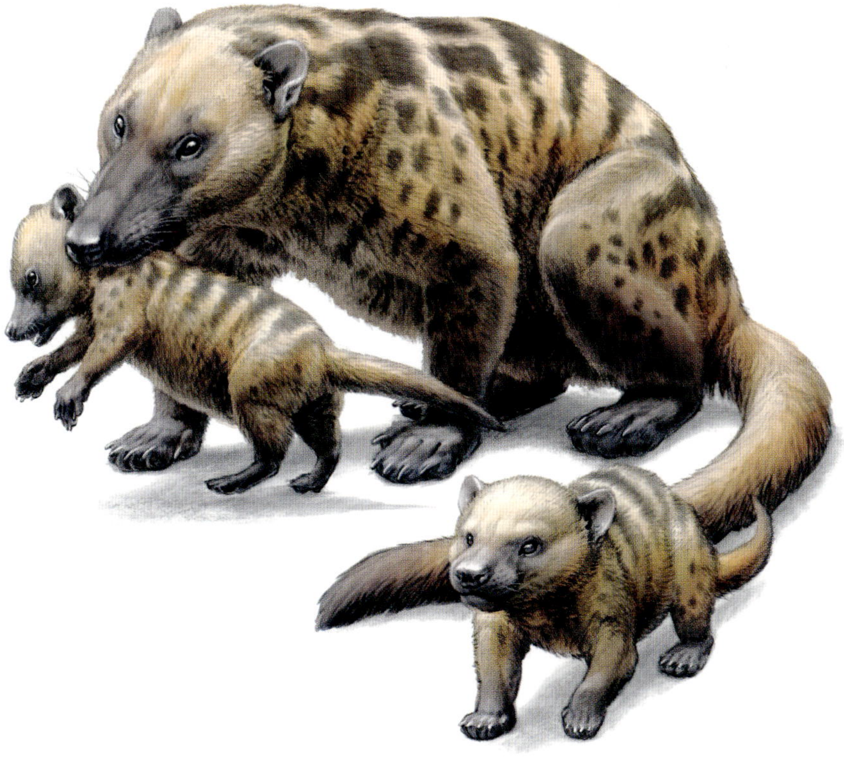

BAIOCONODON ▸

Numerous mammal groups existed during the Paleogene (that is, between 66 and 23 million years ago), most unheard of except to specialists. *Baioconodon* from Colorado, USA, is an arctocyonid, a group of predators or omnivores noted for their massive canine teeth and large skulls. *Baioconodon* was thought to have some adaptation for climbing and so was at home in the forest understorey.

• Digital | 2019

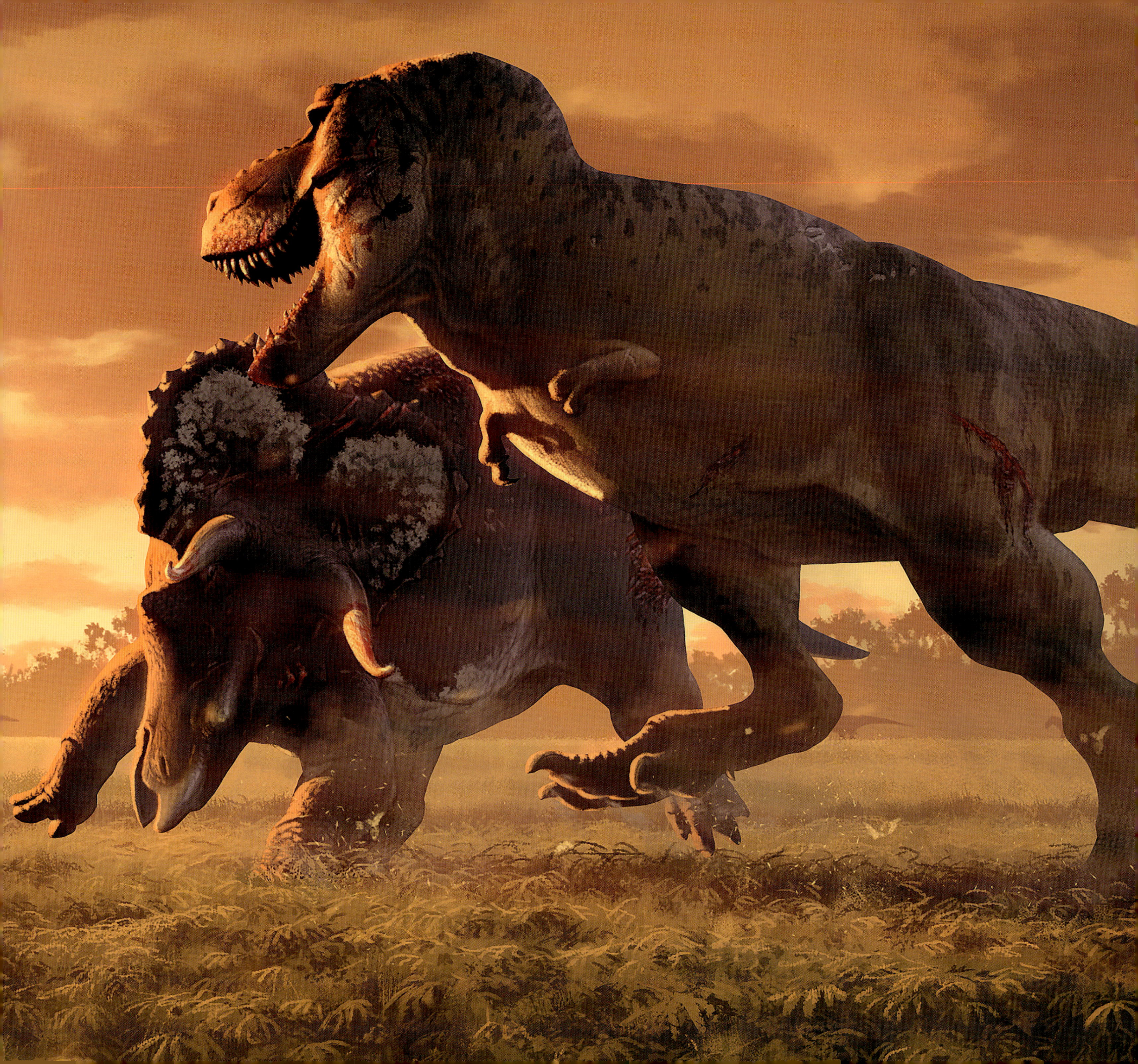

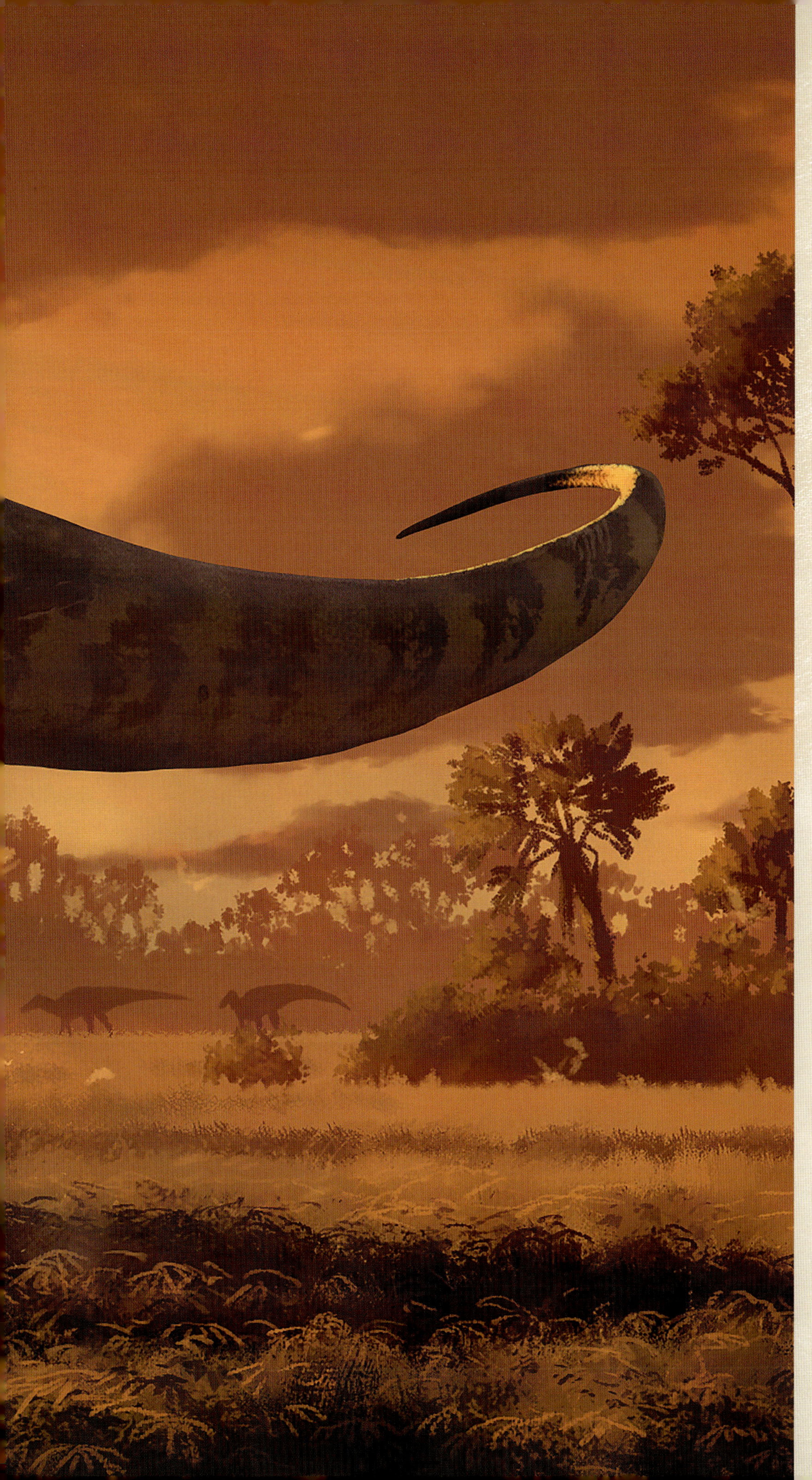

FREDERIC WIERUM

A FAMILY BACKGROUND rooted in the tropics of Belize has given Frederic lifelong immersion in the beauty and complexity of the natural world. It was the powerful believability of the dinosaurs rendered in the BBC's *Walking with Dinosaurs* that galvanised his passion for extinct as well as living animals. From doodling in the elementary classroom to now working as a full-time digital creature and concept artist, Frederic's main focus has always been dinosaurs. Having grown up with encyclopaedias and countless wildlife documentaries, he tries his best to replicate the mystery, brutality, awe and beauty of nature in his palaeoart. He has worked in the gaming industry, on children's books and magazines, a novel, YouTube channels and imagery for Wikipedia. Via the artistic fun of video games and film and the science communication of documentaries and books, all he wants to do is share his love and passion for prehistory with everyone around him, whether in or out of the palaeo-community.

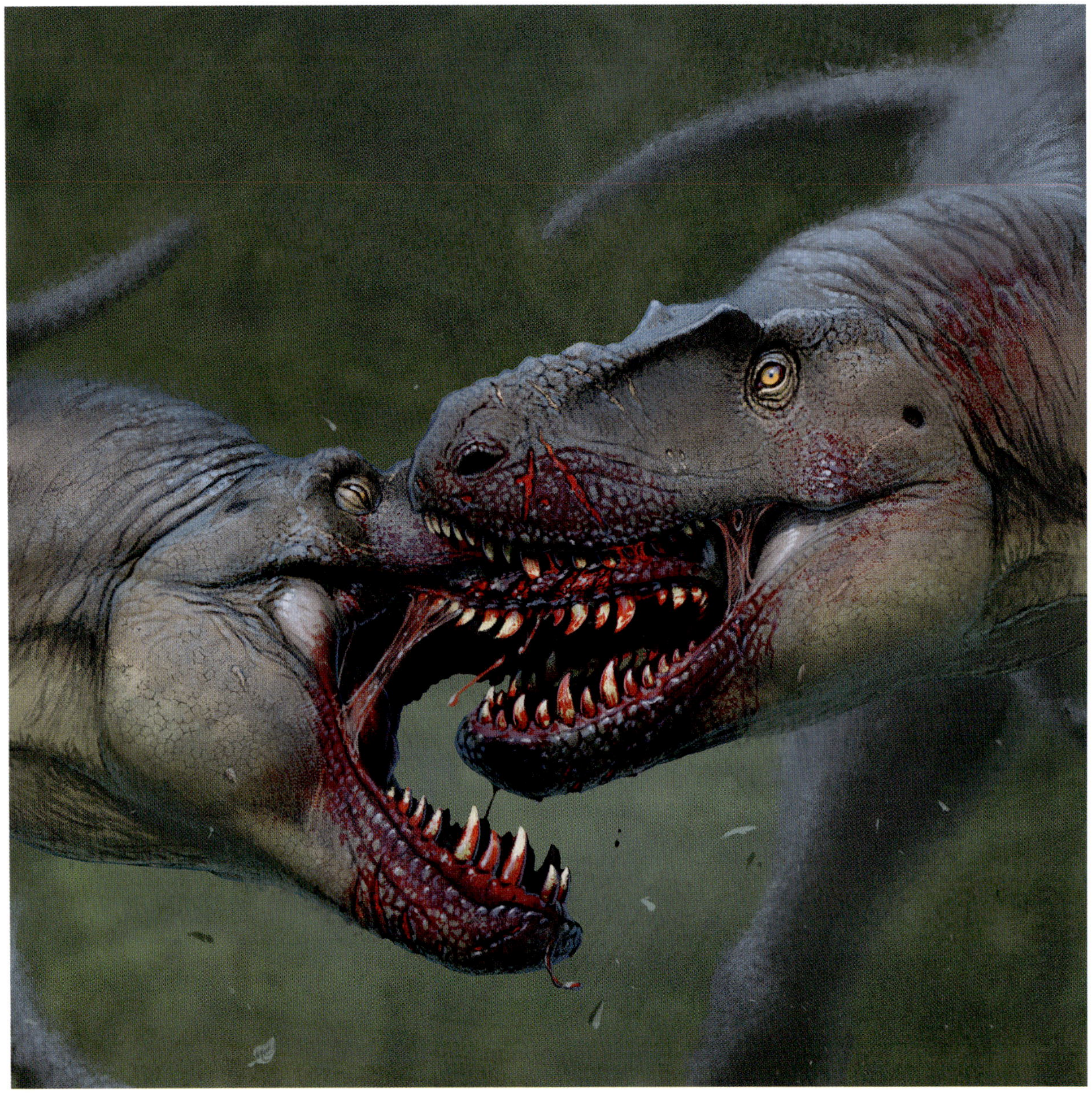

ALLOSAURUS FIGHT ▸

Fights between modern predators like tigers and bears can be brutal, bloody, and even end in death. With that in mind, combat between big theropods like the *Allosaurus* individuals shown here would have been horrific. Teeth and jaws could slice and tear, and the massively hooked hand claws could gouge, rip and punch holes.

• Digital | 2021

▾ *UTAHRAPTOR* STALK

The first remains of *Utahraptor* were found in 1975, but this large dromaeosaurid was not officially described until 1993 – just in time for pop culture stardom due its similarity to the giant dromaeosaurids of a certain popular movie. Reaching 6–7m (20–23ft) and possibly weighing up to 770kg (1,700lbs), *Utahraptor* was a gigantic and unusually heavily built member of its group.

• Pencils & digital | 2018

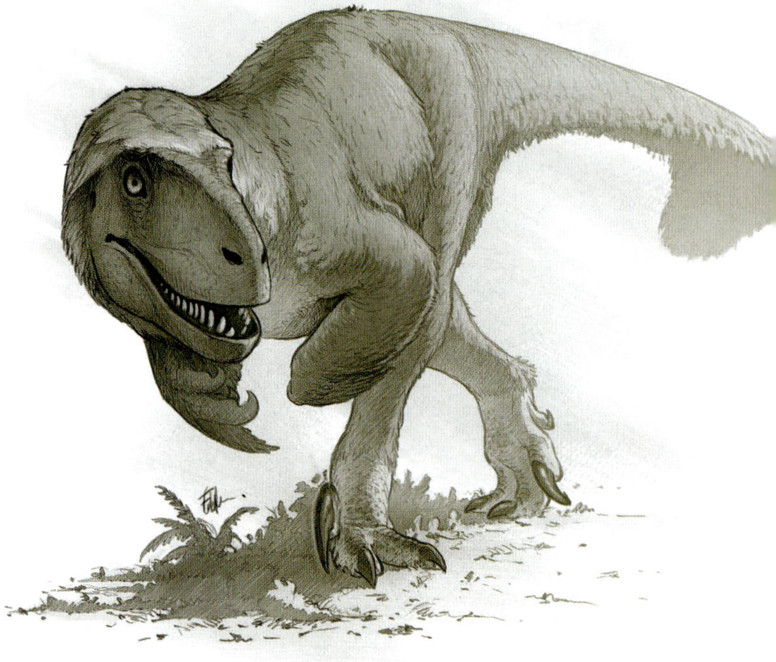

[Previous pages]

TYRANNOSAURUS VS *TRICERATOPS*

It might seem that this iconic battle – involving the most famous of Late Cretaceous beasts – has been illustrated so many times than an artist might struggle to show something new. But that's not so, as shown by this gorgeously lit scene, inspired by legendary artist Jay H. Matternes.

• Digital | 2021

▴ *DASPLETOSAURUS* FIGHT

The fossils of several tyrannosaur specimens preserve bite marks, holes and scores almost certainly made by the teeth of other individuals of the same species, suggesting ritualised facial biting of this sort happened often. These are daspletosaurs, a group of species unique (so far!) to North America's northern regions.

• Digital | 2021

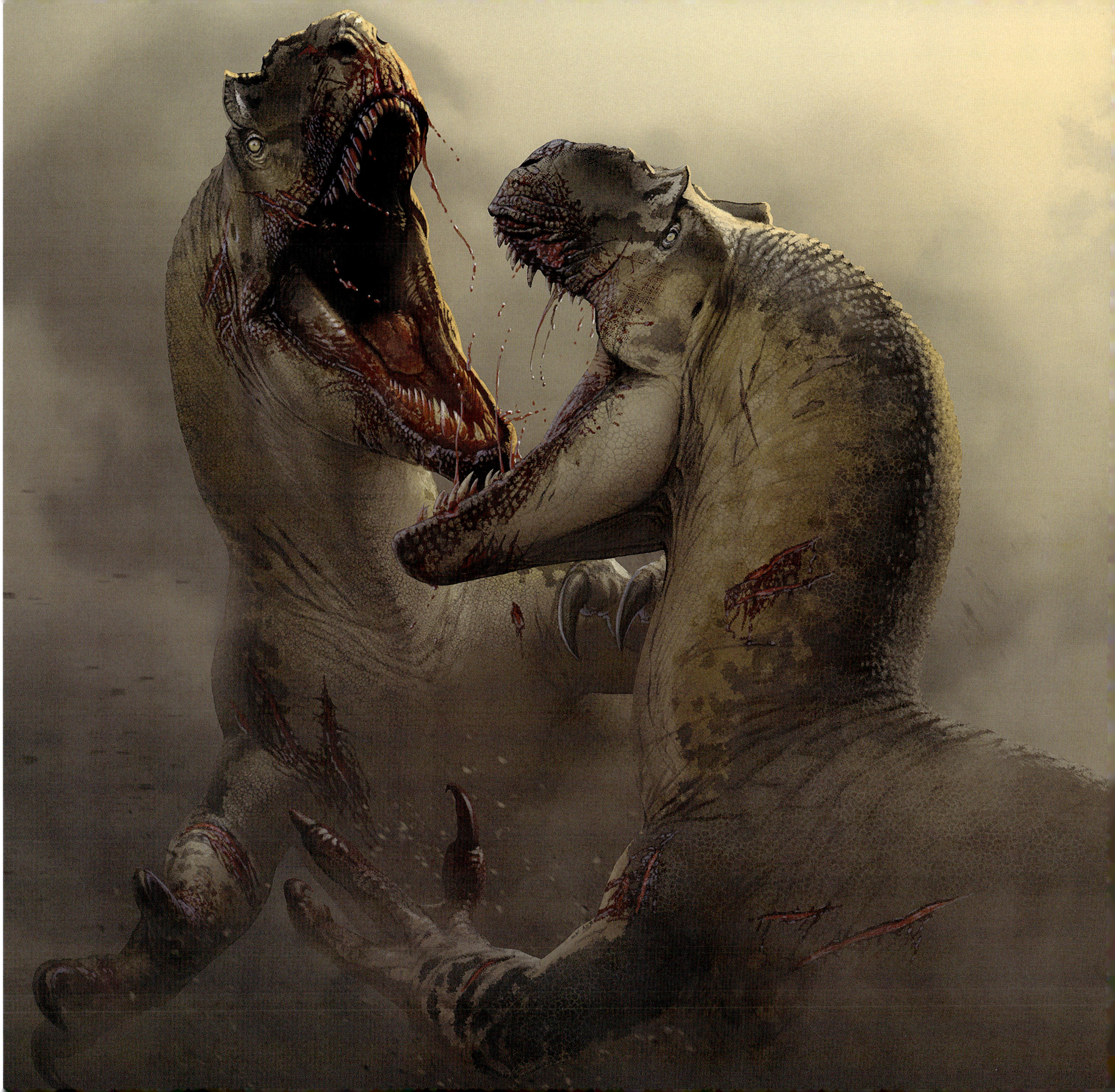

▾ *DEINONYCHUS* SWAMP

The iconic North American dromaeosaurid *Deinonychus* is most typically illustrated as a pack-hunting animal of arid environments. But some of its fossils come from swampy environments not unlike the bayous and deltas of the modern American South. As a part-time swamp dweller, maybe the Deinonychus of some areas were more likely to eat fish, frogs and waterbirds than other dinosaurs, and maybe they were solitary hunters, not social ones.

• Digital | 2024

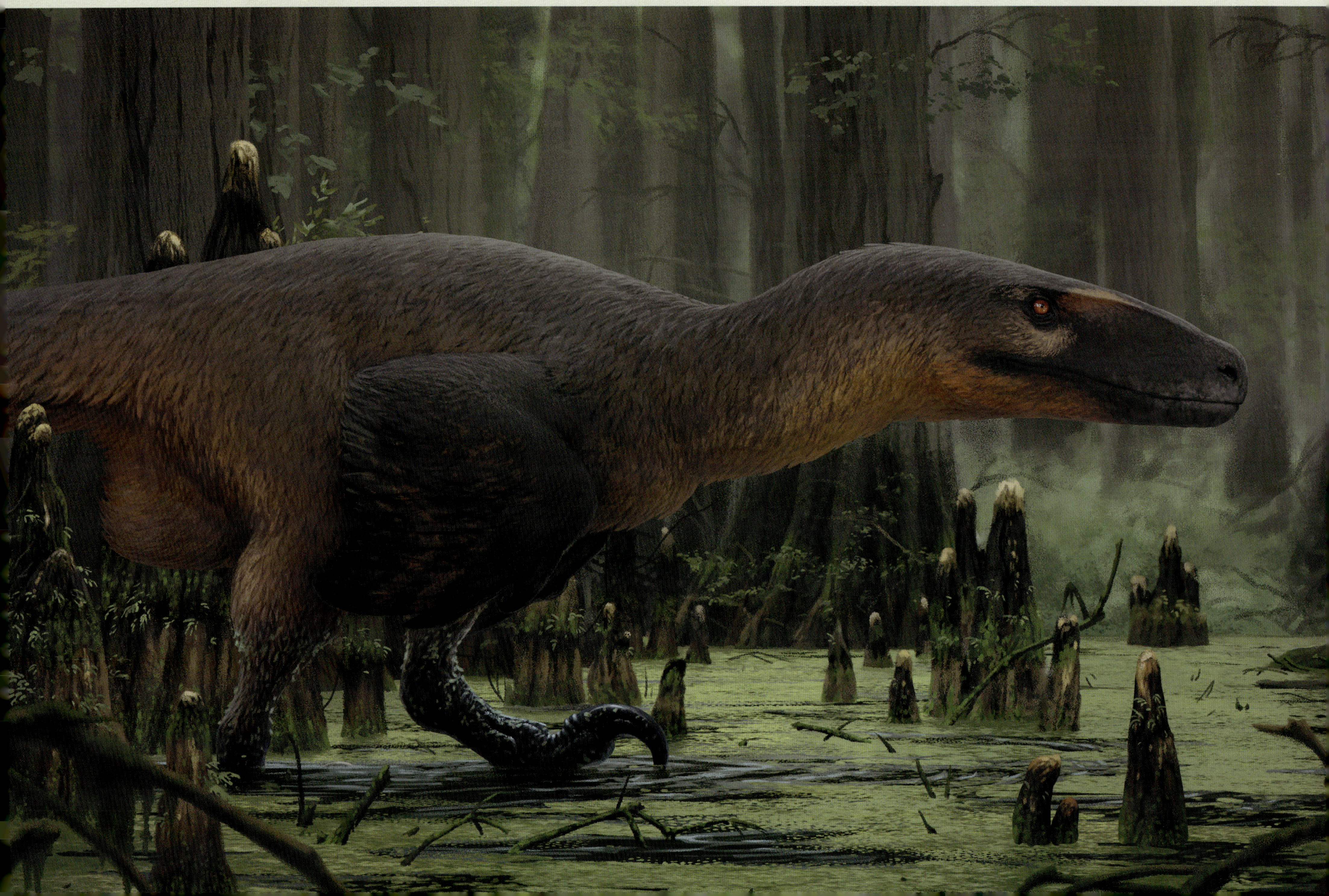

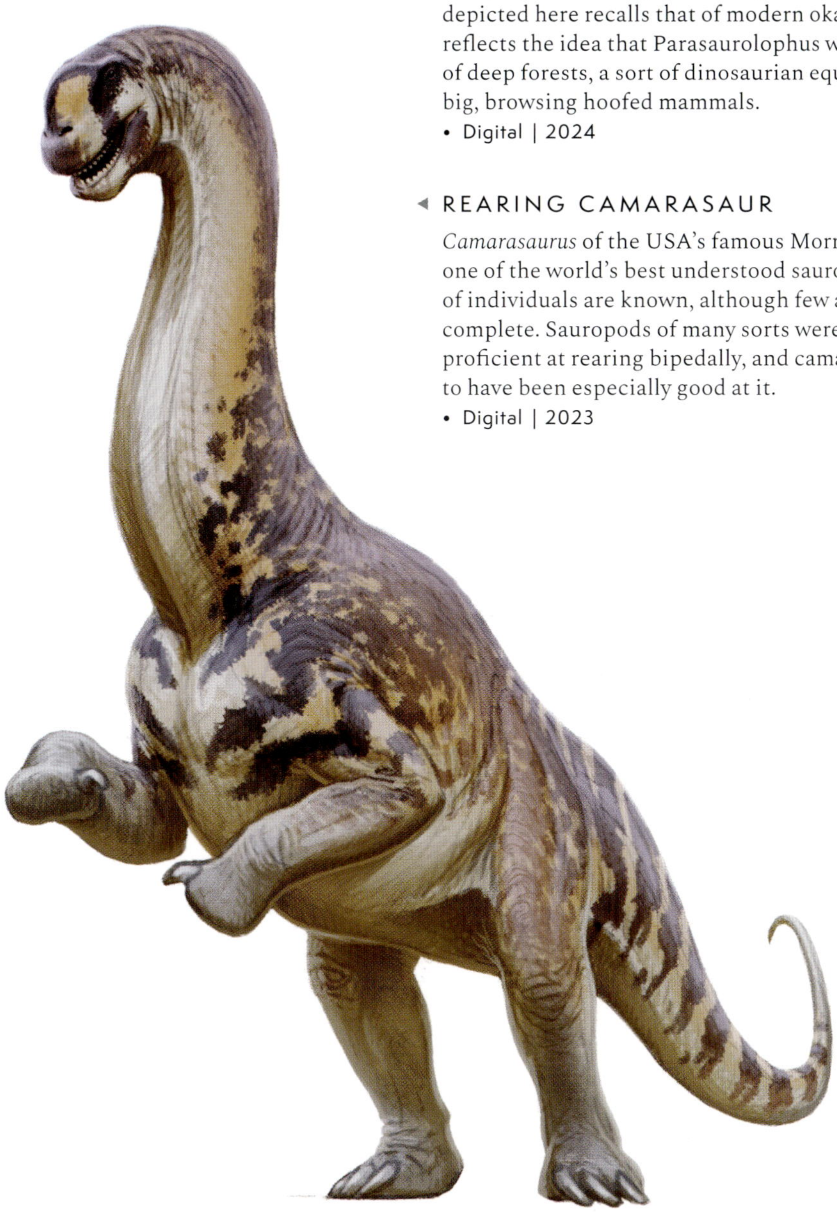

PARASAUROLOPHUS ▸

This Late Cretaceous hadrosaurid is known from three North American species, the best known being *P. walkeri*, with its lengthy, trombone-like head crest. The purplish-brown, white-striped colour scheme depicted here recalls that of modern okapis. This reflects the idea that Parasaurolophus was a denizen of deep forests, a sort of dinosaurian equivalent of big, browsing hoofed mammals.

• Digital | 2024

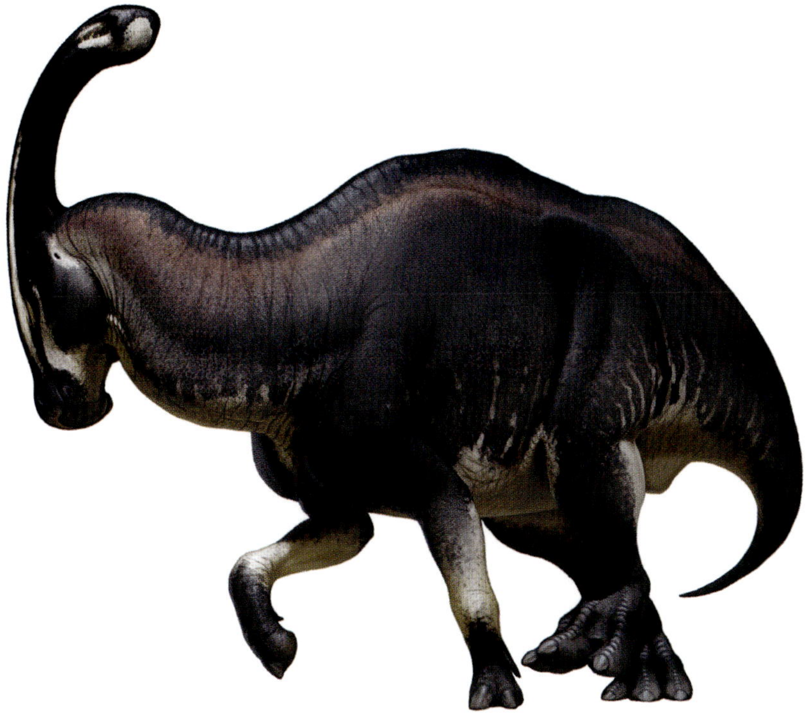

◂ REARING CAMARASAUR

Camarasaurus of the USA's famous Morrison Formation is one of the world's best understood sauropods. Hundreds of individuals are known, although few are skeletally complete. Sauropods of many sorts were probably quite proficient at rearing bipedally, and camarasaurs appear to have been especially good at it.

• Digital | 2023

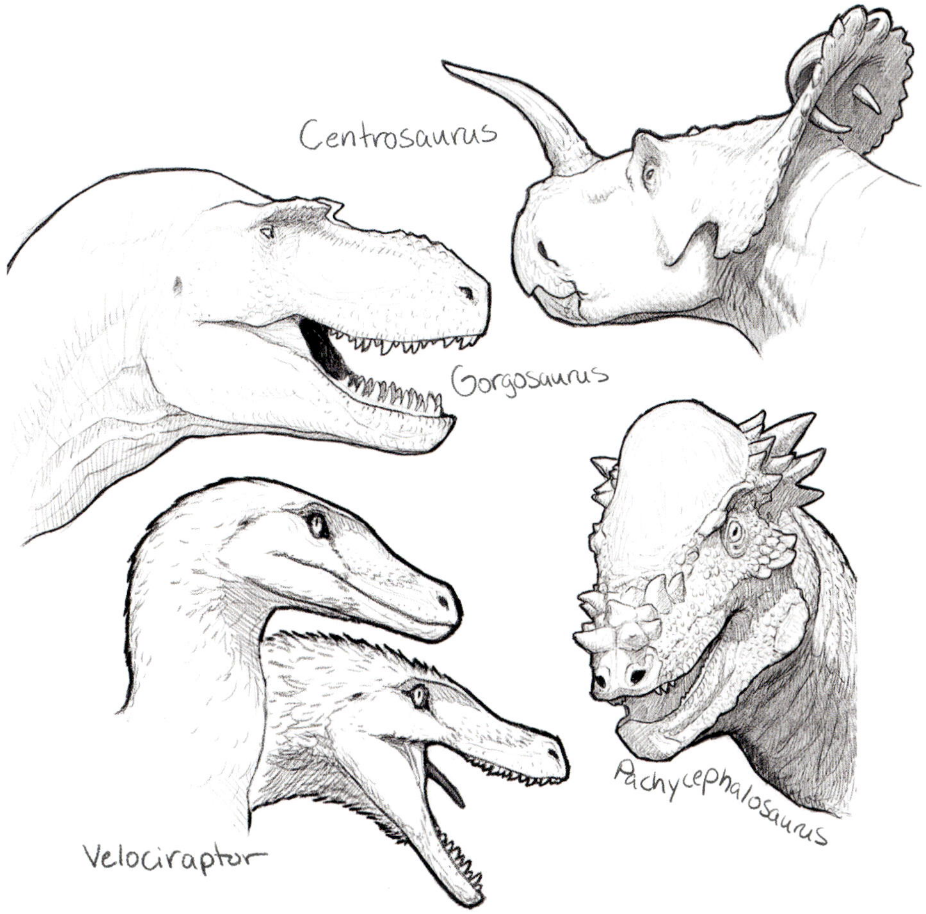

HEAD SKETCHES ▸

For all those claims about dinosaur popularity being some kind of mystery, a big part of their appeal is that they look *so* awesome. Sweeping lines, gnarly edges, horns and immensely detailed textures make them irresistible subjects for artists who like drawing animals.

• Pencils & digital | 2022

JANE ▸

This study of Jane – a relatively intact, teenage specimen of *T. rex* – emphasises the shallow snout, prominent facial crests and dynamic, agile air of this rhino-sized dinosaur. Jane is currently on show at the Burpee Museum of Natural History, Illinois, and is one of several key specimens that have helped in our understanding of tyrannosaur growth and shape change.

• Digital | 2023

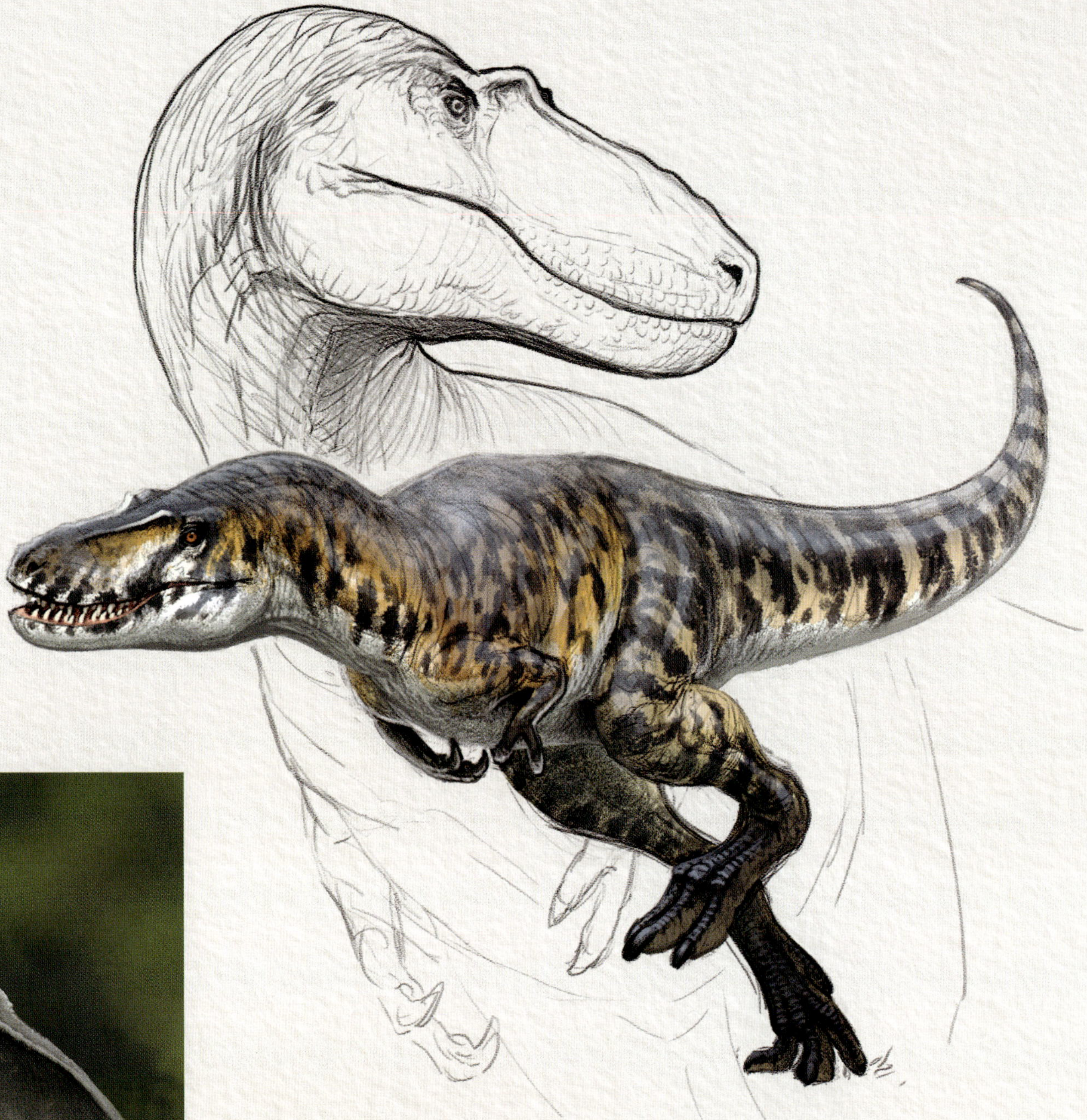

▾ *ACROCANTHOSAURUS*

Known from the Early Cretaceous of the USA, *Acrocanthosaurus*, an allosauroid named in 1950, was among the first fossils to demonstrate that this clade of theropods was not restricted to the Jurassic as previously thought. Tall bony spines along the midline of its back and neck anchored muscles that would have enhanced its strength and power.

• Digital | 2021

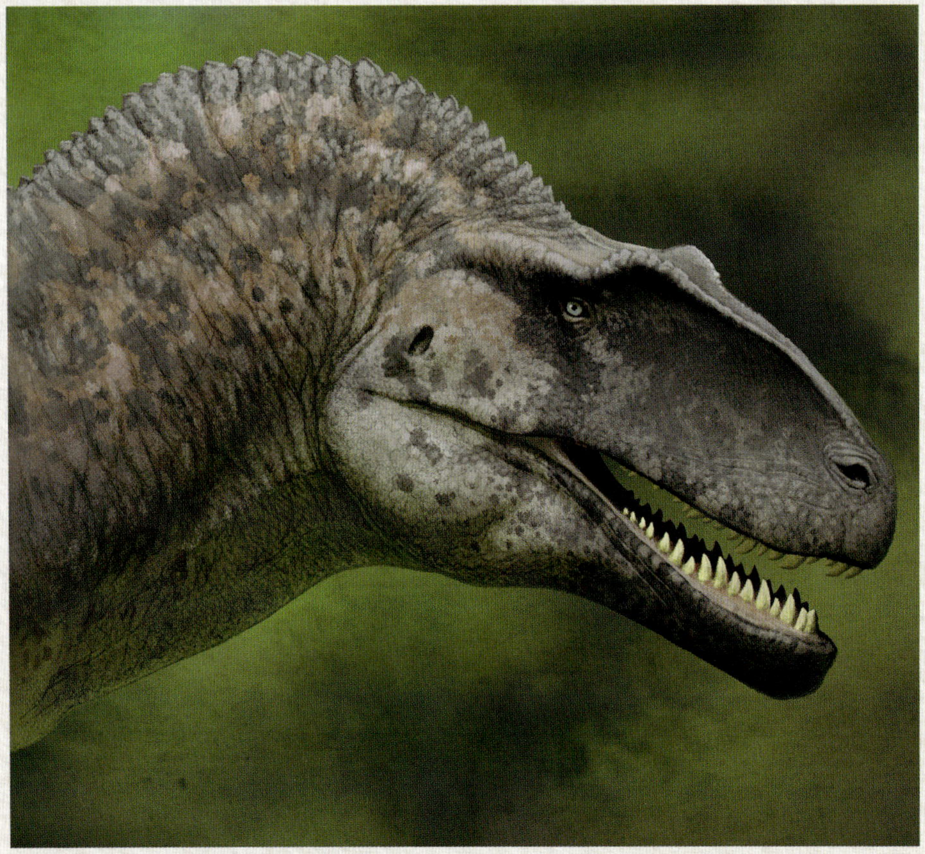

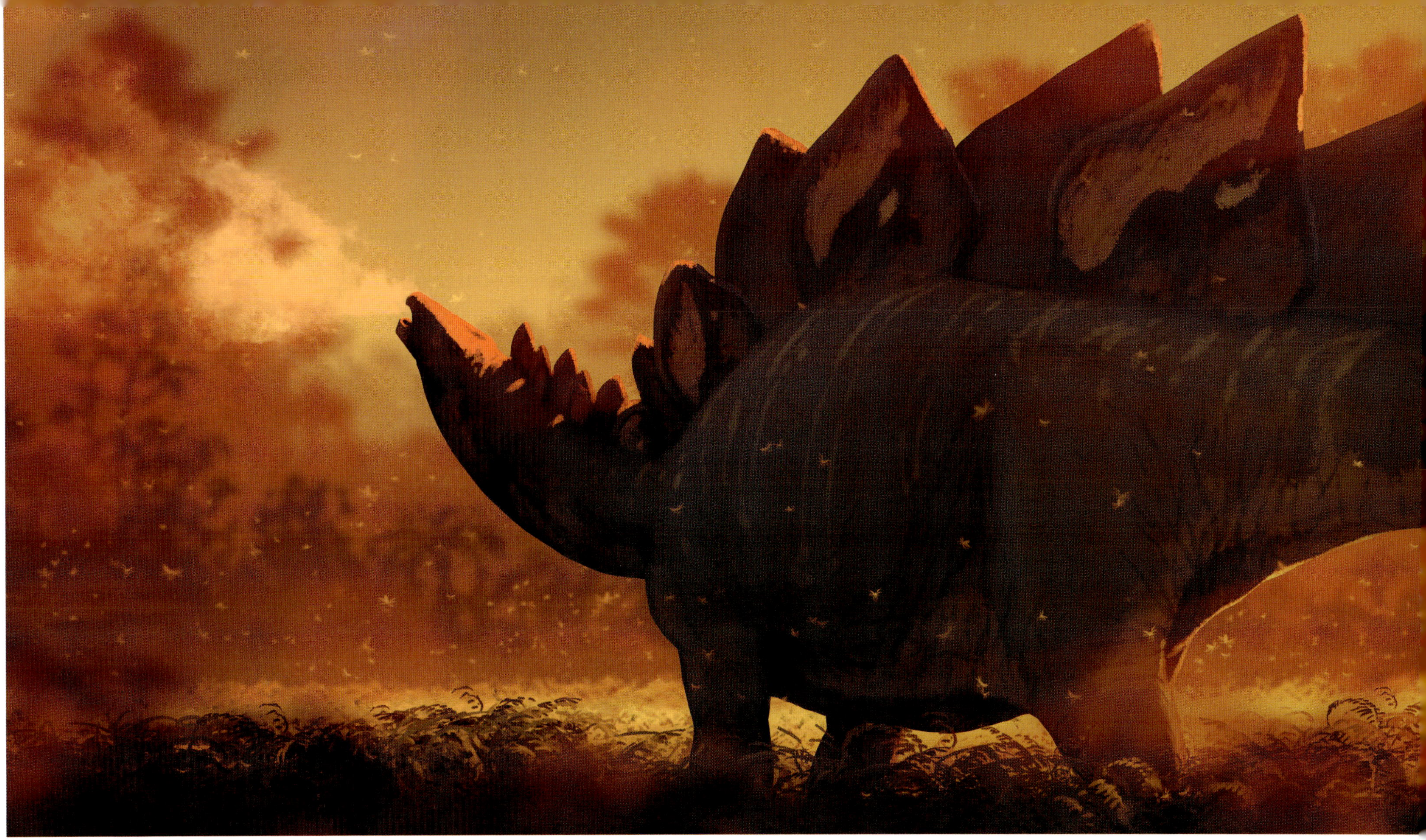

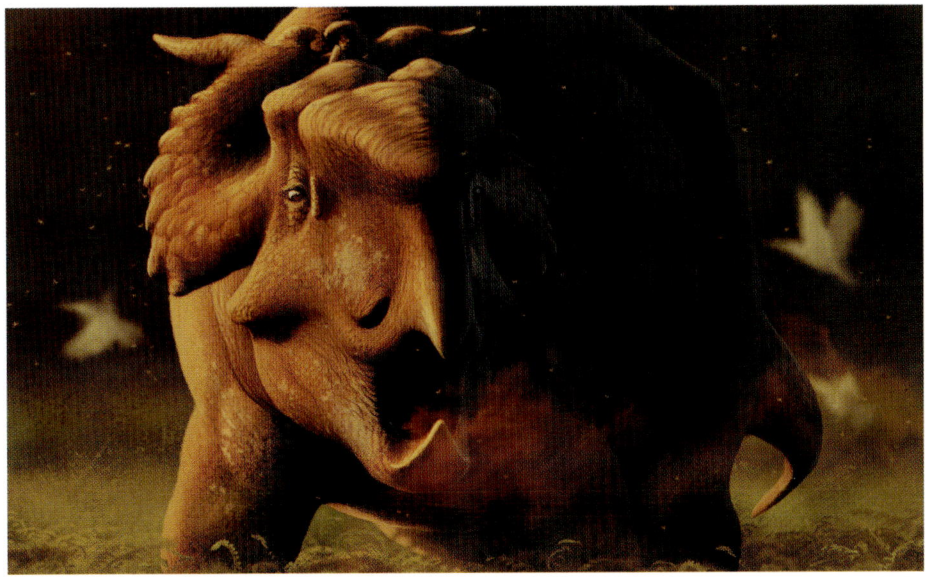

◄ *PACHYRHINOSAURUS*

When first unveiled to science during the 1950s, *Pachyrhinosaurus* – the 'thick-nosed lizard' – was a real surprise. Instead of horns on the nose and over the eyes, it sported a thick, hugely gnarly 'nasal boss'. Decorated with lumps, bumps and a striated texture, this was perhaps used as a battering ram in sexual combat.

• Digital | 2022

▲ *STEGOSAURUS* SUNSET

Stegosaurus is such a familiar dinosaur that we sometimes forget how extraordinary it is, and how bizarre it might have been when alive. And did stegosaurs make loud noises, as implied here? Most dinosaur enthusiasts have some idea of what such dinosaurs as tyrannosaurs and hadrosaurs might have sounded like, but…stegosaurs? Make your best guess.

• Digital | 2022

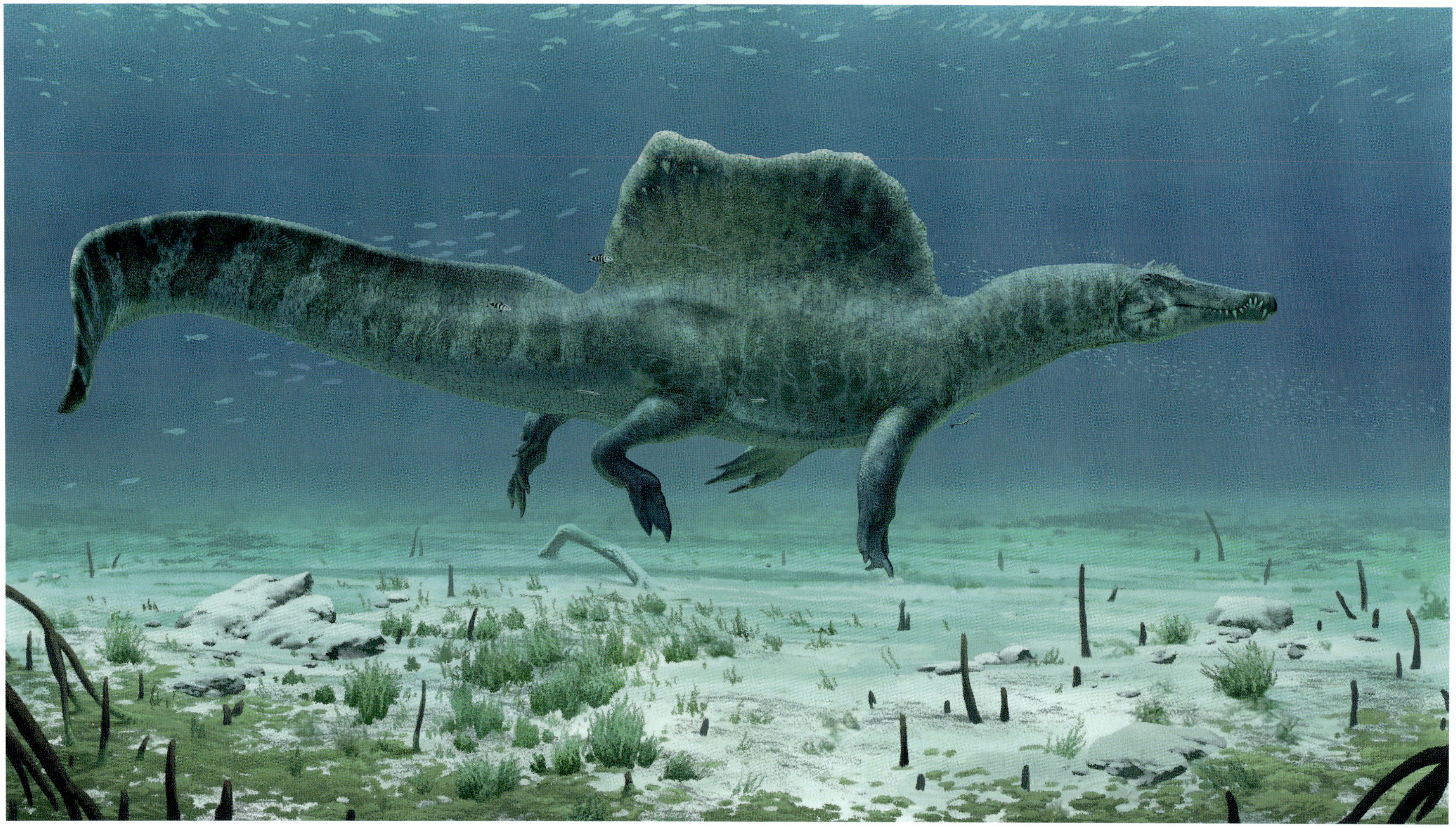

▲ UNDERWATER SPINOSAUR

It's a standard in-joke within palaeoart discourse that *Spinosaurus* has undergone constant changes in appearance over recent years. In reality, experts mostly agree on the appearance of this animal. The disagreement is whether it was highly aquatic – as implied here – or more of a waterside forager. This image incorporates data from a 2018 study that described the tail as deep, laterally compressed and suited for swimming.

• Digital | 2020

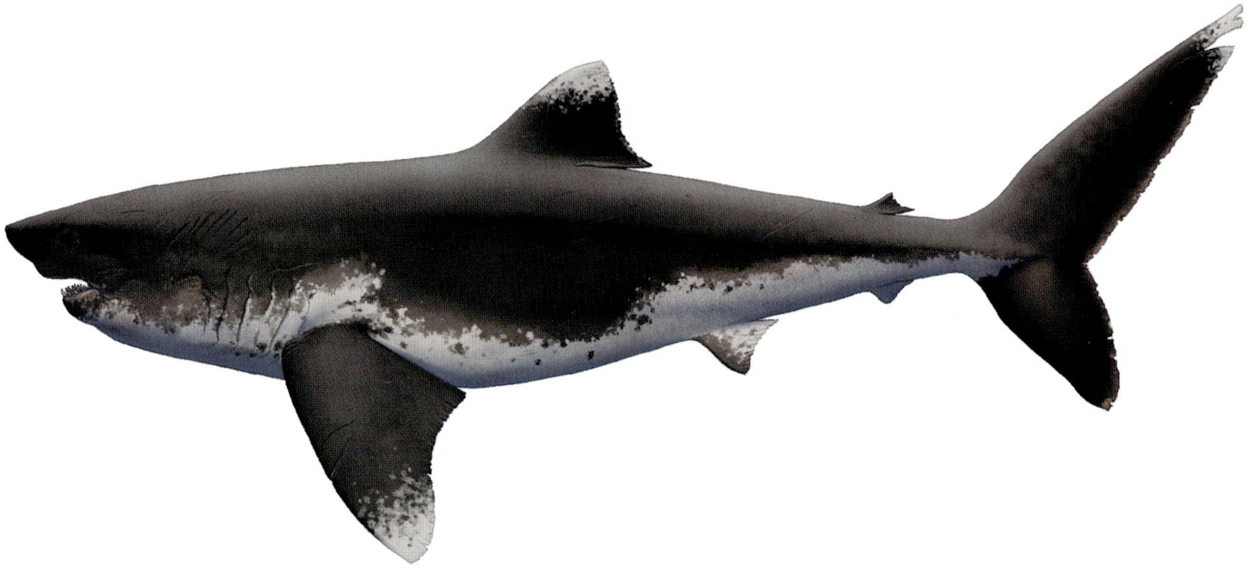

MEG ▶

Few fossil sharks have been reconstructed more often than the mega-toothed species currently known as *Otodus megalodon*, and popularly called Meg. Mild disagreement continues over its precise proportions. A challenge faced by the artist is how to make it look suitably big, since the largest individuals certainly exceeded 15m (49ft) in total.

• Digital | 2023

MORRISON WATERHOLE ▸

The fact that giant dinosaurs of some or many species shared the same Jurassic and Cretaceous environments is hugely exciting for anyone who enjoys imagining the animal life of the past. This image casts an impression of *Brontosaurus* as a heavyset bruiser, more than willing to bully other sauropods (like the *Diplodocus* in the background) and even large theropods (like the *Marshosaurus* shown here) away from a waterhole.

• Digital | 2021

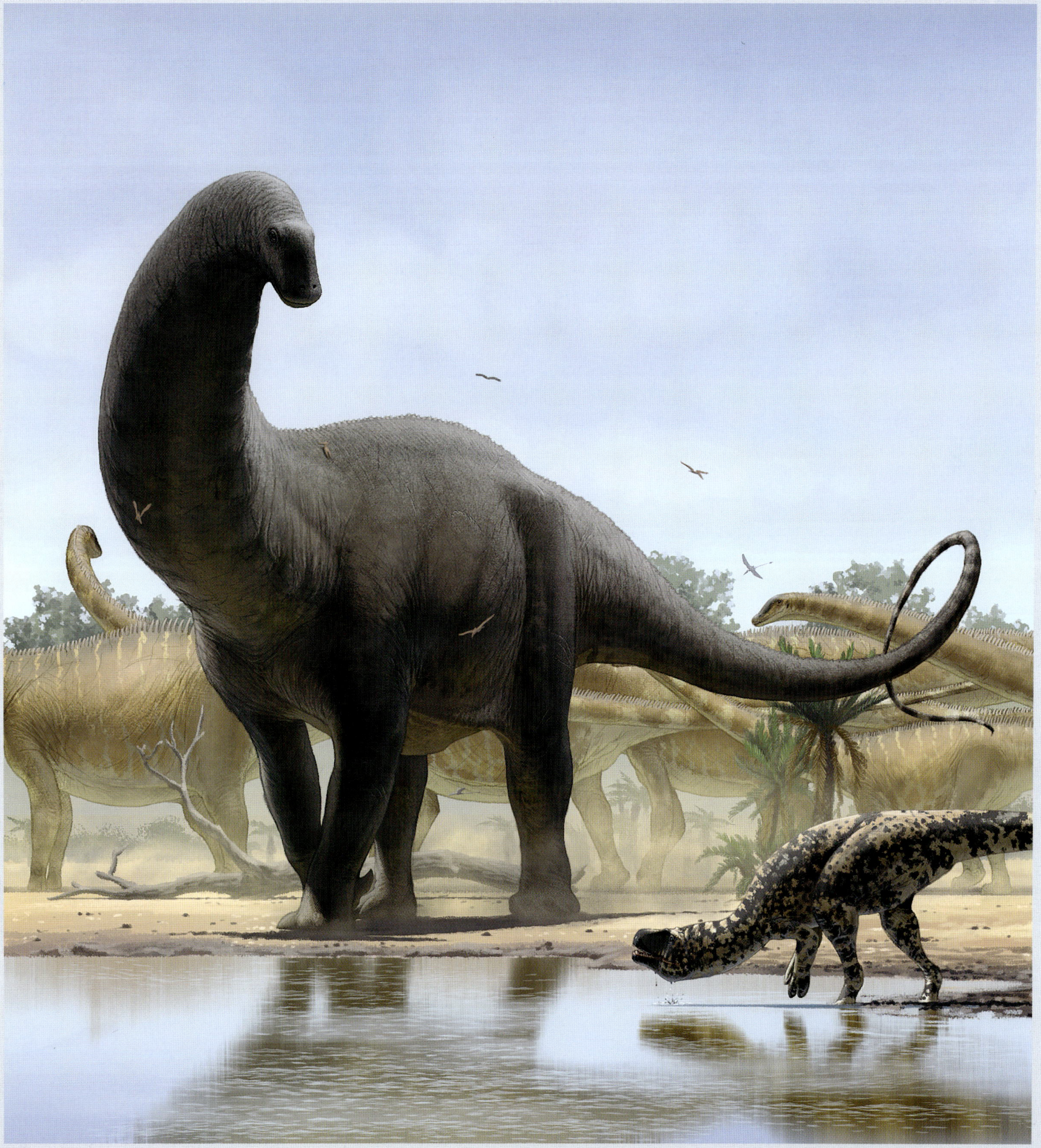

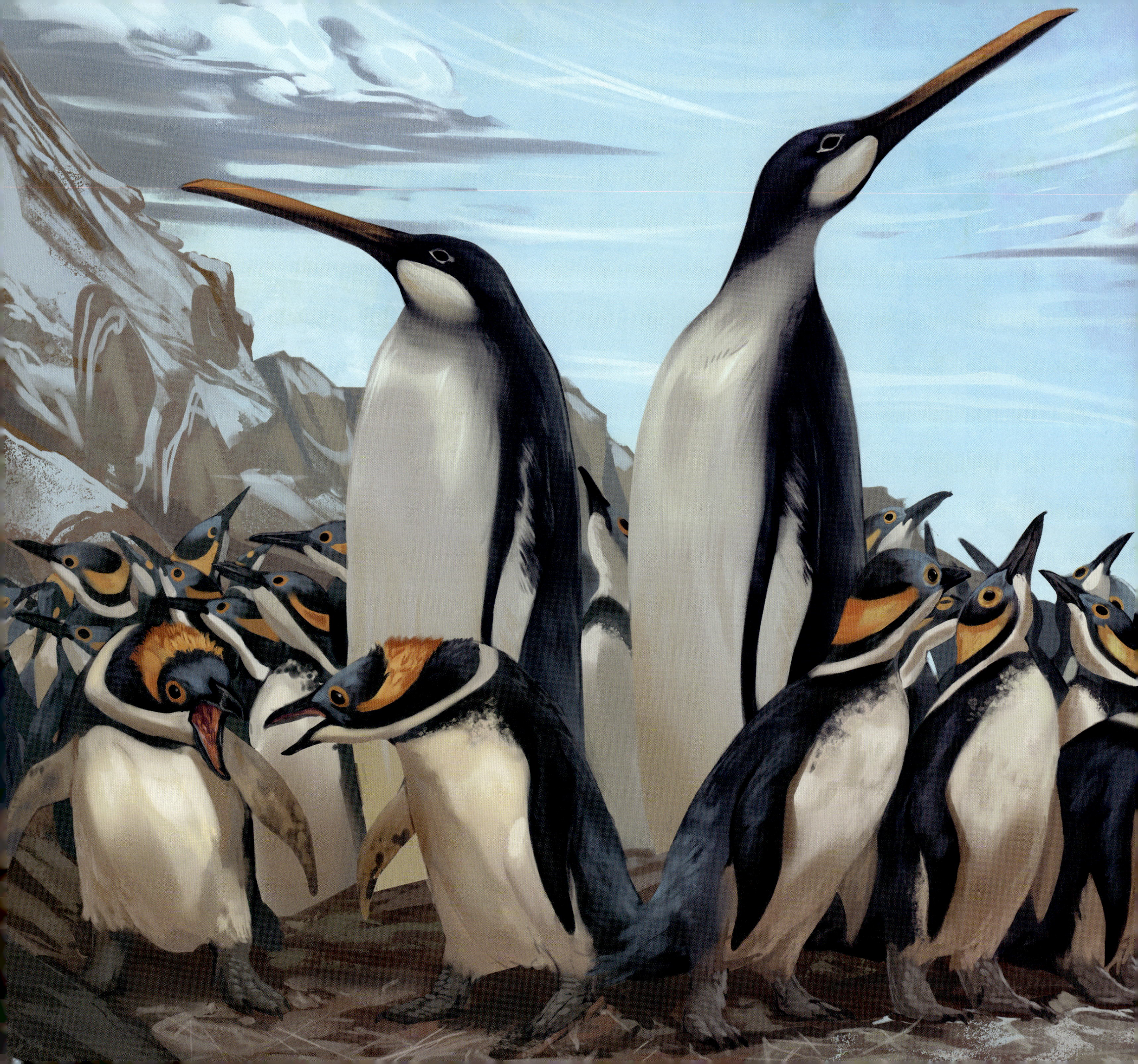

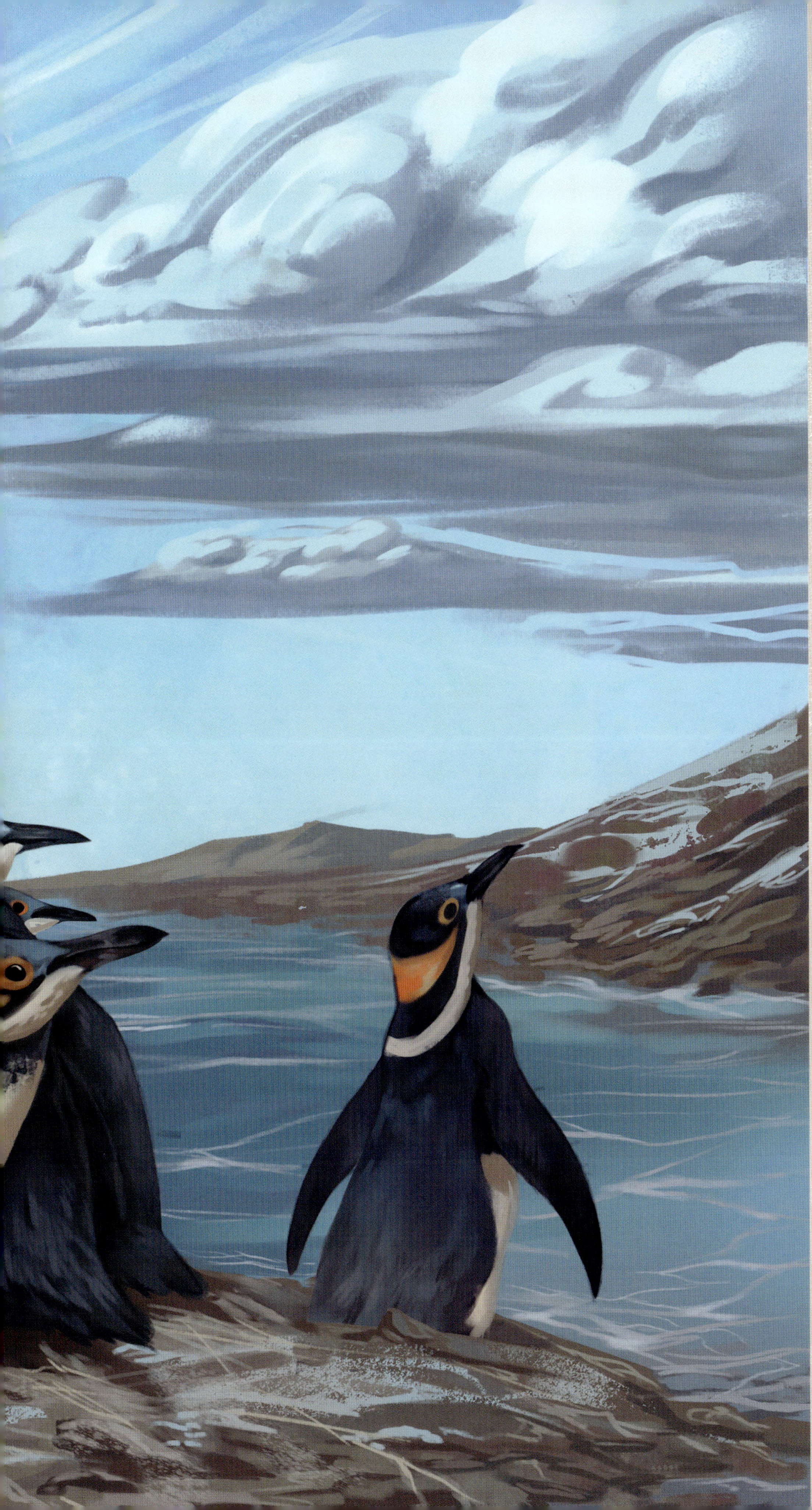

SIMONE GIOVANARDI

BORN IN 1990 IN AOSTA, a small town among the Italian Alps, Simone always loved drawing extinct animals as well as learning about them. He began studying biology and natural sciences at the University of Turin, always keeping his drawing material nearby. He continued his academic journey to New Zealand, studying for a PhD on the evolutionary morphological rates of extinct and extant penguins. There he specialised in illustrating New Zealand's extinct avifauna, particularly his principal study subjects. While penguins remain his favourite subjects to draw, (non-penguin!) dinosaurs and pterosaurs are also special to him.

Simone works mostly with digital media – computers and tablets – and he also likes to work on more stylised and cartoonish interpretations of palaeofauna. He now works as freelancer illustrator and concept artist.

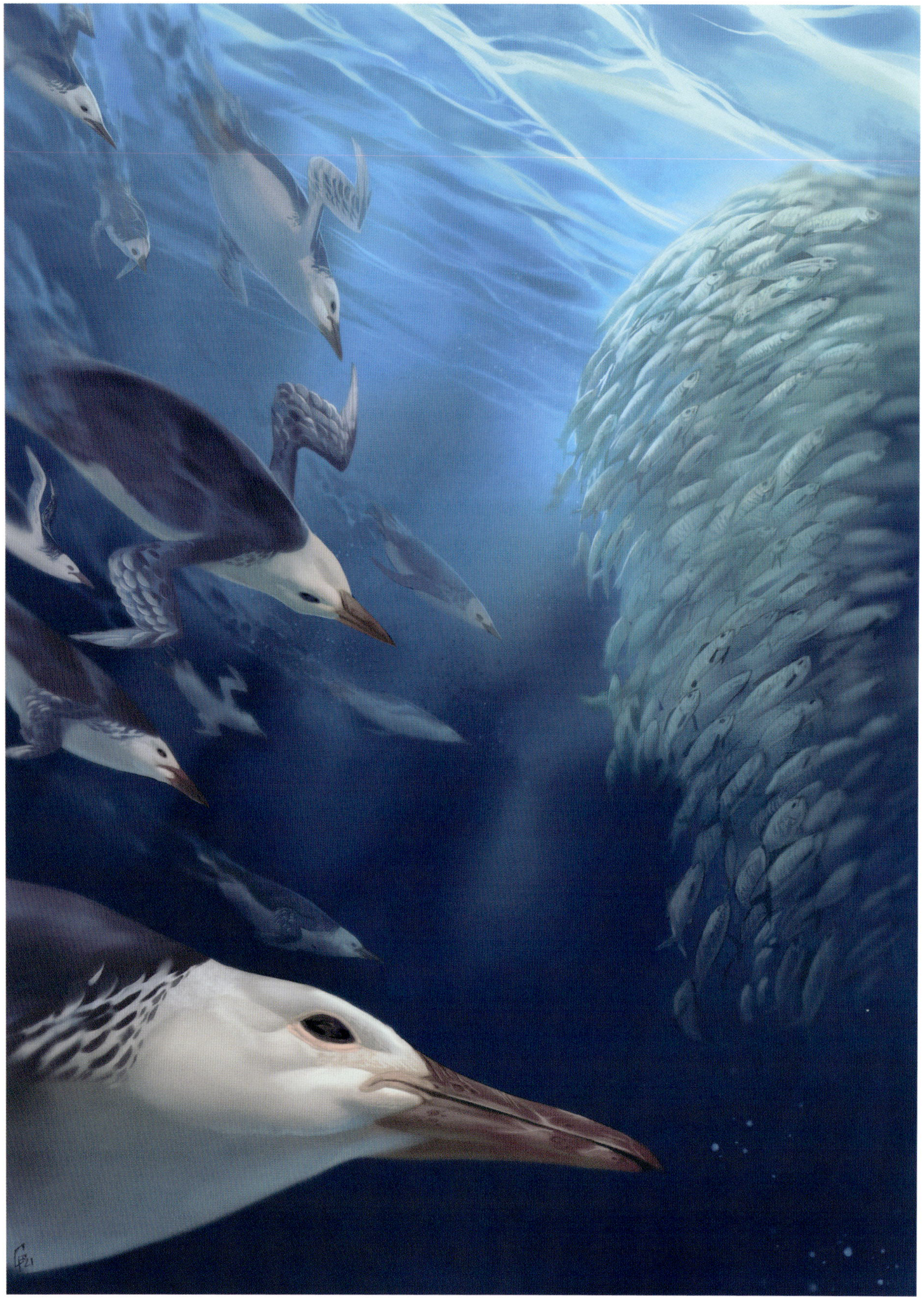

[Previous pages]

ANCIENT PENGUINS OF SEYMOUR ISLAND

Some ancient penguins were gigantic: *Anthropornis grandis*, shown here, was 1.8m (5ft 1in) long from bill tip to tail. Birds like this were different from living penguins in possessing spear-shaped bills that demonstrate an unusual hunting style relative to modern species. Smaller penguins lived alongside these giants, such as *Delphinornis*, which was less than 60cm (23.6in) long.

• Digital | 2021

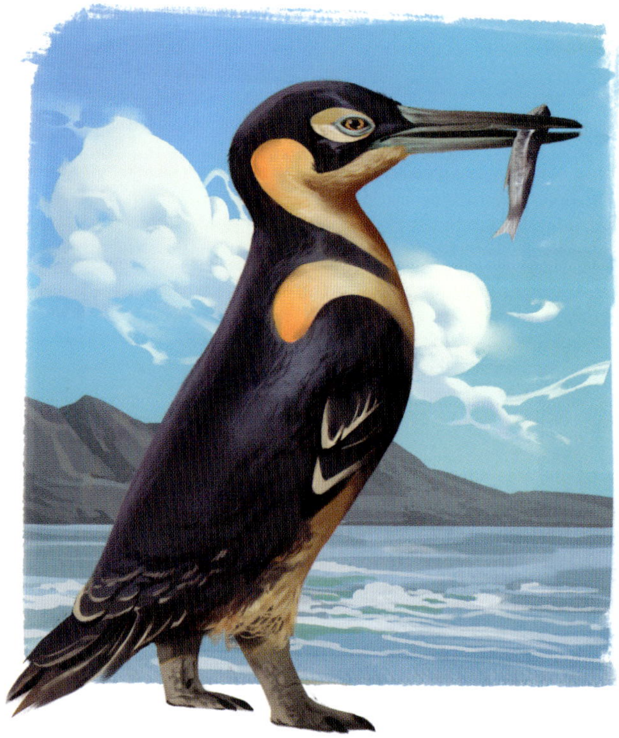

▲ *CROSSVALLIA*

Crossvallia is an especially ancient penguin (known both from Antarctica and New Zealand) and some of its fossils date to the Palaeocene, 55 million years ago. It was large at over 1.4m (4.6ft) long. Modern penguins have stiffened, flipper-like wings. The oldest penguins had more mobile elbow and wrist joints than seen in modern penguins.

• Digital | 2023

◀ *KAIIKA*

Several archaic penguins are currently known from poor remains (most typically their highly distinctive upper wing bones), so their exact appearance in life remains speculative. Simone's art resists the idea that they should be depicted as clones of modern species. This is *Kaiika* of New Zealand, a large Early Eocene penguin.

• Digital | 2021

▼ *KUMIMANU & PETRADYPTES*

The two ancient penguin species shown here – *Kumimanu fordycei* and *Petradyptes stonehousei* – inhabited New Zealand around 60 million years ago. *Kumimanu* was a giant, perhaps weighing around 150kg (331lbs). It has been suggested these spear-billed ancient penguins might have resembled loons, and this illustration captures that idea.

• Digital | 2023

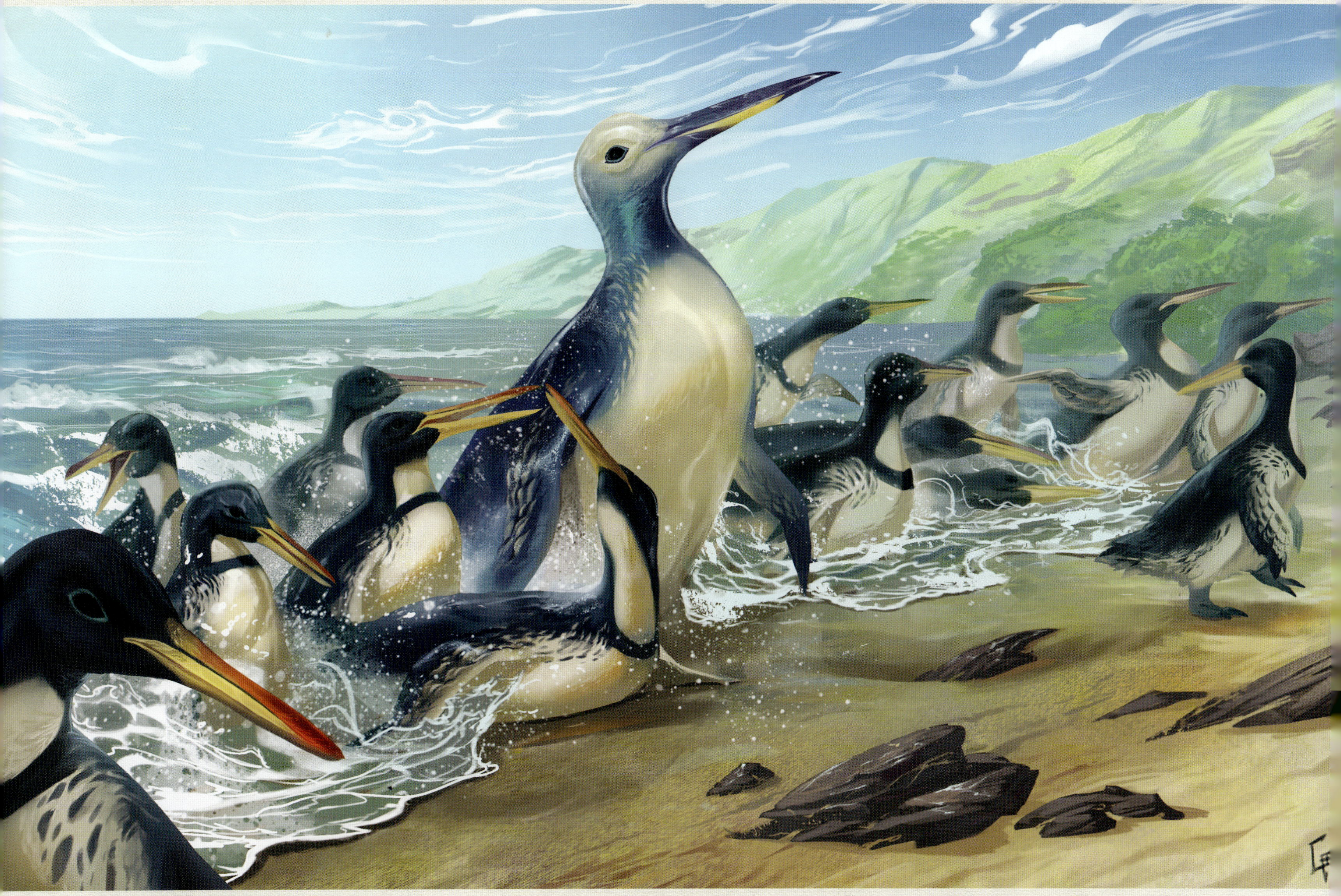

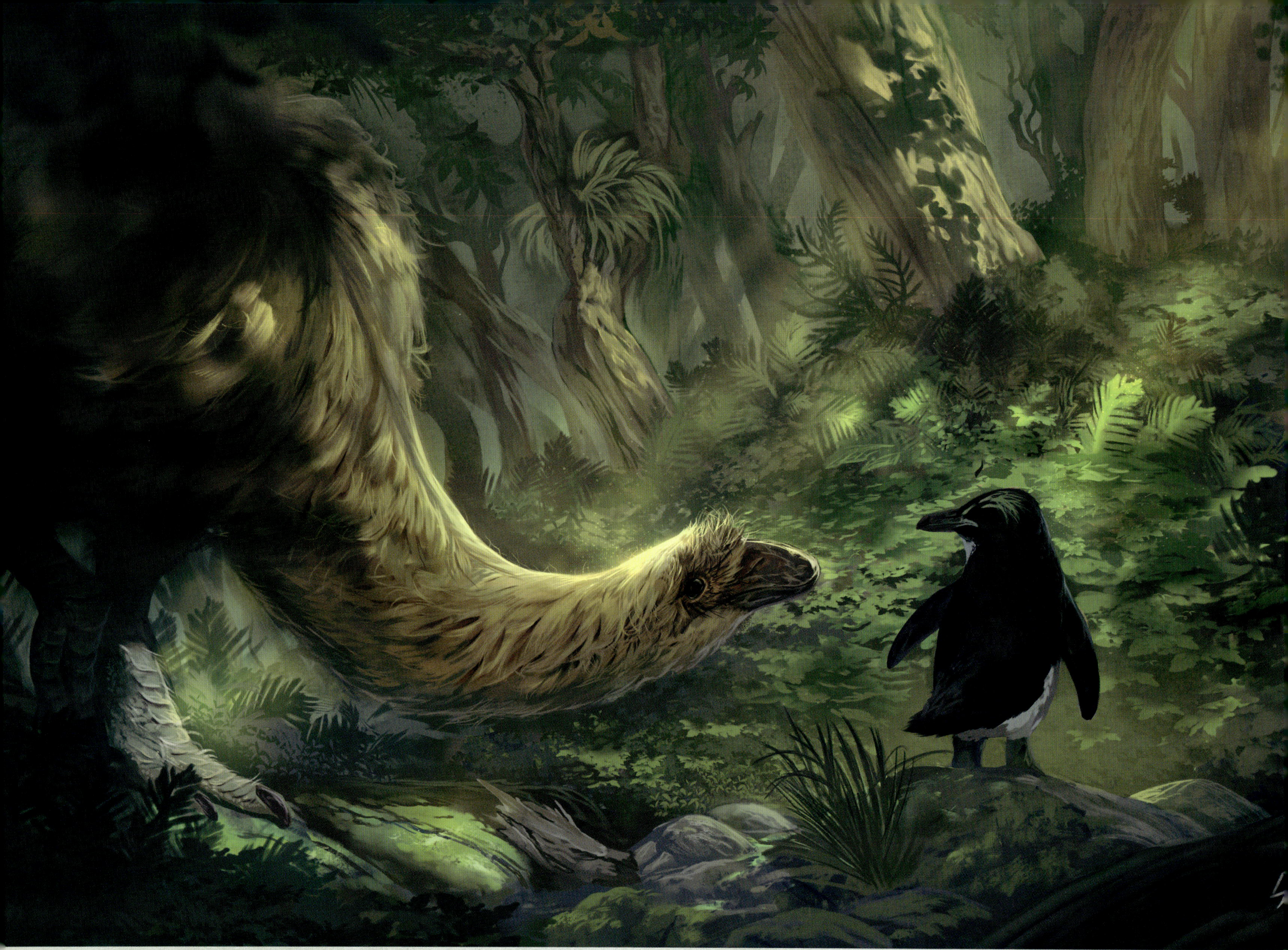

▲ CHATHAM PENGUIN & NORTH ISLAND MOA

Moas are a group of wingless ratite birds, endemic to New Zealand. The North Island giant moa was one of the biggest and best known of the nine or so species. Here, one encounters a Chatham crested penguin, another recently extinct species, named in 2019 but made extinct around 1500 years ago.
• Digital | 2021

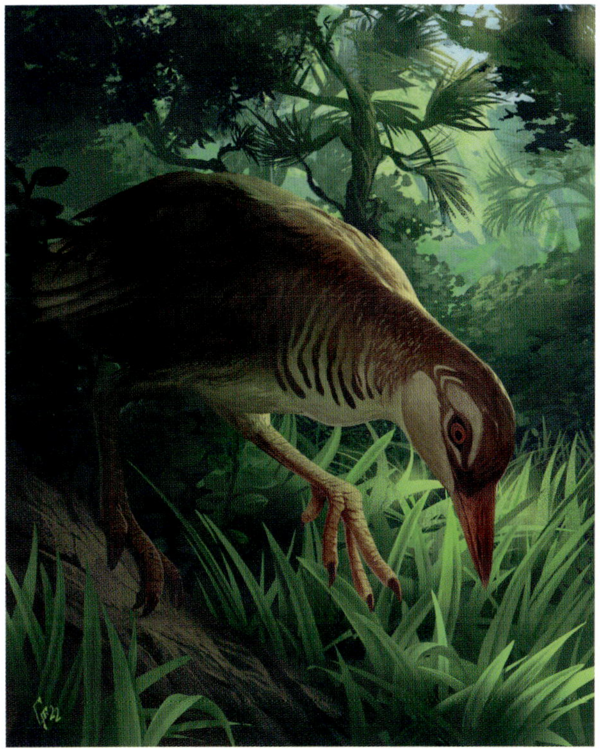

◂ *GALLIRALLUS ASTOLFOI*

New to science as of 2021, Astolfo's rail is a recently extinct species from the Bass Islands of French Polynesia in the South Pacific. It's one of hundreds of rail species made extinct by human hunting. Like other *Gallirallus* species, it was likely a robust-billed, omnivorous rail that was either flightless or with poor flight abilities.

• Digital | 2022

OWLET-NIGHTJAR ▸

Owlet-nightjars are a group of relatively obscure Australasian birds with a wide bill, large eyes and a long tail. They recall nightjars but are actually more closely related to swifts. New Zealand was once home to both especially large and especially small members of this group.

• Digital | 2022

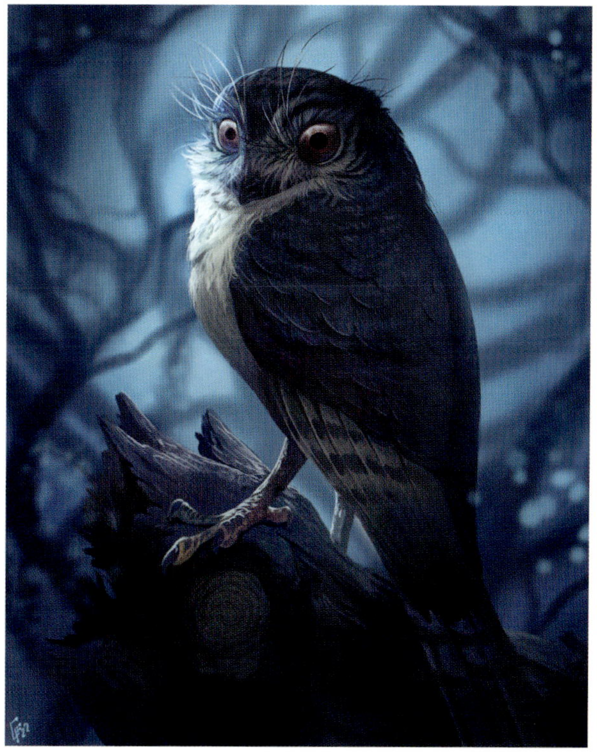

MIOTADORNA ▸

New Zealand is famous for extinct birds like moa, but many more familiar species used to live there too. The large duck *Miotadorna* is from the Miocene, about 20 million years ago. Little is known of its appearance, but it is thought to be a member of the shelduck group (the tadornines), so was likely long-legged and heavily built.

• Digital | 2022

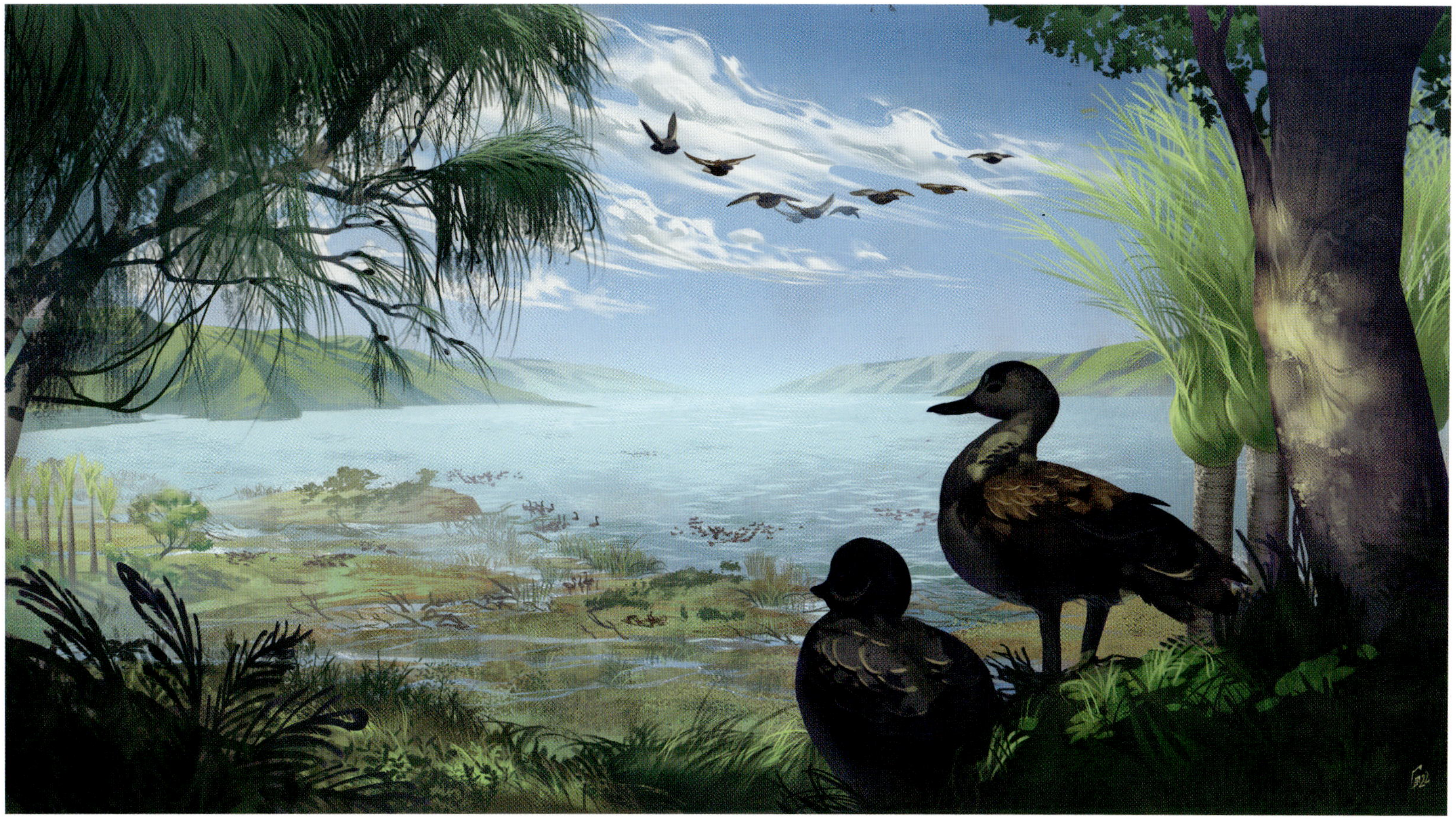

POSTOSUCHUS ▸

The existence of a great empire of so-called 'crocodile-line archosaurs' during the Triassic remains very much unexplored by palaeoartists. These images illustrate one of the few that has been depicted in popular media: the giant rauisuchid *Postosuchus* from the USA. Big individuals of this top predator reached 6m (20ft) at least.

• Digital | 2022

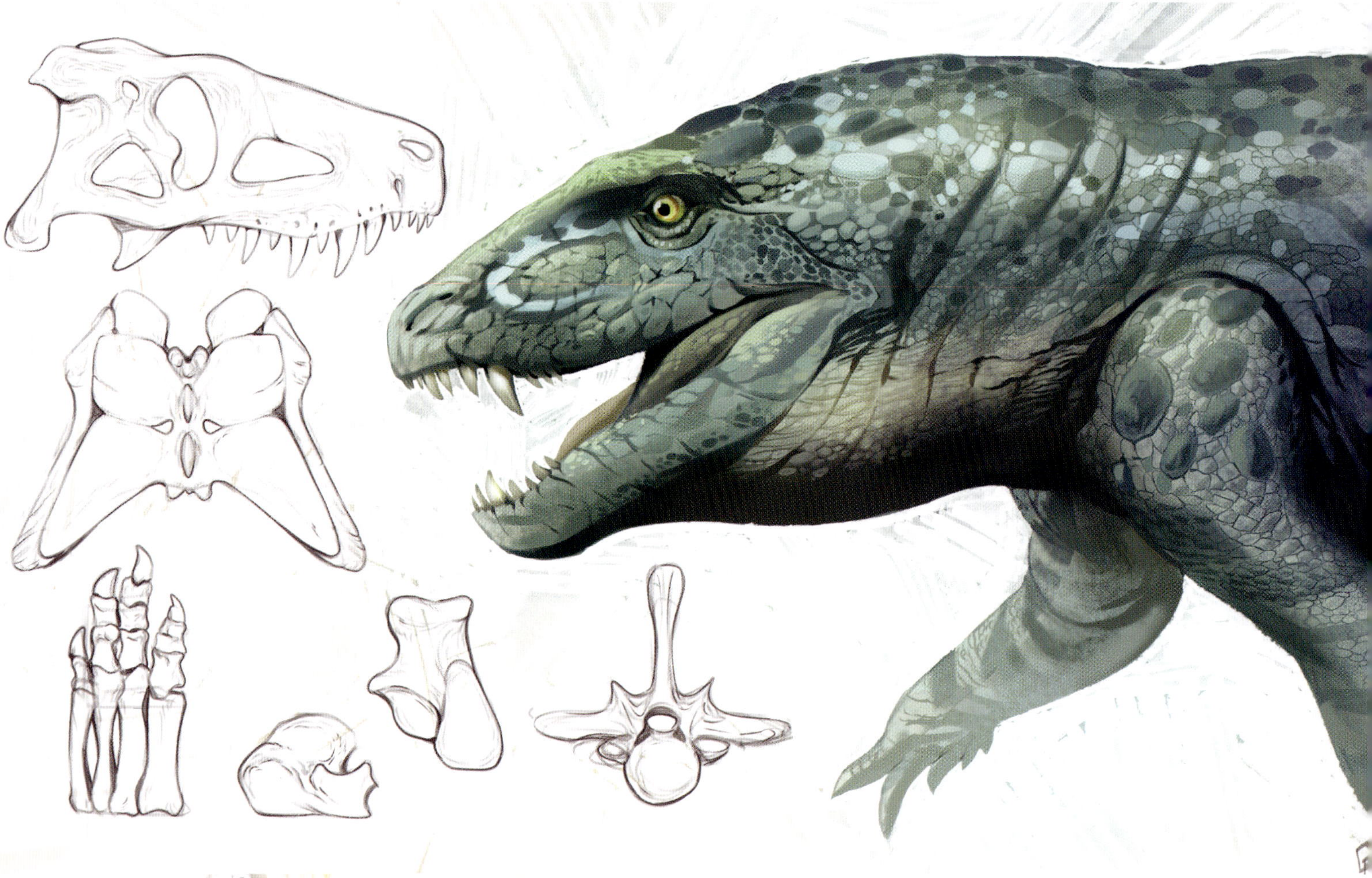

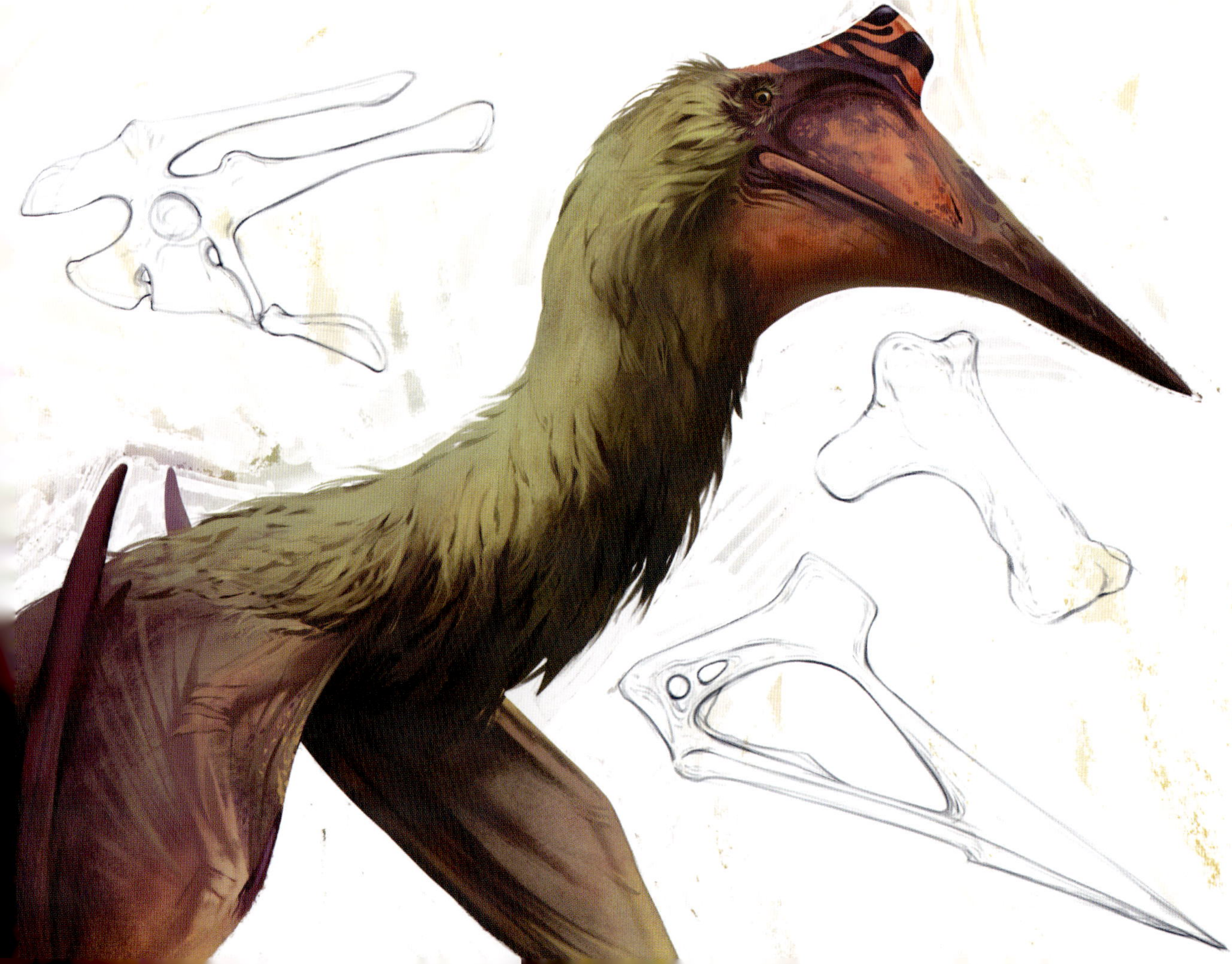

◂ *QUETZALCOATLUS*

Views differ on how the very biggest of the azhdarchid pterosaurs – like North America's *Quetzalcoatlus*, shown here – should be portrayed. The long-standing tradition of making them as slender and thin-necked as possible might be wrong: a shaggy coat of hair-like fibres would allow for a very different look.

• Digital | 2022

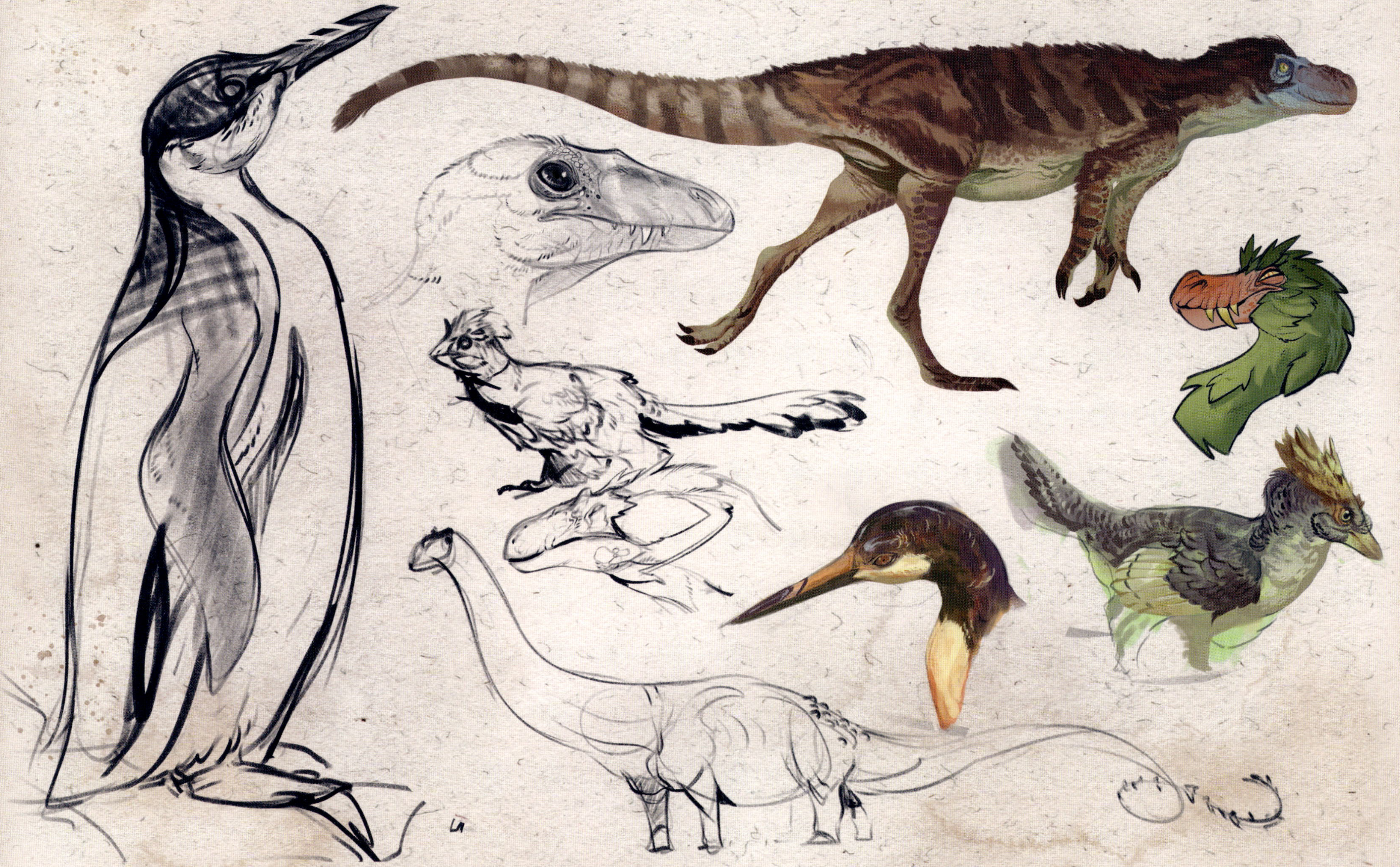

▲ SKETCHES

Penguins, Mesozoic dinosaurs like sauropods and maniraptorans… here is an insight into some of Simone's favourite subjects. While many palaeoartists rightly emphasise the need to understand the most nuanced details of anatomy, experimenting with form and style when learning to illustrate animals is key as well.

• Mixed media | 2021–2022

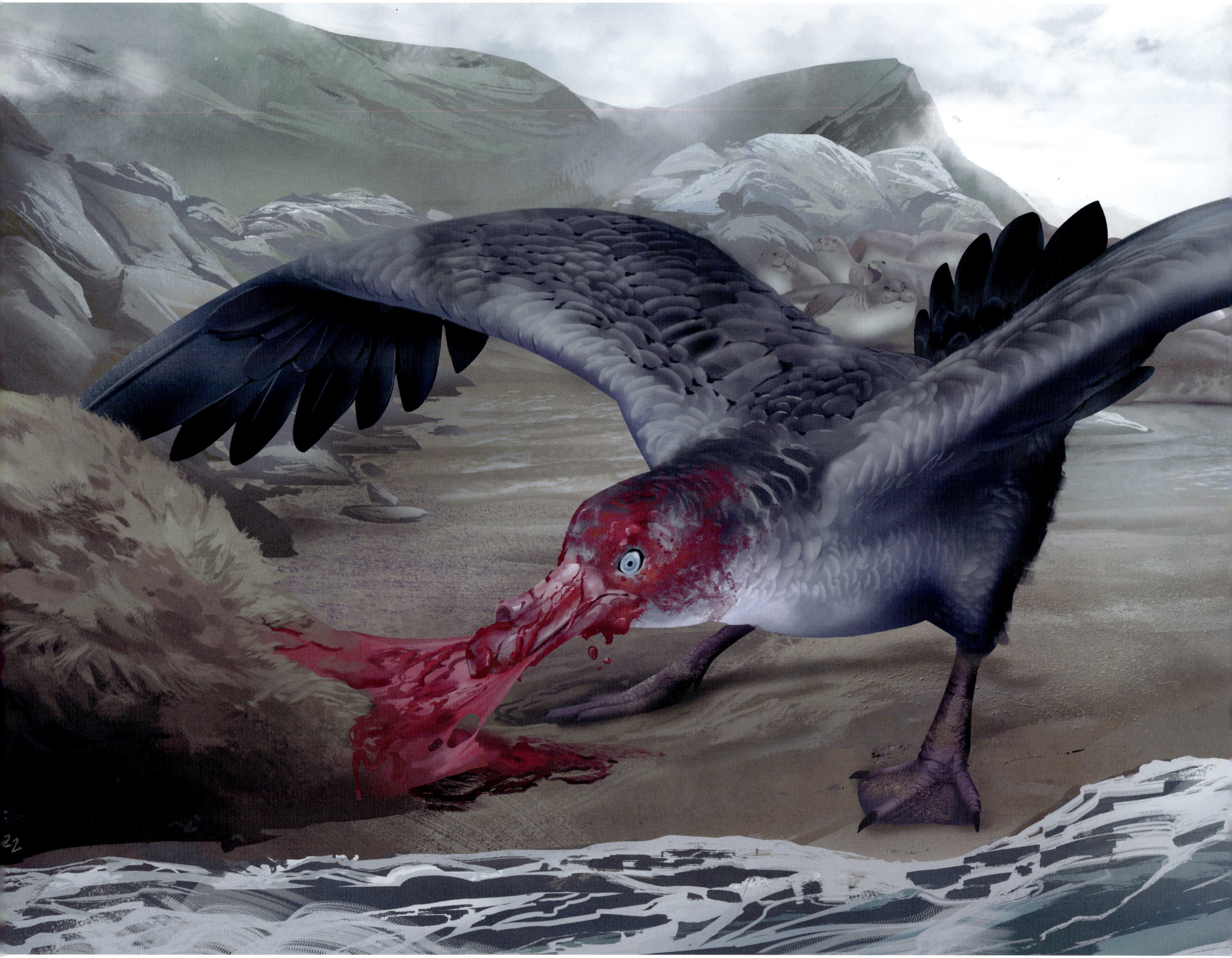

◂ *MACRONECTES TINAE*

Named in 2023, *Macronectes tinae* is an extinct species of giant petrel, a group of formidable, hook-billed scavengers and predators of the Southern Hemisphere. It is known from a three-million-year-old skull and upper wing bone from New Zealand's North Island. The wing bone indicates a wingspan of about 1.5m (5ft), shorter than the 2m (6.5ft) wings of living giant petrels.

• Digital | 2023

▾ *SEQUIWAIMANU*

Another archaic penguin from New Zealand, *Sequiwaimanu* is about 58 million years old and among the oldest of its group. Experts have long hoped to discover 'proto-penguins': fossils of avifauna that possessed key penguin traits but were still able to fly. *Sequiwaimanu* was not flight-capable, but Simone's reconstruction here makes it look less penguin-like than its relatives.

• Digital | 2020

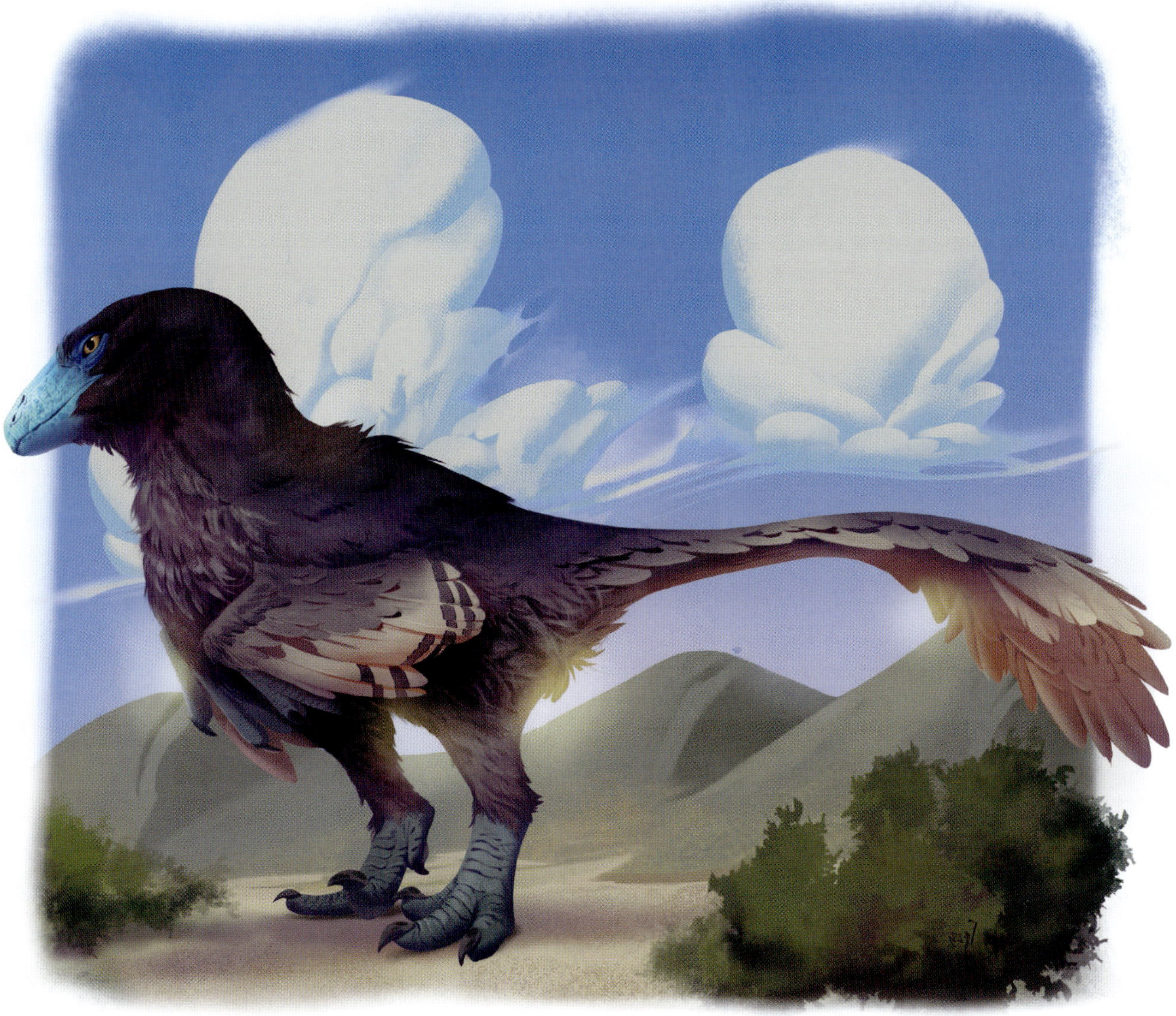

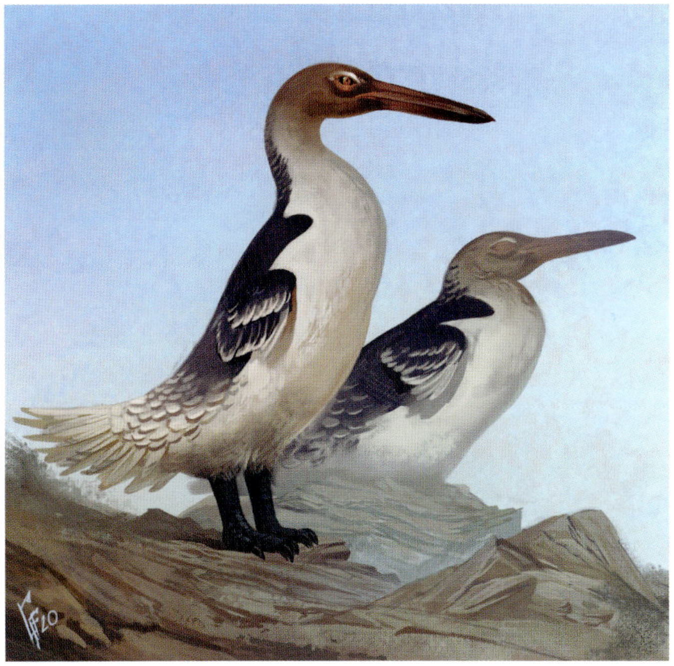

▴ *UTAHRAPTOR*

Utahraptor exceeded 6m (20ft) in length and is one of several giant dromaeosaurids – the family of sickle-clawed, feathered and very bird-like theropod dinosaurs. Many smaller dromaeosaurs such as *Deinonychus* and *Velociraptor* were likely fully feathered, and it's equally likely their giant relatives were similarly attired.

• Digital | 2023

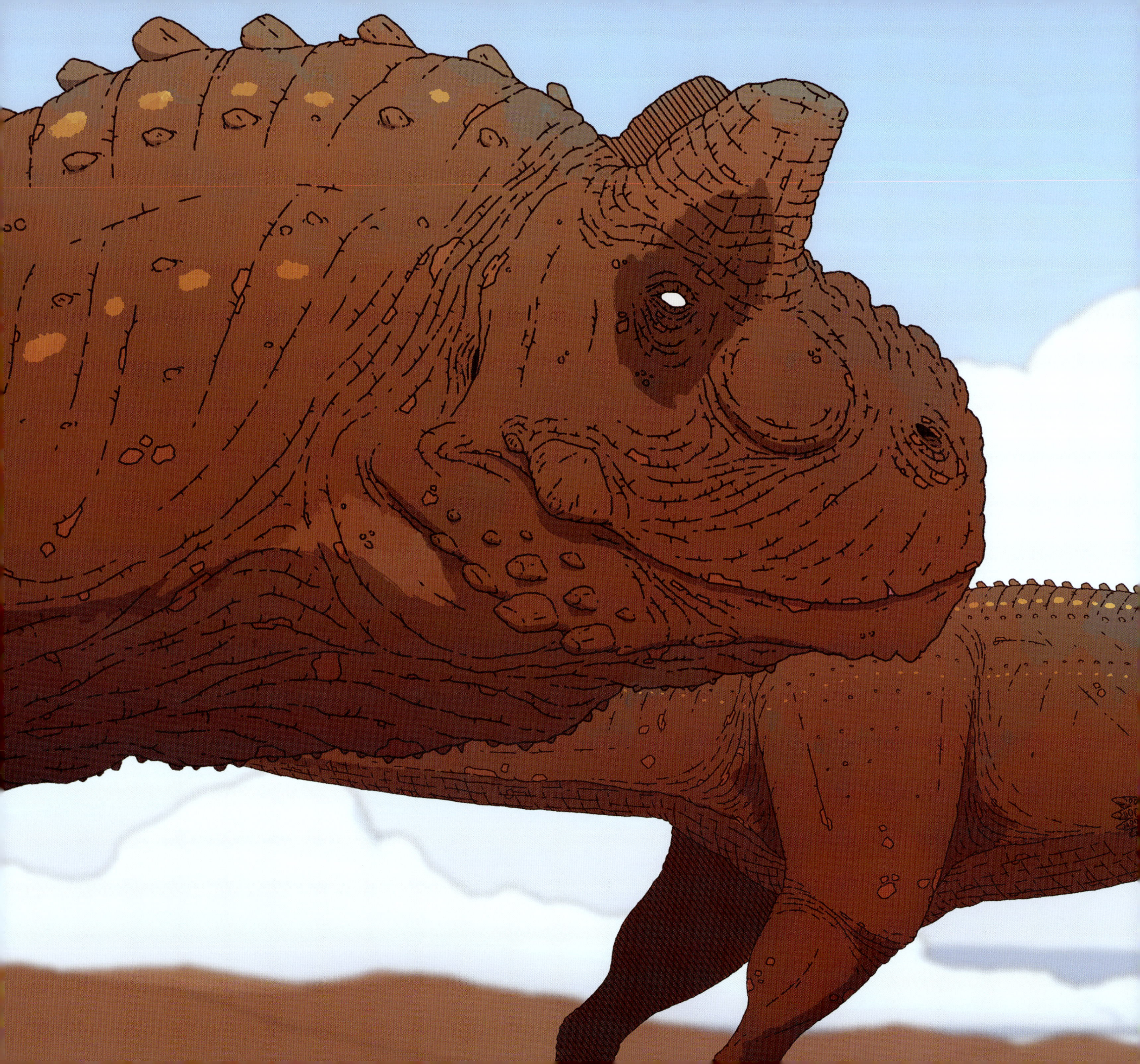

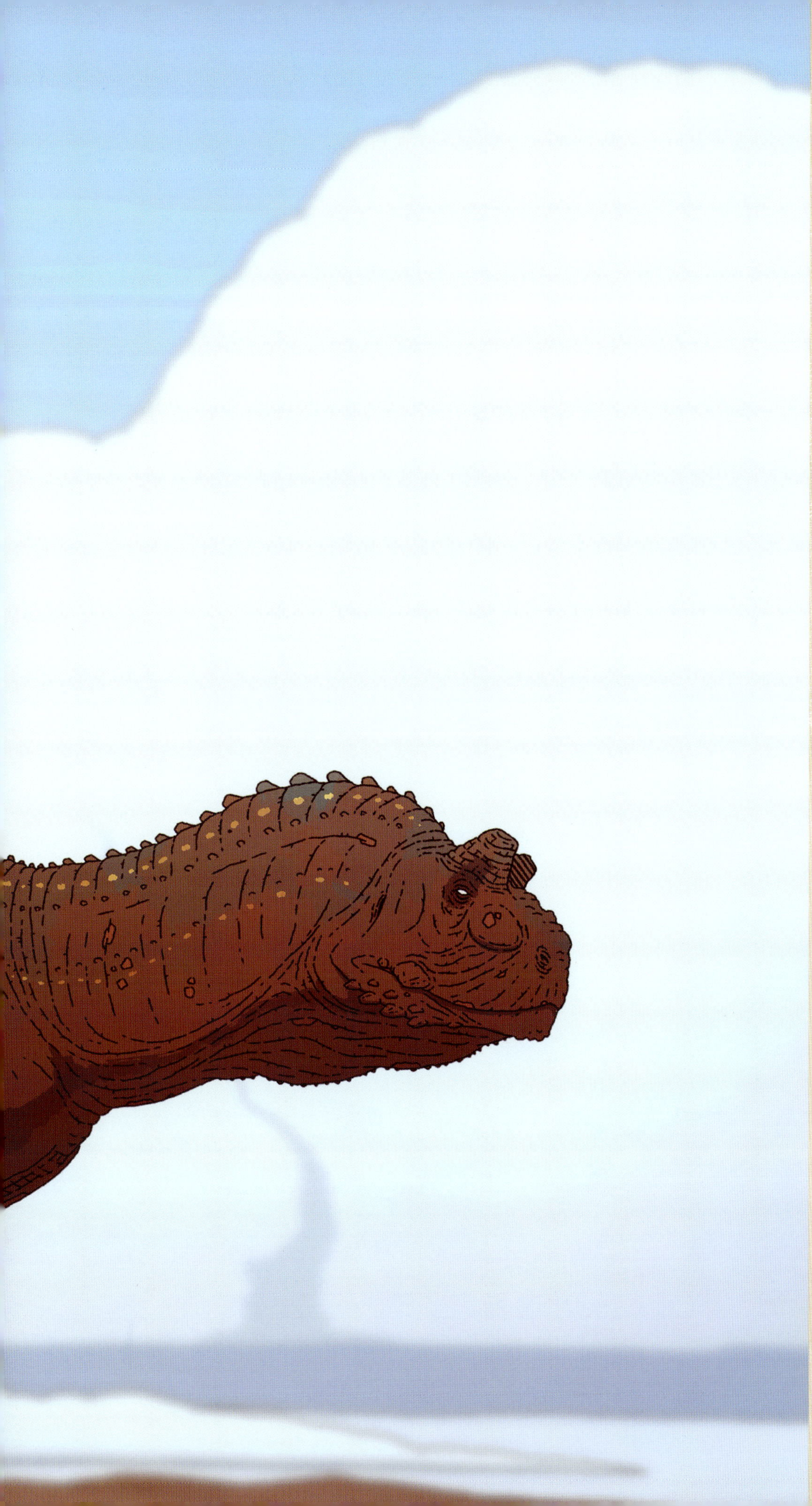

CURTIS LANAGHAN

CURTIS HAS BEEN DRAWING dinosaurs as far back as he can remember. As he grew older, that love of art and nature never went away, and after debating whether to study art or palaeontology in college, he chose the former, obtaining his Bachelor of Fine Arts in Media Arts & Animation in 2010. He has been working primarily as a 3D artist since, but in the last few years has begun drawing on a regular basis again, branching out into illustration. Dinosaurs and other prehistoric creatures are his regular subject matter once more. Heavily influenced by series like James Gurney's *Dinotopia*, the acclaimed comic series *Age of Reptiles*, and the hugely influential *Walking with Dinosaurs*, his aim has been to capture dinosaurs as living animals going about their own lives.

He currently resides near the west coast of Washington in the United States, with his wife and their slowly increasing number of animals.

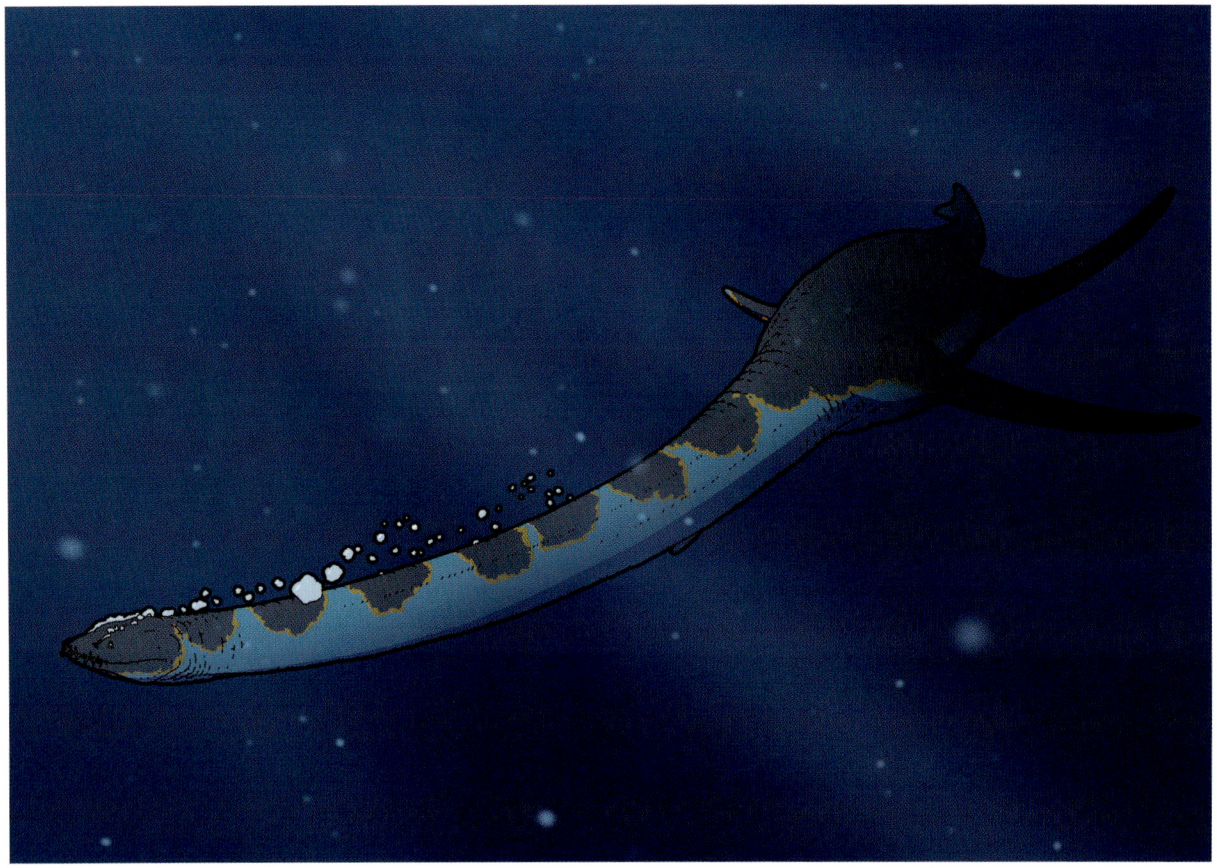

[Previous pages]

CARNOTAURUS

Named in 1985, the Argentinian abelisaurid theropod *Carnotaurus* is one of South America's best understood Mesozoic theropods, and certainly its most famous. Indeed, it is today a global superstar, and one of the most frequently illustrated of all Cretaceous dinosaurs. The key features of this animal are depicted here: its prominent horns, short face and remarkably odd arms and hands.

• Digital | 2022

▼ MOSASAUR

Mosasaurs foraged in shallow, continental seas right across the Cretaceous world but may also have made voyages across massive tracts of open ocean. Some species frequented estuaries and rivers too. The massive, thickset frame of some especially big, powerfully built mosasaurs is nicely conveyed in this image.

• Digital | 2024

▲ *STYXOSAURUS*

This image shows the elasmosaurid plesiosaur *Styxosaurus*, a North American member of the group from the very end of the Cretaceous. We assume that plesiosaurs had a 'marine livery' with blueish and/or greyish colour schemes and patterns, and obvious countershading. The idea that the necks of long-necked species were marked with stripes, bars or blotches is a popular speculation.

• Digital | 2023

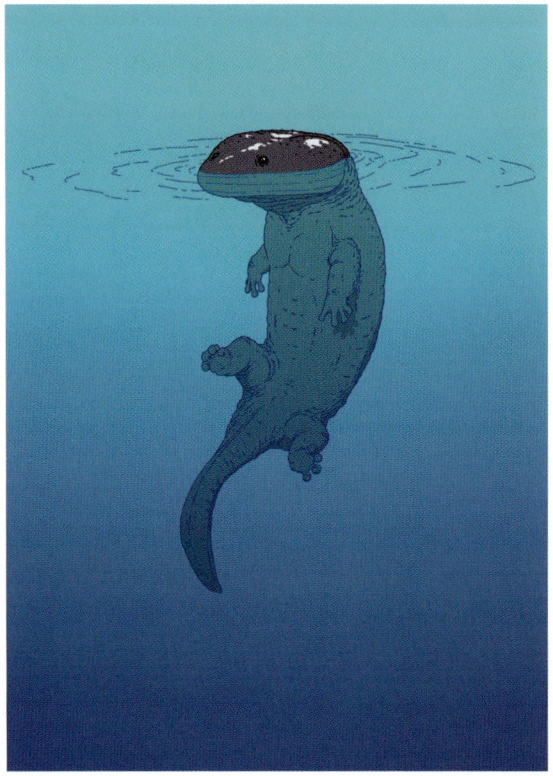

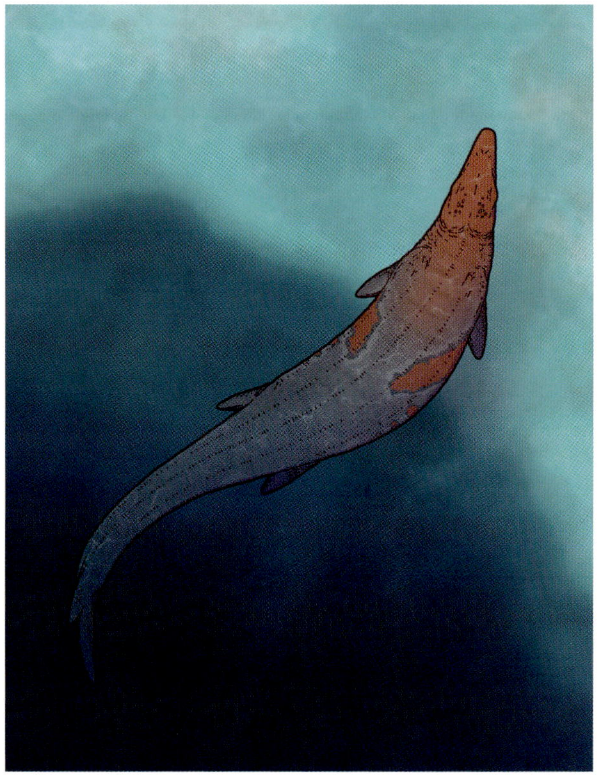

KOOLASUCHUS ▶

Superficially resembling a gargantuan salamander, the temnospondyl amphibian *Koolasuchus*, from the Early Cretaceous of southern Australia, inhabited cool freshwater pools, eating fishes, other animals and perhaps dinosaurs that were careless at the water's edge. It has a modicum of fame thanks to an appearance in the 1999 TV series *Walking With Dinosaurs*.

• Digital | 2024

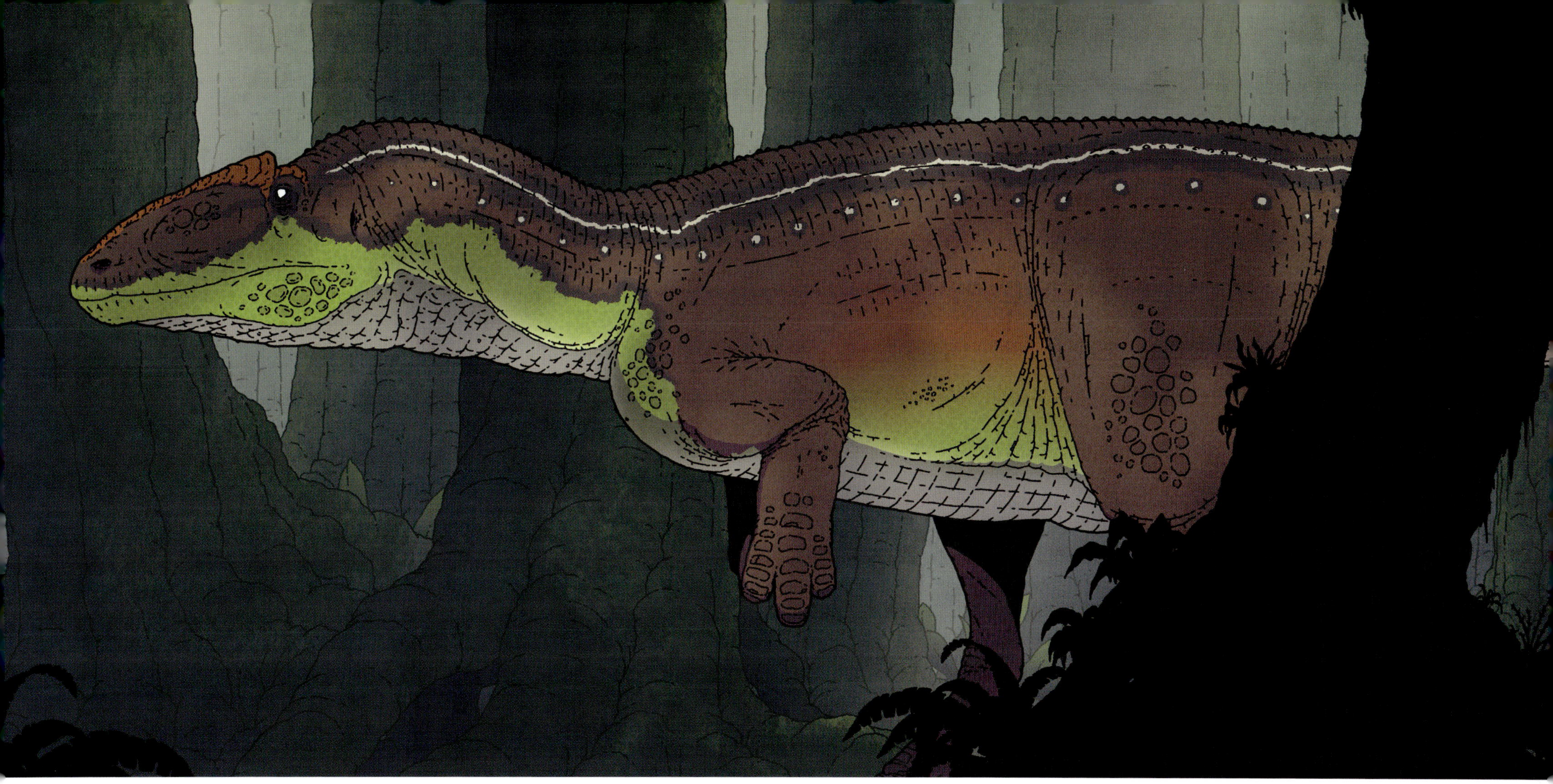

▲ *NEOVENATOR*

The English theropod *Neovenator* – a member of one of the famous Early Cretaceous Wealden dinosaur assemblages – stalks through a woodland dominated by massive trees. *Neovenator* fossils reported so far come from environments where shrubby trees and small bushes were typical. As ever, however, the possibility exists that it ranged far and wide when alive.
• Digital | 2023

STEGOSAURUS ▶

This image invites questions of whether large, formidably armed dinosaurs like stegosaurs were brightly coloured (because predation was simply not a concern), or did they have stripes or dapples to offer a degree of camouflage? Traditionally, palaeoartists make the plates different colours to the body. That's a reasonable idea, but we don't know whether it was true.
• Digital | 2024

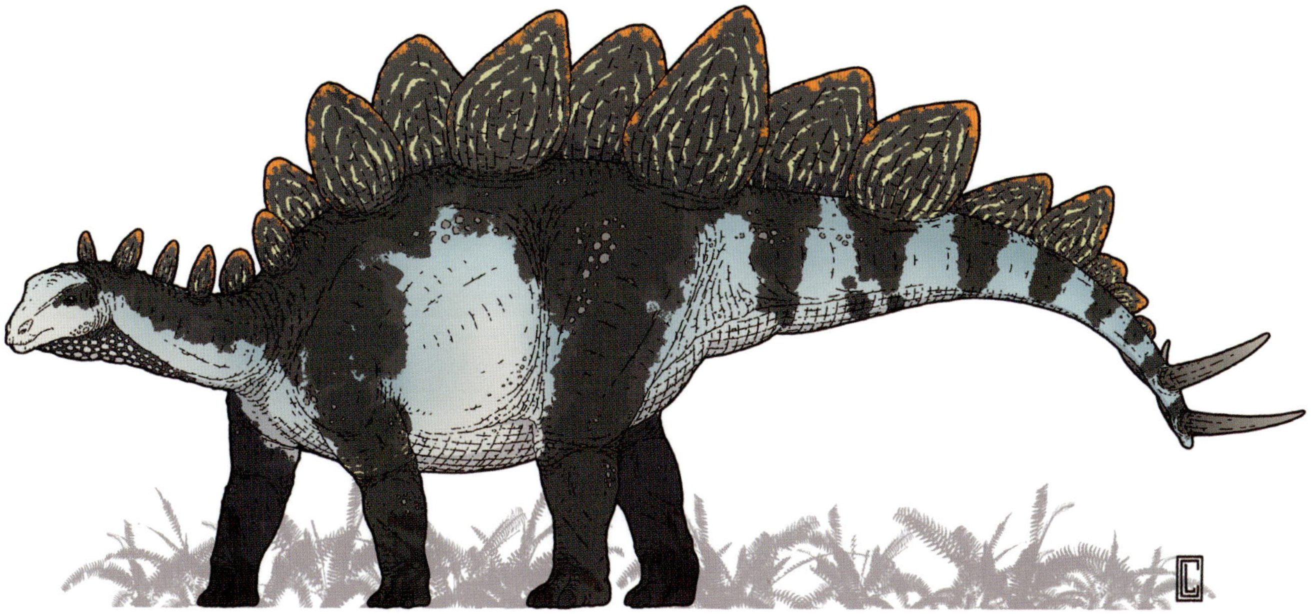

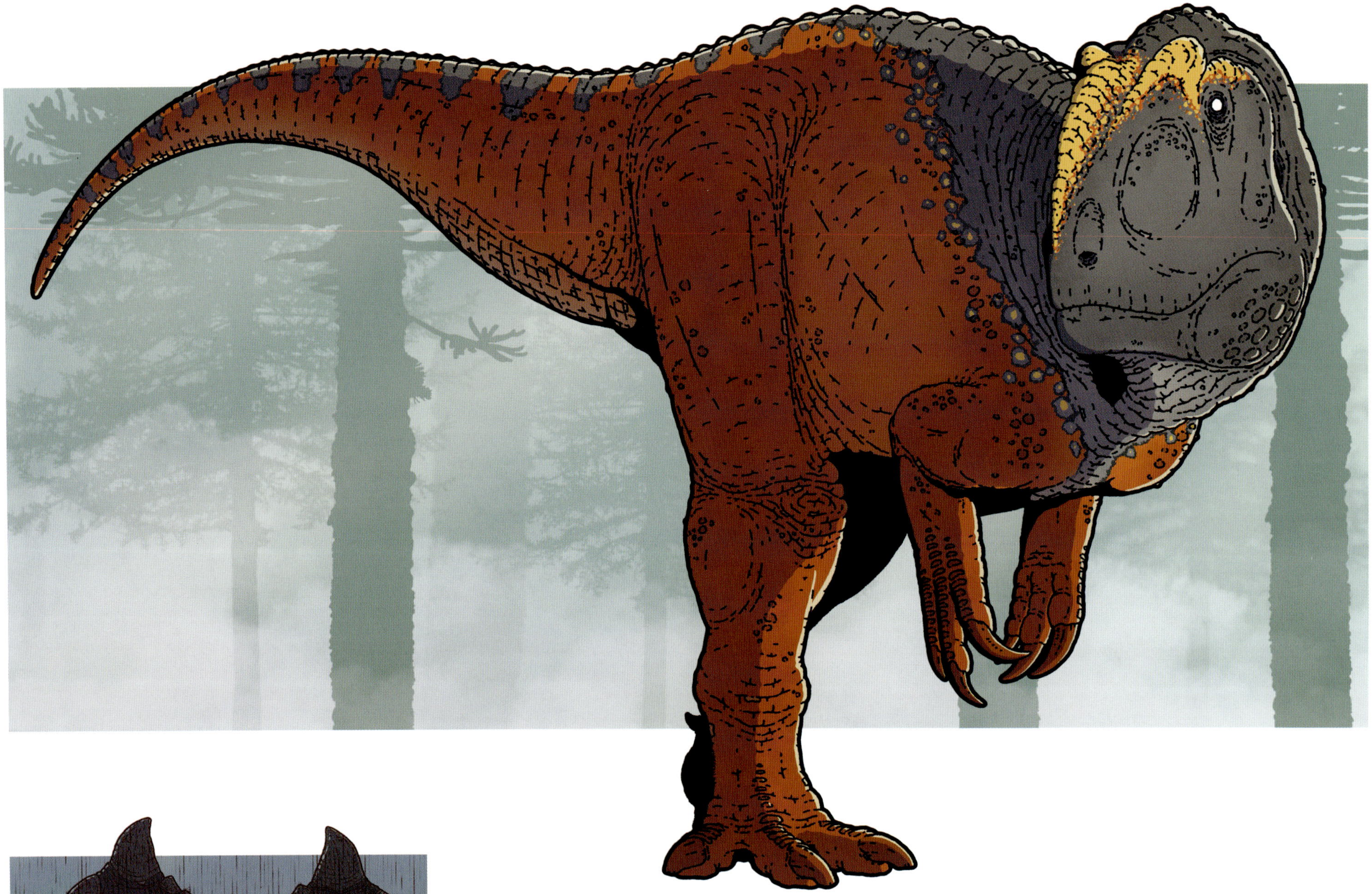

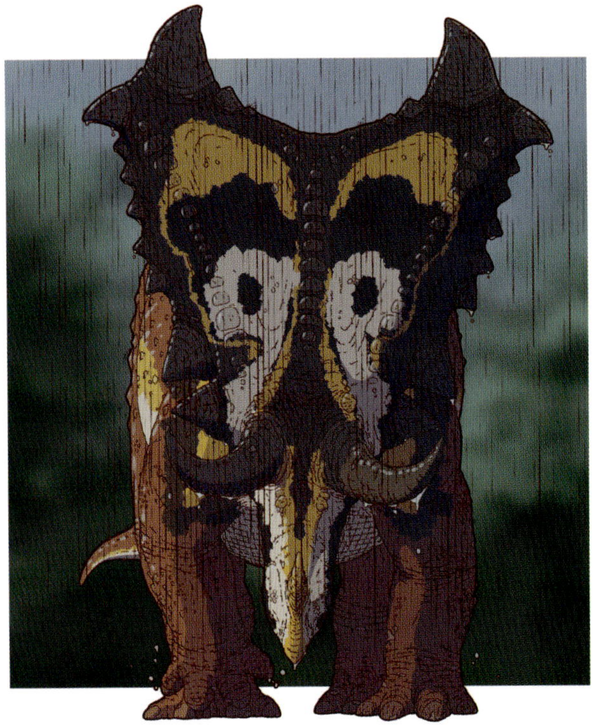

◄ *CHASMOSAURUS*

The massive size of the horned dinosaur frill – this one belongs to *Chasmosaurus*, specifically the Canadian species *C. belli* – makes it likely that the frill's surface was prominently marked and coloured, and used in sending signals to rivals and potential predators. Eyespots, bright flashes of colour and areas of obvious visual contrast offer exciting possibilities for the artist to explore.

• Digital | 2022

▲ *YANGCHUANOSAURUS*

Two or three species of *Yangchuanosaurus* – originally named for a city in Shanxi Province, China – inhabited Late Jurassic Asian dinosaur communities. Some reached 8m (26ft) – as long as their famous cousin, *Allosaurus*. Yangchuanosaurs had an especially deep, narrow snout, with hornlets along its upper margin and a bumpy midline crest.

• Digital | 2024

▼ *CERATOSAURUS*

A spectacular predator best known from the famous Upper Jurassic Morrison Formation, *Ceratosaurus* has nose and eyebrow horns, armour plates along the midline of its back and especially long teeth in the upper jaw. The colour scheme is based on the bright breeding colours of the male common green forest lizard, *Calotes calotes*.

• Digital | 2023

DREPANOSAURUS ▶

Drepanosaurus is one of several small, pointy-snouted, climbing reptiles known from the Late Triassic. Curved claws, powerful limbs and a prehensile tail are typical of these animals, grouped together as the drepanosauromorphs. Most were climbers, but some appear to have been burrowers and even gliders. Where in the reptile family tree they belong remains the topic of debate.

• Digital | 2023

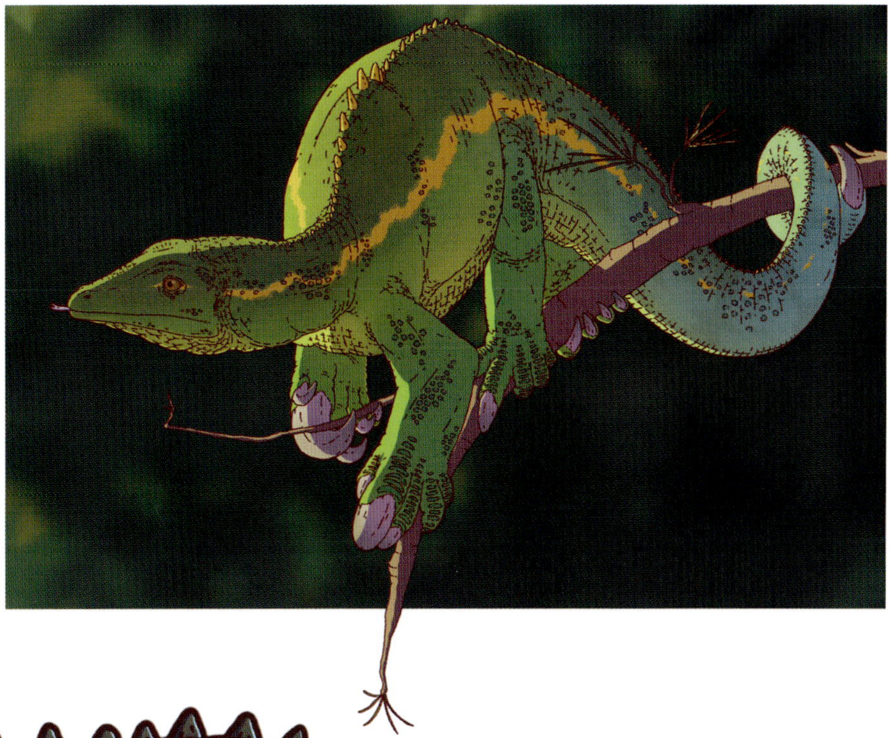

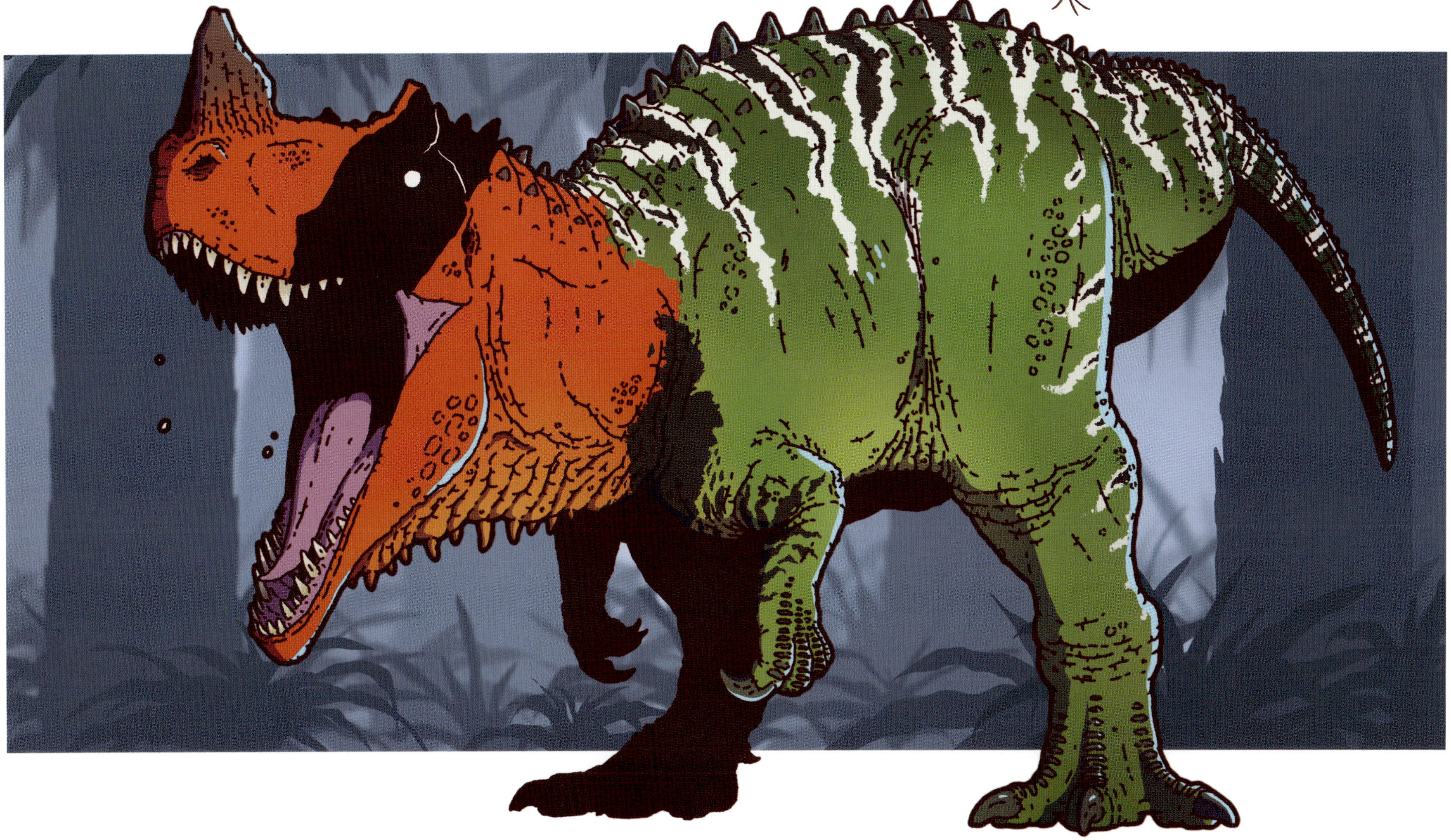

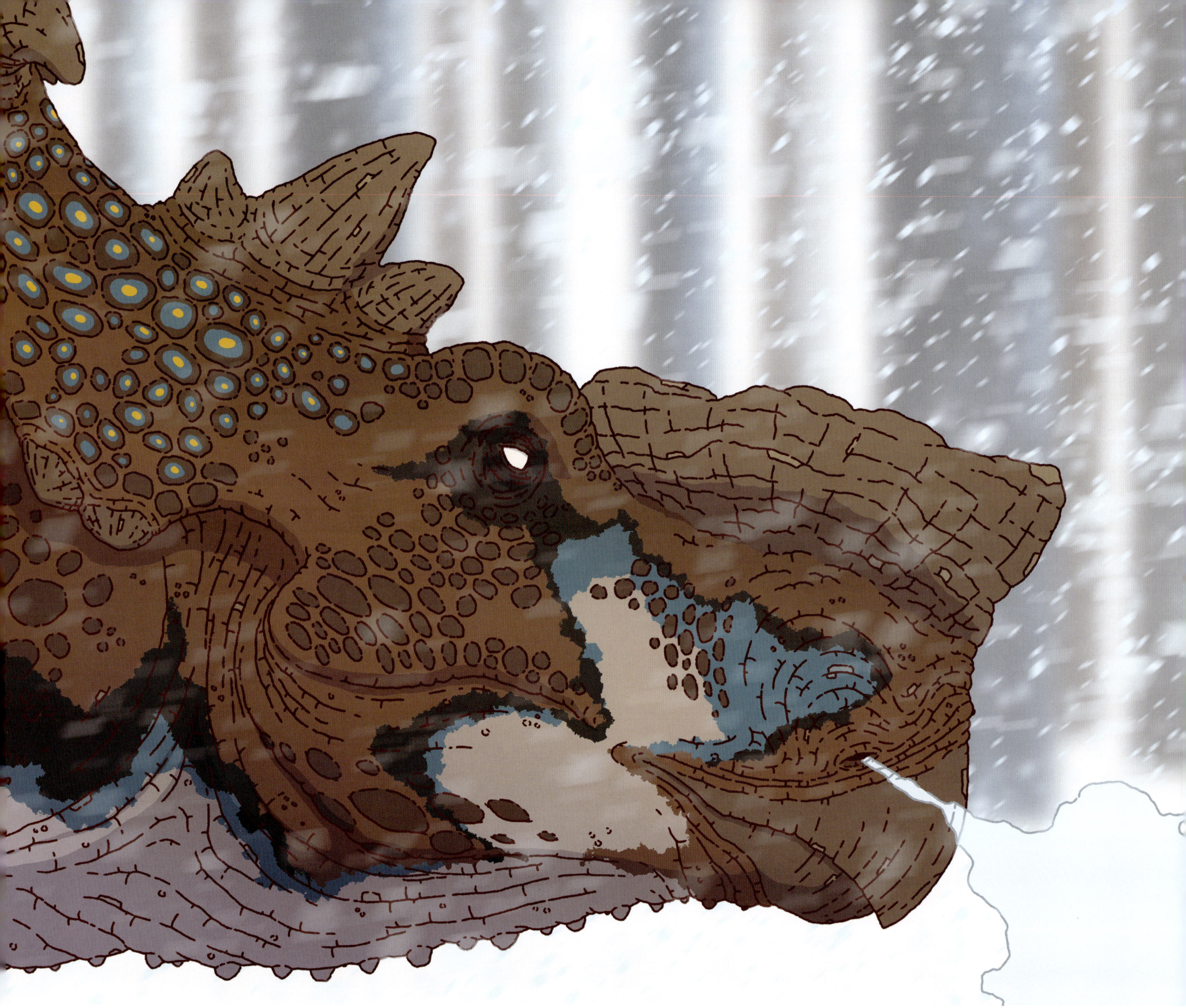

◂ *PACHYRHINOSAURUS*

Pachyrhinosaurus has transitioned from obscure and scarcely mentioned to being among the most frequently portrayed horned dinosaur. The cluster of spikes along the frill midline mark this species as *P. lakustai*, named from Canada in 2008. These animals didn't live in perpetually snowy landscapes, but did have to deal with snow, ice and freezing temperatures on occasion.

• Digital | 2022

▾ *TRICERATOPS*

A long-running debate in palaeoart concerns the application of colour to giant animals. Is a reconstruction like this too gaudy, or should we consider it plausible that even dinosaurs like *Triceratops* could be coloured this way, perhaps only on occasion (like, in the breeding season)? And were horn-covered structures always grey…could they be colourful too?

• Digital | 2023

RHAMPHORHYNCHUS ▸

Rhamphorhynchus of Late Jurassic Europe is among the most well-studied of all pterosaurs, and numerous fossils show that it inhabited lagoons, coasts and deltas. The sharply pointed, interlocking teeth of this pterosaur were used to grab fishes, crustaceans and other prey. It remains uncertain whether the jaws were lined with horny beak tissue, as shown here.

• Digital | 2023

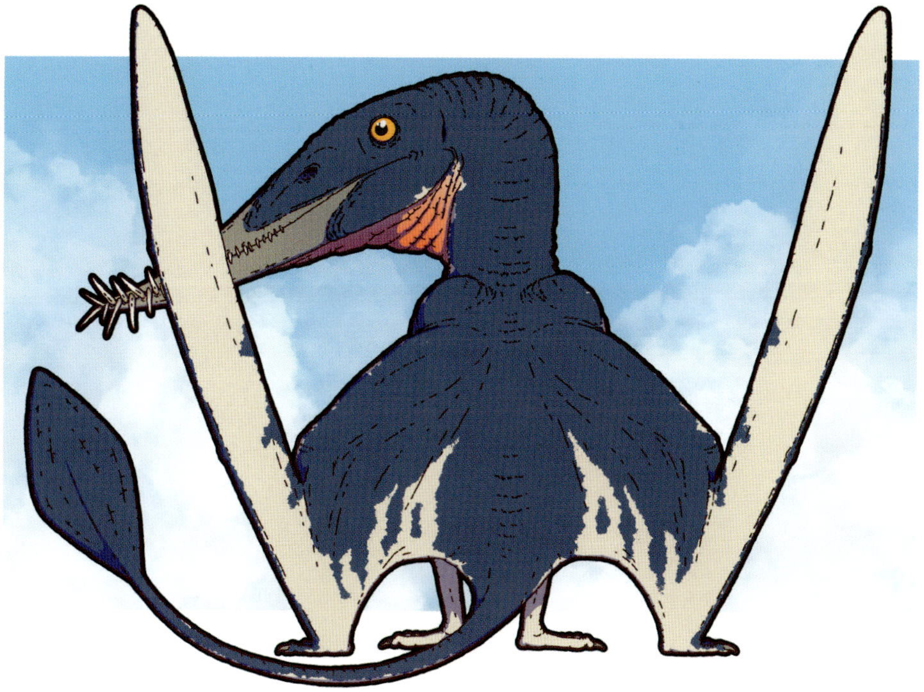

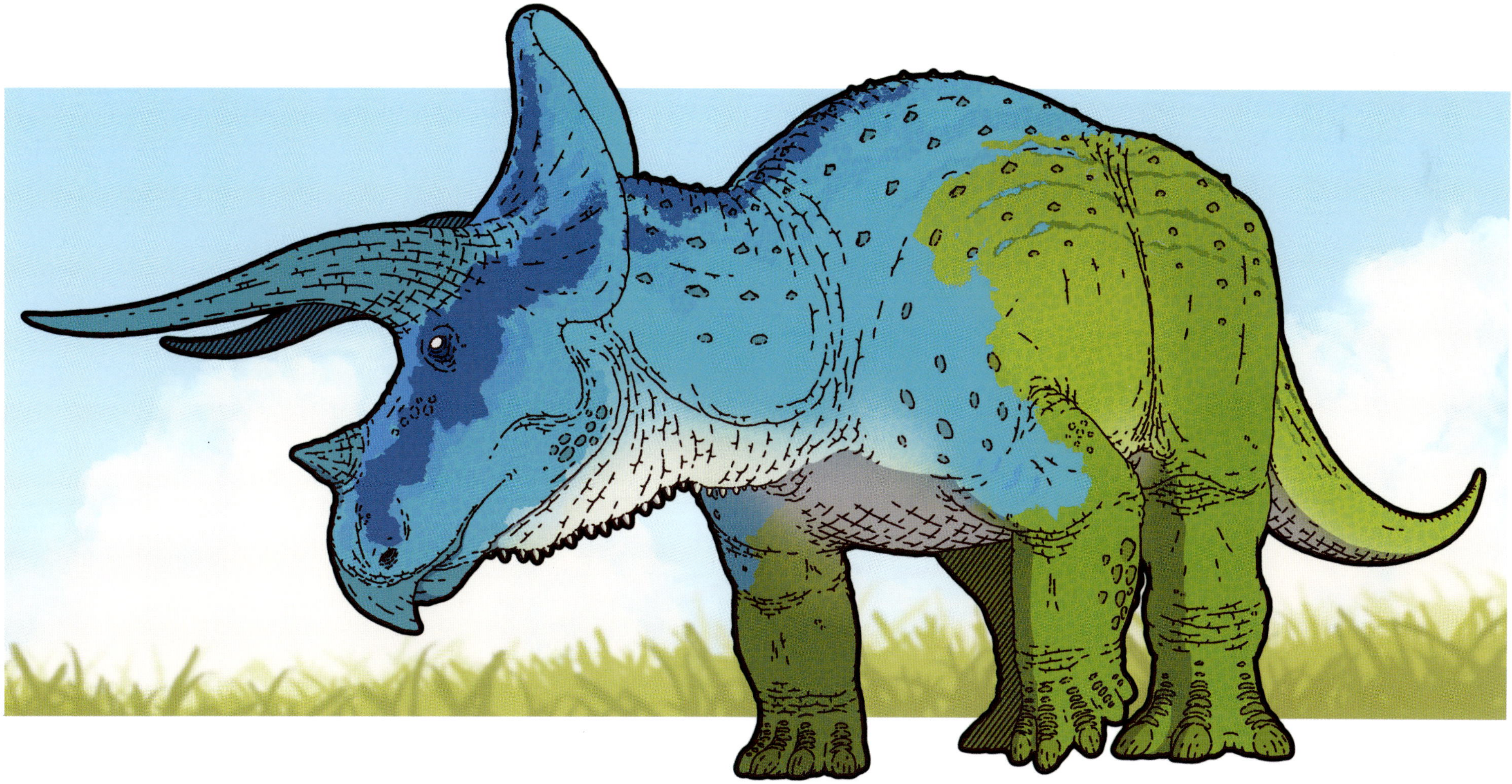

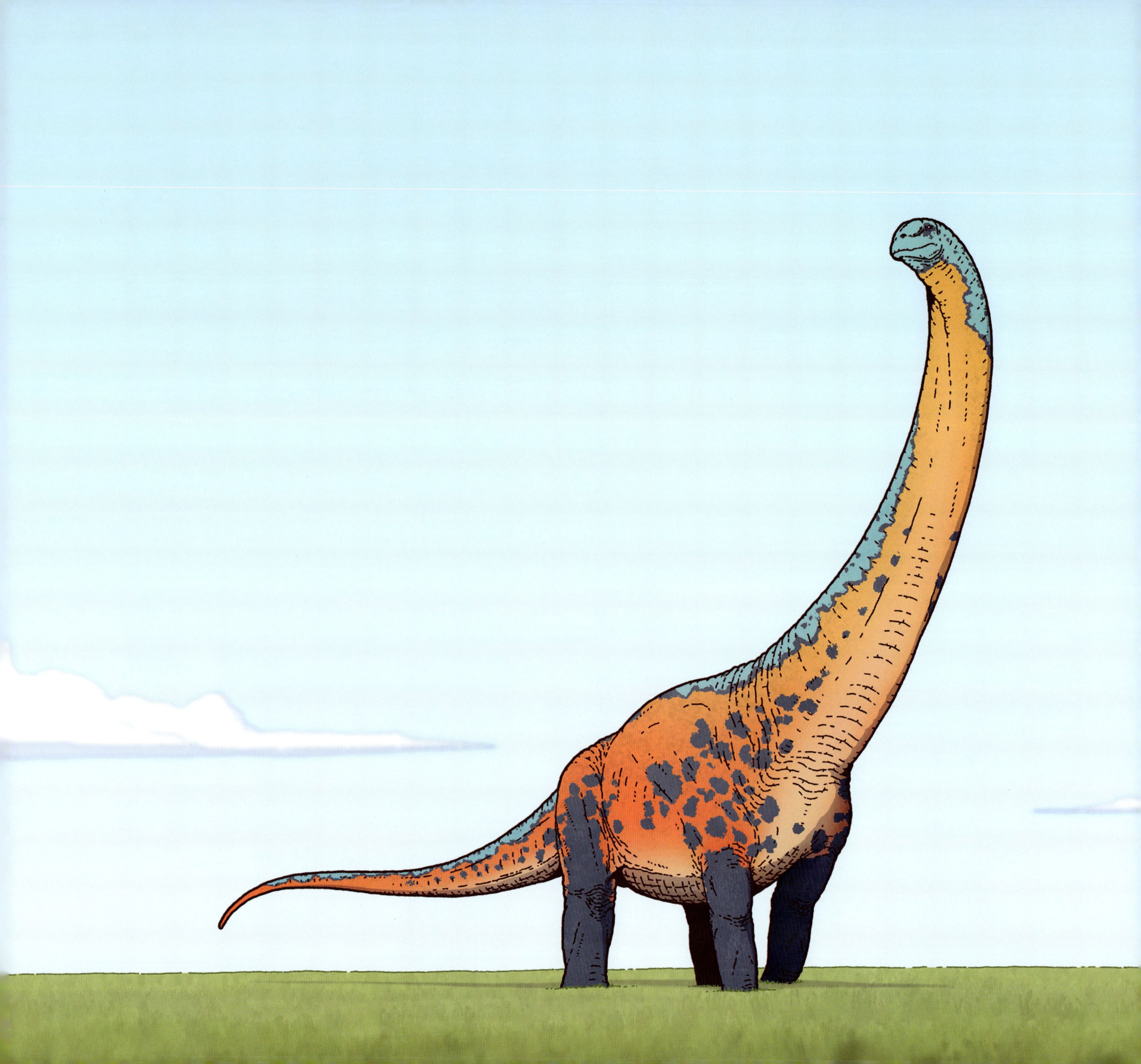

◂ TITANOSAUR

The massive size of giant sauropods like this titanosaur invites all kinds of questions about their biology, but also their impact – visual as well as ecological – on the landscape. The titanosaurs were a diverse group of sauropods known from across the globe and including some of the largest land animals ever known, including *Argentinosaurus* and *Patagotitan*.

• Digital | 2024

▾ *APATOSAURUS*

In illustrating the Late Jurassic sauropod *Apatosaurus*, Curtis has once again been provocative on the issue of colouration. Perhaps sauropods were like this during the early years of their lives – when camouflage was essential – and gradually transitioned to something plainer as adult size was achieved. Changes of this sort are typical of living reptile groups.

• Digital | 2024

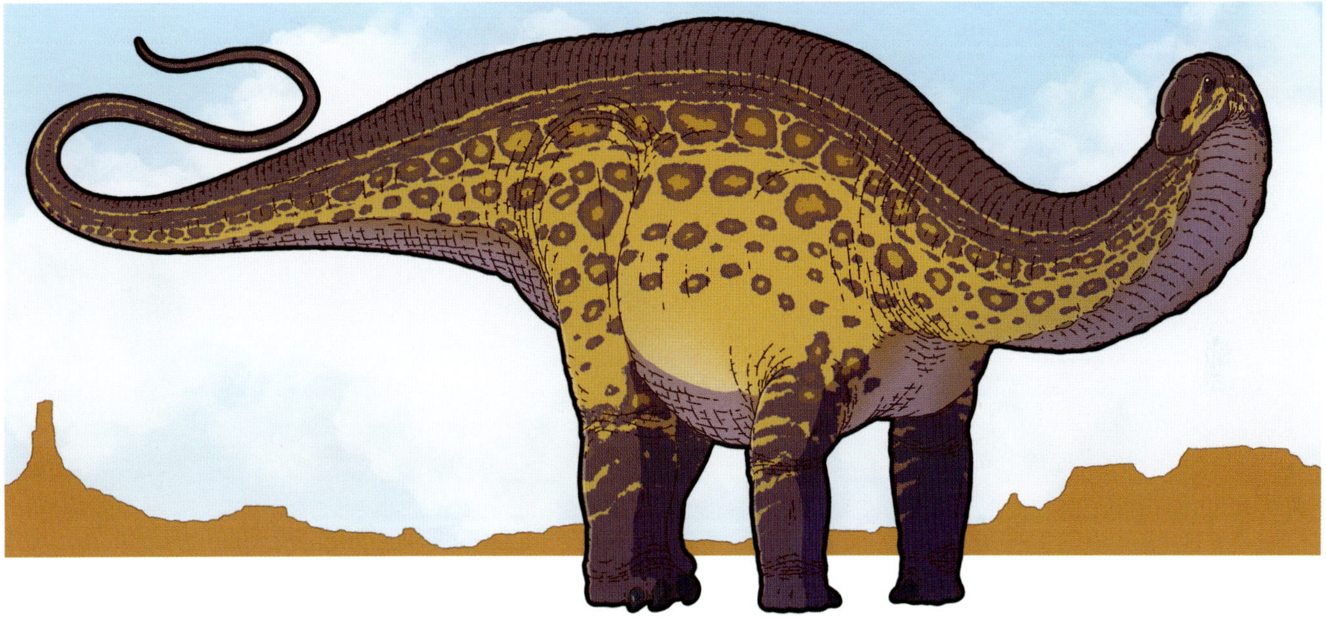

SPINOSAURUS ▸

There was a time when *Spinosaurus* was consistently depicted as an animal of hot deserts. Those days have mostly gone, since modern renditions show it as an animal of rivers and deltas. The North African animals lumped together as *Spinosaurus aegyptiacus* by some authors likely represent several distinct spinosaurid species.

• Digital | 2023

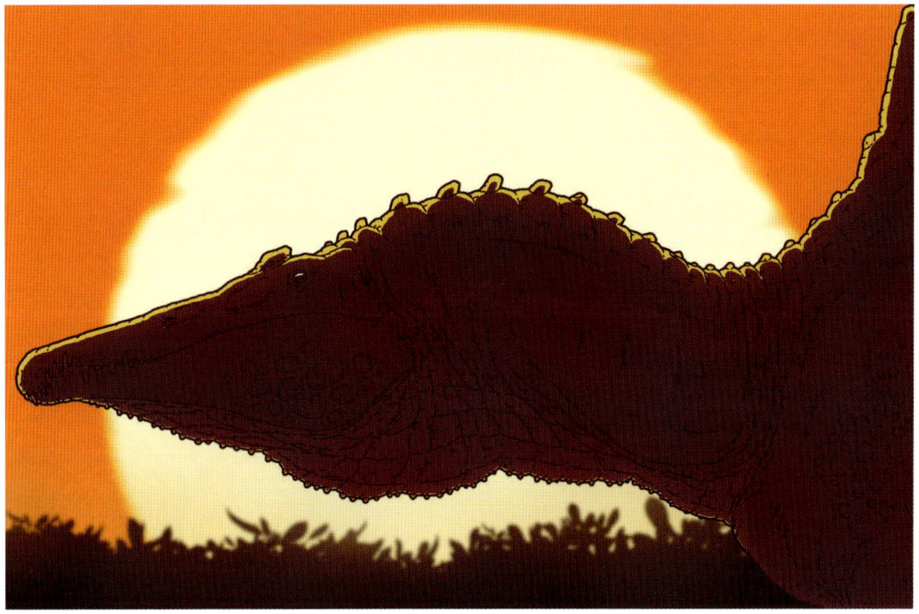

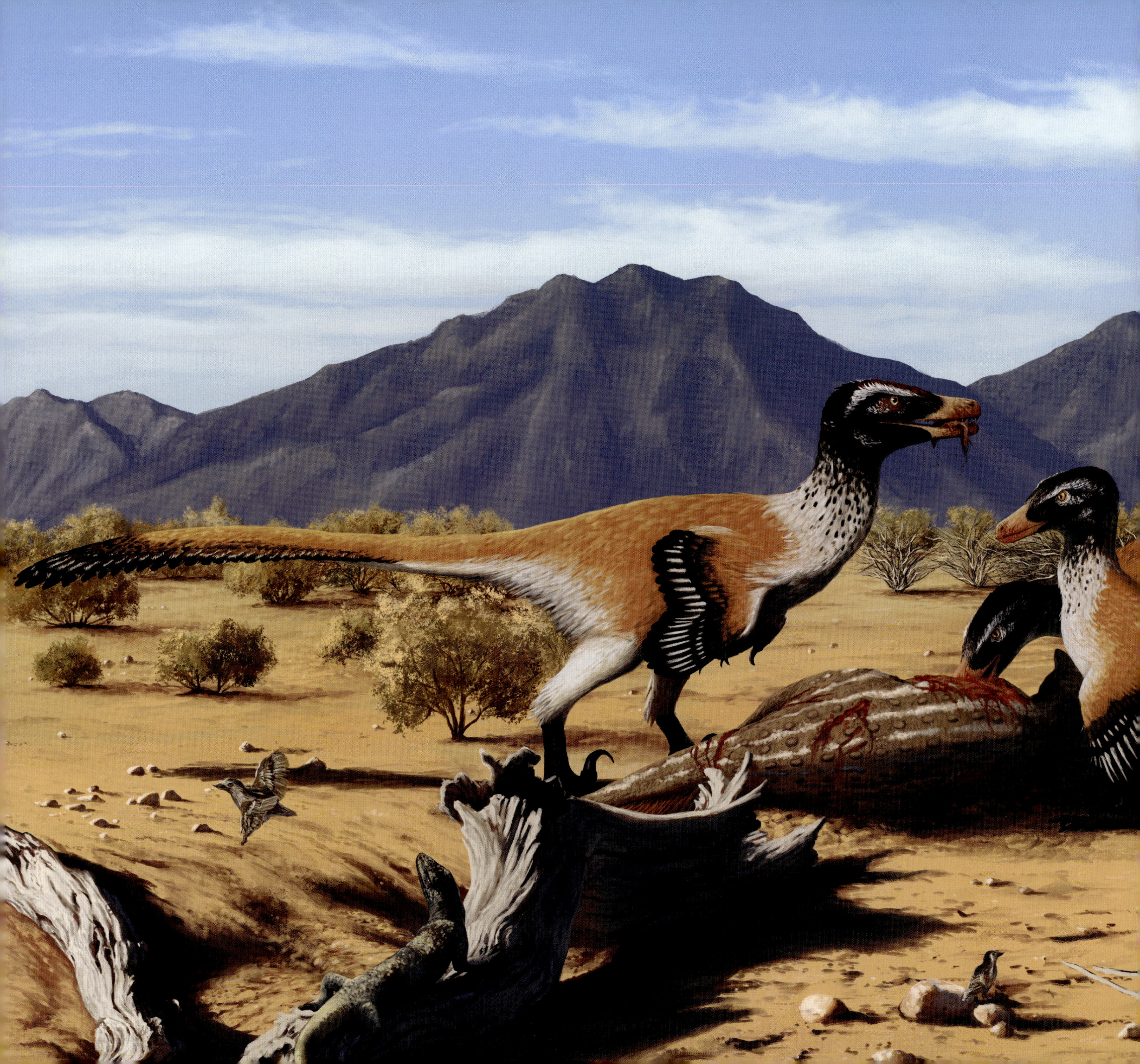

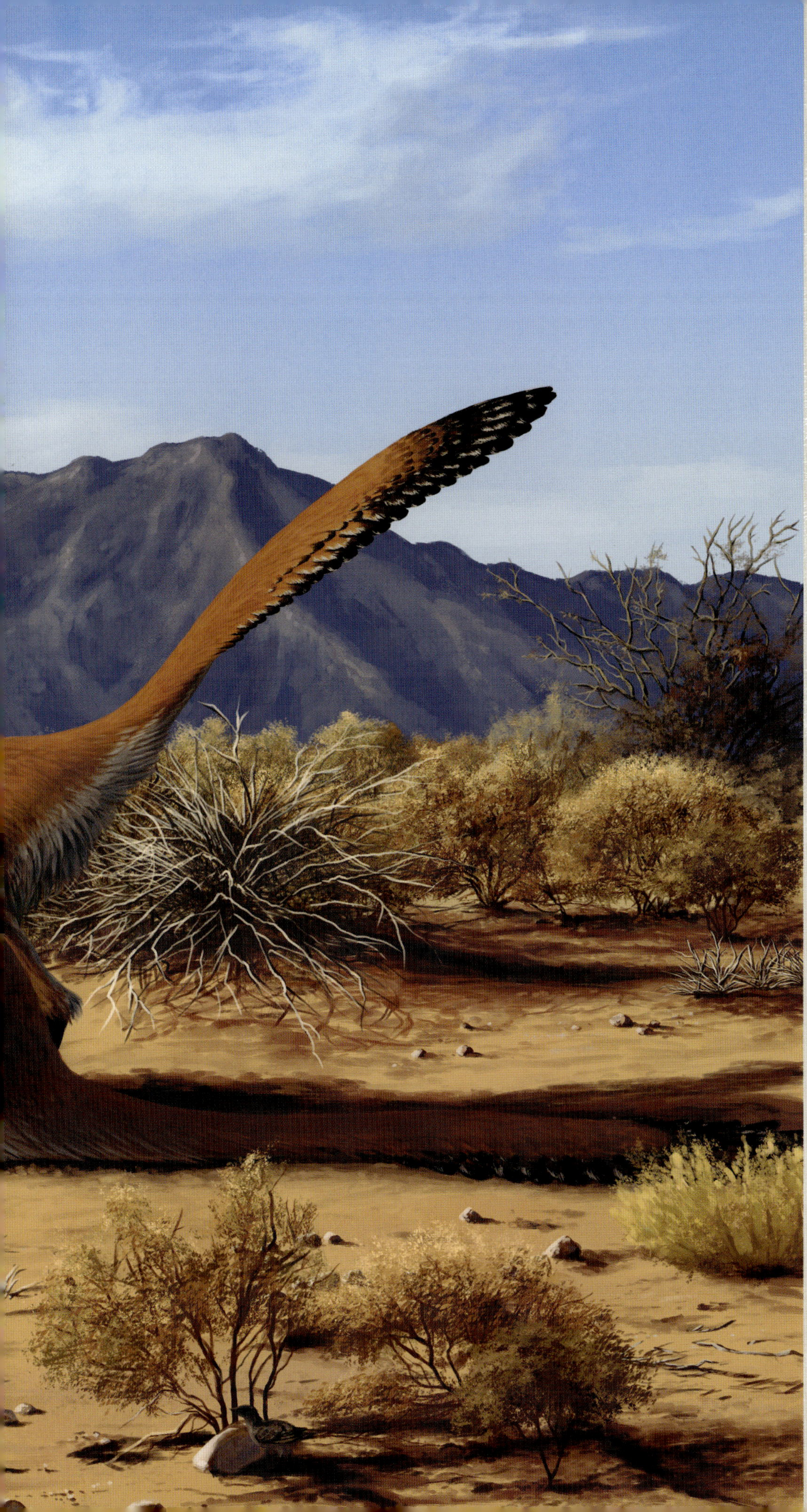

EDYTA FELCYN-KOWALSKA

EDYTA WAS BORN IN 1987 in Bydgoszcz, Poland. At the age of four, she was given a few books about animals, among which was a book about dinosaurs. This was the one that interested her the most. She began copying illustrations from the book, learning how to draw in the process.

She initially drew only for her own entertainment until the early 2000s, when the Internet became available to her. She began publishing her work online and learned digital painting, her new art bringing her to the attention of the Polish palaeontological community. As a result, she produced palaeoart illustrations for the Polish Geological Institute, the National Research Institute and Institute of Paleobiology, part of the Polish Academy of Sciences. However, she never intended to be a professional artist, graduating instead in medical biotechnology.

She now lives in Poznań, working full time as a translator. Palaeoart remains one of her hobbies: she regularly produces pencil sketches and digital art, and strives to blend realism with a painterly style.

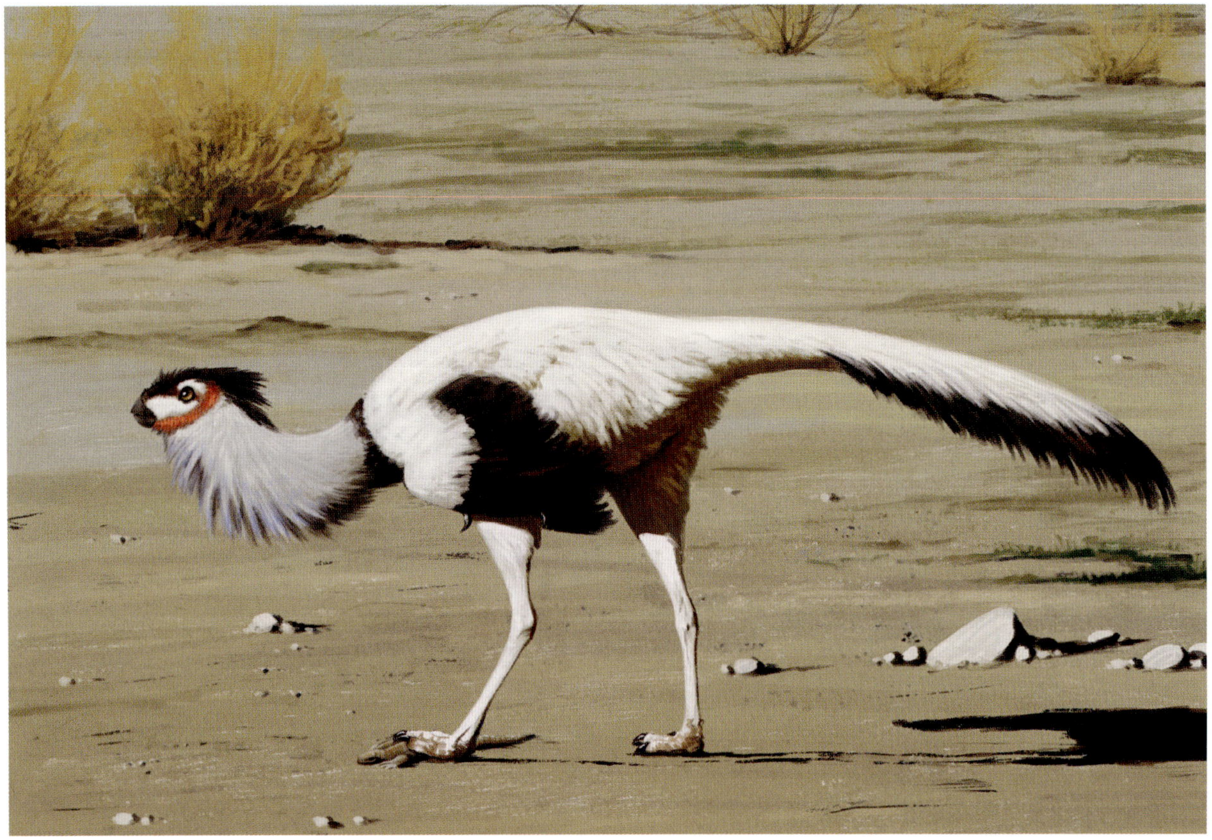

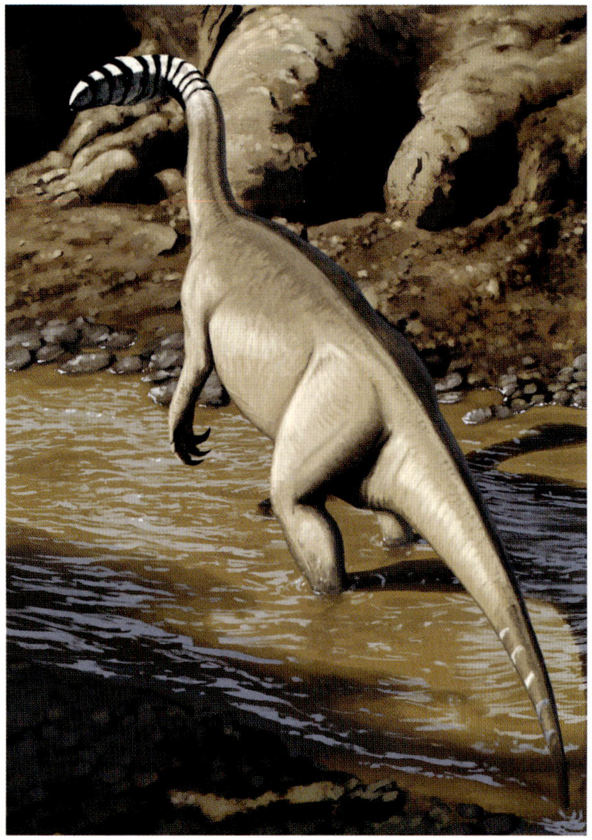

▲ AVIMIMUS

Originally identified as very close to the ancestry of modern birds, *Avimimus* (of Late Cretaceous Mongolia) is currently thought to be an oviraptorosaur, a feathery theropod group whose members are typically mostly toothless, beaked and often crested. *Avimimus* was about 1.5m (5ft) long, short-snouted and with long, slender legs.
• Digital | 2012

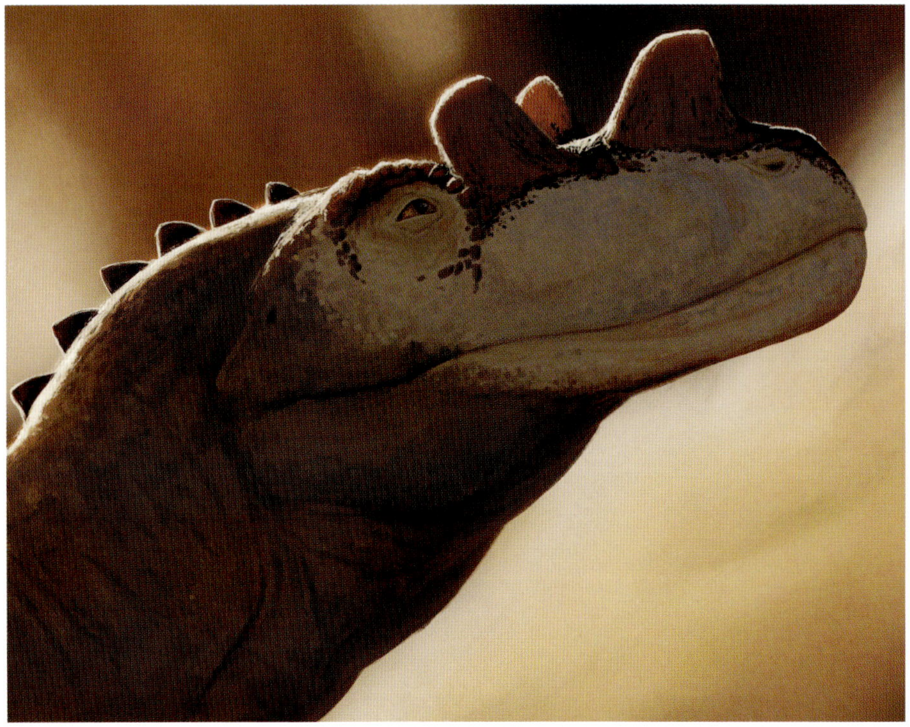

CERATOSAURUS ▶

Ceratosaurus – the fantastic horned theropod of Late Jurassic USA (and perhaps Portugal and elsewhere) – is generally shown these days as having horns that are prominently coloured relative to the rest of the face. This image also emphasises the fact that the dentition would likely have been concealed when the jaws were closed.
• Digital | 2021

[Previous pages]

VELOCIRAPTOR & GOBIPTERYX

This vista sets a *Velociraptor* group within the semi-desert habitat that formed part of their East Asian range. This is reflected in their colour scheme, which recalls that of modern desert birds. Other animals known from the same fossil beds as *Velociraptor* include lizards of many sorts and small birds like the finch-sized *Gobipteryx*.
• Digital | 2024

◂ *PLATEOSAURUS*

A 'classic' dinosaur of the European Late Triassic, *Plateosaurus* has long been portrayed as bland and greyish. However, if the remarkably long necks of the great sauropods had a role in visual display, maybe the same was true of earlier long-necked dinosaurs, a possibility explored here. Studies of forelimb posture and function show that the traditional view that plateosaurs were quadrupedal may now be incorrect. Like theropods, their hands were kept palms-inward.
• Digital | 2014

LEAELLYNASAURA ▸

Measuring 90cm (3ft) long, this little Early Cretaceous dinosaur is known from the famous Australian fossil site — Dinosaur Cove — in Victoria. During this period, Victoria was likely within the Antarctic Circle, thereby exposing the region to periods of intense cold and darkness. It is likely that *Leaellynasaura* was insulated by a coat of hair-like feathers, structures preserved in other, similar bipedal ornithischian dinosaurs.
• Digital | 2013

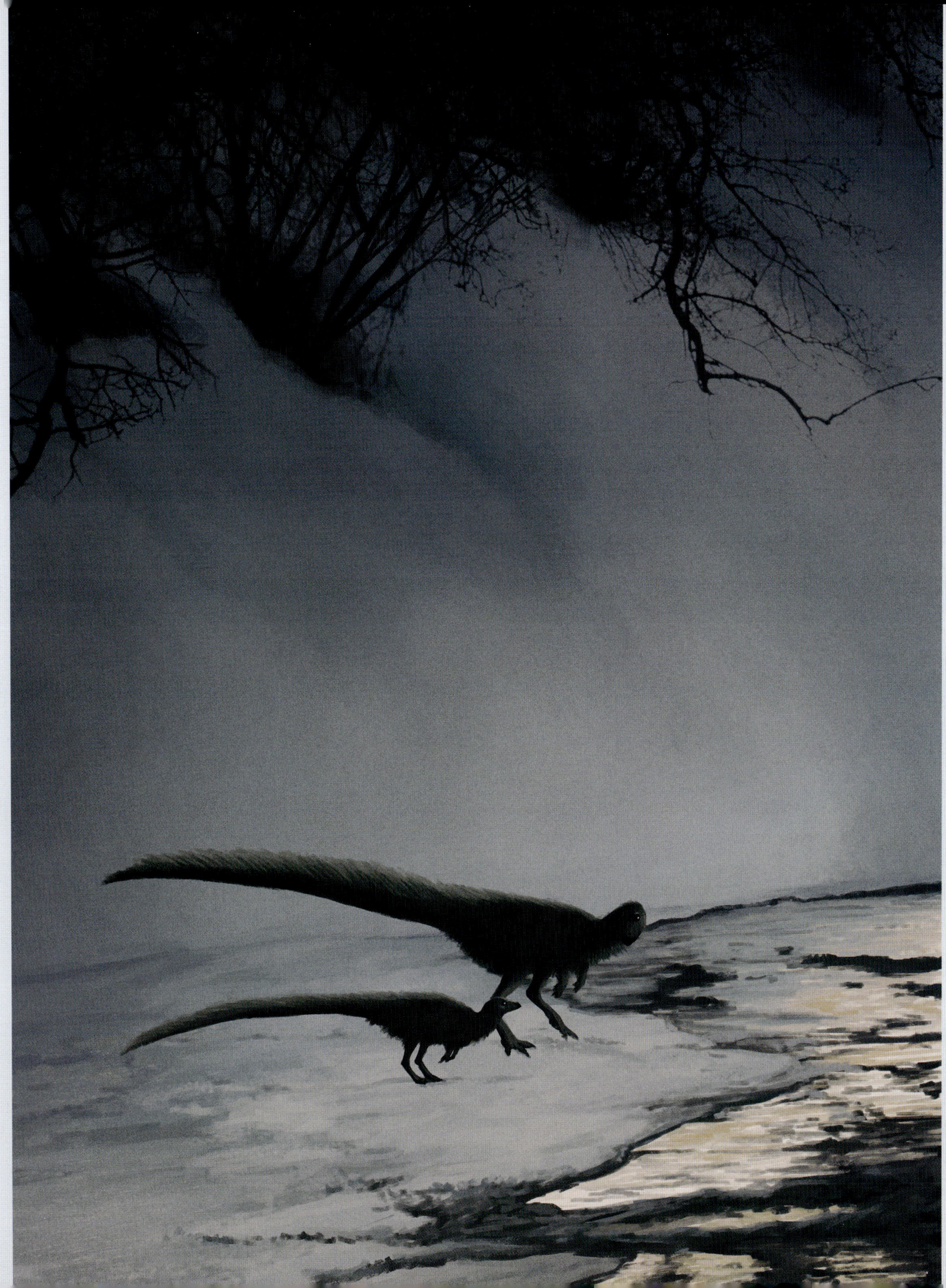

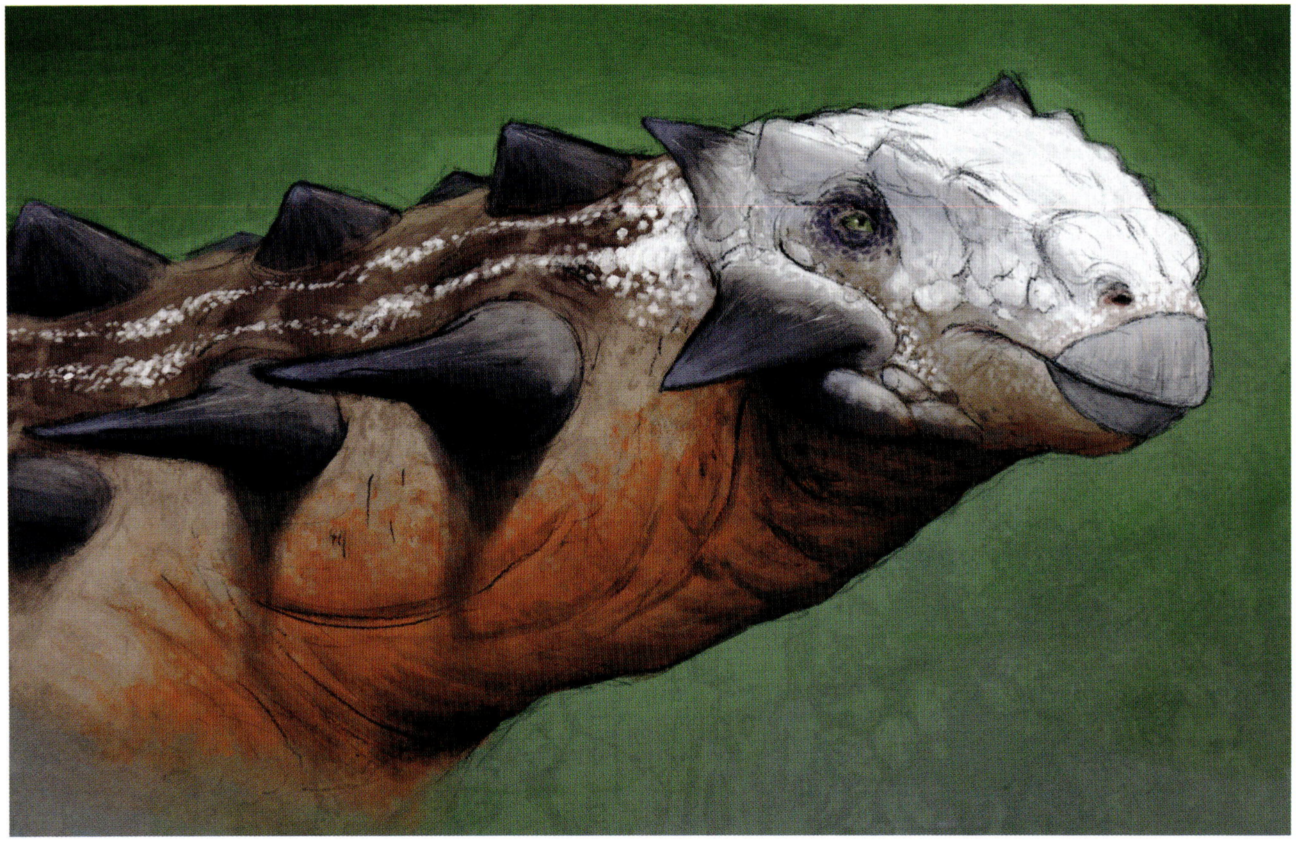

◂ *PINACOSAURUS*

A vast amount of speculative prose could be penned on which colours and patterns might best fit ankylosaurids, like this *Pinacosaurus* of Late Cretaceous eastern Asia. Were they earth-toned for camouflage, or could their armour and defensive capabilities allow for prominent markings that could warn off a predator?

• Pencil, watercolour and digital | 2019

▾ *SIMOSUCHUS*

It's gradually becoming better known that crocodyliforms – the huge archosaur group known traditionally as 'crocodilians' – evolved a massive diversity throughout the Mesozoic. The very weird *Simosuchus* – Late Cretaceous Madagascar's 'pug crocodile' –had leaf-shaped teeth and wide jaws that suggest it was a herbivore.

• Pencil, watercolour and digital | 2018

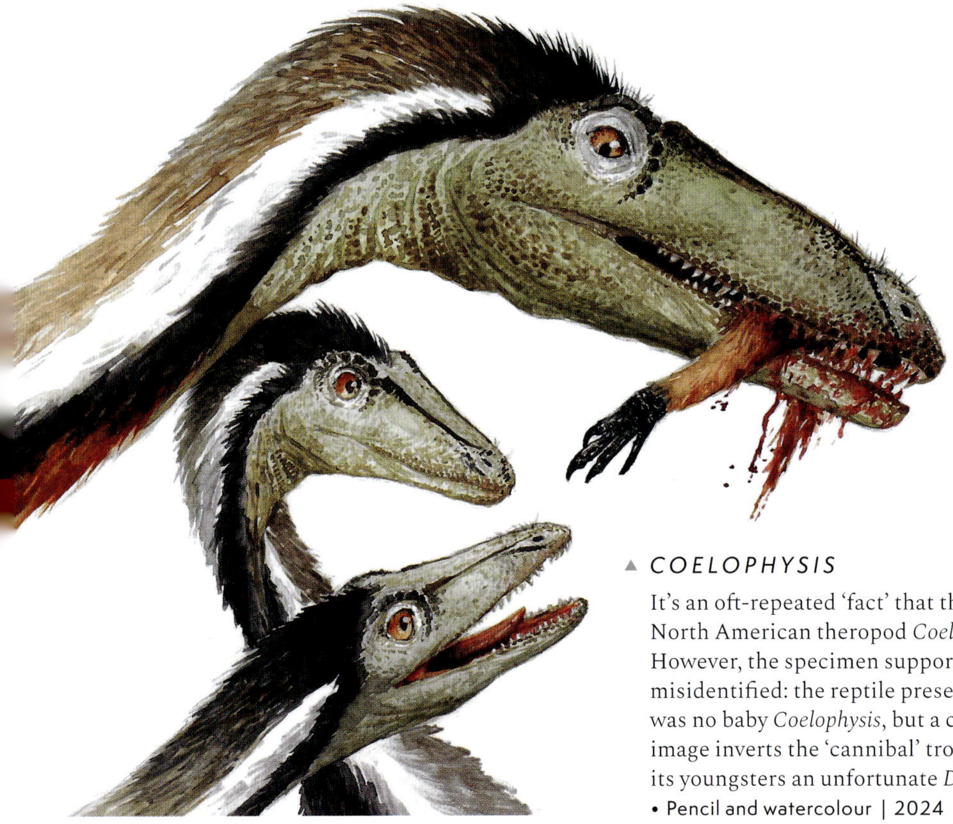

▴ *COELOPHYSIS*

It's an oft-repeated 'fact' that the Early Jurassic North American theropod *Coelophysis* was a cannibal. However, the specimen supporting this contention was misidentified: the reptile preserved as stomach contents was no baby *Coelophysis*, but a crocodylomorph. This image inverts the 'cannibal' trope: the parent is feeding its youngsters an unfortunate *Dromomeron*.

• Pencil and watercolour | 2024

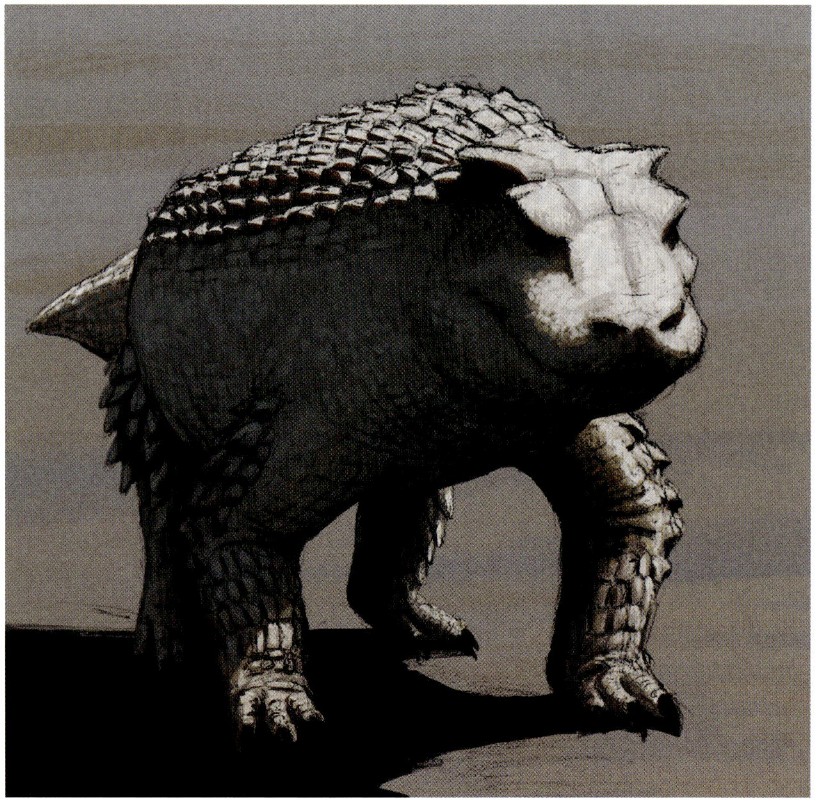

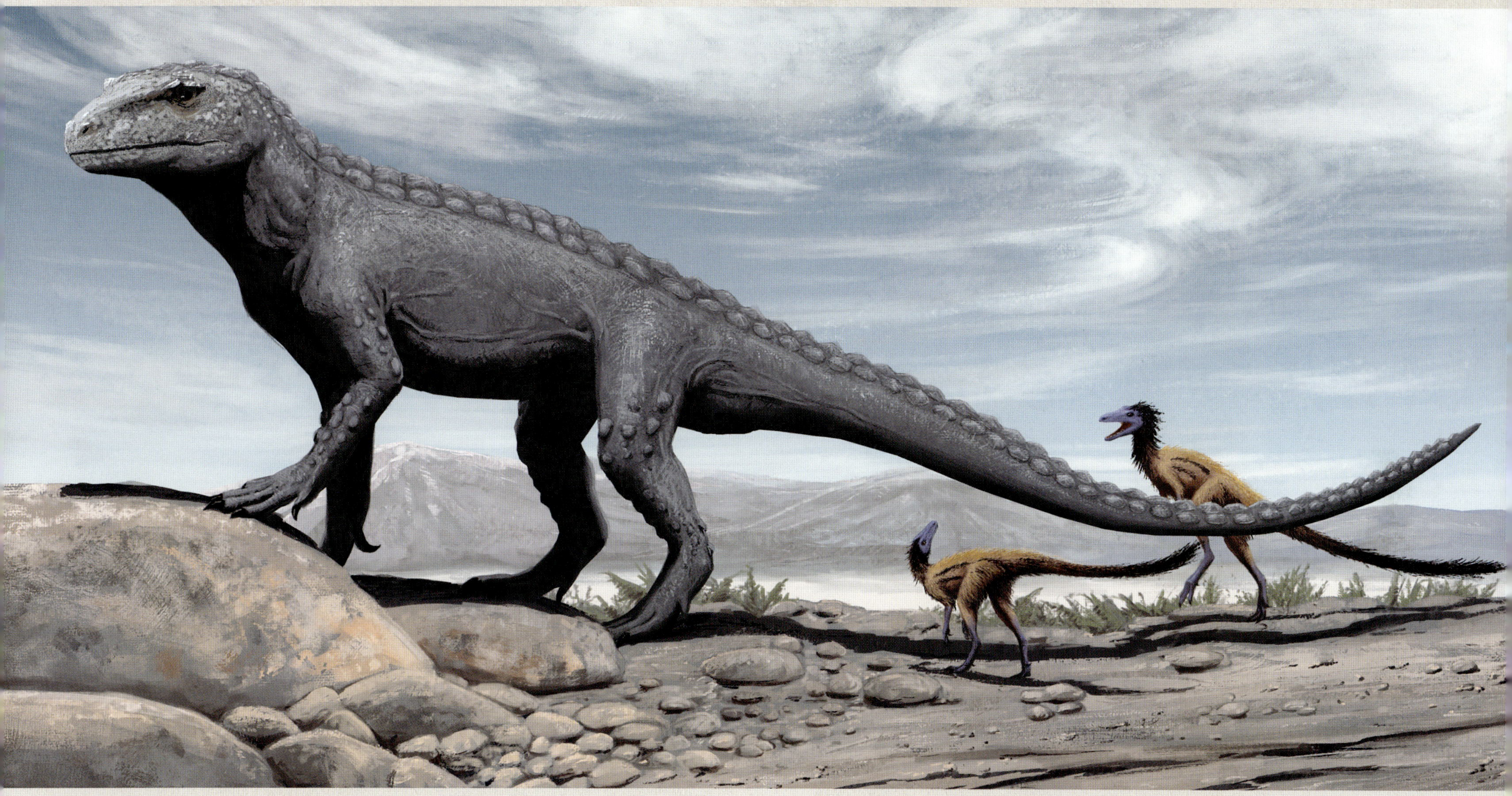

▲ *TARJADIA & MARASUCHUS*

For at least part of the Mesozoic, small, lightly built, probably fuzzy-coated members of the bird lineage – like the two *Marasuchus* shown here – lived alongside gnarly, reptilian beasts that had a far more 'old school' look. *Tarjadia*'s armour plates (termed osteoderms) show that it was a distant cousin of crocodylians. It appears to be part of a poorly known group called the erpetosuchids. The setting is Late Triassic Argentina.

• Digital | 2019

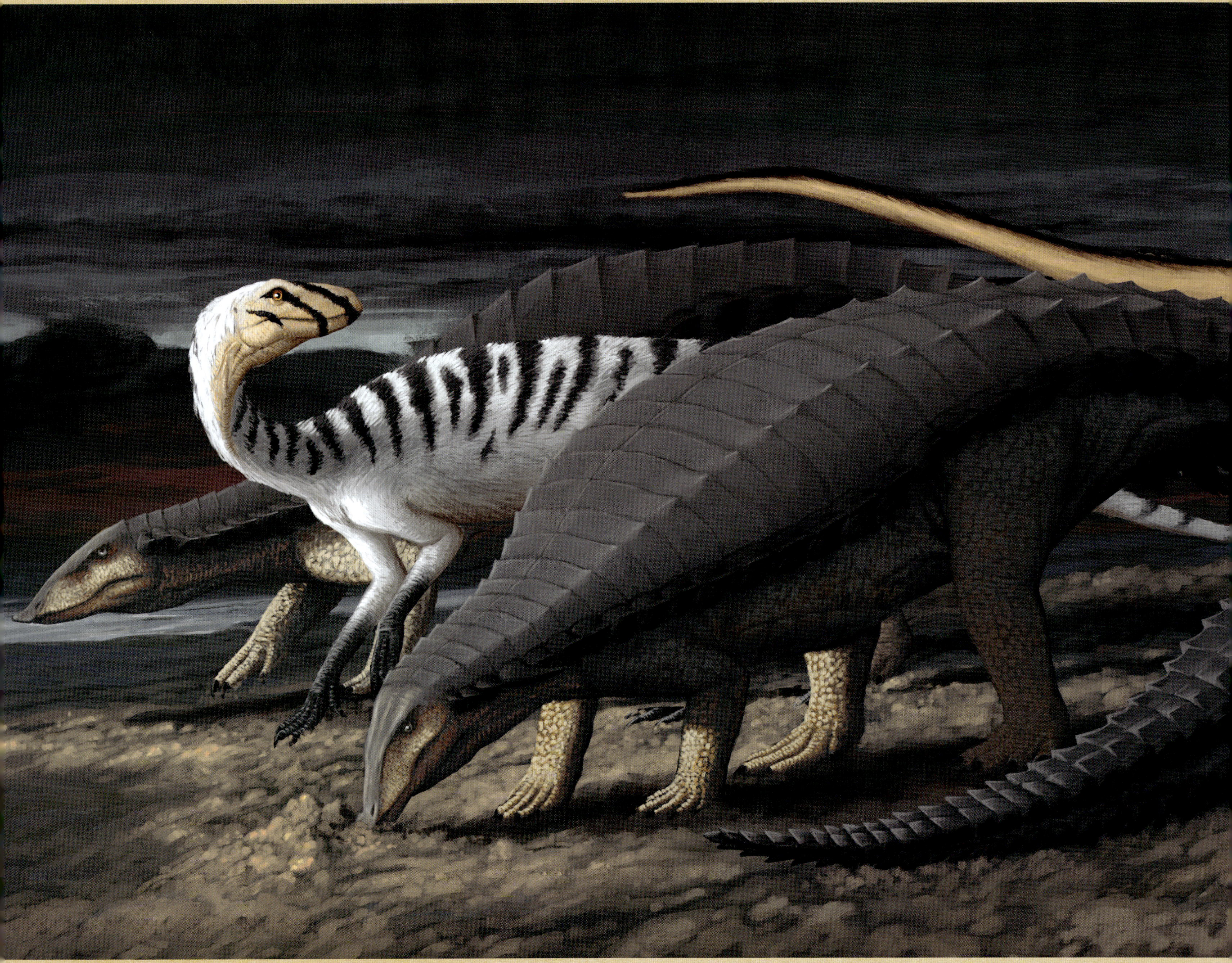

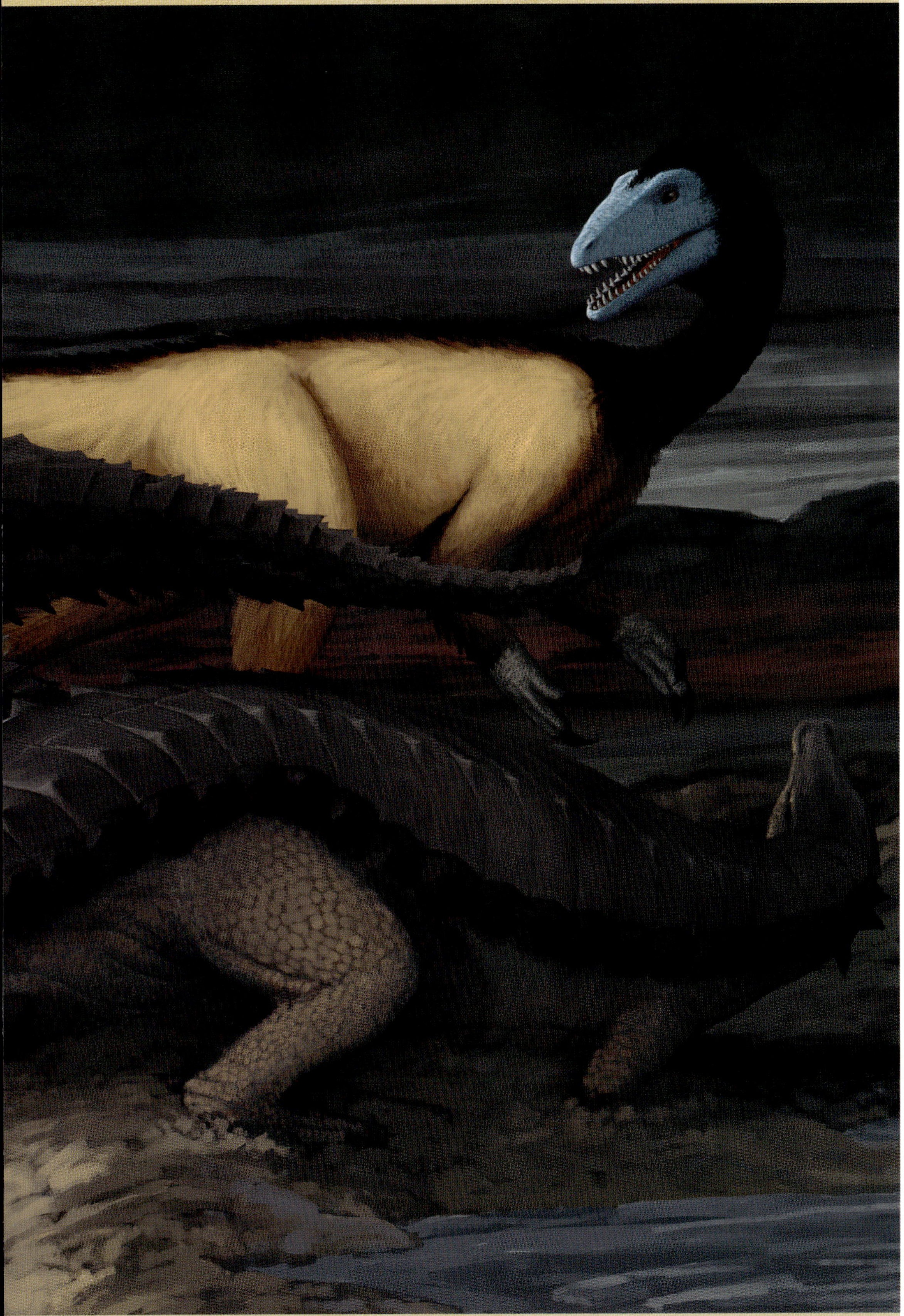

◂ *EUCOELOPHYSIS, TYPOTHORAX & TAWA*

Living alongside the dinosaurs of the Late Triassic were diverse members of the crocodile lineage, technically (but unfortunately) named Pseudosuchia ('unfortunate' because it confusingly means 'false crocodiles'). Here, *Tawa* (the blue-faced theropod) and *Eucoelophysis*, a silesaurid, are surrounded by several aetosaurs, *Typothorax*. Aetosaurs are armour-plated members of crocodile lineage, with jaws and teeth adapted for omnivory and perhaps for rooting in soil.

• Digital | 2016

▾ *STYRACOSAURUS*

Many fossil animals have a 'traditional' colour scheme whose use was established by a pioneering palaeoartist. For the ceratopsian *Styracosaurus*, that colour scheme is grey, grey and grey. This bold black-and-white colouration is a prominent departure and suggests an interesting relationship between the animal and its pale (perhaps ash-covered) surroundings.

• Digital | 2021

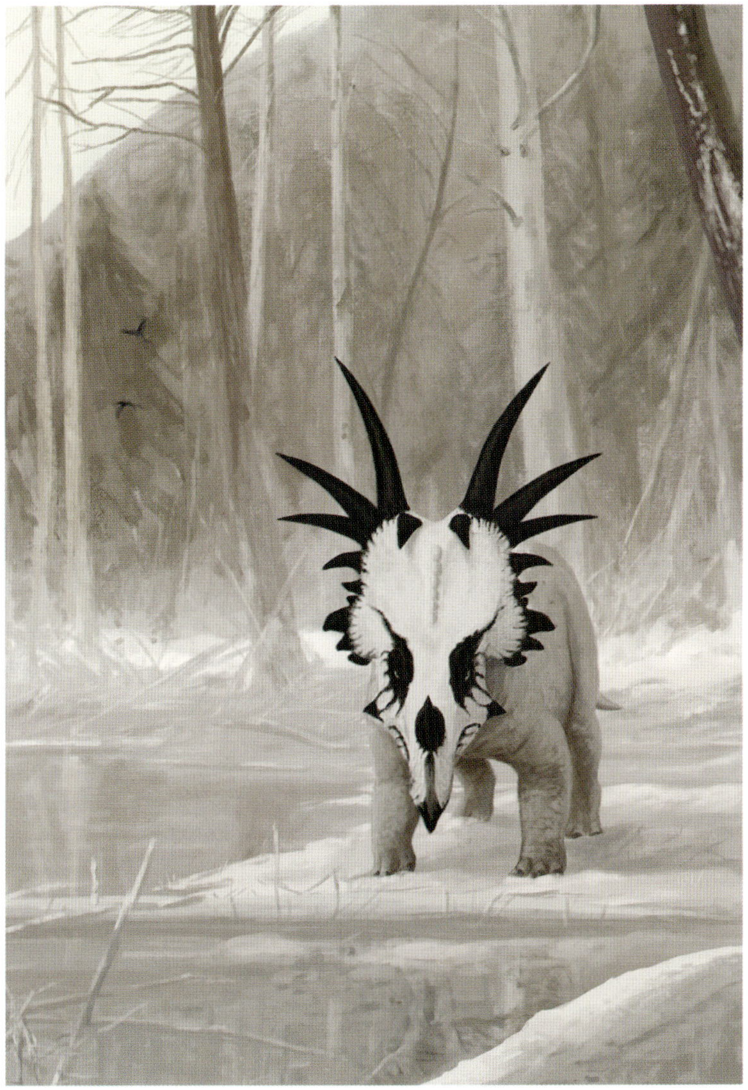

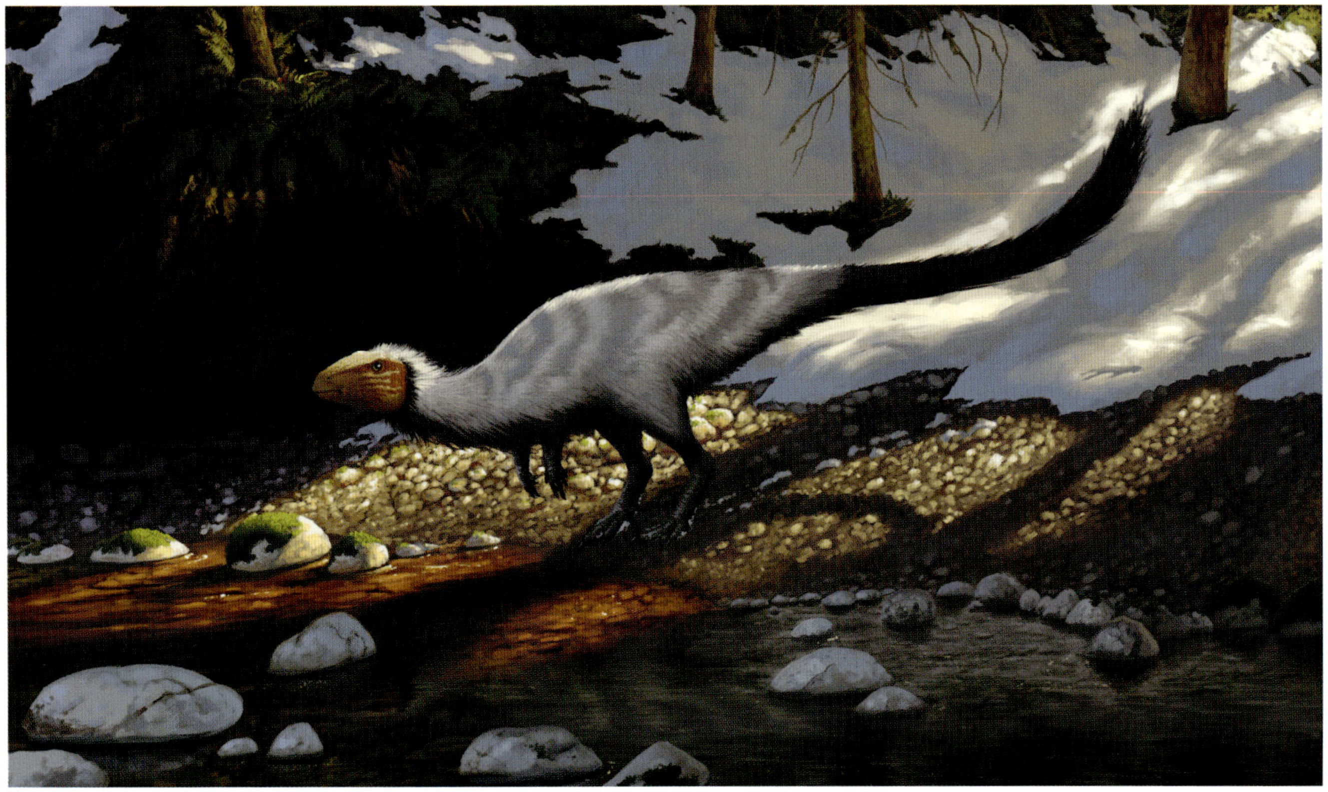

◂ *TRINISAURA*

Trinisaura, from Late Cretaceous Antarctica, is one of several non-bird dinosaurs that inhabited the cool, seasonally dark, sometimes snowy habitats of the Cretaceous Far South. *Trinisaura* is an ornithopod – specifically part of the Southern Hemisphere group Elasmaria. Based on information from other related dinosaurs, it is plausible that it had a filamentous covering.
• Digital | 2020

▾ *EUCOELOPHYSIS & REDONDASAURUS*

In this Late Triassic scene, a *Eucoelophysis* falls prey to *Redondasaurus*, a phytosaur. *Eucoelophysis* is a member of the Silesauridae, a dinosaur-like, mostly quadrupedal group. Some experts argue that silesaurids aren't just dinosaur-*like* – they're actual members of Dinosauria. Several phytosaurs such as *Redondasaurus* were thick-jawed enough, and big enough, to behave like certain crocodiles today.
• Digital | 2013

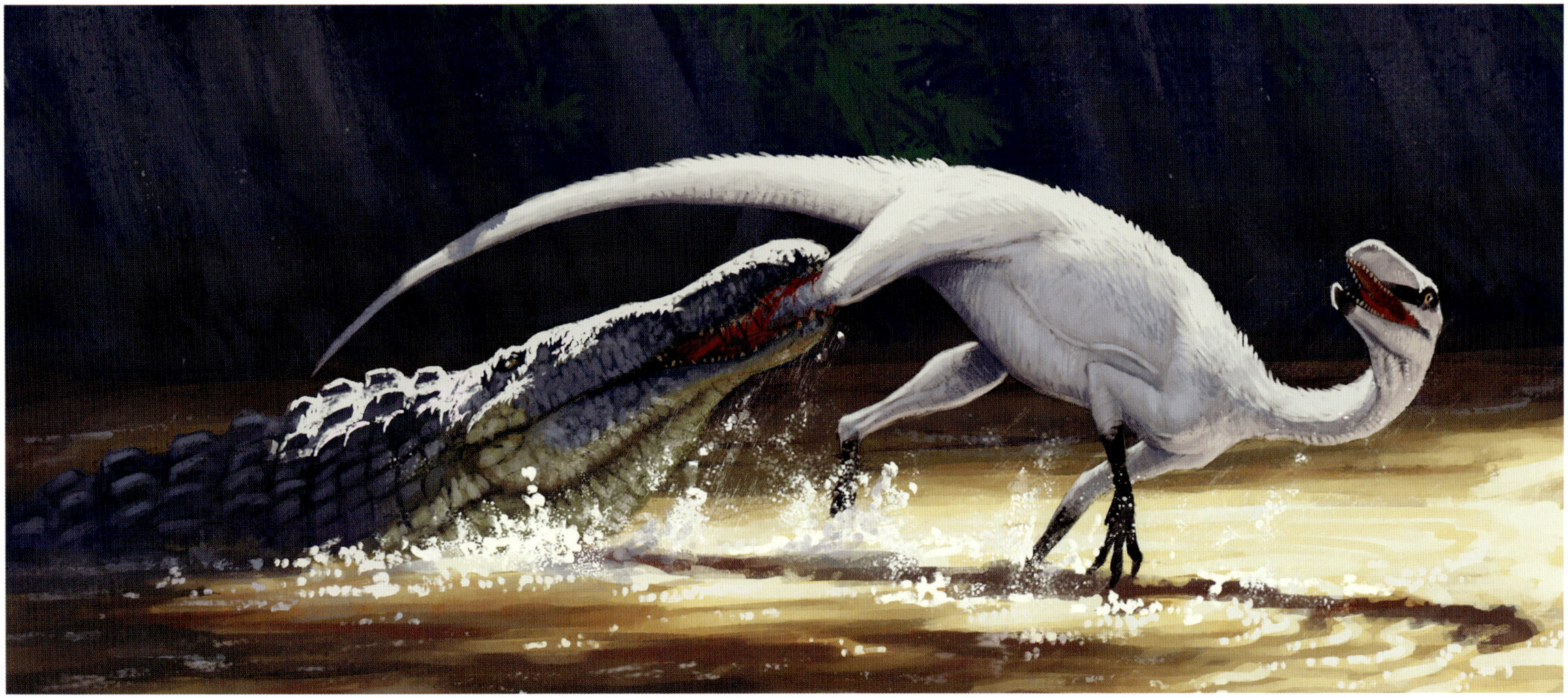

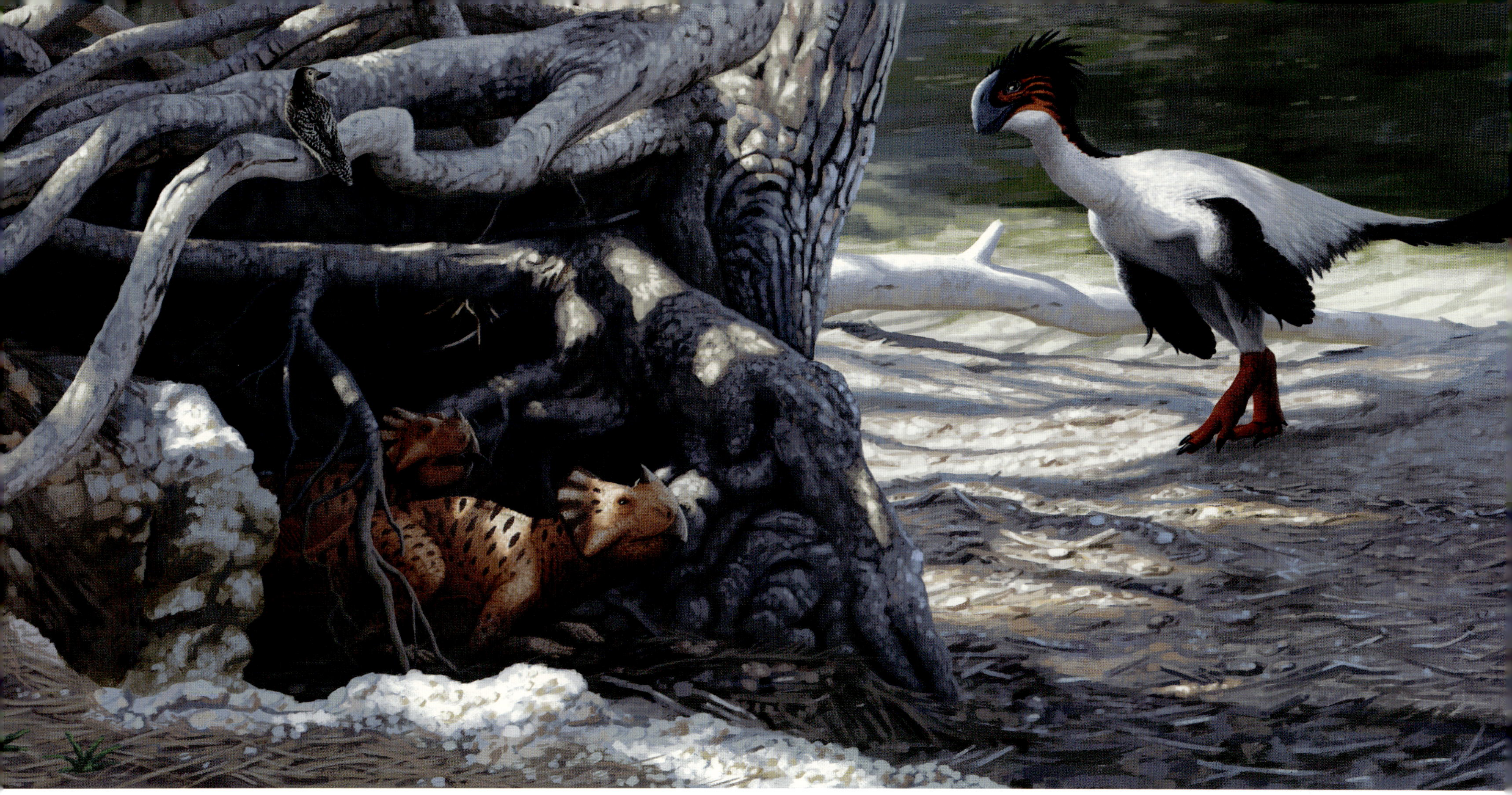

▲ *PLATYCERATOPS, AJANCINGENIA & GOBITERYX*

In this Late Cretaceous Mongolian scene, the oviraptorosaur *Ajancingenia* approaches a tree being used by a pair of the ceratopsian *Platyceratops* as a den. Small ceratopsians are known to have used burrows and dens like this. The enantiornithine bird *Gobipteryx* perches nearby. *Ajancingenia* was formerly known as '*Ingenia*' and is included by most experts in the genus *Heyuannia*.

• Digital | 2016

EFFIGIA ▶

Effigia, of Late Triassic New Mexico, USA, caused quite the stir when announced in 2006. It turned out that certain specimens long assumed to represent the theropod *Coelophysis* were something else entirely. They were not theropods, nor even dinosaurs, but long-necked, small-armed, toothless, bipedal members of the crocodile lineage. Today, *Effigia* is but one of several archosaurs of this kind, grouped together as the shuvosaurids.

• Digital | 2016

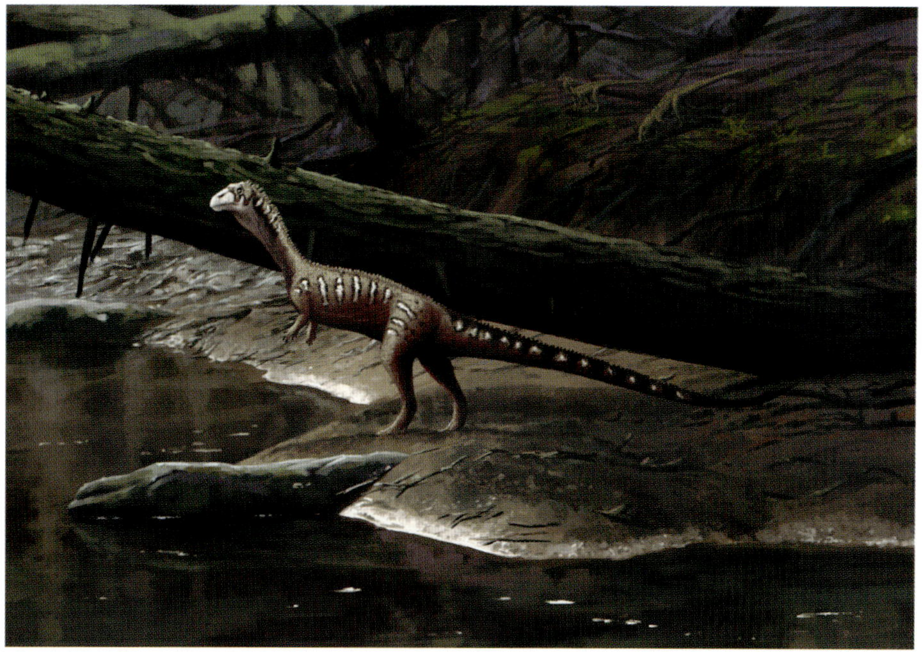

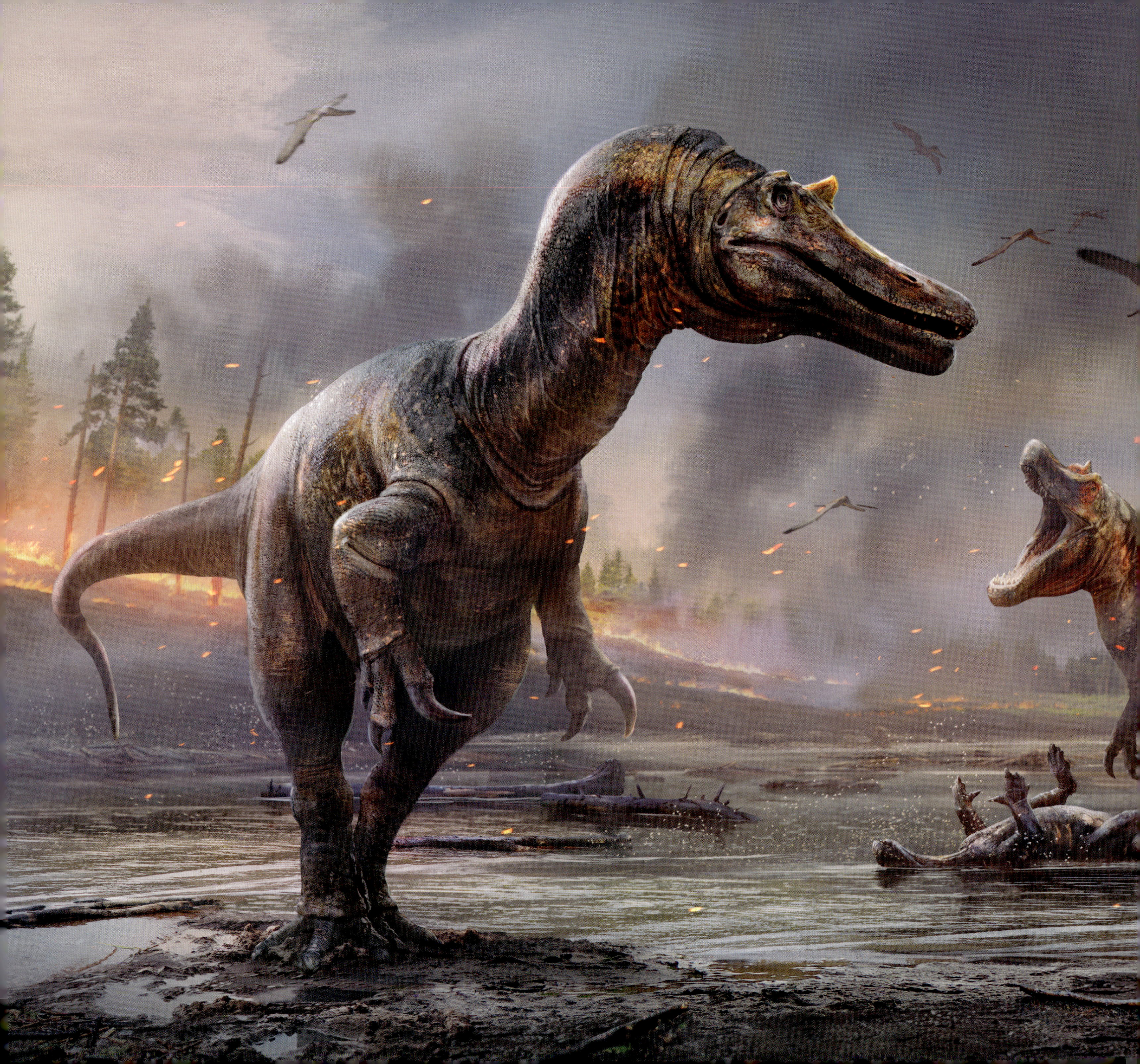

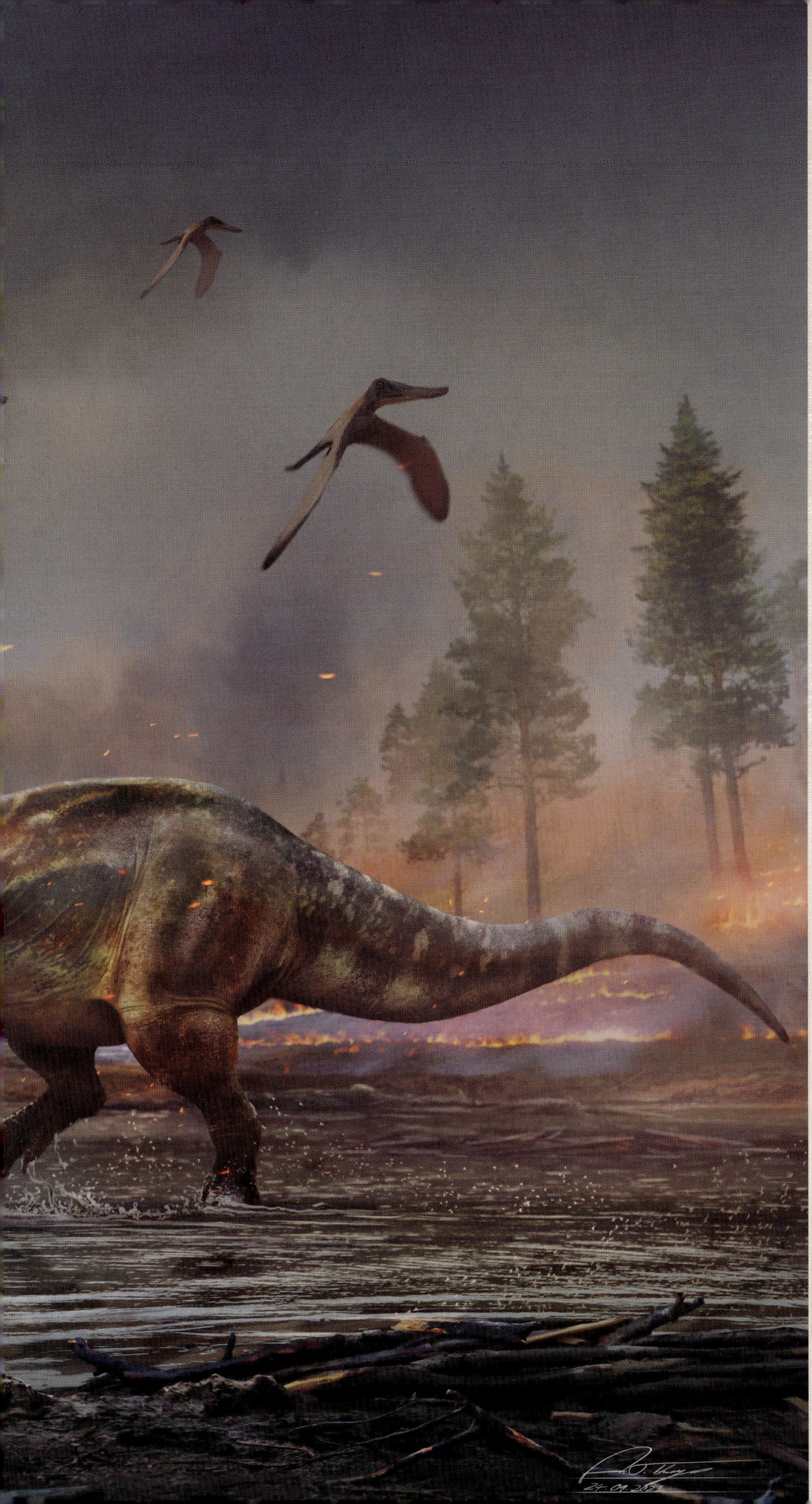

ANTHONY J HUTCHINGS

ANTHONY J HUTCHINGS is a digital artist based in London. Originally from Birmingham in the UK, Anthony began his art journey as a young child, where he quickly developed a fascination for prehistoric life. After graduating with a bachelor's degree in Illustration, he went on to become a concept artist for the theming and entertainment industry. Although themed design is his primary profession, he always retained a deep passion for prehistory and has been steadily growing his palaeoart portfolio over recent years.

When depicting extinct animals, his main goal is to create realistically believable and lifelike images, as if they were living nature shots. His preferred method to achieve this look is by physically sculpting maquettes and using these as heavy reference for digital paintings and enhancement.

In recent years he has had the privilege of collaborating on projects such as *Duelling Dinosaurs* at the North Carolina Museum of Natural Sciences, along with creating several pieces for press releases announcing the discovery of new spinosaurids on the Isle of Wight.

So far he has thoroughly enjoyed his palaeoart journey and is looking forward excitedly to creating more artwork in the foreseeable future, ever striving to bring our planet's distant past back to life.

[Previous pages]

ISLE OF WIGHT SPINOSAURS

The Lower Cretaceous sediments of southern England have proved important in the discovery of new spinosaurid theropods. This scene depicts two members of the group – *Ceratosuchops* (left) and *Riparovenator* – named from the famous dinosaur-bearing sediments of the Isle of Wight in 2021. The animals are shown in a floodplain environment where wildfires were a frequent occurrence. Istiodactylid pterosaurs fly nearby.

With thanks to Southampton University.

• Digital | 2021

▼ *PERUCETUS* MOTHER AND CALF

The gigantic fossil whale *Perucetus*, named in 2023, lived in Peru around 38 million years ago. Based on the anatomy of its vertebrae (the only parts known so far), its describers interpreted it as hugely thick-bodied, with a mass somewhere between 77 to 308 metric tonnes (85 to 340 tons). The lower end of that range is considered most likely by experts.

• Digital | 2023

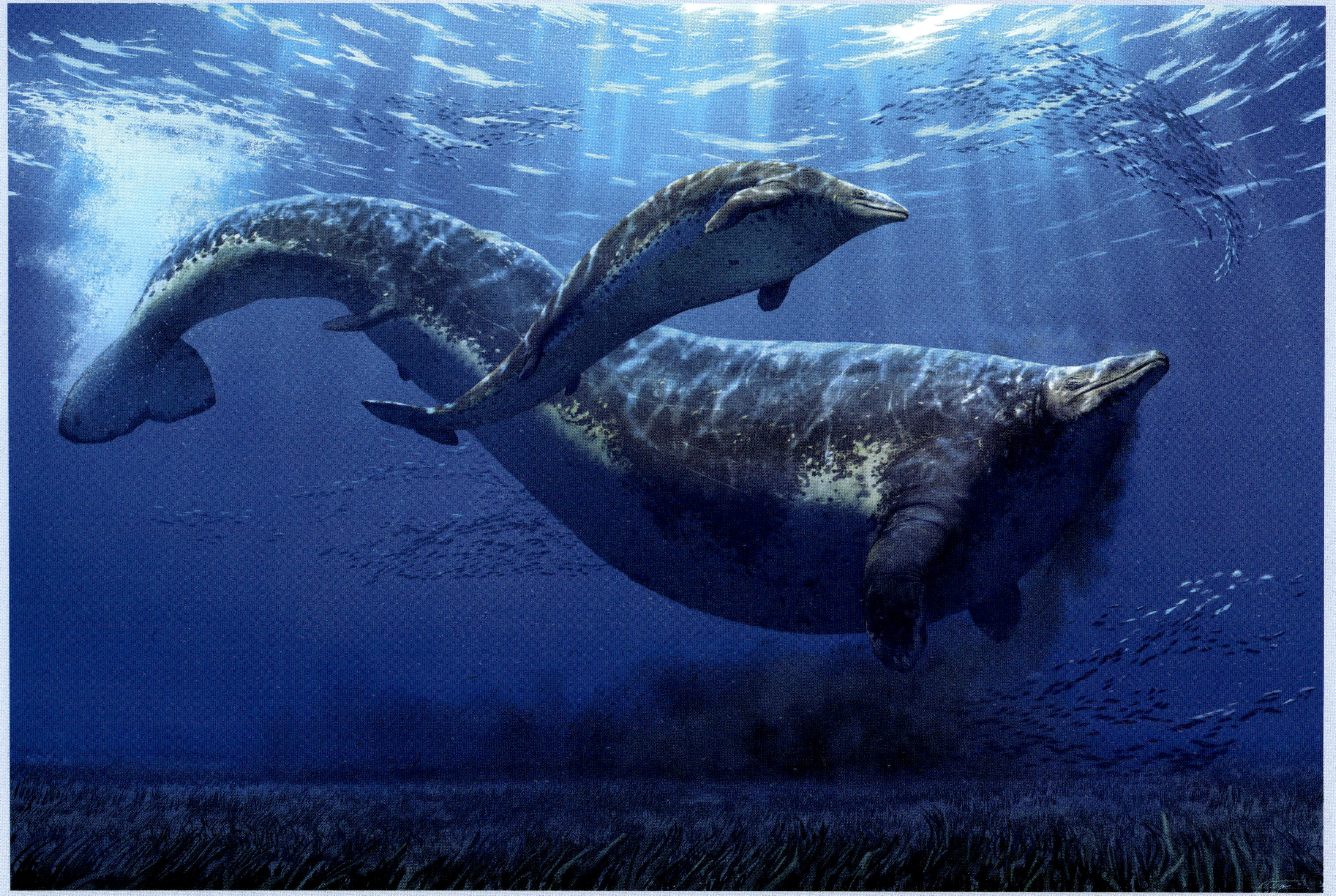

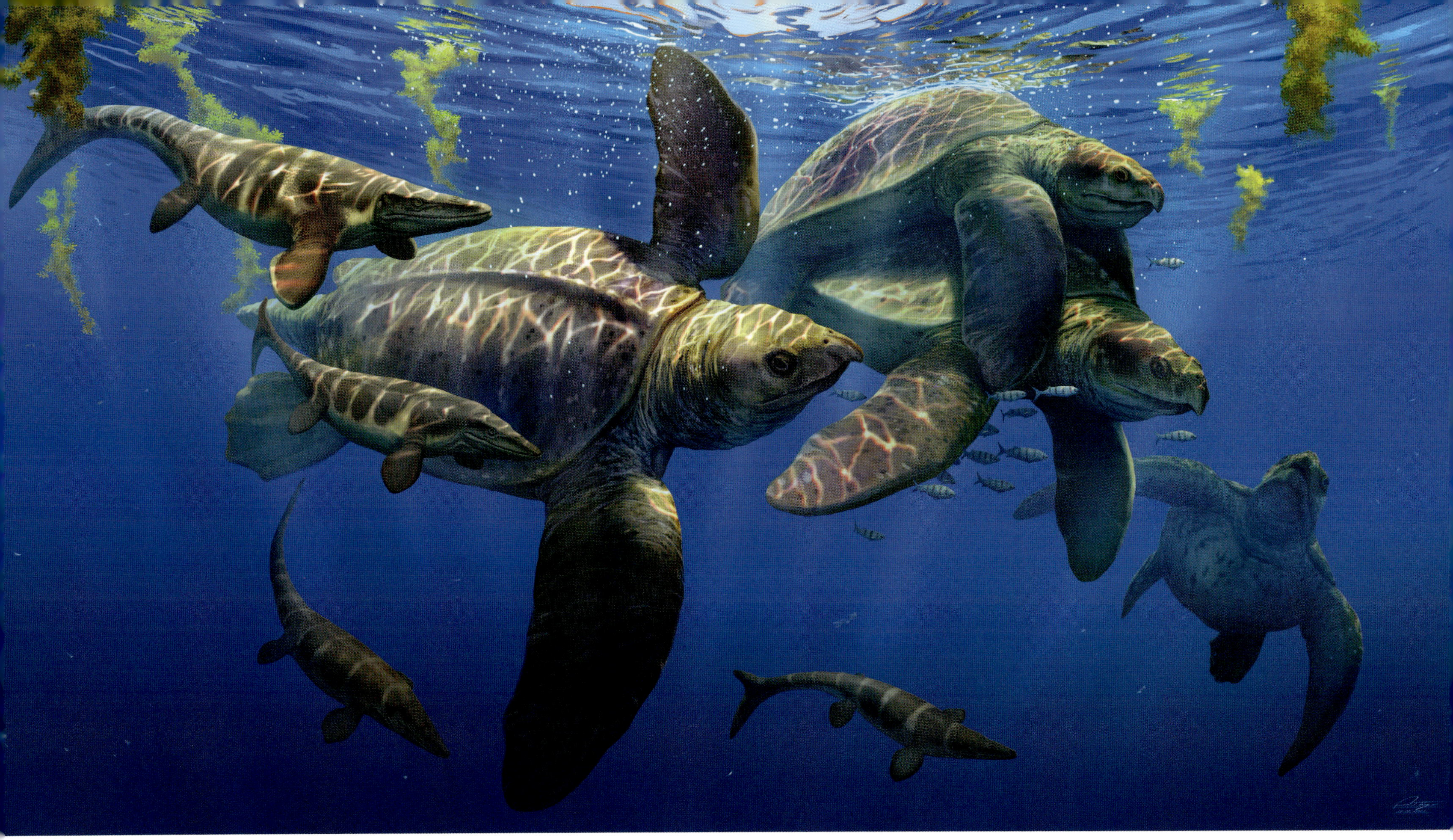

▲ *ARCHELON* MATING

Among the most famous of all fossil turtles is *Archelon* of Late Cretaceous North America's Western Interior Seaway, giant specimens of which were 4.6m (15ft) long, with a flipper-span of 4m (13ft). Almost nothing is known about the behaviour of these animals, but they surely mated, fought with rivals, and laid eggs on beaches as sea turtles do today.
• Digital | 2021

RHOMALEOSAURUS ▶

Several plesiosaur groups included 'intermediate-necked' species that were neither truly short-necked, nor properly long-necked. *Rhomaleosaurus* is an Early Jurassic plesiosaur from England that reached 7m (23ft) in length. Reinforced jaws and stout teeth show that it was a predator of moderately large prey. Social behaviour in plesiosaurs is likely, as are colour schemes that enhanced their camouflage.
• Digital | 2020

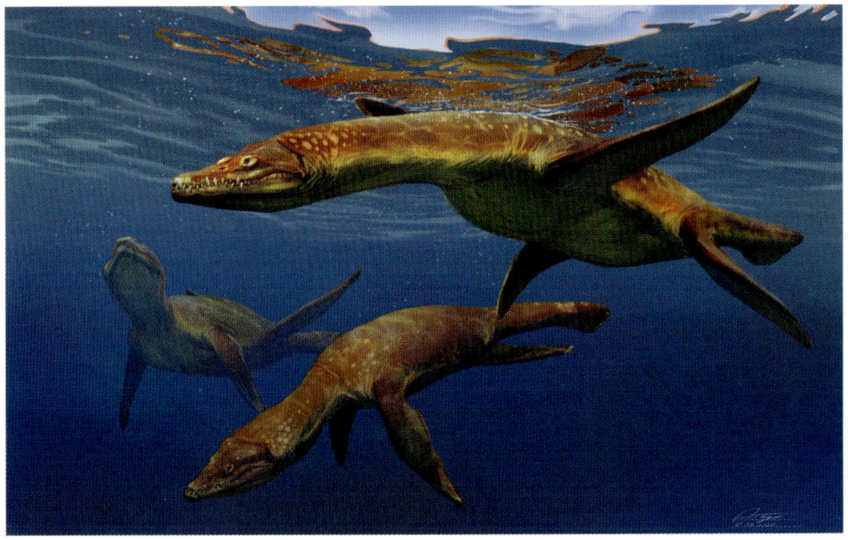

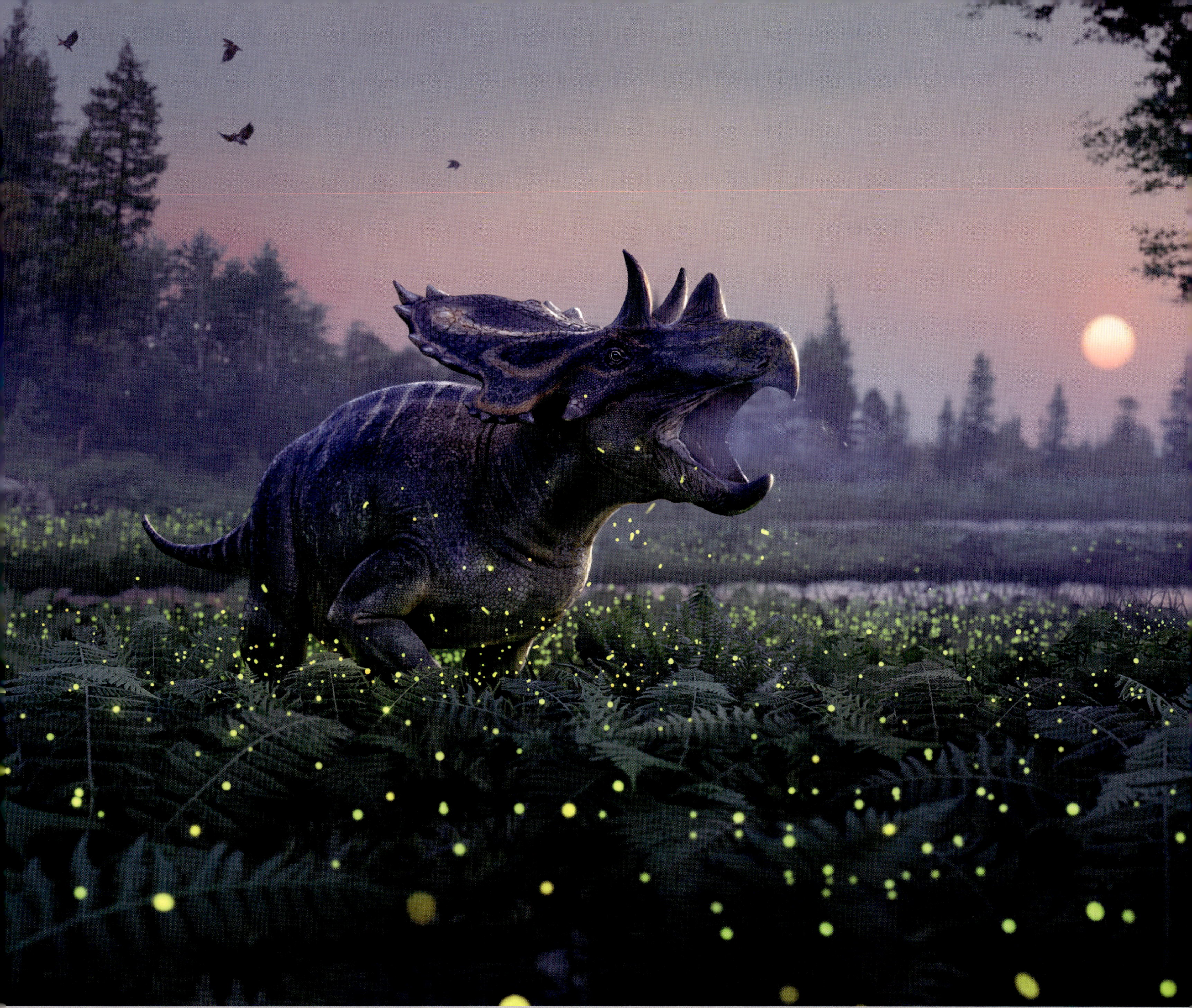

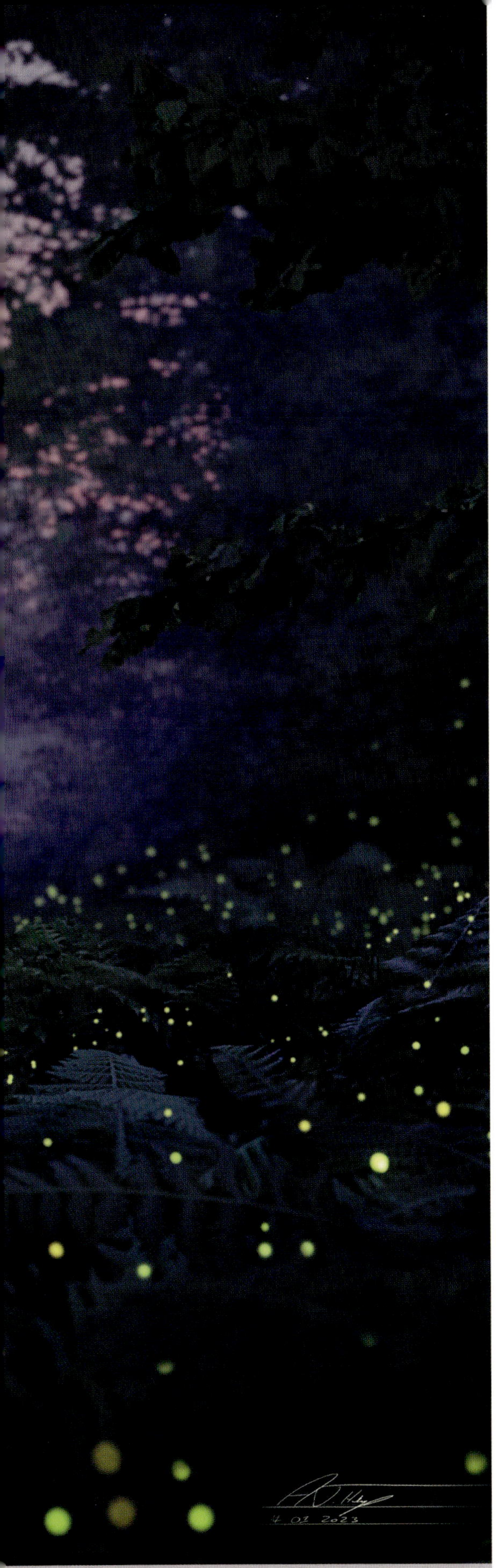

◄ ***MERCURICERATOPS* & FIREFLIES**

In a Late Cretaceous North American woodland, the mid-sized chasmosaurine horned dinosaur *Mercuriceratops* calls for its herd. *Mercuriceratops* was first described in 2014 from Alberta, Canada, but later reported from Montana, USA as well. The group of beetles that includes fireflies has a fossil record that extends back into the Cretaceous, and even includes amber specimens with their light-emitting organs present.
• Digital | 2023

LLUKALKAN ►

New to science as of 2021, *Llukalkan* from the Late Cretaceous of Argentina is an abelisaurid, seemingly part of the same specific abelisaurid group (termed Furilesauria) as the famous *Carnotaurus*. The single known *Llukalkan* specimen – a partial skull – represents a juvenile around 5m (16.4ft) long. Adults were probably a few metres longer.
• Digital | 2021

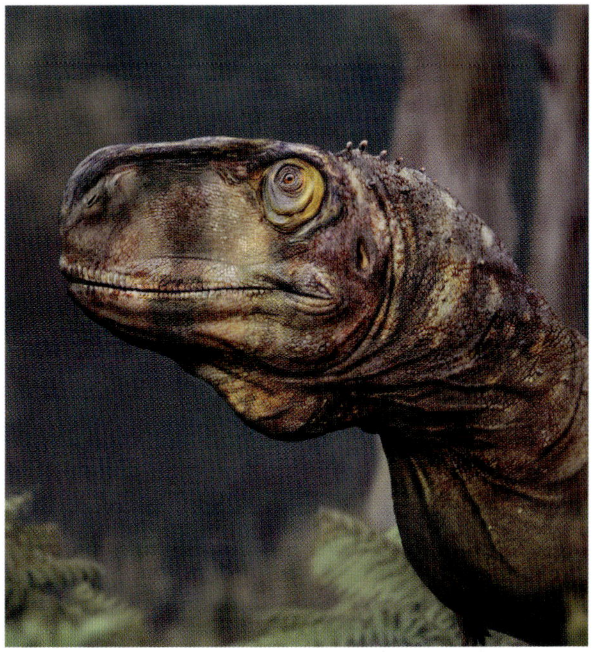

▼ **PSITTACOSAUR RIVALRY**

Fossils show that *Psittacosaurus* lived in groups for at least part of their lives. Cheek horns, a short, powerful beak and bipedal abilities suggest that they fought face to face while standing tall. This image suggests that squabbles for dominance occurred within their social groups.
• Digital | 2020

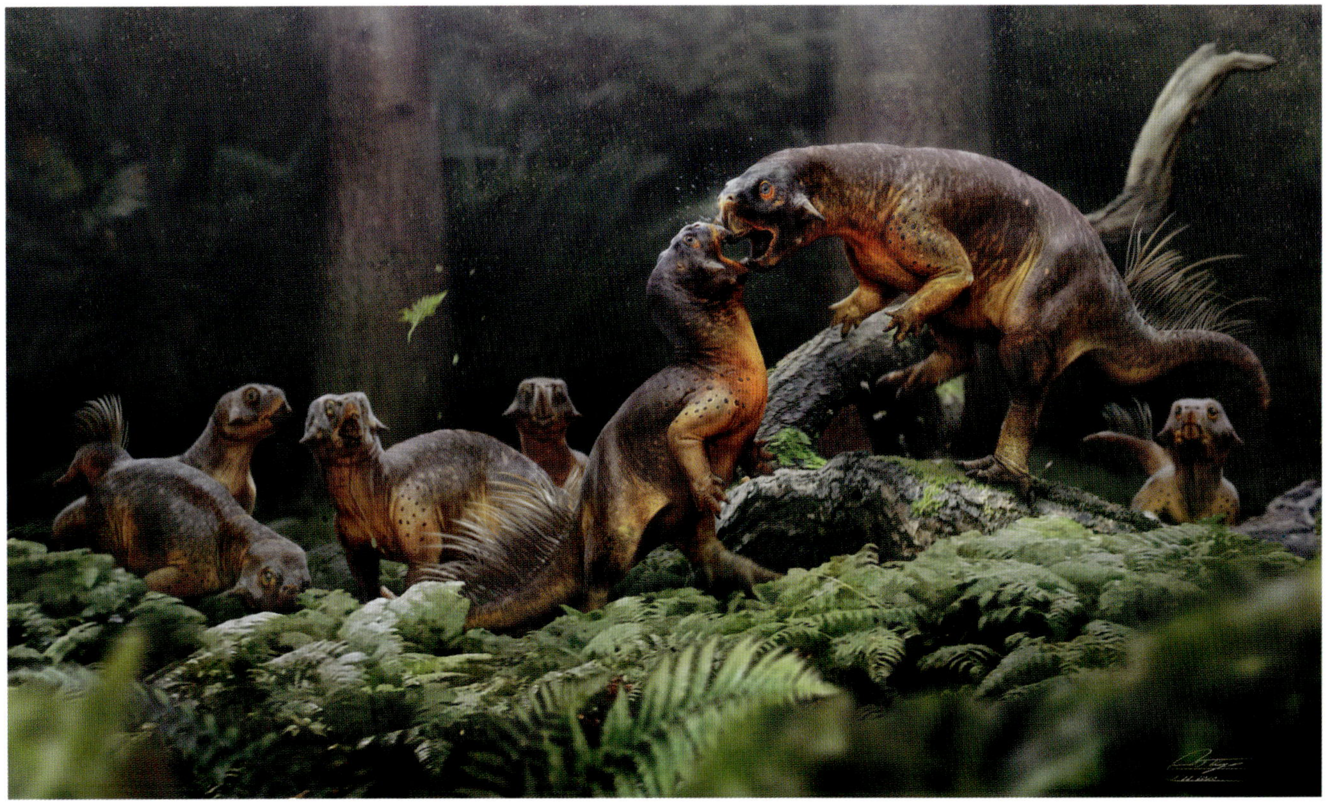

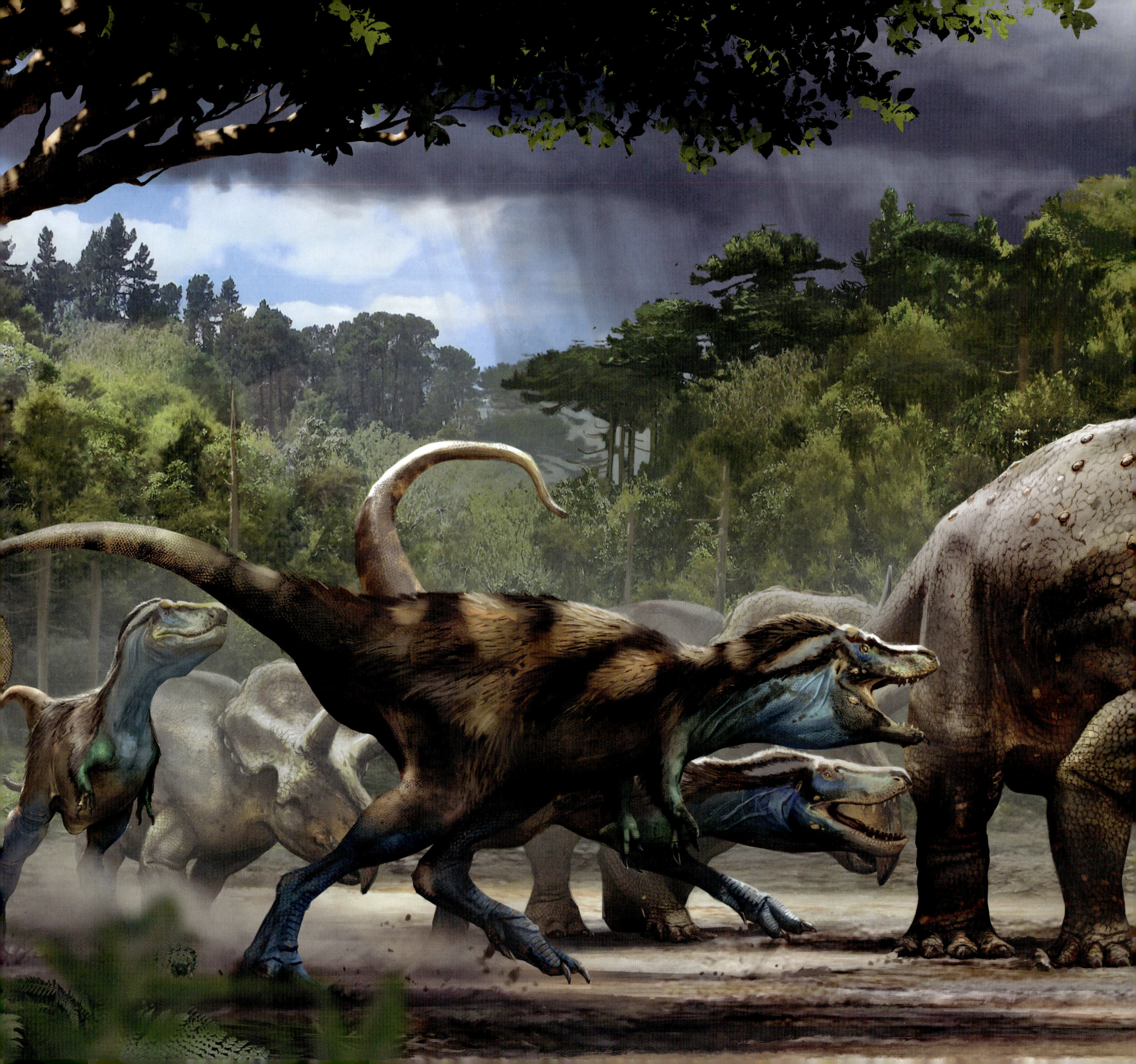

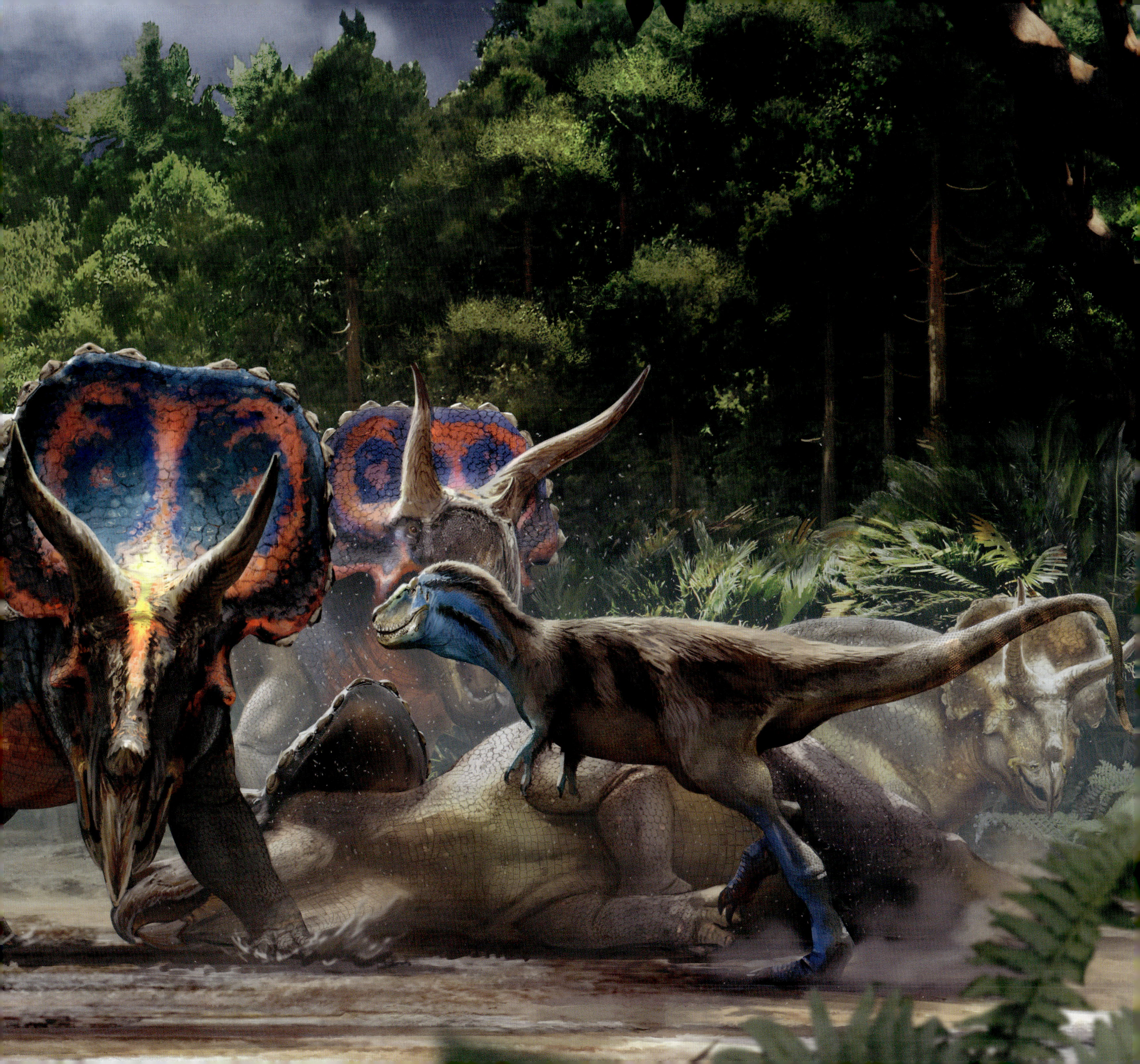

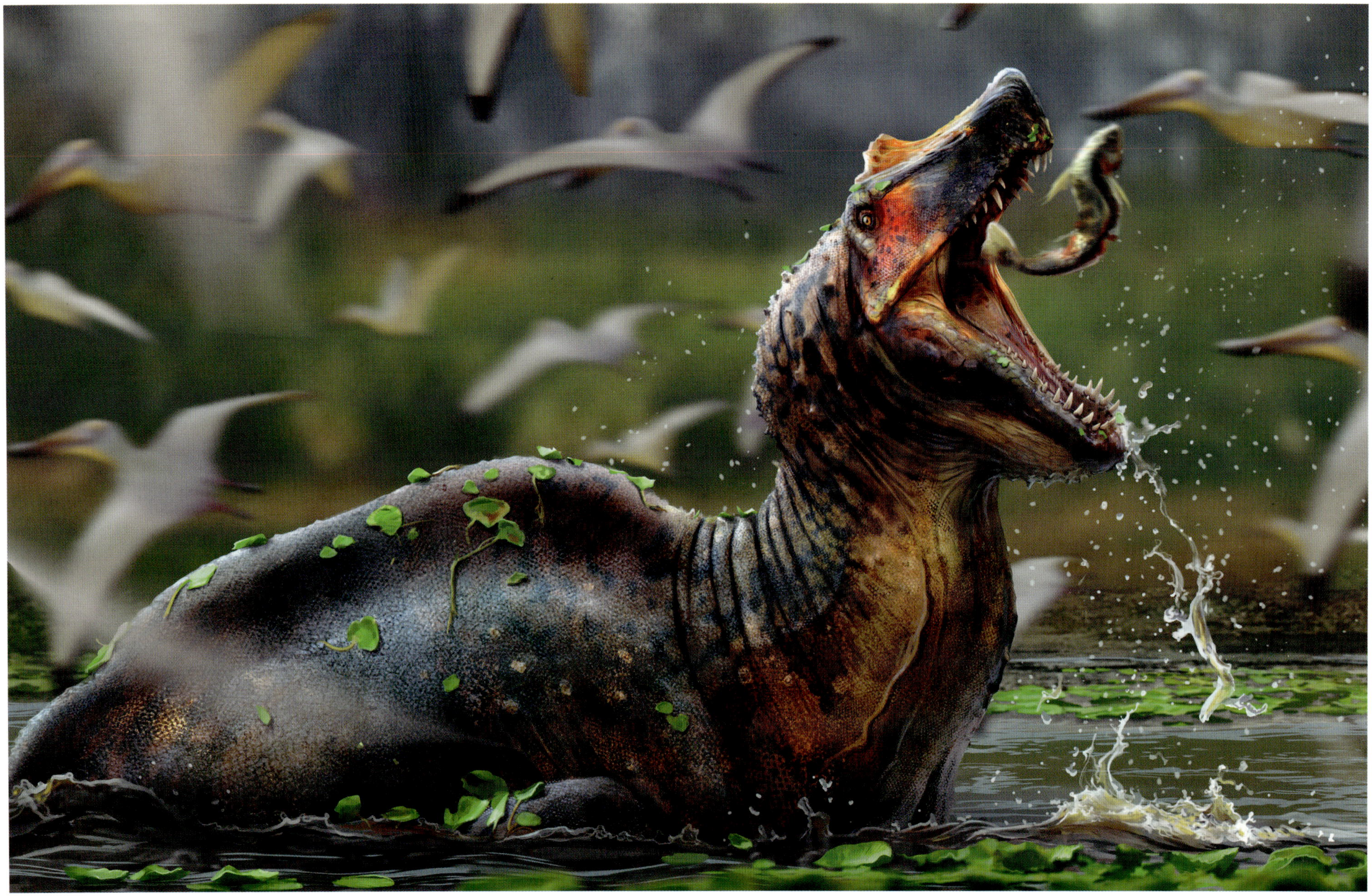

[Previous pages]

THE DUELLING DINOSAURS

A remarkable fossil from Montana, discovered in 2006, preserves a juvenile *T. rex* very close to an adult *Triceratops*. It has become known as the 'duelling dinosaurs'. This scene, inspired by that discovery, shows a *Triceratops* guarding the recently dead. This might seem like a distinctly mammalian piece of behaviour, but it's known that social lizards, turtles and birds will sometimes remain with the body of a fallen family member.

• Digital | 2020

▲ *IRRITATOR*

A surprising theropod with a complex backstory (it was initially misidentified as a giant pterosaur!), Brazil's *Irritator* is a narrow-snouted spinosaurid with jaws and teeth that look suited for the fish-catching. The more we learn, the more unusual *Irritator* becomes. A 2023 study showed how it was able to widen the rear parts of its jaws as they were opened.

• Digital | 2020

TAPEJARA ▶

Tapejarid pterosaurs are noted for their impressive head crests. In *Tapejara* from Brazil, a bony flange extended upwards from the snout, a bony spar projected from the skull's rear, and a sheet of keratin connected the two. As implied here, this sheet was thin enough to be semi-translucent. The crest was almost certainly a display structure, and very likely gaudily coloured.

• Digital | 2023

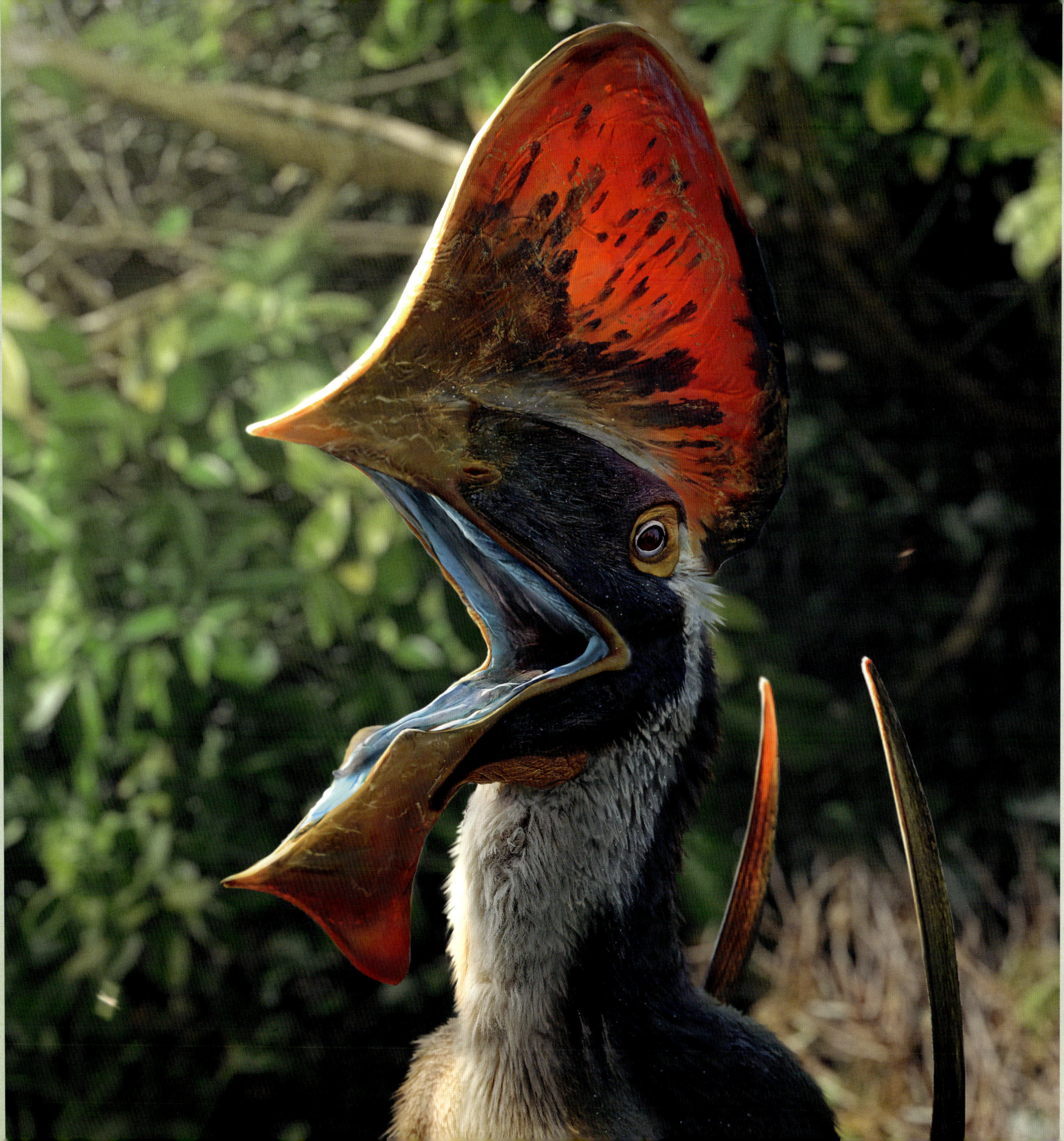

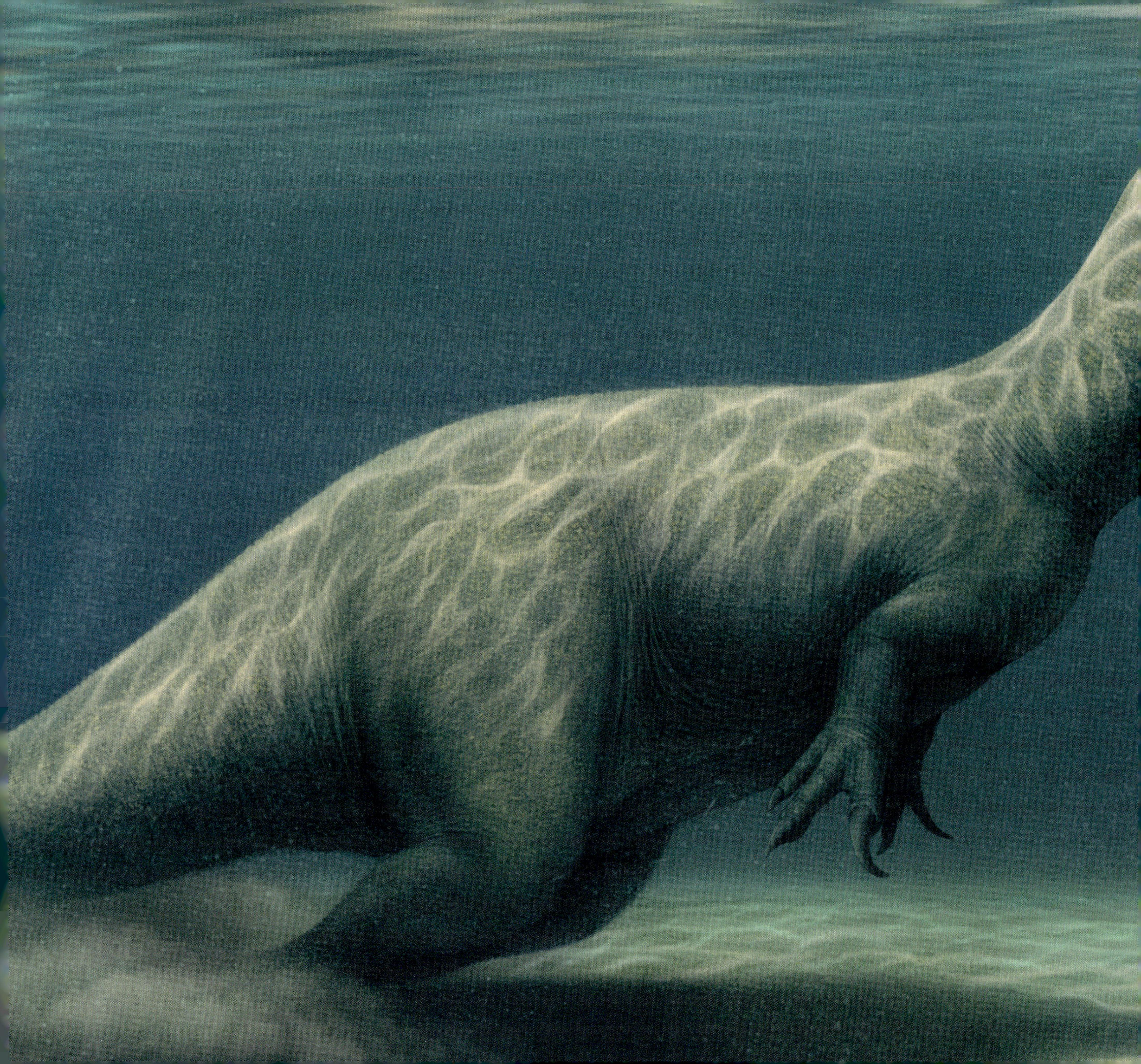

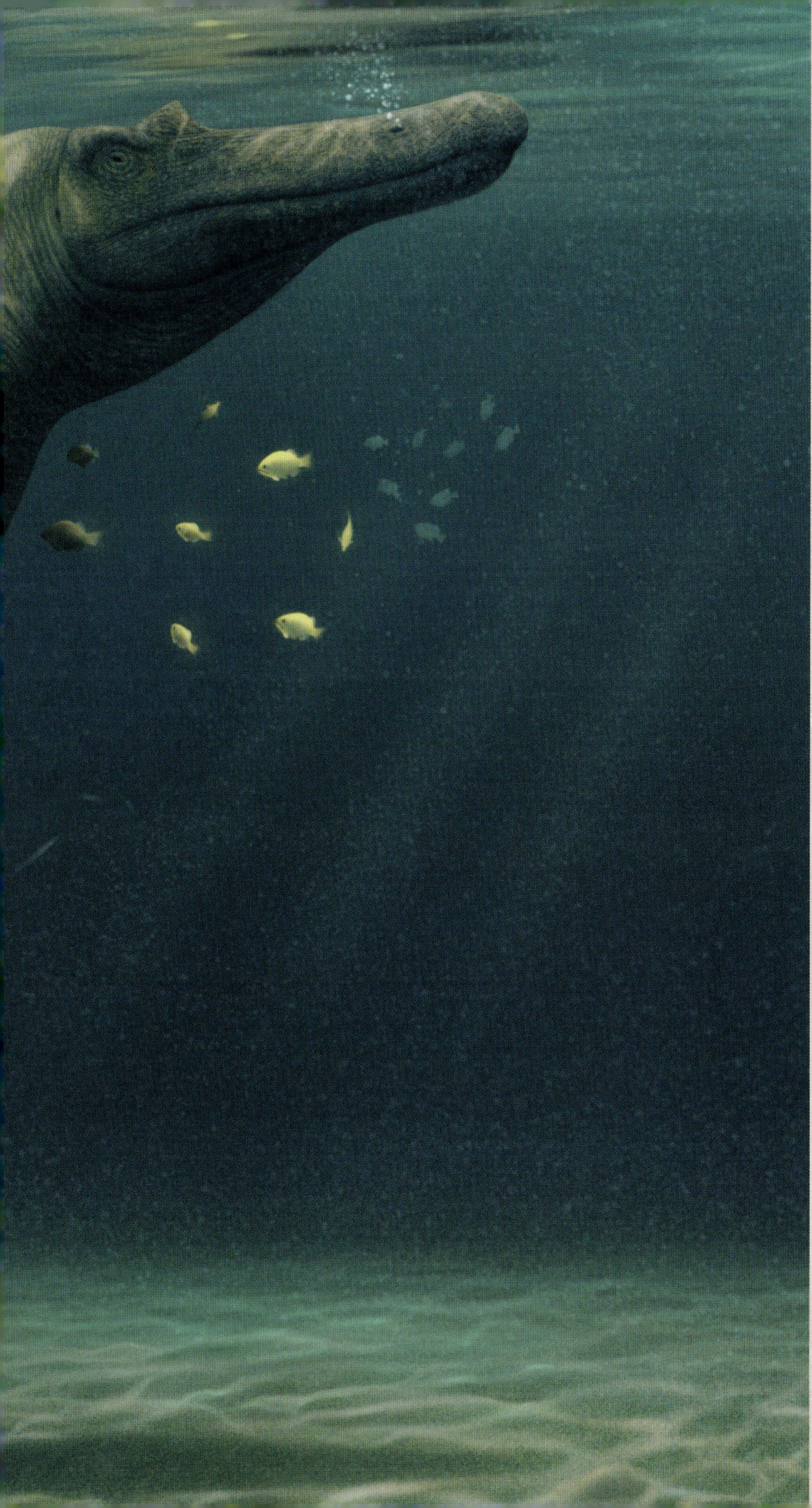

HAIDER JAFFRI

HAIDER SYED JAFFRI is an artist, palaeoartist and dentist living in the UK with a deep fascination with life both extant and extinct. Art has been his passion from an early age and his belief in Islam has helped him appreciate the beauty of nature. His father first noticed his fondness for drawing at the age of seven and has, since then, been Haider's biggest supporter, inspiring his son to draw as much as possible.

Most of Haider's art features landscapes and modern animals in their habitats, as well as palaeoart. His interest in palaeontology also began at an early age, inspired by classic TV shows such as *Walking with Beasts* and *Sea Monsters*, which amazed him with their unique approach in realistically conveying extinct life. These series would provide the catalyst for him to try and replicate something similar through his own work, and since 2020, at the age of 17, he has drawn many extinct species in their long-vanished environments. He also provided the cover art for Darren Naish's *Ancient Sea Reptiles* (Natural History Museum, 2023). He continues to develop his art to depict the prehistoric world, utilising modern sciences and research to make the distant past as real as the living world.

[Previous pages]

***BARYONYX* UNDERWATER**

Discussion about the behaviour of *Spinosaurus* rumbles on. One exciting idea that has emerged from this debate is that several spinosaurids might have been occasional 'subaqueous foragers'. Here, the spinosaurid *Baryonyx* from the Early Cretaceous of England submerges itself while hunting. Many modern animals today can do this, even those lacking aquatic specialisations.
• Digital | 2022

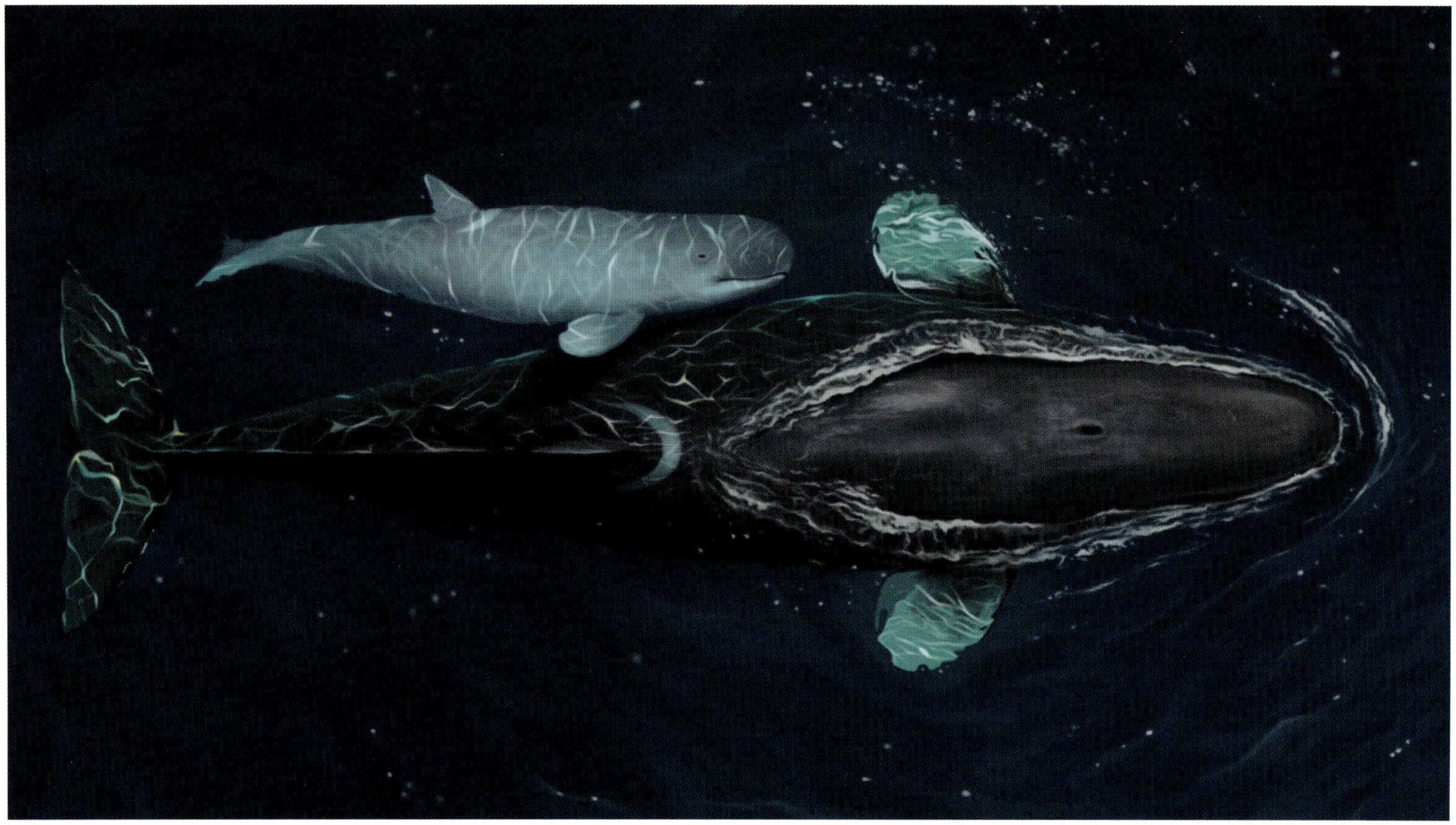

***LIVYATAN* MOTHER & CALF ▸**

Superficially similar to the living sperm whale, *Livyatan*, from the Miocene about nine million years ago, was far more predatory. It had shearing teeth in both jaws – the sperm whale has teeth in the lower jaw only – and likely hunted the smaller cetaceans that swam the Miocene oceans. Here, a calf swims under the protection of its mother.
• Digital | 2022

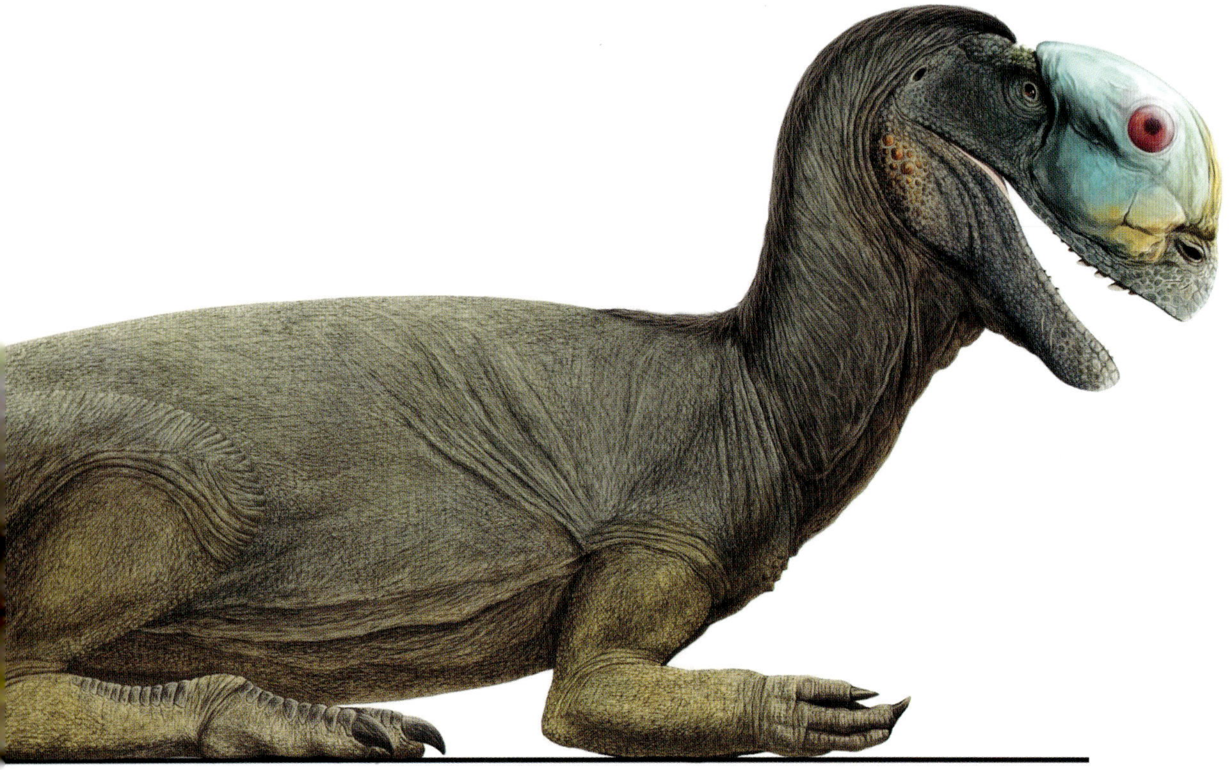

◂ *SINOSAURUS*

Sinosaurus of Early Jurassic China is related to the twin-crested *Dilophosaurus* of North America, and indeed was initially regarded as an Asian member of that genus. Recent work has shown that the plate-like crests of these dinosaurs were probably part of a large casque-like structure, perhaps brightly coloured, boldly patterned, or even functioning as a scary 'pseudoface'!
• Digital | 2023

***HATZEGOPTERYX* IN FLIGHT ▸**

Among the largest of the famous azhdarchid pterosaurs, *Hatzegopteryx* of Late Cretaceous Romania was a giant, and – relative to other members of its group – a heavily built one at that. An idea suggested several times is that such giants were flightless. Haider would seem to disagree…A substantial amount of material almost certainly belonging to *Hatzegopteryx* is known but remains unpublished.
• Digital | 2022

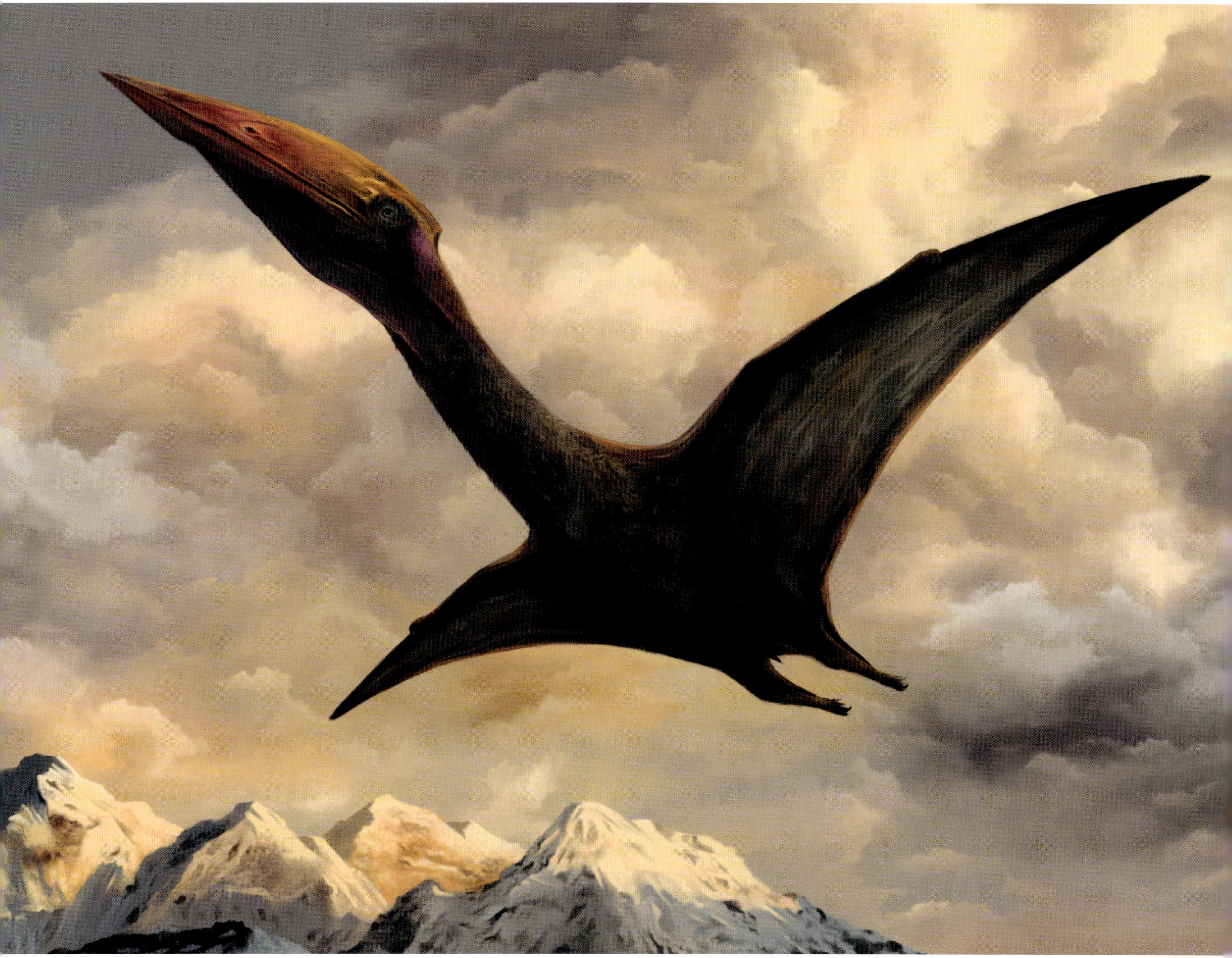

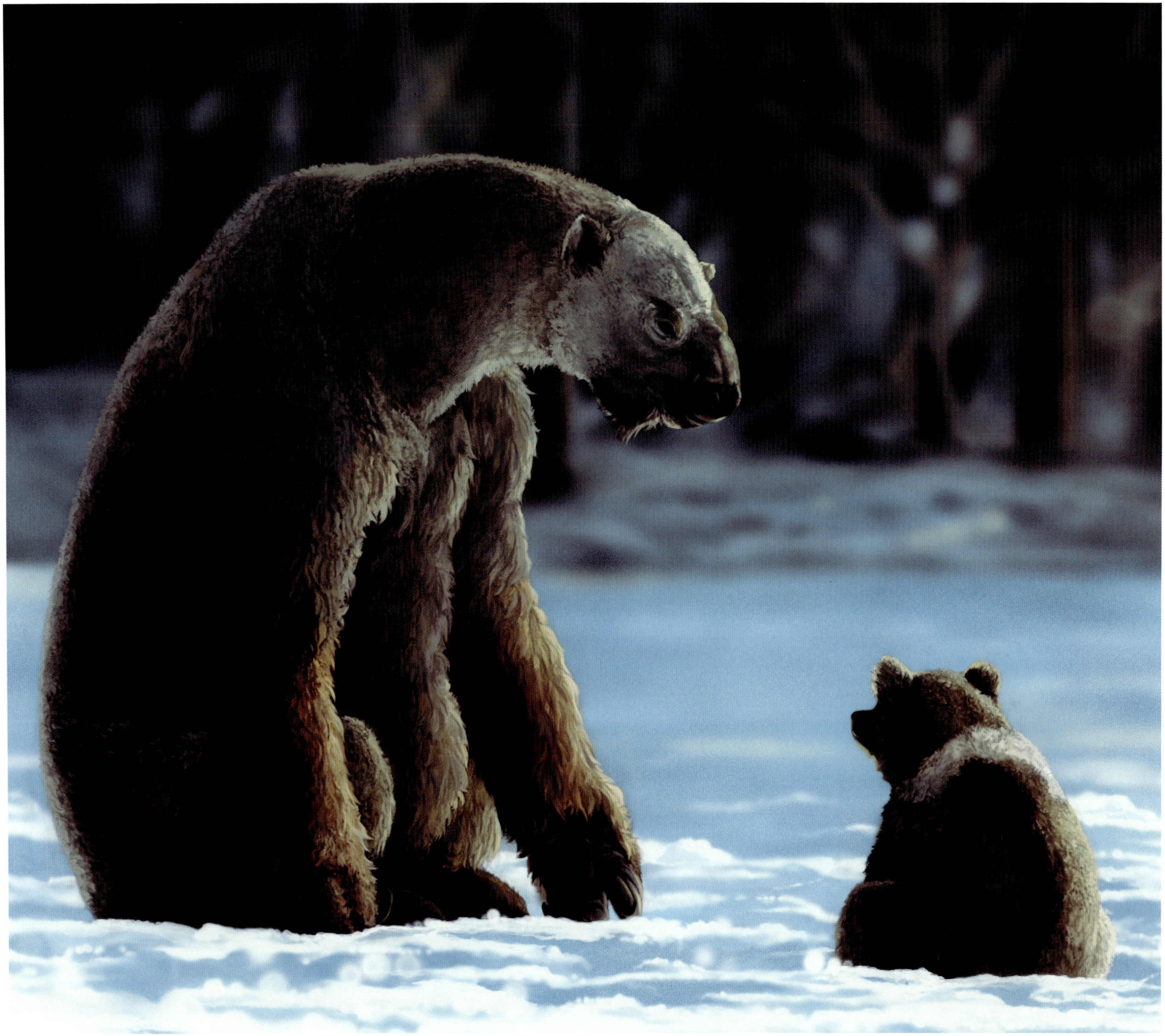

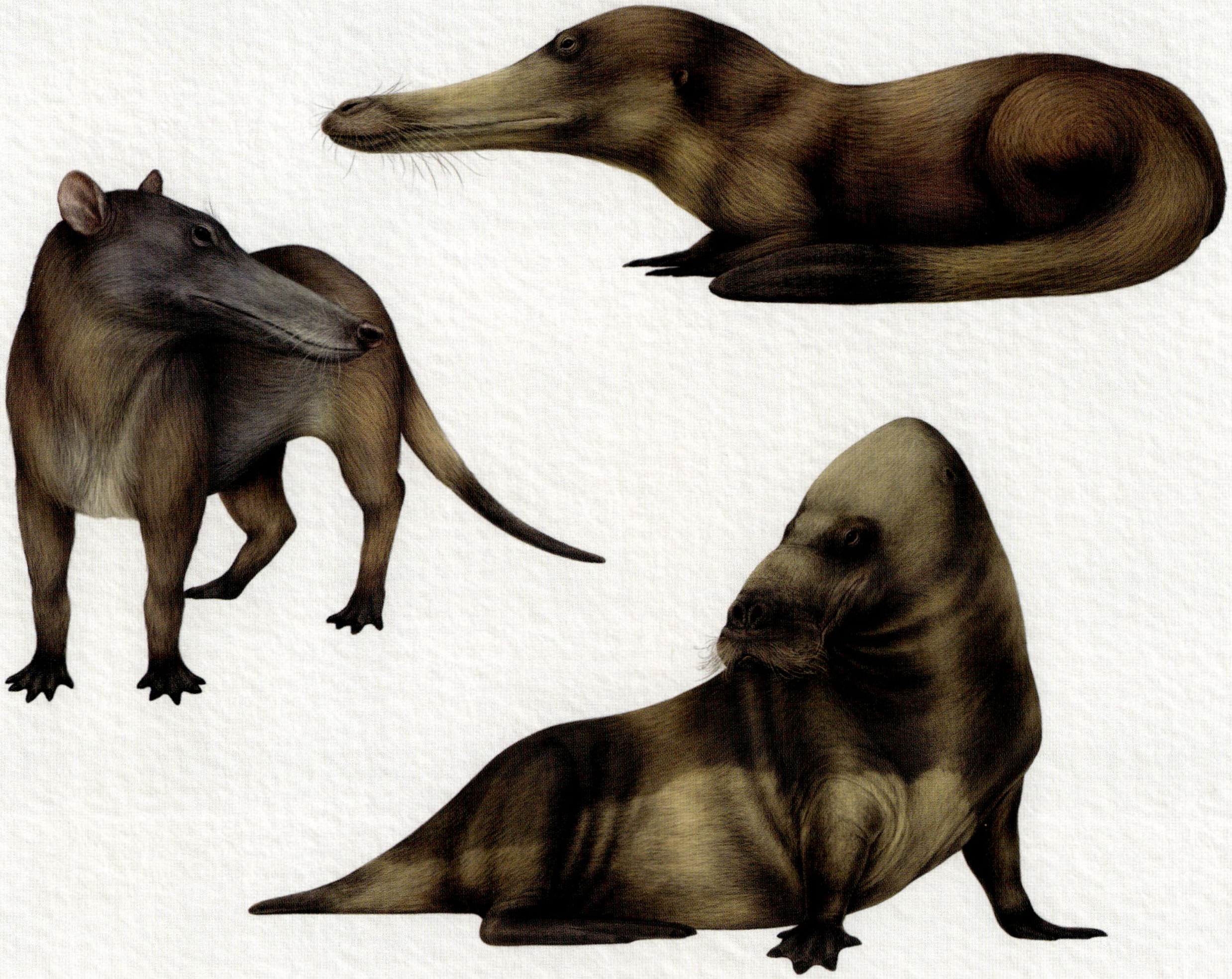

▼ *ANUROGNATHUS*

Despite their recent popularity in modern palaeoart, the anurognathid pterosaurs of the Jurassic and Early Cretaceous are not newly discovered. *Anurognathus* from Germany, shown here, was named in 1923. The anurognathids' (generally) small size, broad, muppet-like faces, and bat- or nightjar-like demeanour makes them entertaining artistic subjects, and the topic of some speculation and discussion among scientists.

• Digital | 2023

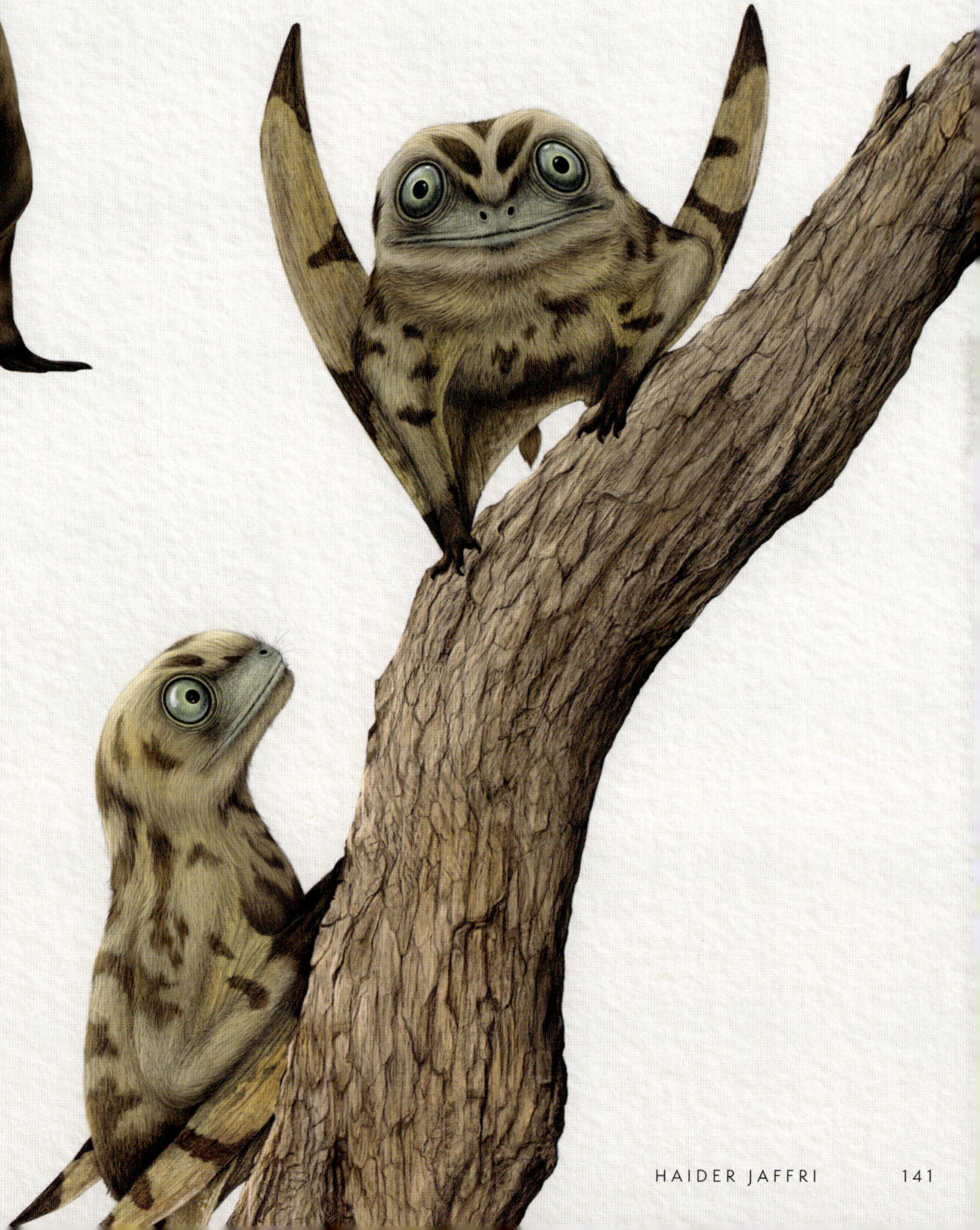

▲ STEM-WHALE DIVERSITY

The great burst of tropical stem-whale discoveries that occurred in Pakistan and India between the 1980s and early 2000s revealed an exciting diversity of early, amphibious members of the whale lineage. As revealed here, various of these animals looked like long-faced, amphibious pseudo-dogs. Here, we see (at left) *Ichthyolestes*, the notably long-jawed *Kutchicetus* (at top), and the droopy-nosed *Makaracetus* (at bottom).

• Digital | 2023

◀ *HESPEROTHERIUM* & ETRUSCAN BEAR CUB

In a Pleistocene winter somewhere in eastern Asia, an Etruscan bear cub sits to observe *Hesperotherium*, a late-surviving chalicothere, among the last of its kind. Part of the same group as horses and rhinos, chalicotheres combined long arms, clawed hands and short hindlimbs with a vaguely horse-like head. Best known from Europe, Etruscan bears occurred across Asia as far east as Tajikistan.

• Digital | 2022

LYCAENOPS ORNATUS ▸

Named from South Africa in 1925, *Lycaenops* is one of the best known of the gorgonopsids, a group of short-tailed, sabre-toothed stem-mammals often dubbed 'gorgons' for short. These animals were important predators during the Permian (between 270 and 260 million years ago) and some species likely inhabited the vast Pangaean deserts of the Permian world.

• Digital | 2022

▾ *TORVOSAURUS* PORTRAIT

Torvosaurus of the Late Jurassic is a gigantic megalosaurid theropod, 9m (30ft) long, known from the USA and Portugal, and possibly elsewhere. This portrait raises questions faced by many palaeoartists: how much wrinkling should there be on the neck? How keratinous should the lumps around the eyes and along the snout be? And were the jaws lined with skin texture differing from that elsewhere on the face?

• Digital | 2021

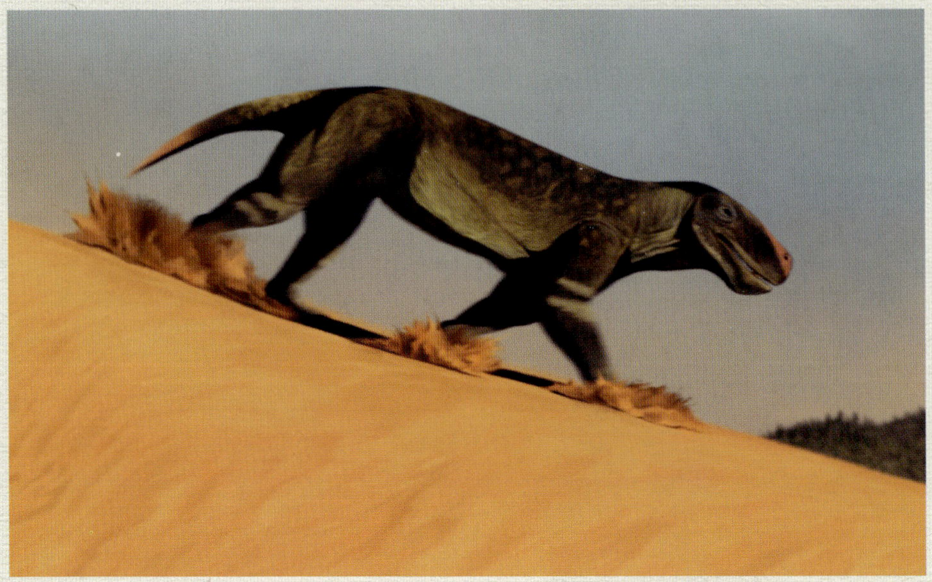

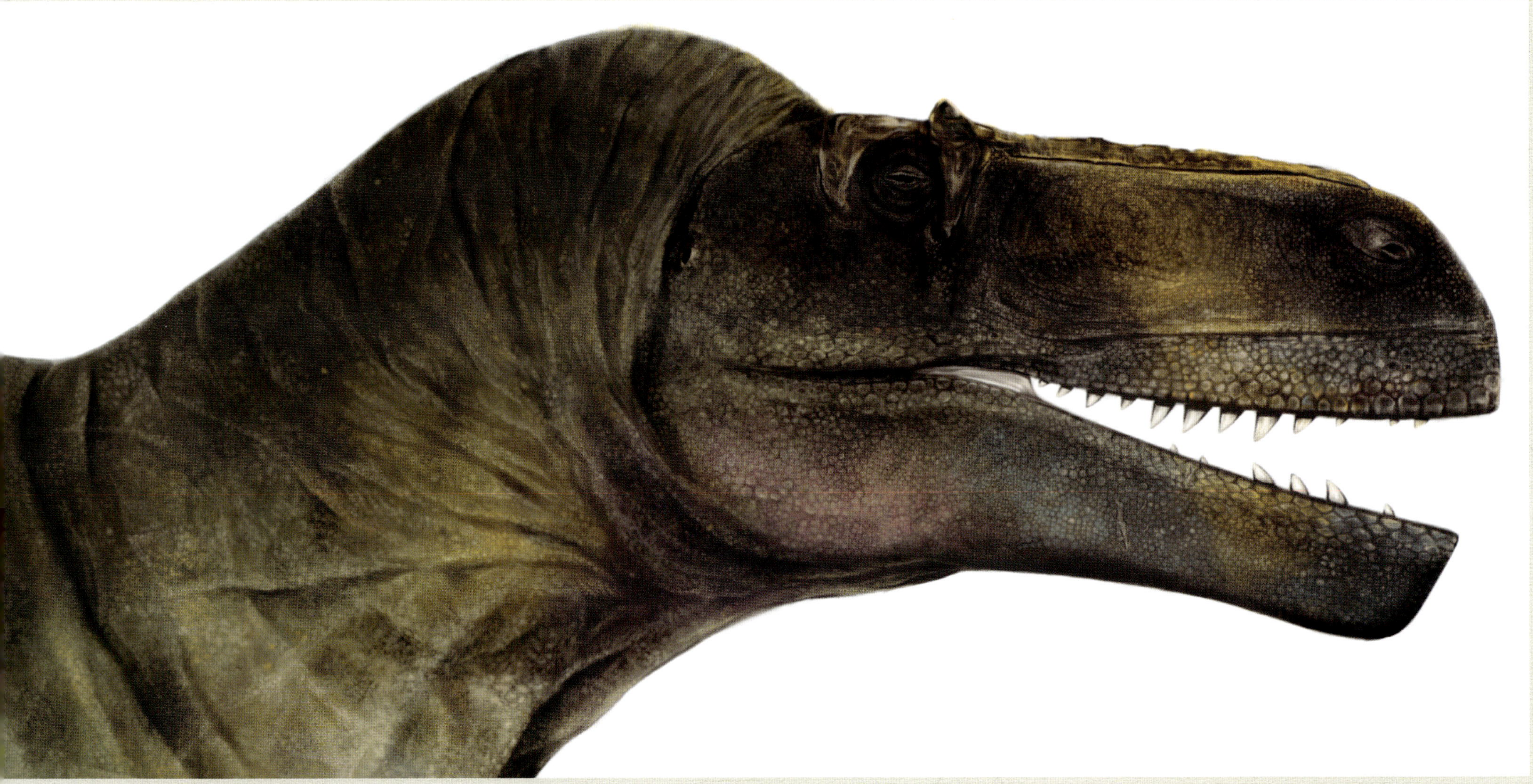

DINOSAUR TRIO ▸

Juxtaposing dinosaurs from different lineages is always fun and can be used as a way of highlighting differences, but also similarities. *Dromornis* the bird – alive in Australia around 3 million years ago – is both phylogenetically and temporally very distinct from the Cretaceous ornithischians *Iguanodon* and *Chasmosaurus*. But it's also similar to them in being a browsing herbivore that used a massive and powerful beak.
• Digital | 2022

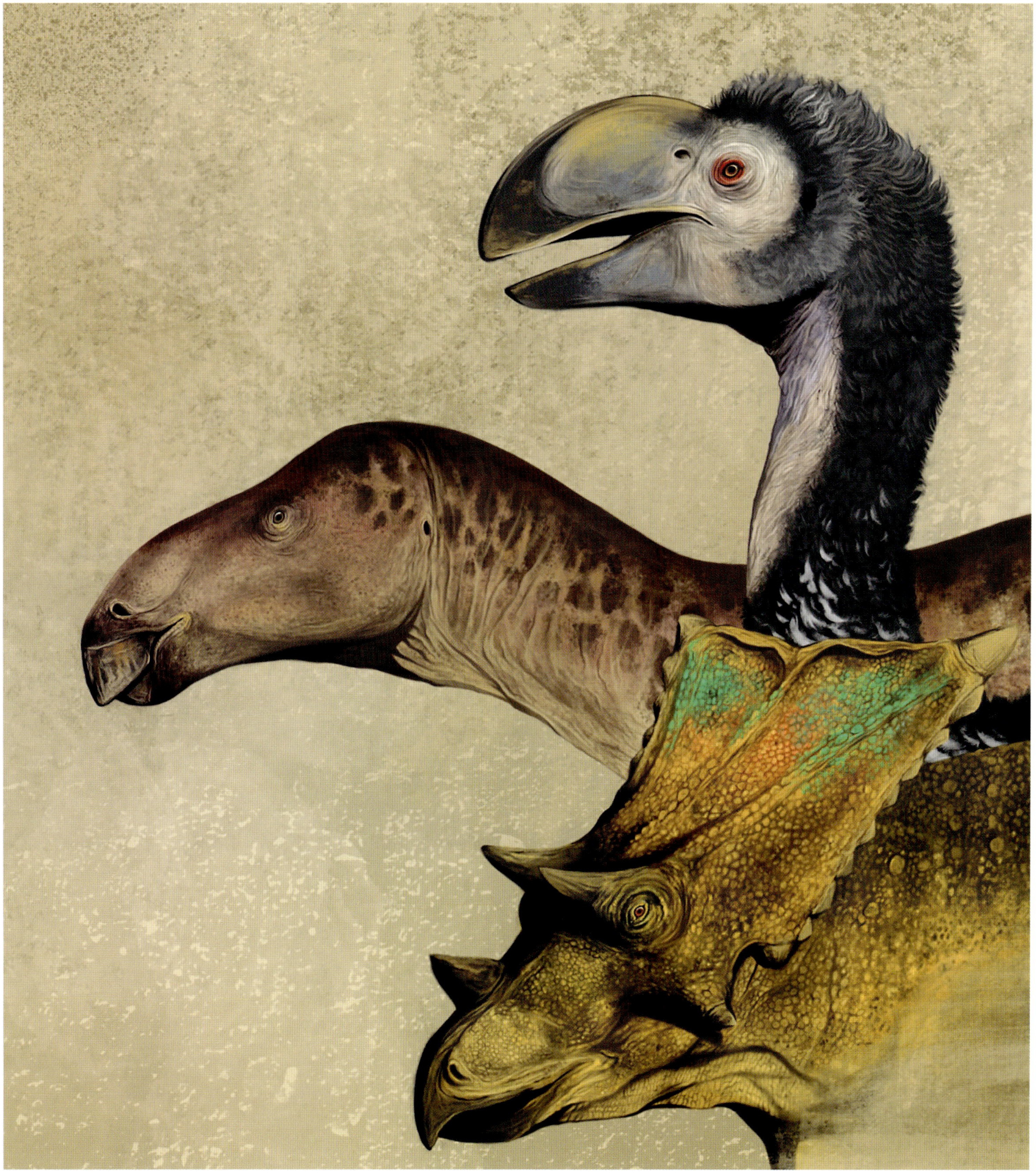

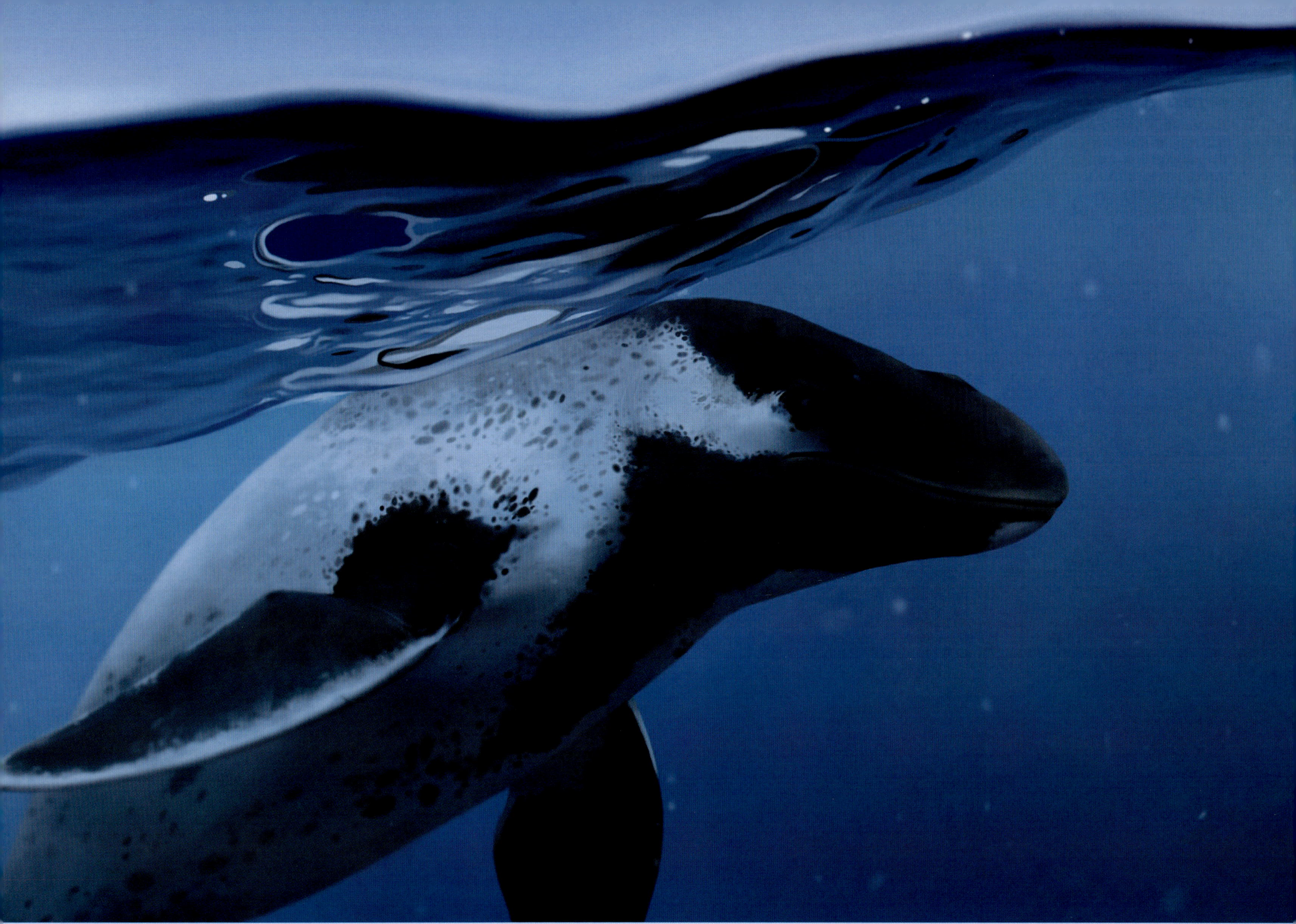

▲ *JANJUCETUS*

Among the many new fossil whales discovered in recent years are early members of the baleen whale lineage. Named in 2006, *Janjucetus*, from the Late Oligocene of Australia – and thus around 25 million years old – was relatively short-snouted and small overall (at 3.5m or 11ft). It was intermediate in many respects between more archaic whales (like the giant *Basilosaurus*) and more modern-looking baleen whales. It was probably an opportunistic predator that used its large eyes in locating prey.
• Digital | 2021

VELOCIRAPTOR ▶

It's increasingly agreed that *Velociraptor* and related maniraptoran theropods would have looked like big, weird birds when alive, as conveyed in this reconstruction. Here, *Velociraptor* recalls some kind of heavy-jawed seabird, the limbs drooping at its side looking like clawed wings rather than feathered arms and hands. White streaks of uric acid show that this is a favoured resting spot.
• Digital | 2022

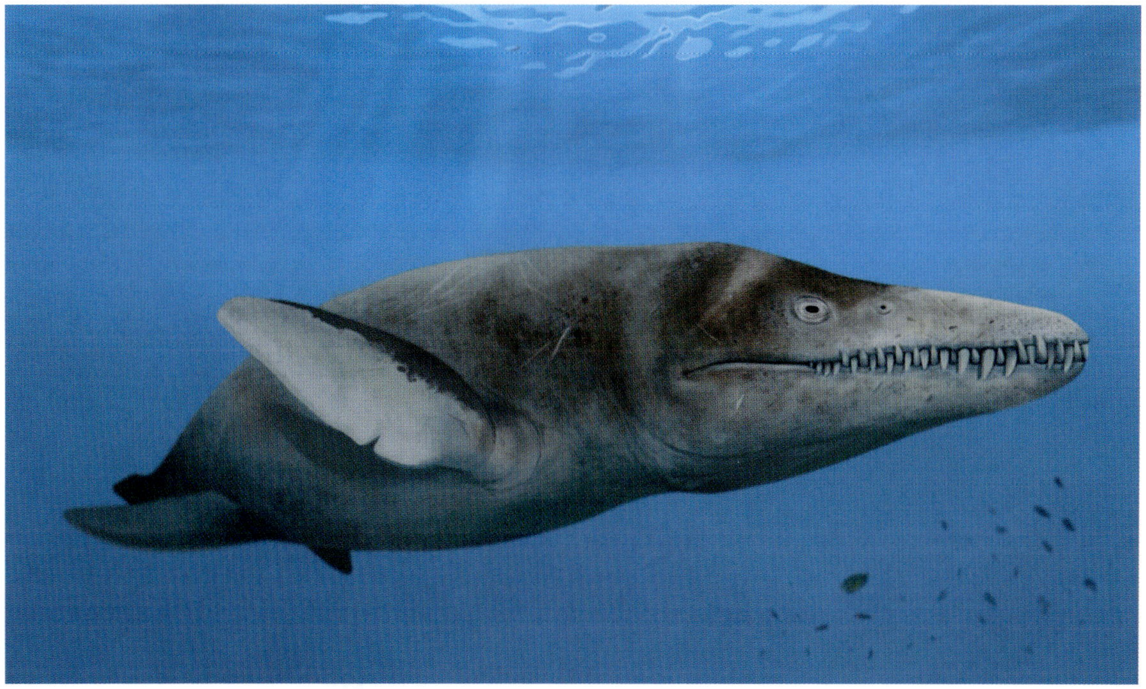

◂ *SACHICASAURUS*

In recent years, several new pliosaurids – a specific group of plesiosaurs, the most familiar of which are short-necked and with a massive skull – have been discovered in the Lower Cretaceous sediments of northern South America. The proportions of certain of these animals – *Sachicasaurus*, shown here, is the most extreme – were different from those more conventionally depicted in images of these animals.
• Digital | 2021

▾ *ENDOCERAS*

In an Ordovician ocean 460 million years ago, the shelled cephalopod *Endoceras* forages for prey. The vertical shaft on the horizon is part of a ring that might have existed around the planet at the time, a consequence of asteroid fragmentation. Evidence of this possible ring system was announced in 2024.
• Digital | 2024

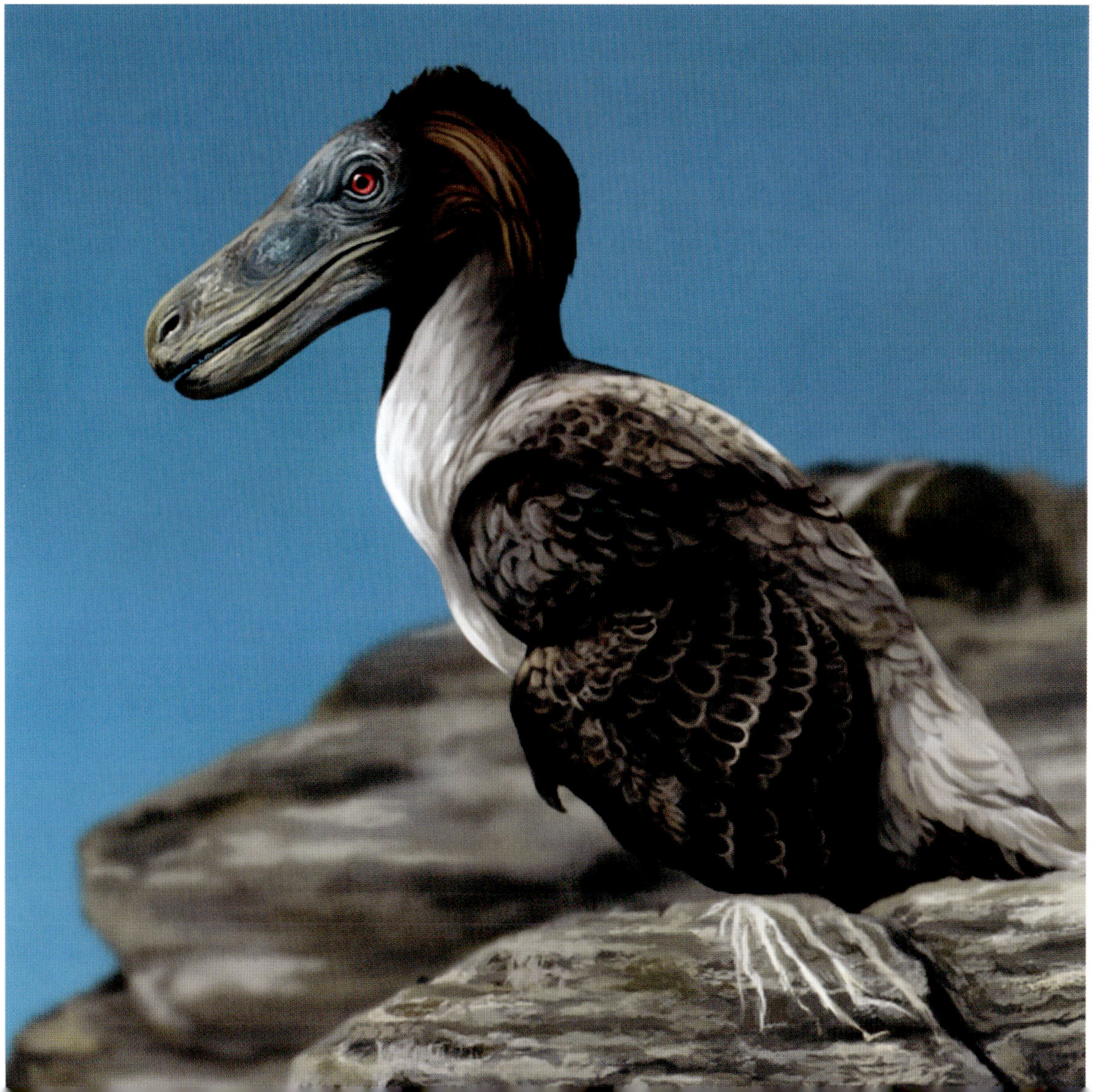

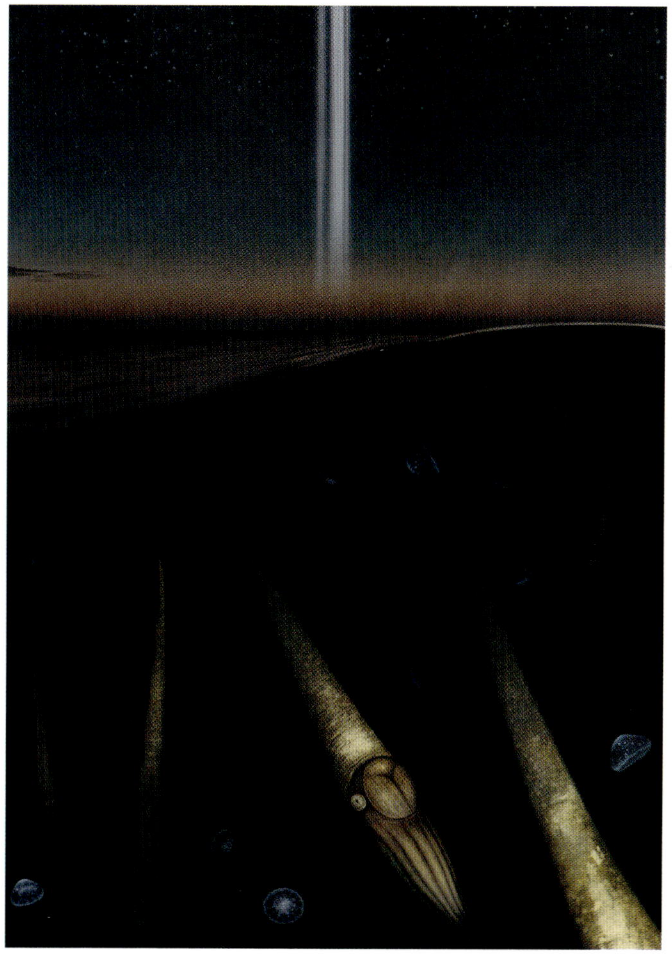

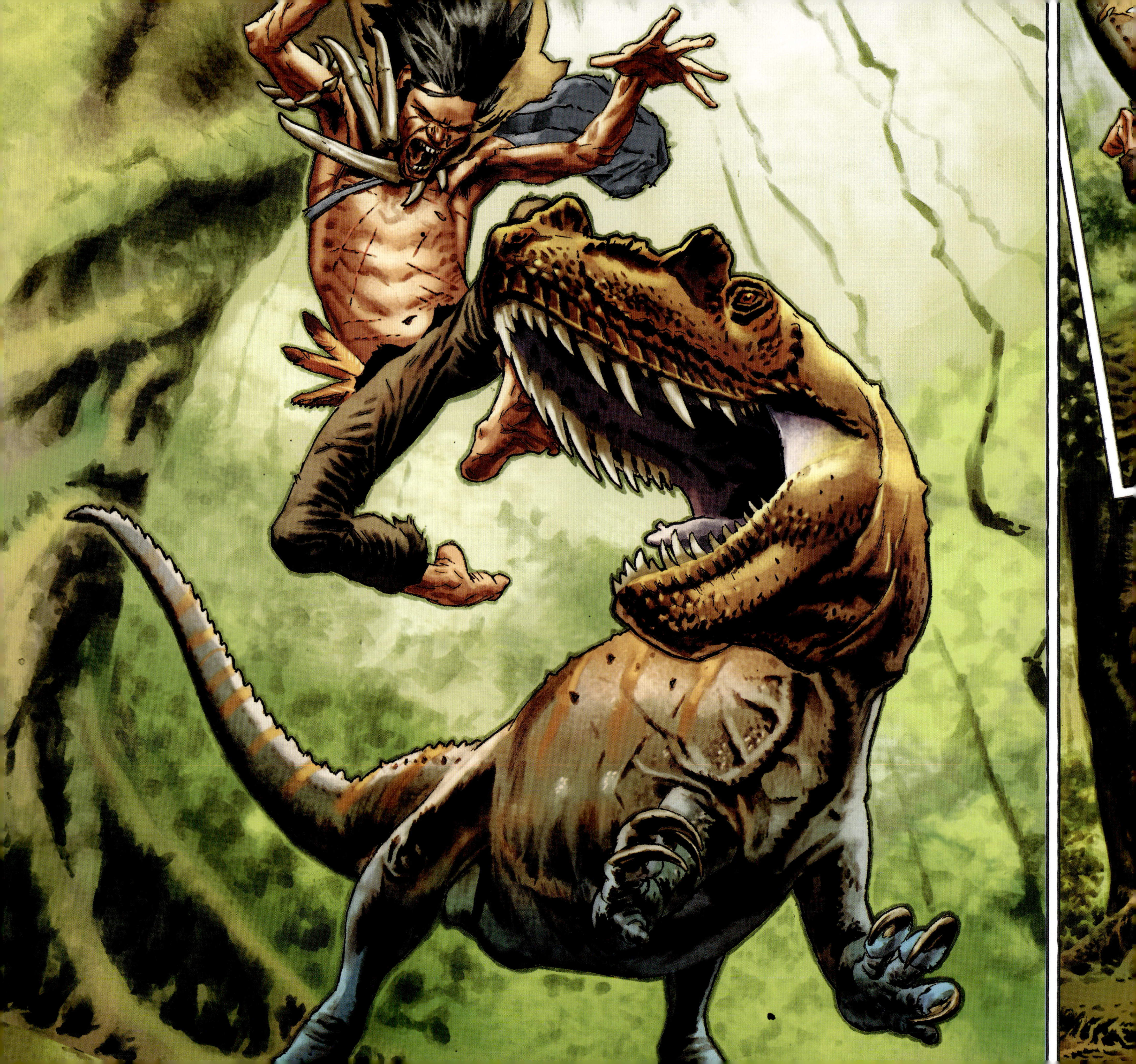

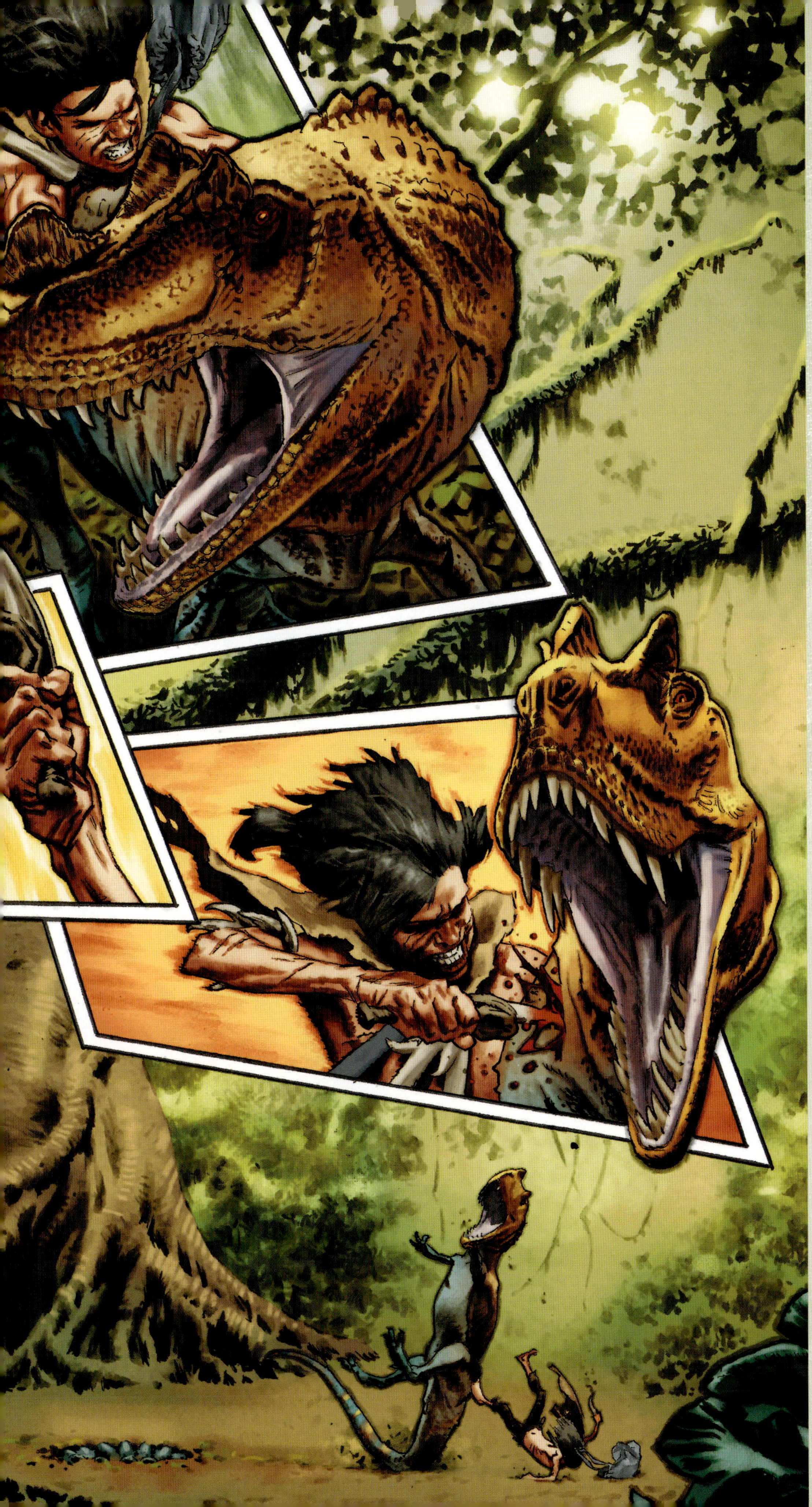

LEWIS LAROSA

LEWIS LAROSA IS AN AMERICAN artist and illustrator living in Marietta, Georgia, with his wife and two children. He specialises in comic book and sequential art, character design and concept art, and his favourite, palaeoart. Lewis believes himself something of a dinosaur, in the sense that his work is strictly traditional, encompassing a mix of pencil, pen, marker, ink wash and paint on board. He earned a bachelor's degree in Studio Art from Kennesaw State University, Georgia, where he spent a considerable amount of time drawing and painting from life. Published internationally for over 20 years, his notable projects projects include reinventing *The Punisher* for Marvel Comics, his rendition heavily influencing the character's movies and television shows; *Bloodshot* for Valiant Comics (which earned him a credit in the motion picture); and *Fortnite* for Epic Games.

Lewis's earliest memories include drawing dinosaurs with crayons at the kitchen table, his imagination ignited by his ever-growing collection of books on prehistoric life and the beautiful palaeoart within. He has nursed a lifelong obsession with palaeontology, always staying current on the newest research and keeping connected with the palaeontology and palaeoart community through social media. Lewis brings his skills as a visual storyteller – honed through his years as a comic book artist – to the world of palaeoart, bringing prehistoric animals and their environments to life through thoroughly researched sequential graphic art.

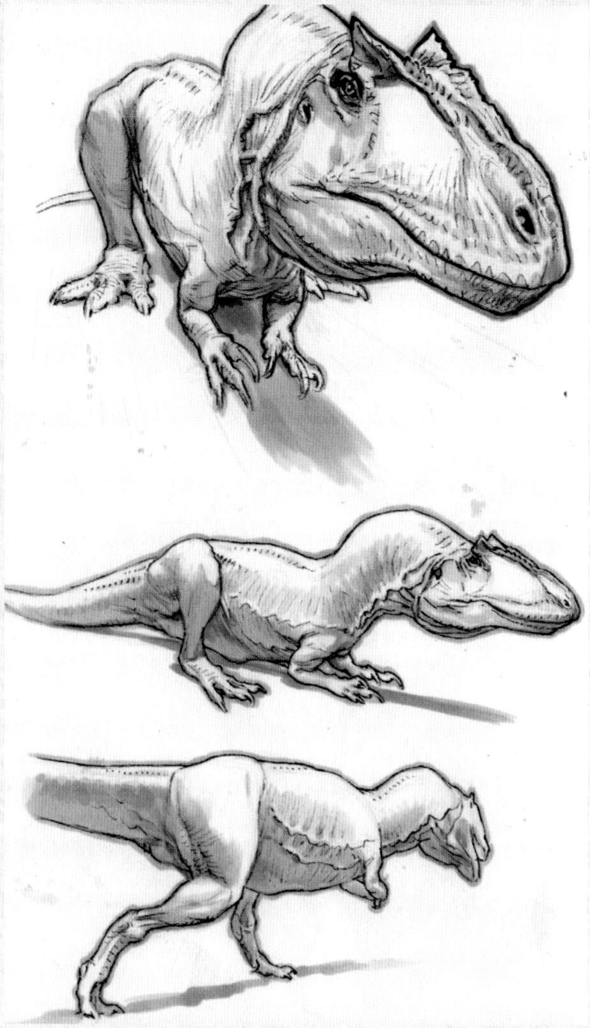

ETERNAL ENEMIES ▶

This piece, is a tribute to the famous 1929 painting of *Triceratops* and *T. rex* by legendary palaeoartist Charles R. Knight. Lewis used 2B, 4B and 8B pencils, and 0.5mm and 0.3mm fineliner pens as well as an ink wash and white paint to create the effects he was after.

• Pencil, ink, ink wash & white paint | 2024

▼ *PARASAUROLOPHUS* STUDIES ▶

Lewis's studies of this famous hadrosaur emphasise a problem particular to palaeoart. Many dinosaur skeletal reconstructions – the jumping-off point for many artists to start their artwork – tend to be simple plan views. This can make visualising the animal in three dimensions – obliquely, from above, even head on – problematic. As such, many palaeoartists build up huge collections of references and photos to overcome this potential problem.

• Pen & marker | 2022

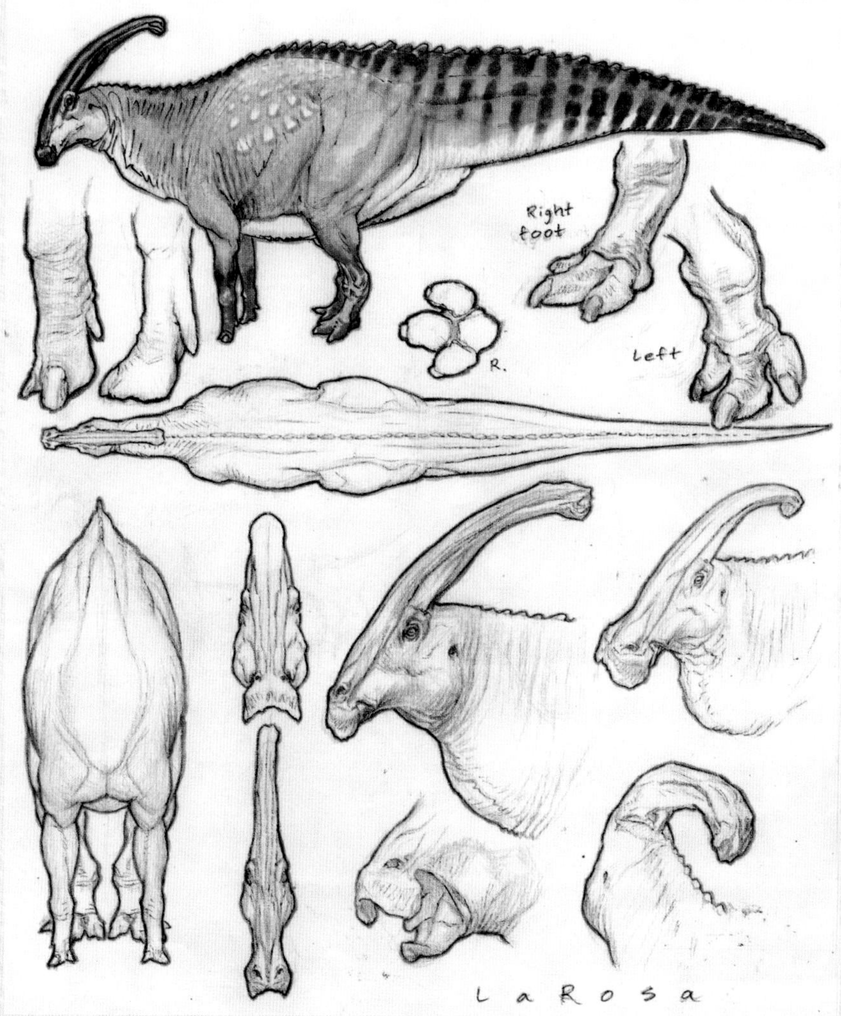

▲ *ALLOSAURUS* SKETCHES

These warm-up sketches of *Allosaurus* are fleshed-out versions based on the skeletal mounts at the Smithsonian Museum of Natural History in Washington, D.C. Many artists start their day with warm-ups as a way to ready their hands and eyes for the more serious work ahead.

• Pen & marker | 2019

[Previous pages]

SAVAGE & *CERATOSAURUS*

These pages from the first issue of *Savage* prove what any pop culture enthusiast has always known: that the Venn diagram between comics creators and dinosaur enthusiasts has considerable overlap. Lewis's art not only has the trademarks of a fine comic artist – dynamism, excellent figure work and superb storytelling – but also includes an accurate rendition of *Ceratosaurus*.

• Pencil, ink, ink wash & white paint | Colours by Brian Reber | 2016

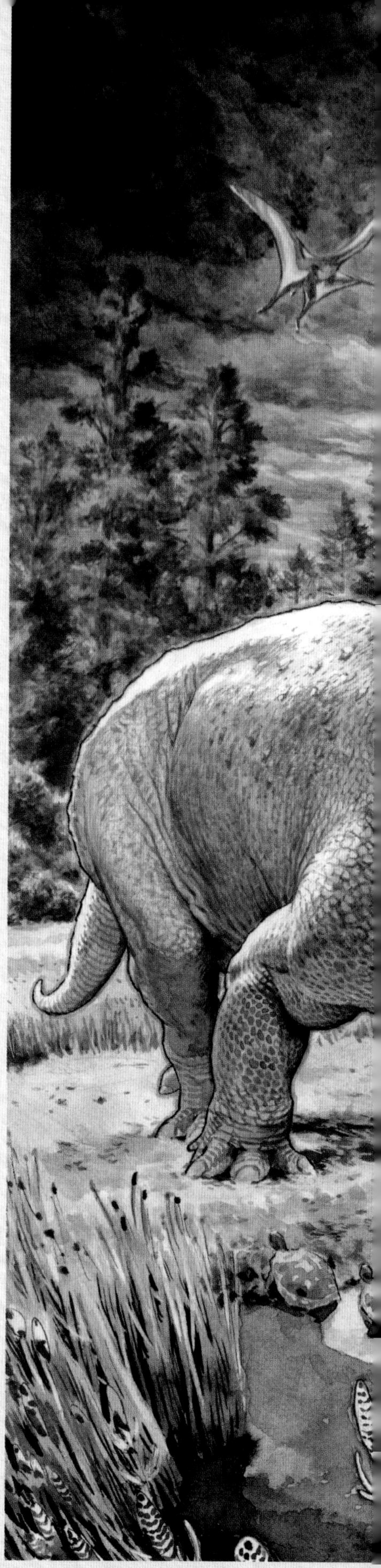

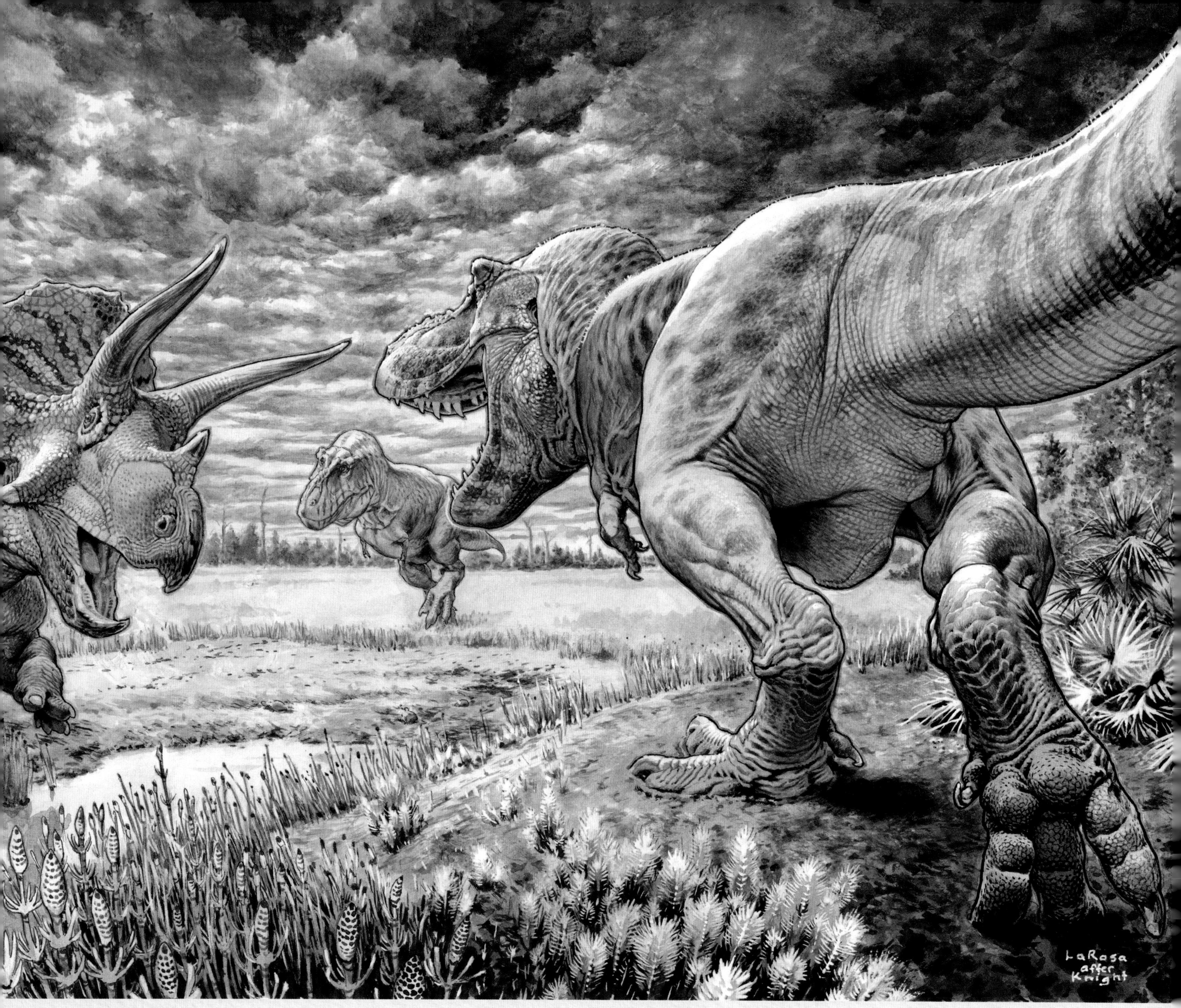
LaRosa
after
Knight

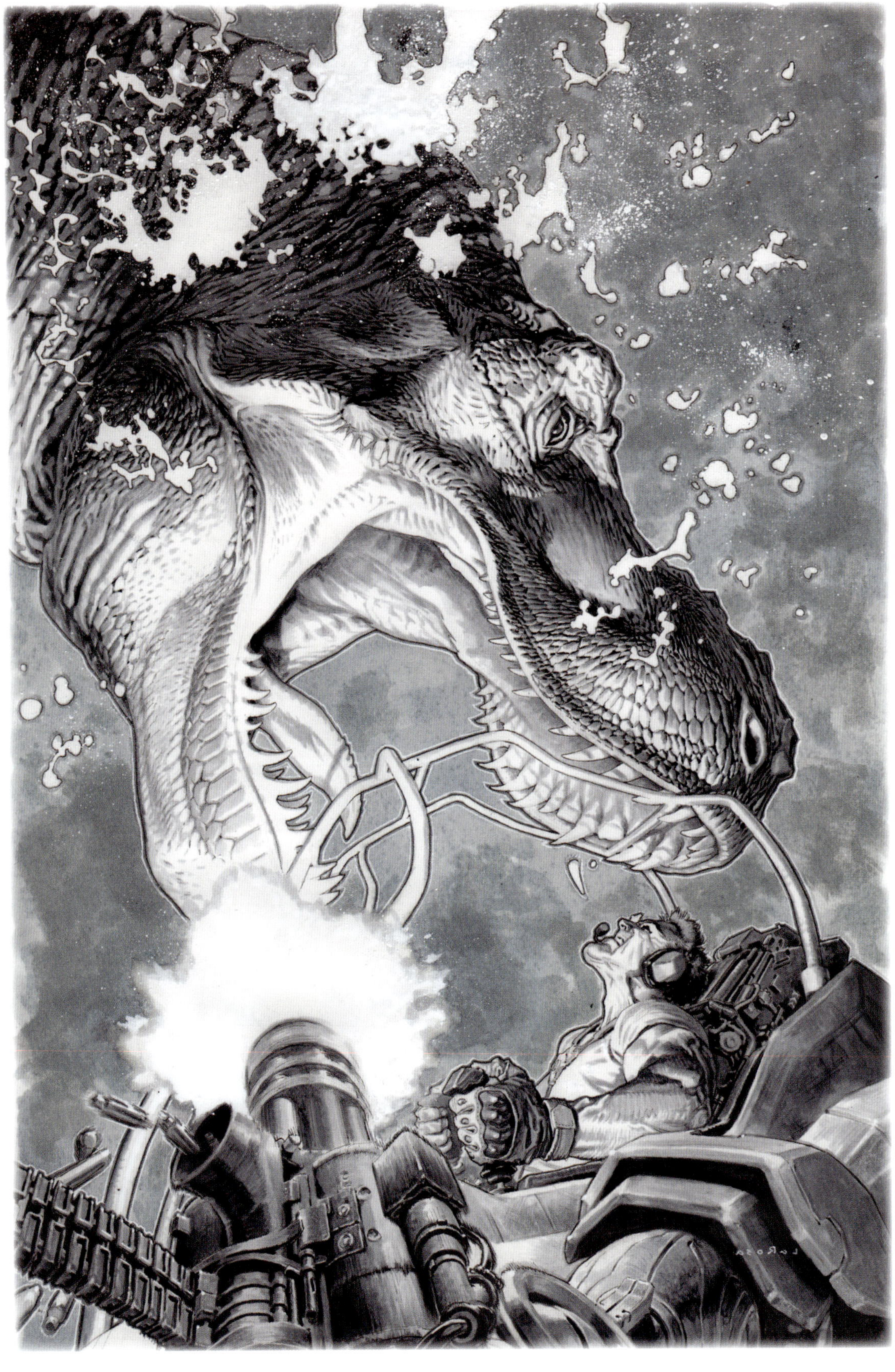

◄ *TANKERS* #2 COVER

A *T. rex* battles a mech-suit! Lewis's rendition of the Most Famous Dinosaur of All Time addresses one of the many contentious issues in palaeoart – that of theropod lips. A recent study of theropod tooth anatomy concluded that extinct theropods did indeed have lizard-like lips that enclosed the tooth rows: they did not possess a configuration where the upper teeth protrude over the lower jaw, as has often been depicted.

• Pencil, ink, ink wash & white paint | 2021

SAVAGE #4 ►

This page is a composite of the cover (on the left) and pages from *Savage* #4. They also illustrate that prehistoric animals are just as exposed to the vagaries of fashion as anything else. In 20th century pop culture, the pterosaur *de jour* was the strikingly crested *Pteranodon*, but the new century saw this classic flying reptile become *passe* and supplanted by the azhdarchids, in particular *Quetzalcoatlus*.

SAVAGE is a trademark of Valiant Entertainment LLC

• Pencil, ink, ink wash & white paint | Colours by Brian Reber | 2016

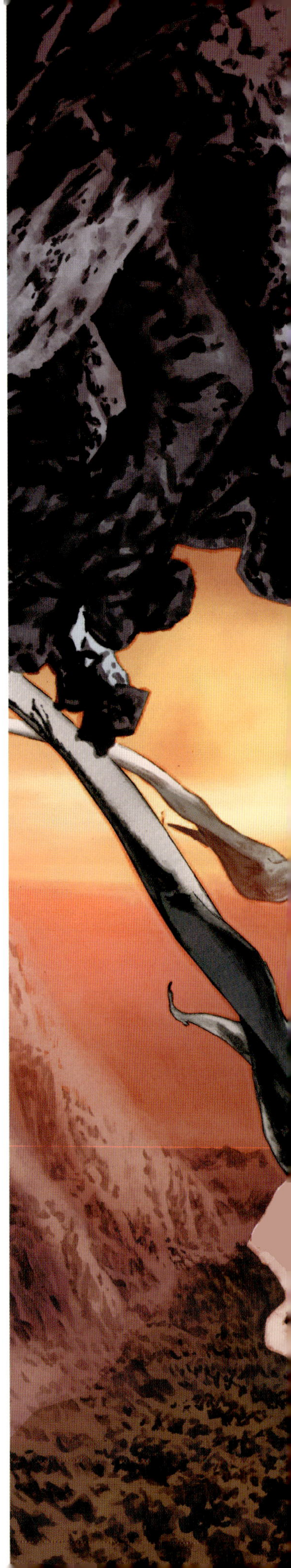

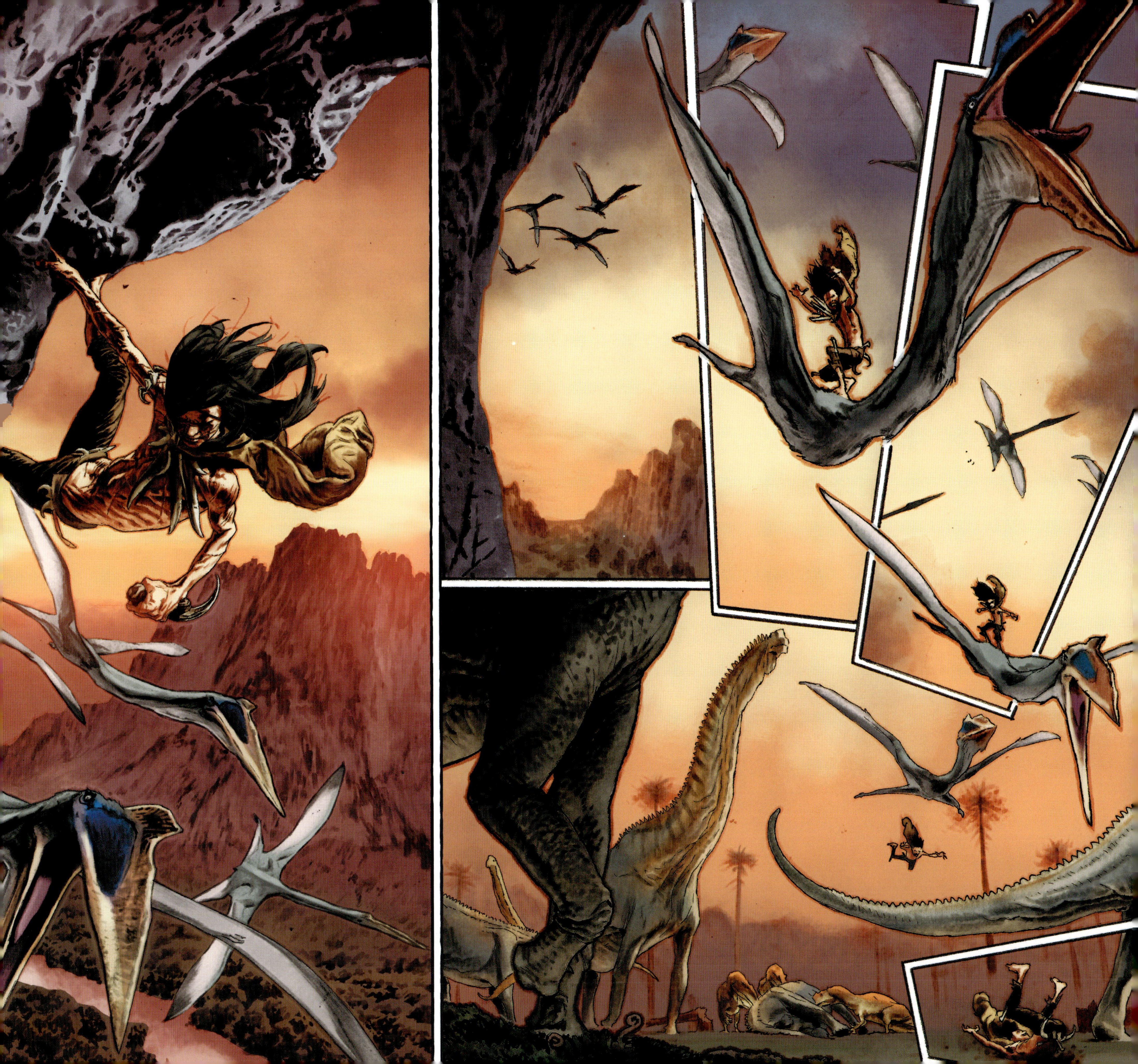

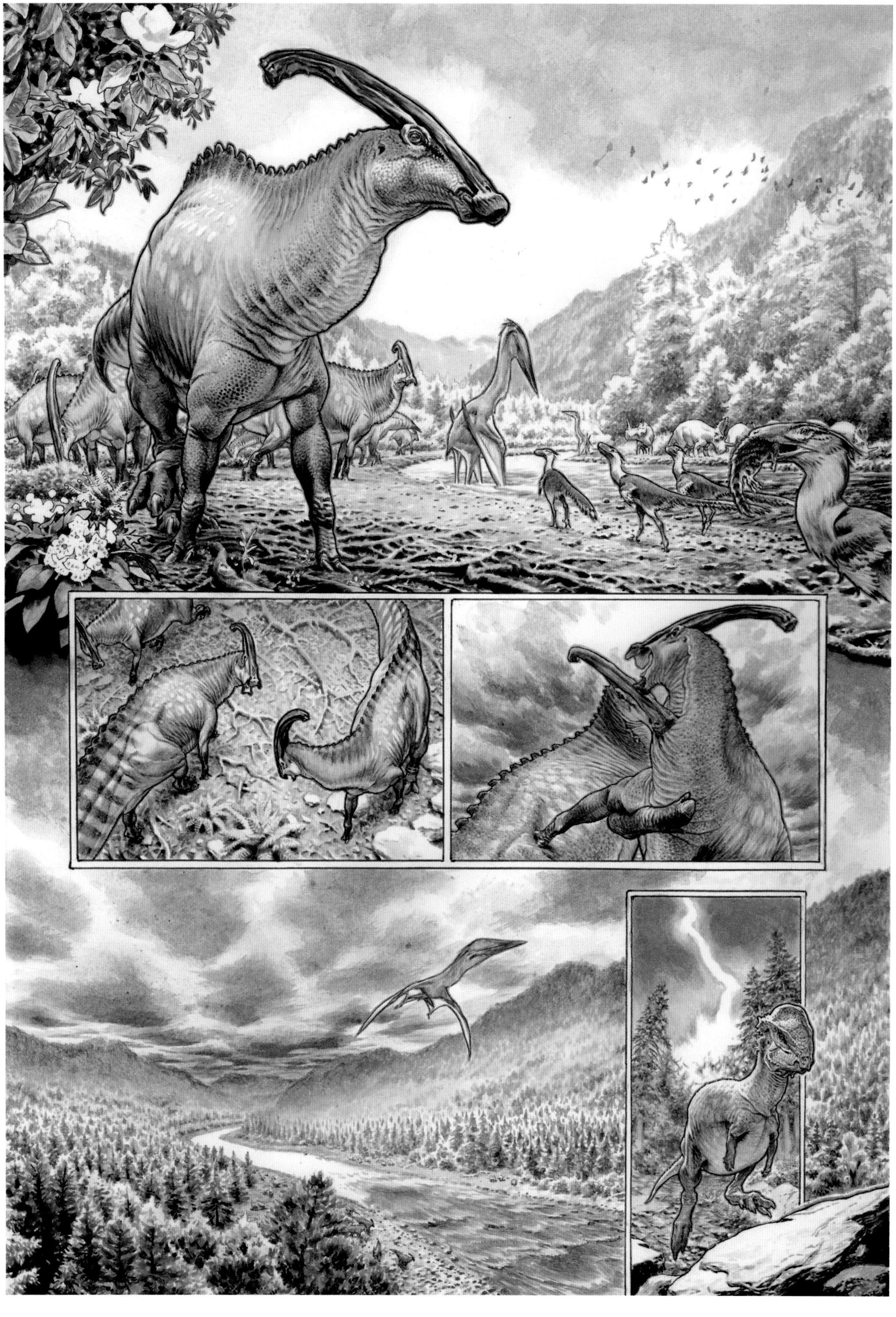

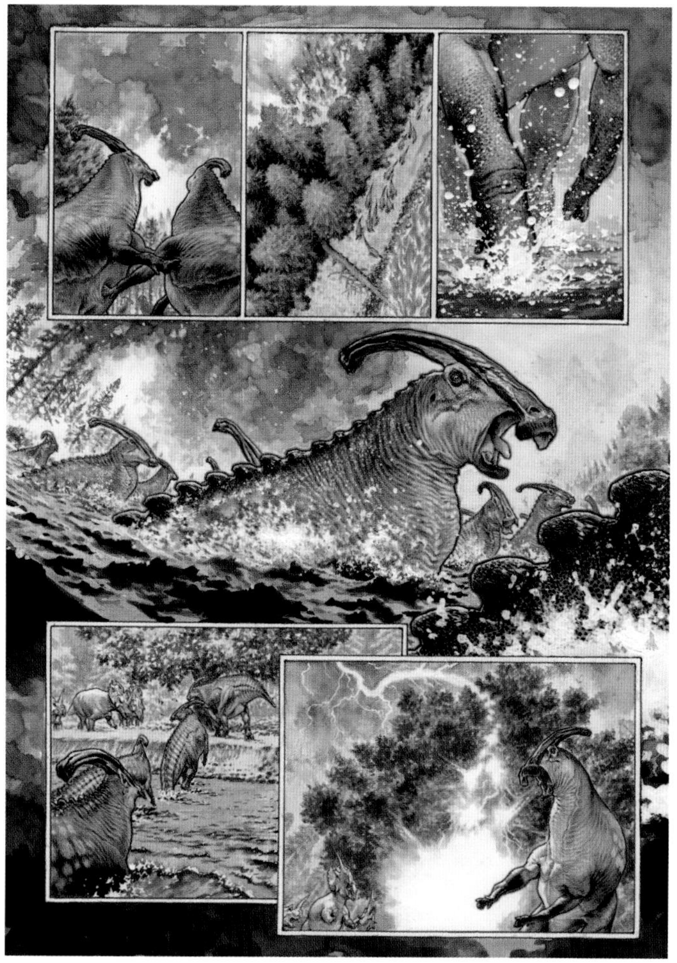

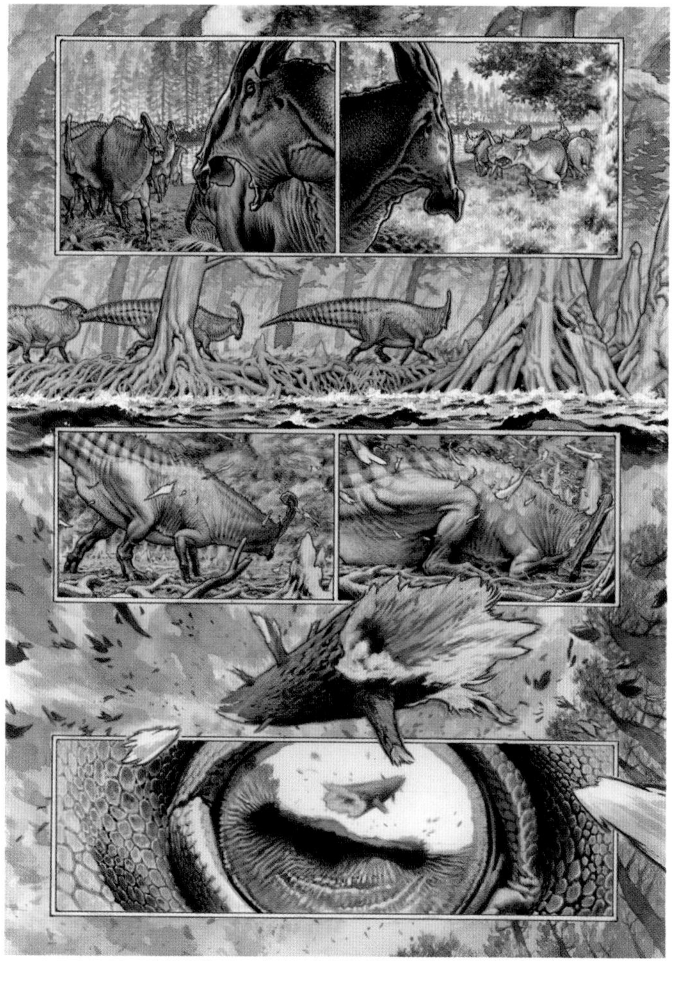

◂ ***PARASAUROLOPHUS* STORY**

These beautiful pages were for a short comic story written by palaeontologist Dr Filippo Bertozzo, for the *Tankers* comic collection. The story was based on a study by Dr Bertozzo and others of the holotype specimen – the original specimen of an organism which is used to describe and name it. This study, published in 2020, suggested that an injury to the animal's shoulders and ribs was caused by a falling tree.

• Pencil, ink, ink wash & white paint | 2022

***PARASAUROLOPHUS* COVER ART** ▸

Cover for the *Parasaurolophus* short story published in the *Tankers* collection. *P. walkeri*, the species featured in the story, is known from the famous Dinosaur Park Formation of southern Alberta, although it is rare compared to other hadrosaurs from the area. The other two species, *P. tubicen* and *P. cyrtocristatus*, are known from New Mexico and Utah, and have shorter crests.

• Pencil, ink, ink wash & white paint | 2022

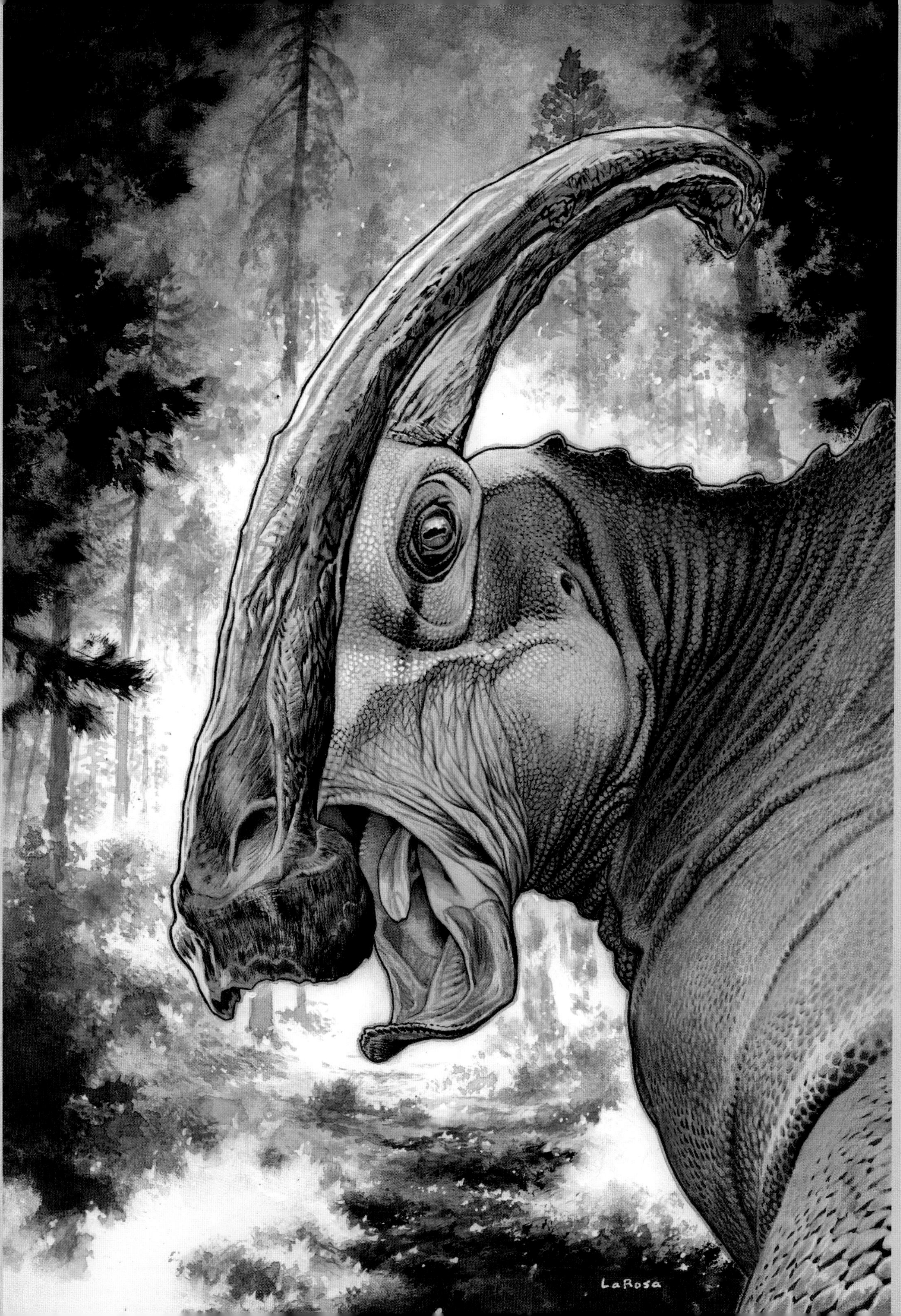

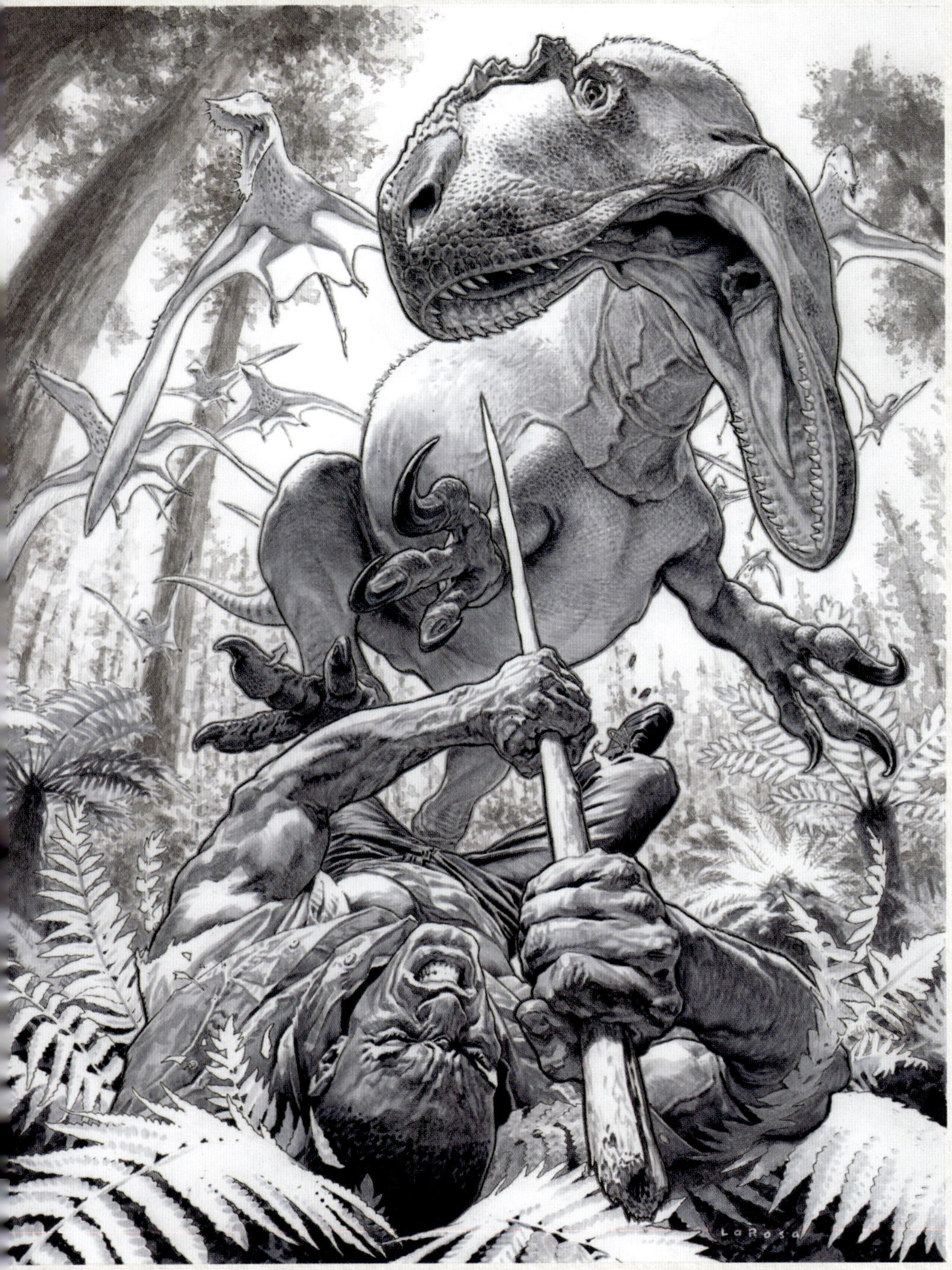

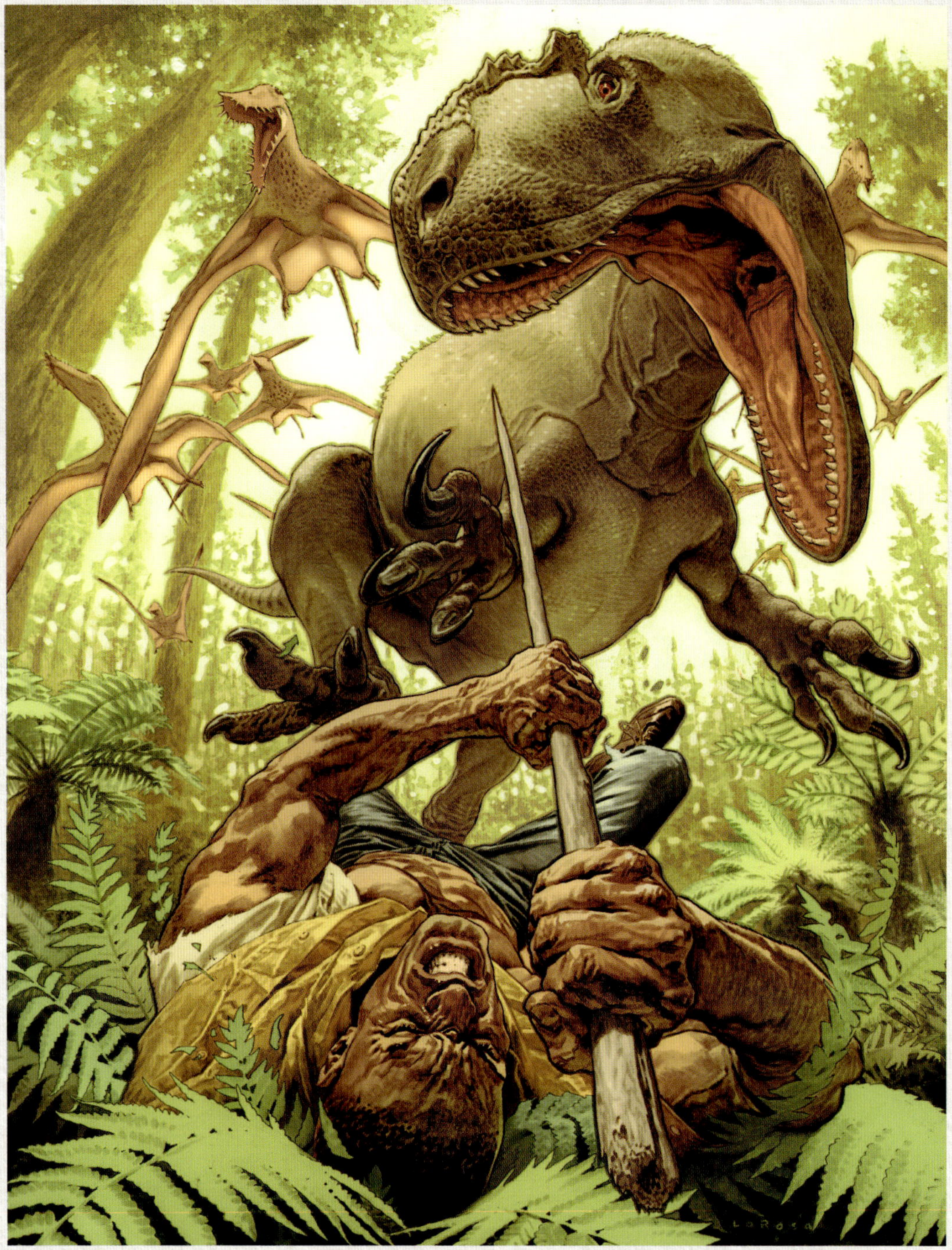

▲ *A VICIOUS CIRCLE* #2 COVER

In any story of a time-travelling assassin out for revenge, the protagonist is inevitably going to end up in the Mesozoic and *A Vicious Circle*, published by BOOM! Studios, is no exception. Unsurprisingly, Lewis was asked to do the cover for #3, featuring an *Allosaurus*. He wanted to emphasise the animal's enormous gape, powerful arms and massive claws. The pterosaurs are *Harpactognathus*.

Published courtesy of BOOM! Studios.

• Pencil, ink, ink wash & white paint | Colours by Matheus Lopes | 2023

▼ ***DEINONYCHUS* SKETCH**

Whilst *Velociraptor* is the most famous member of the dromaeosaurid family, *Deinonychus* is, historically, far more significant. When first described in 1969, the sickle claw on its foot, large brain, rigid tail for balance, and avian characteristics helped revolutionise our understanding of dinosaurs, triggering what would become known as the 'Dinosaur Renaissance'.
• Watercolours | 2023

***T. REX* SKETCH** ▶

Painted the same day as the *Deinonychus* piece below, Lewis produced these while making art with his young daughter one Saturday afternoon, on the same day as the*Deinonychus* piece below. Both paintings were done without the use of references but are still very nice representations of their subjects.
• Watercolours | 2023

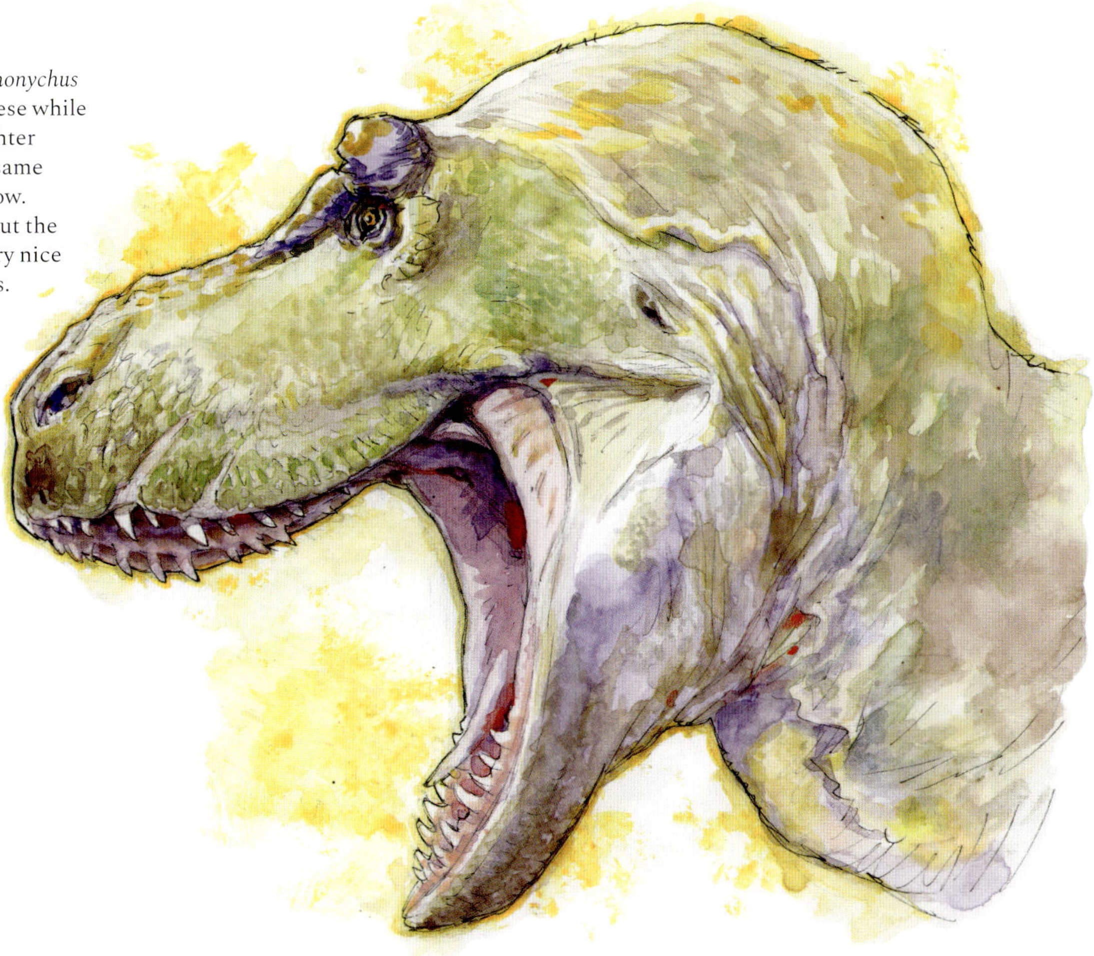

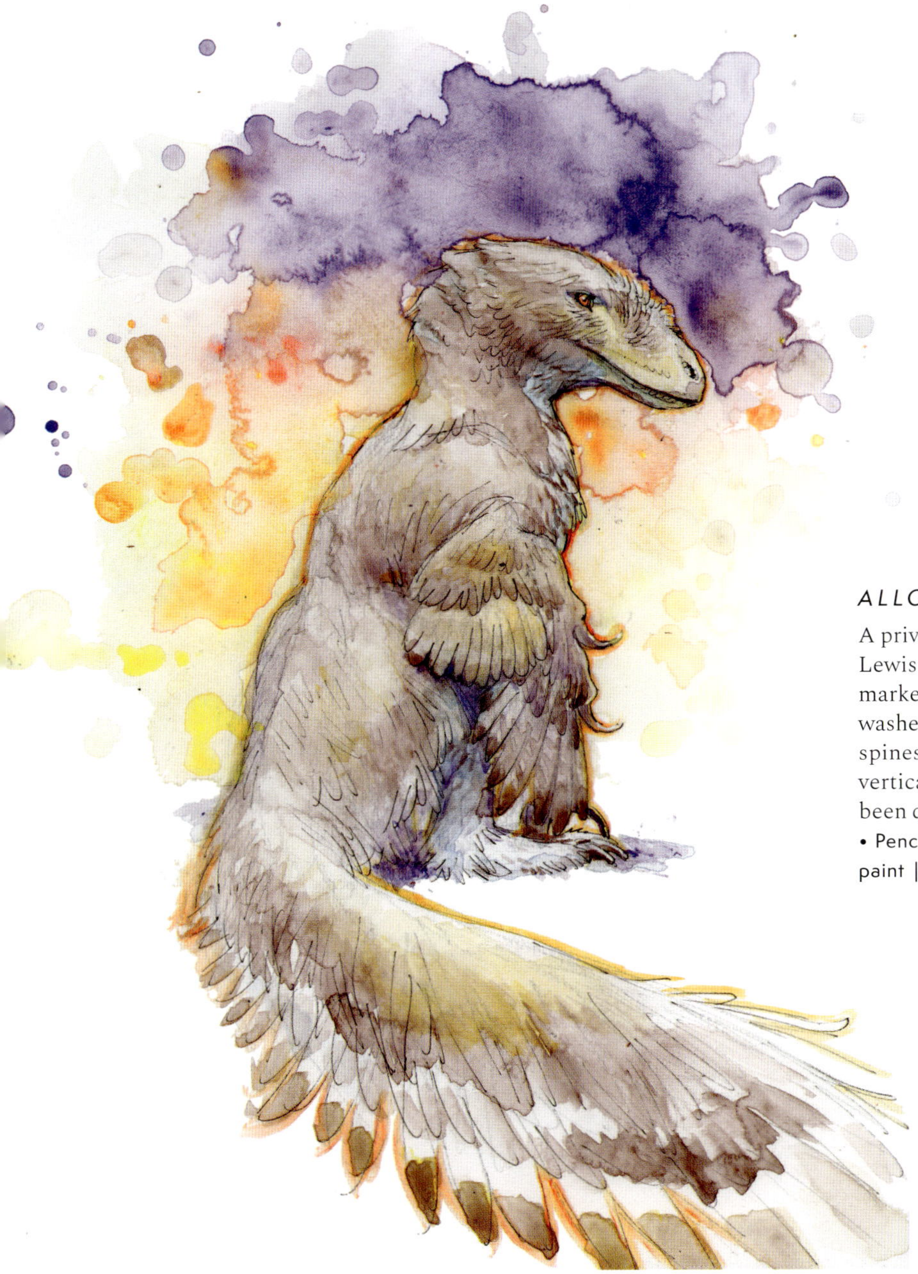

ALLOSAURUS & SAVAGE ▶

A private commission that saw Lewis move away from the use of markers, relying instead on ink washes. It doesn't look like theropod spines moved very much in the vertical plane so this would have been quite a comfortable ride.
• Pencil, ink, ink wash & white paint | 2023

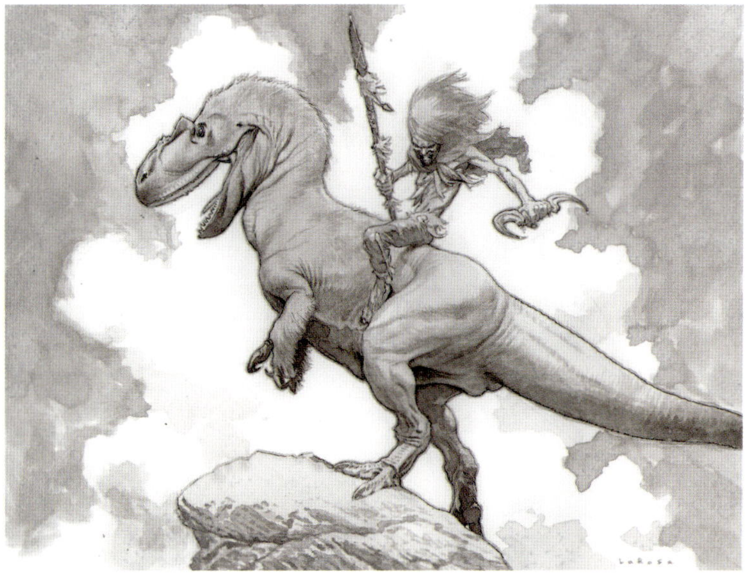

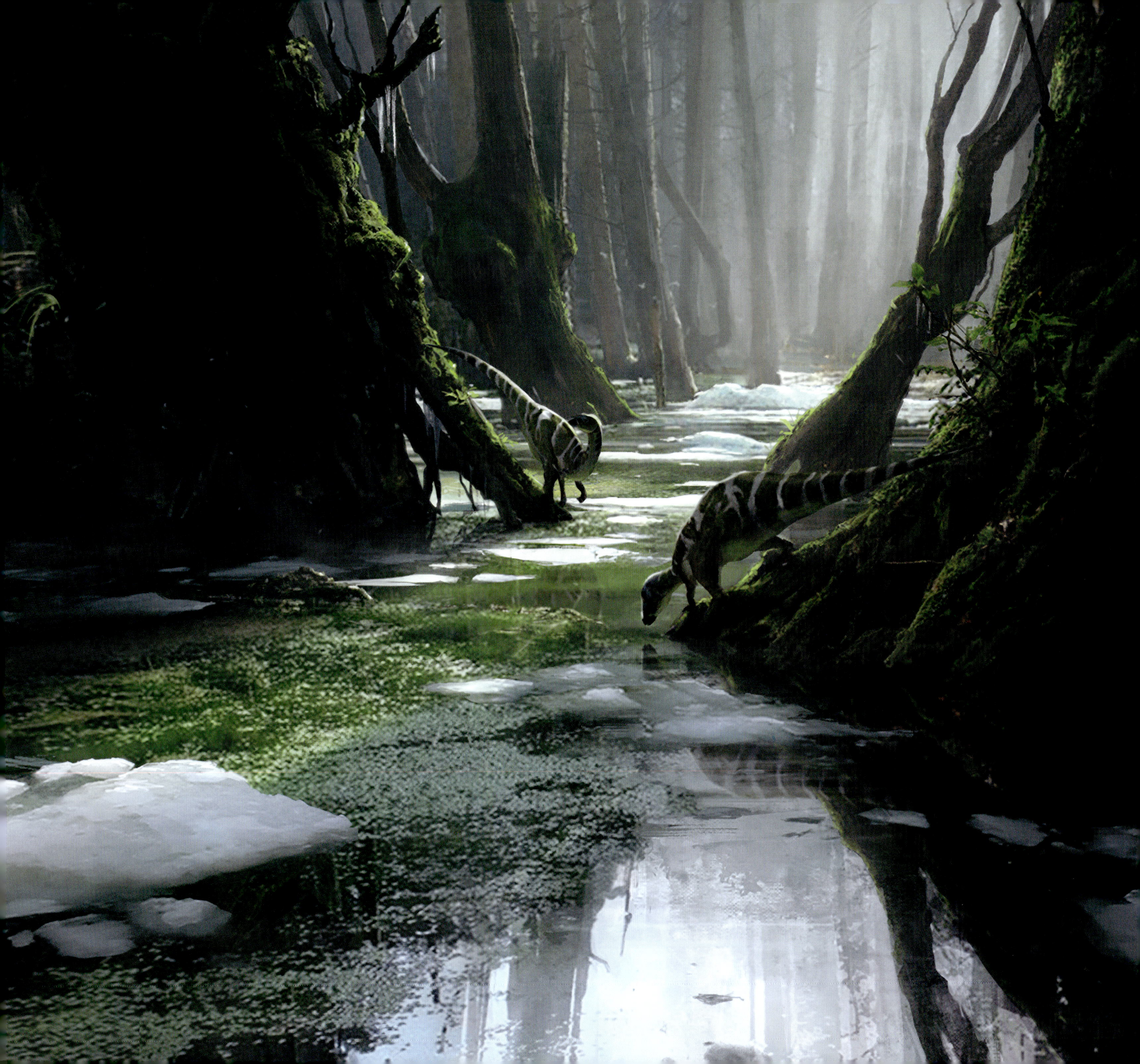

GAËLLE SEGUILLON

GAËLLE SEGUILLON IS a concept artist and palaeoartist with over 10 years of experience in the movie and video game industries. Her lifelong passion for art, movies and dinosaurs led her to study Visual Effects in France at ArtFx after which she began working as a digital matte painter and concept artist with such companies as Industrial Light & Magic, MPC, National Geographic and Quantic Dream, where she specialised in the creation of creatures and digital environments. Working on Apple TV+ BBC Studio's *Prehistoric Planet* rekindled her love for palaeoart and she continues to enjoy natural history projects. Some of her recent work includes contributions to the *Star Wars: Eclipse* video game, and the TV shows *Jurassic World: Fallen Kingdom* and *Prehistoric Planet* season 2.

Gaëlle's passion for travel and photography remains a primary source of inspiration, profoundly influencing her distinctive style. In addition to her professional work, Gaëlle is also dedicated to mentoring aspiring artists and sharing her knowledge through workshops and online tutorials. She constantly explores new techniques and mediums, blending scientific accuracy with creative storytelling.

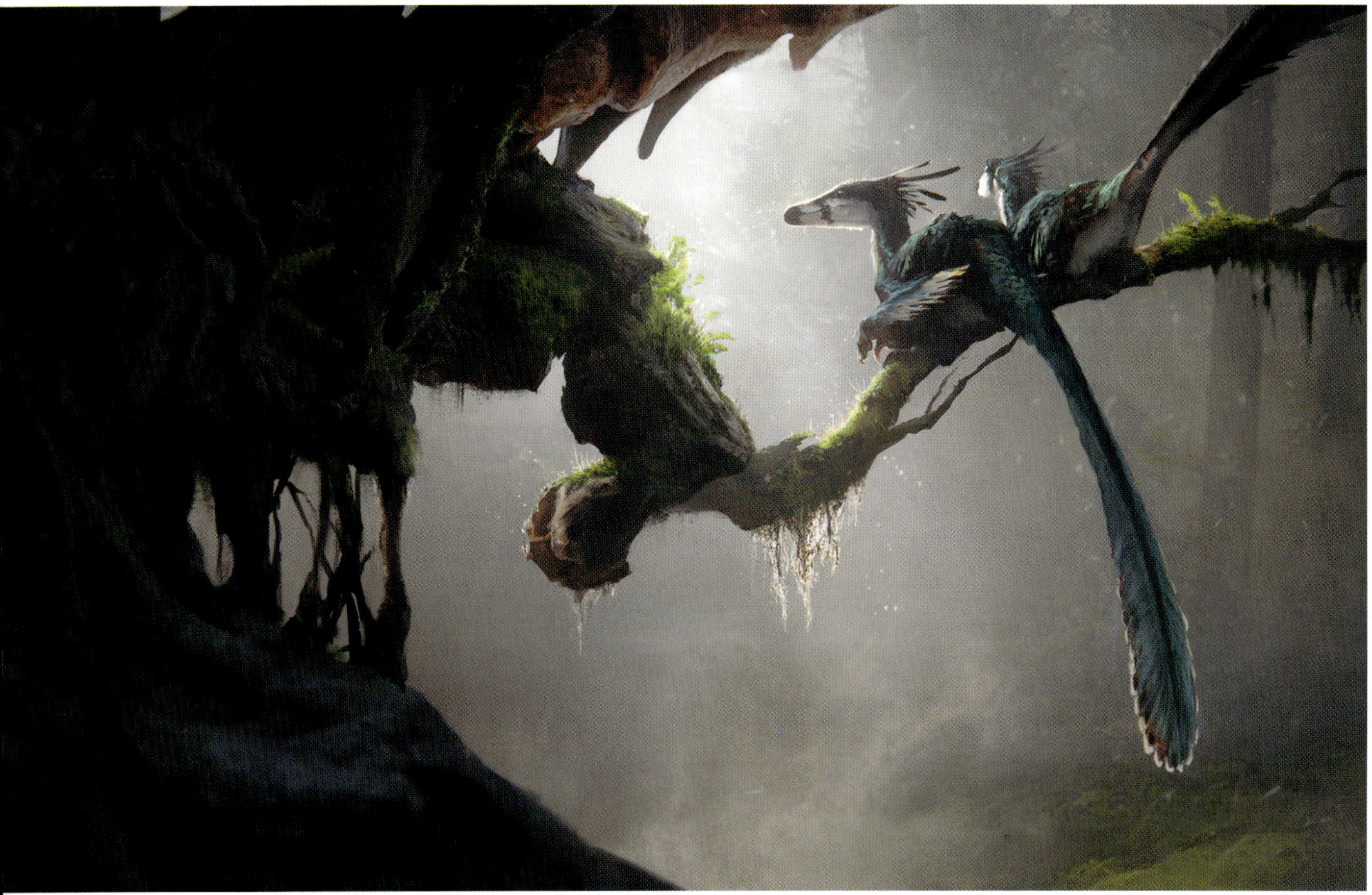

▲ FOREST DROMAEOSAURIDS

Velociraptor and its relatives – the dromaeosaurids – were feathered and bird-like when alive, and some species likely had a subdued, camouflaged appearance. But this was a group that persisted for over 100 million years across many environments. The possibility exists that elaborate plumes, bright colours and even iridescence was present, perhaps only during the breeding season. This forest scene depicts a mid-sized species that would have been similar in mass to a small eagle.

• Digital | 2022

[Previous pages]

FROZEN CRETACEOUS AUSTRALIA

Polar regions during the Cretaceous featured a mix of familiar and unfamiliar conditions. The climate was cool enough for there to be snow and ice, but lush forests existed and there were periods where ice caps did not exist. Climate modelling indicates that these places were extremely misty. Here, small Cretaceous ornithischians forage in a world that is green and verdant, cool and icy, misty but also bathed in perpetual sunlight, six months of the year.

• Digital | 2018

MEET ME AT THE *TRICERATOPS* SKULL ▶

In a Late Cretaceous North American forest, dromaeosaurids make use of an unusual structure as a resting spot. This image raises the interesting question of how long the skeletal remains of a giant dinosaur (like this *Triceratops*) might persist if not destroyed by scavengers or the elements. At least some dromaeosaurids – certain of the crow-sized microraptorine species in particular – could glide.

• Digital | 2022

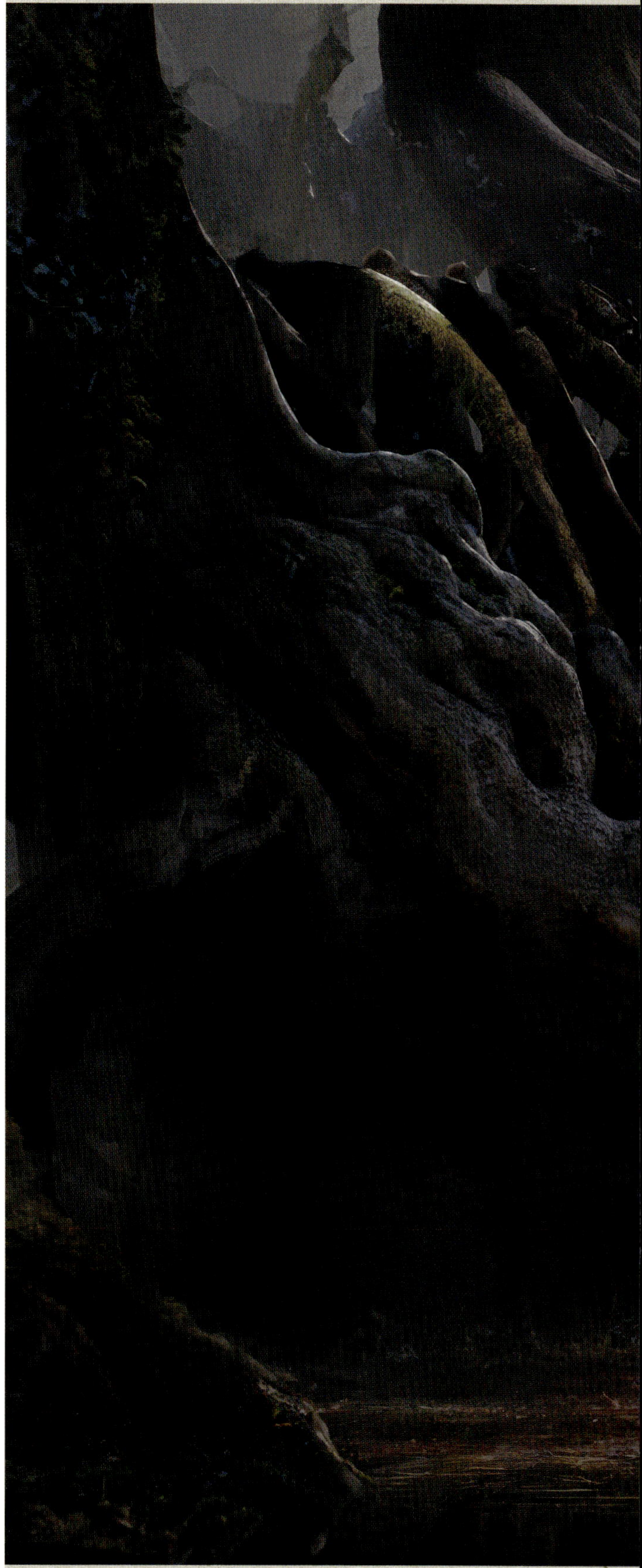

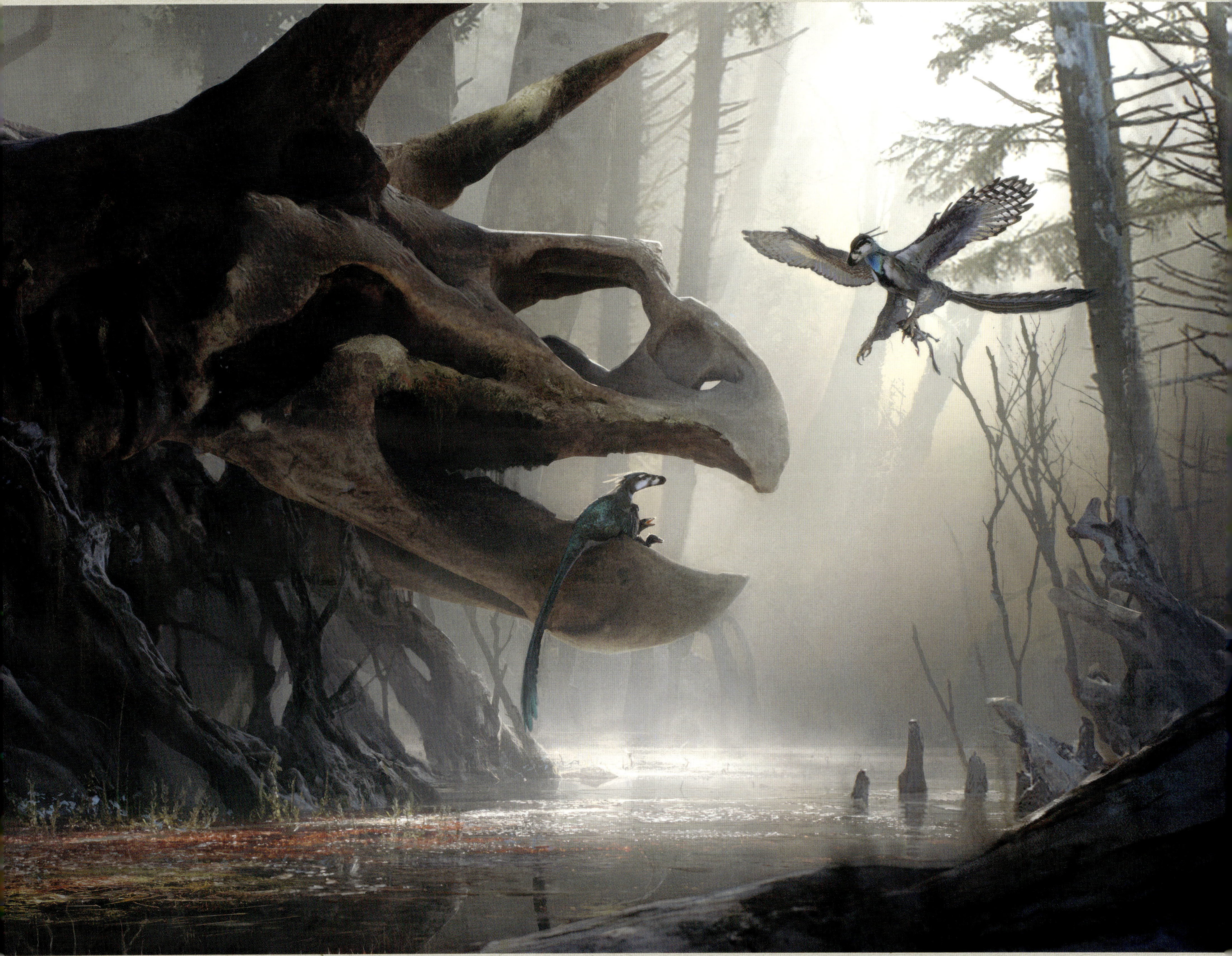

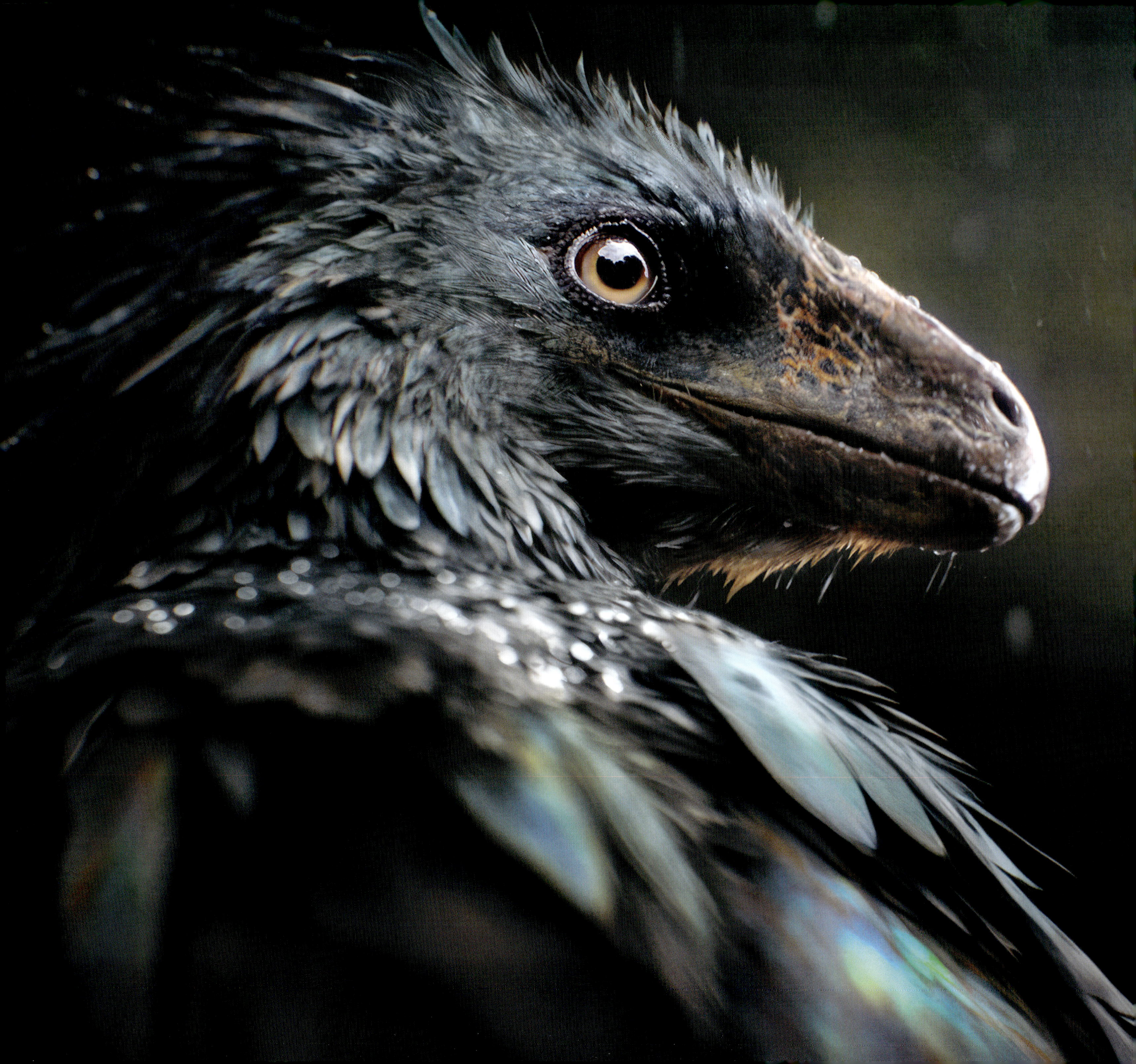

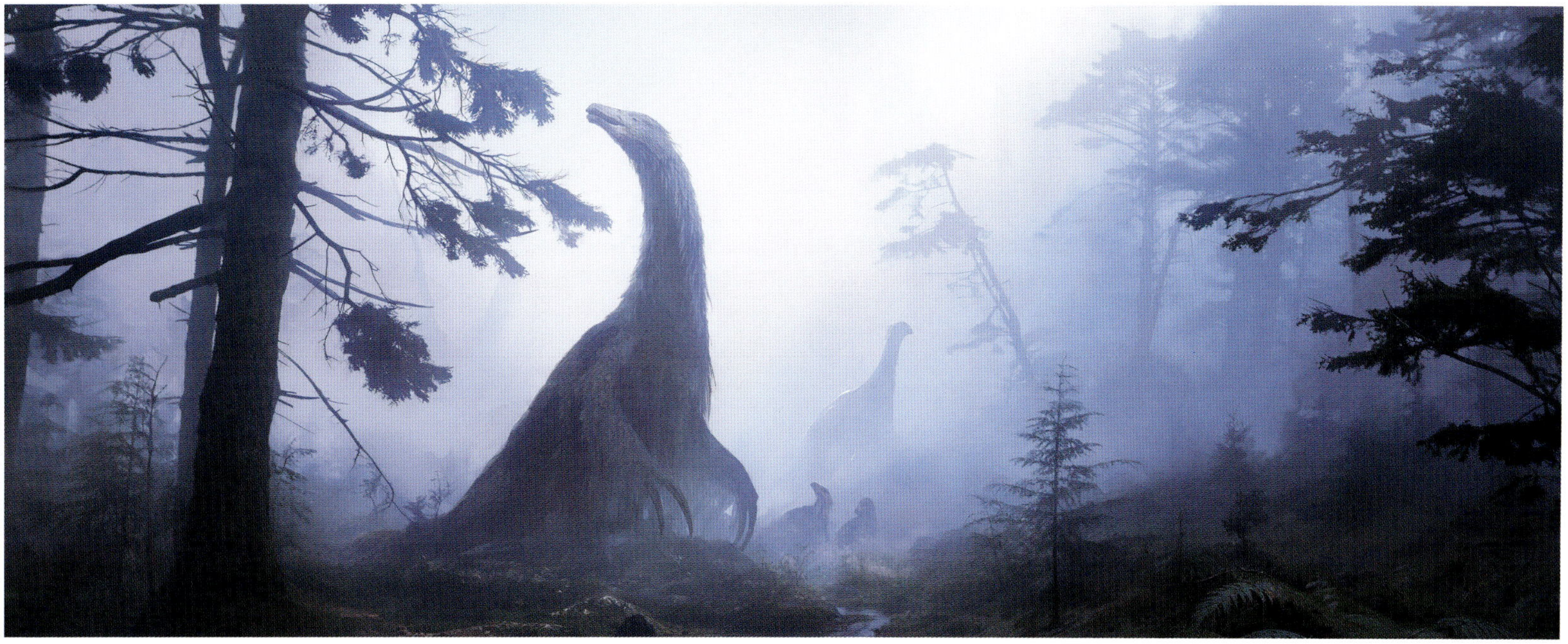

◄ ***BALAUR* PORTRAIT**

Balaur is a Late Cretaceous dinosaur from Romania, initially described as a close relative of *Velociraptor*. However, several details suggest that *Balaur* might be a large, flightless, archaic member of the bird lineage. Uncertainty surrounds the facial covering of archaic birds and other maniraptorans: it seems that they had flexible, naked skin and feathers in places where modern birds have horny beak tissue.

• Digital | 2022

▲ ***THERIZINOSAURUS***

Known from fragmentary remains found in 1948, *Therizinosaurus*, from Late Cretaceous Mongolia, gives its name to one of the weirder dinosaur families. Most are large to gigantic herbivores, long necked with barrel-shaped bodies, short hindlegs but massive forearms equipped with three very long claws. They may also have been feathered. So weird.

• Digital | 2024

▼ **WATERFALL BATH**

Feathered dinosaurs of the Mesozoic – like this *Balaur* – must have cared for their plumage in the same way that living birds do, preening, dustbathing and using pools and even running water to wash. An unusual feature of *Balaur*, visible here, is the presence of a large claw on its hallux (its innermost toe), meaning that it had *two* notably large, curved claws on each foot.

• Digital | 2022

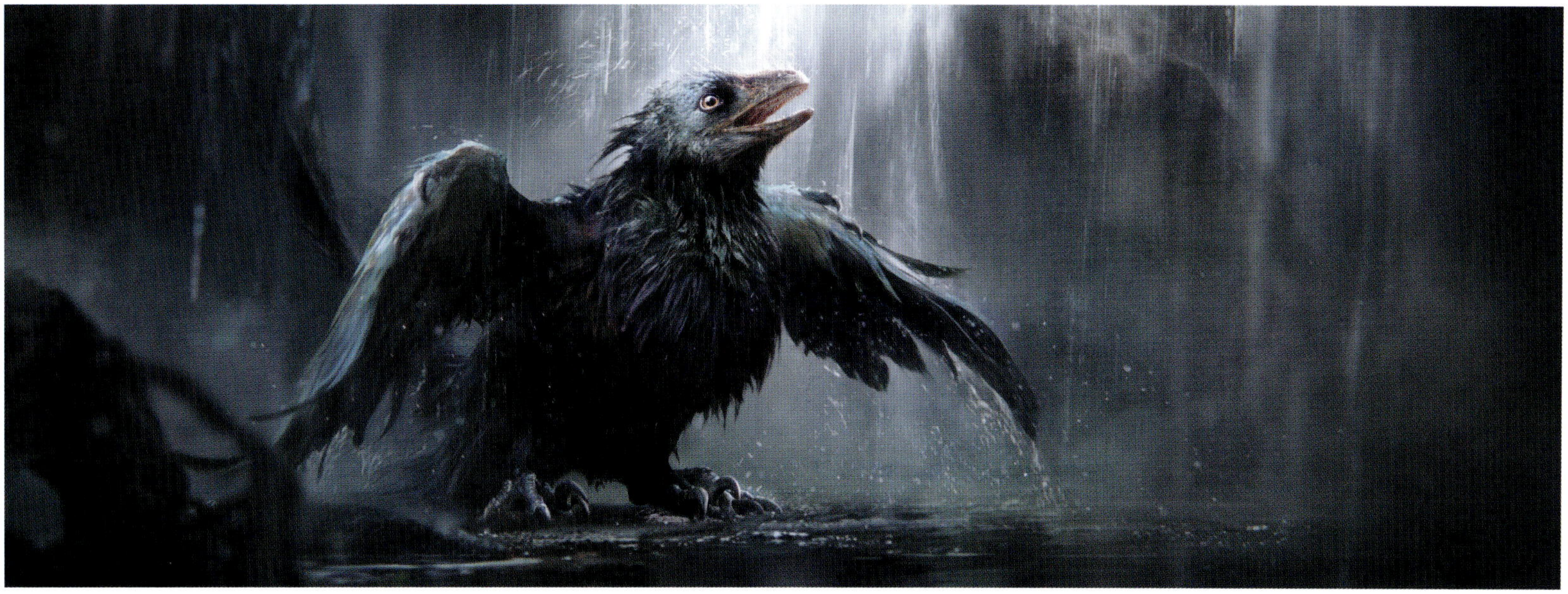

NANUQSAURUS ▸

The great landmass that includes the far north of eastern Asia, western North America and the land in between – known collectively as Beringia – was home to many dinosaurs during the final days of the Cretaceous. In 2014, Beringian tyrannosaurid remains from Alaska were recognised as a distinct form, *Nanuqsaurus* (meaning 'polar bear lizard'). As an animal of cool places, *Nanuqsaurus* might have been insulated with a shaggy coat, a feature known for other cool-climate members of the tyrannosaur family.

• Digital | 2019

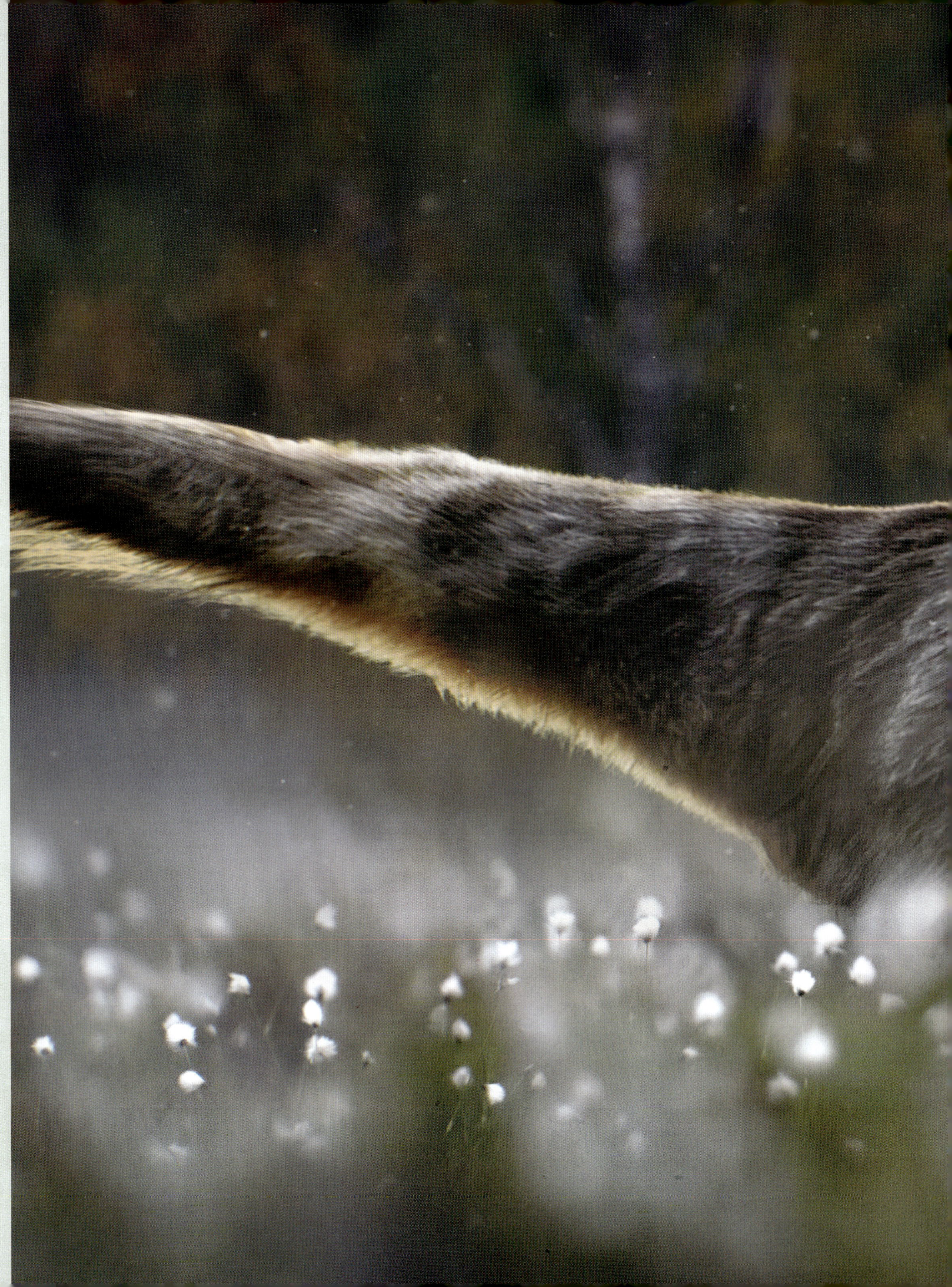

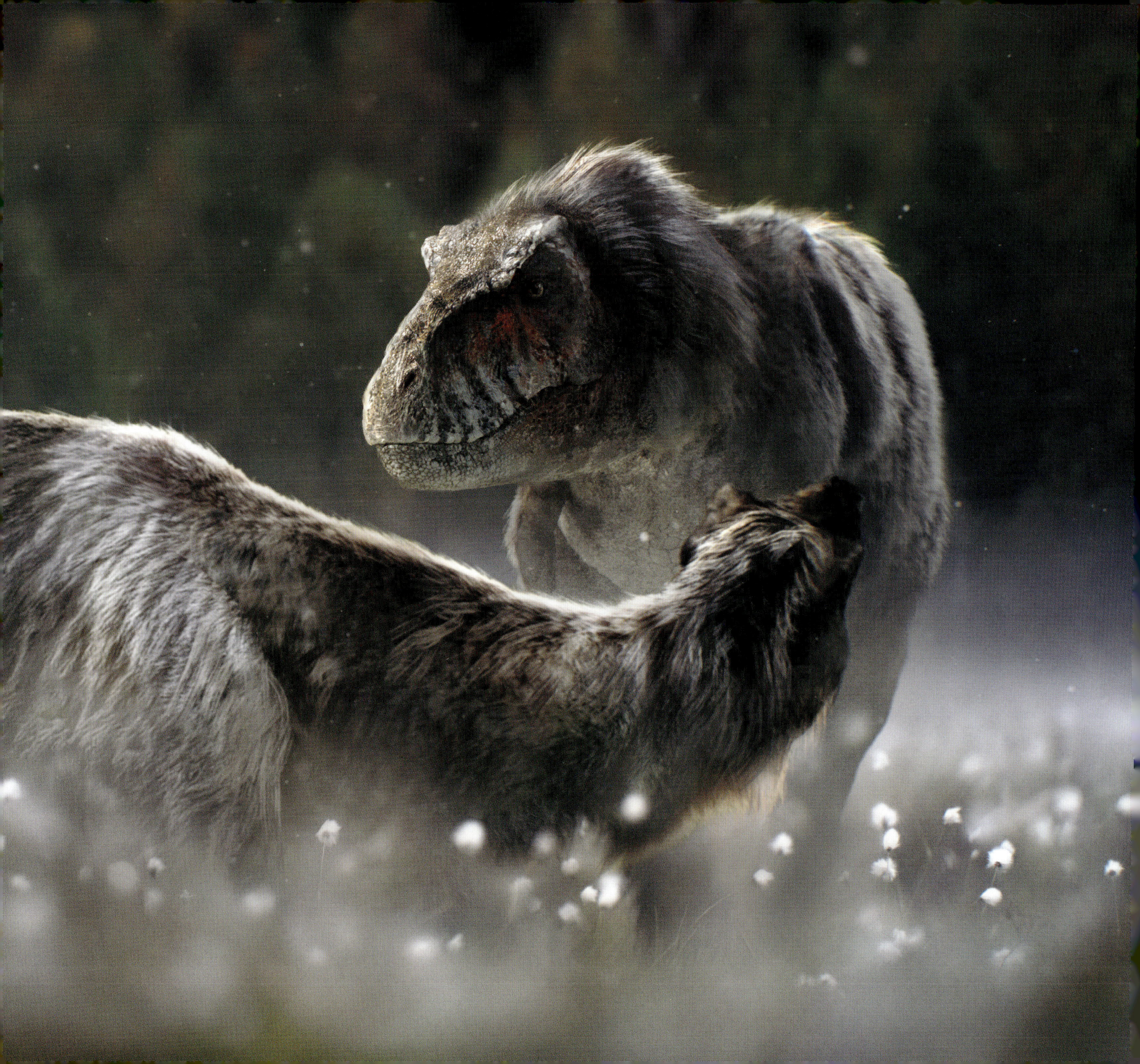

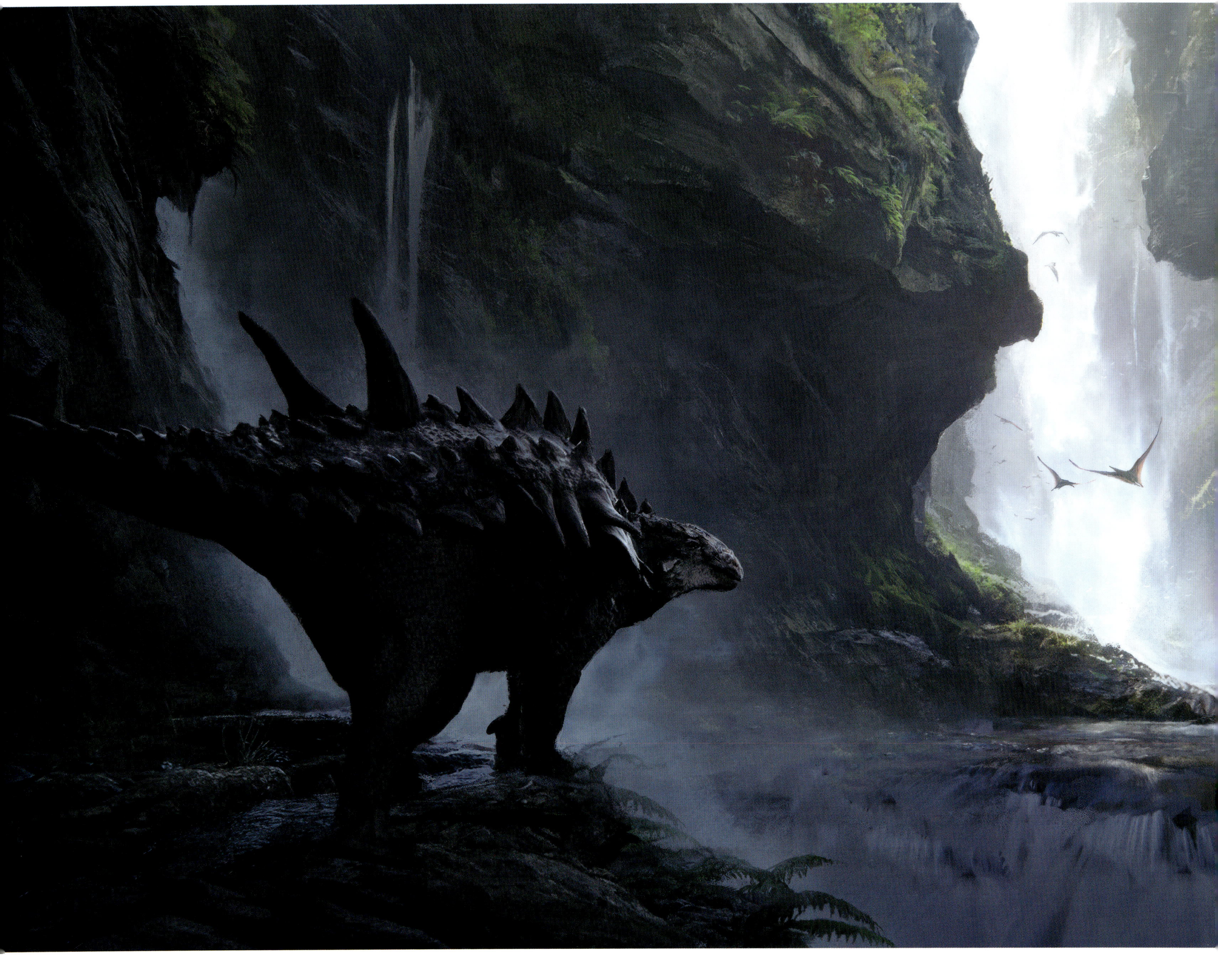

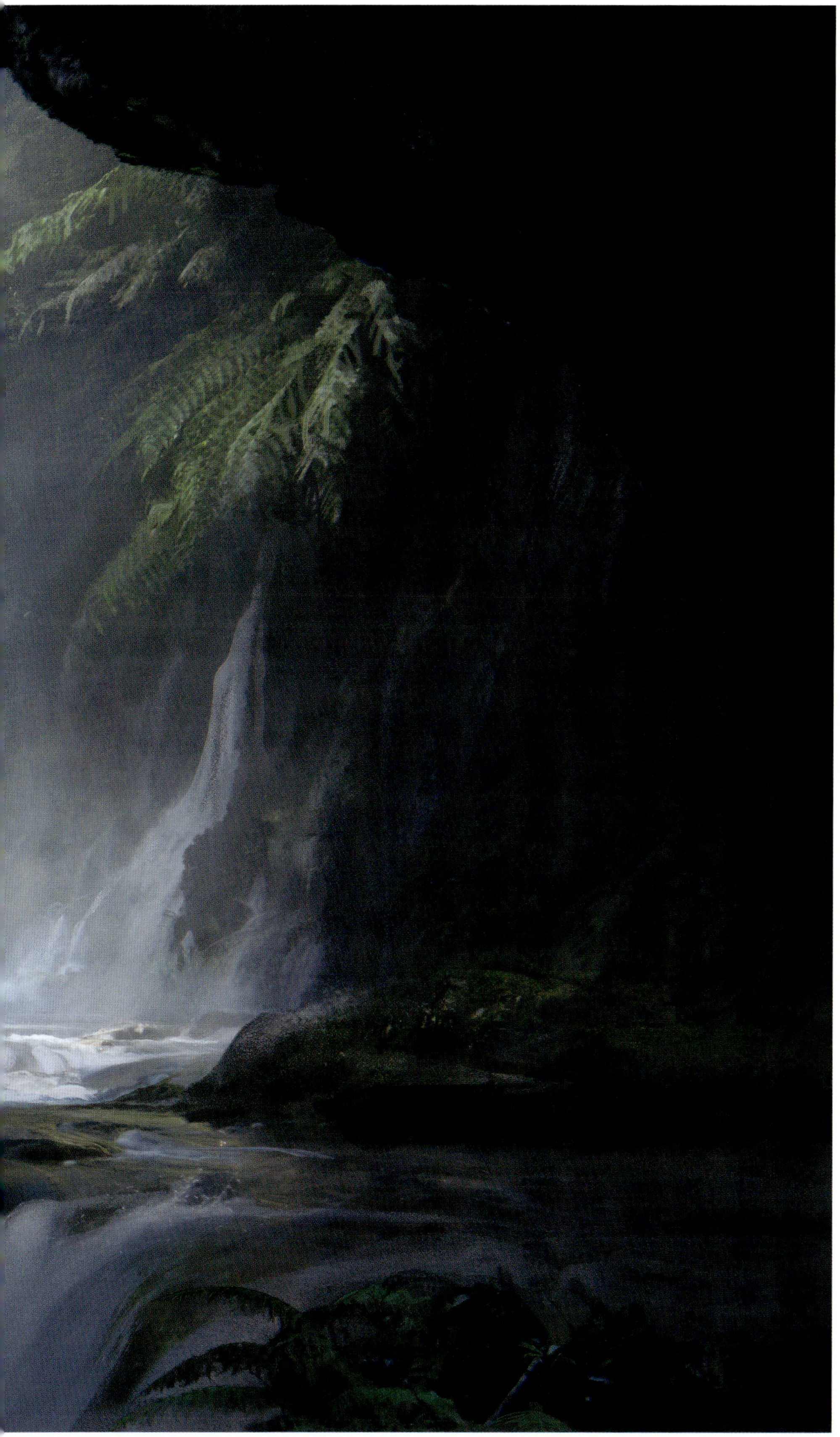

◂ *STRUTHIOSAURUS*

Struthiosaurus is a small Late Cretaceous ankylosaur, species of which are known from France, Austria, Hungary and Romania. Ankylosaurs have generally been portrayed as animals of forests and plains, but like large living mammalian herbivores of today, they may have explored or even inhabited uplands, gulleys and cliffs. Their short, muscular limbs and low centre of gravity might even have made them unusually sure-footed.
• Digital | 2024

▾ A SPRING NIGHT DURING THE K/PG EVENT

Uncertainty remains over the exact look of the few seconds when a giant rock from space raced through our atmosphere and crashed into our planet 66 million years ago, but here's one possibility. This scene makes deliberate reference to research showing that the impact happened during the northern spring.
• Digital | 2023

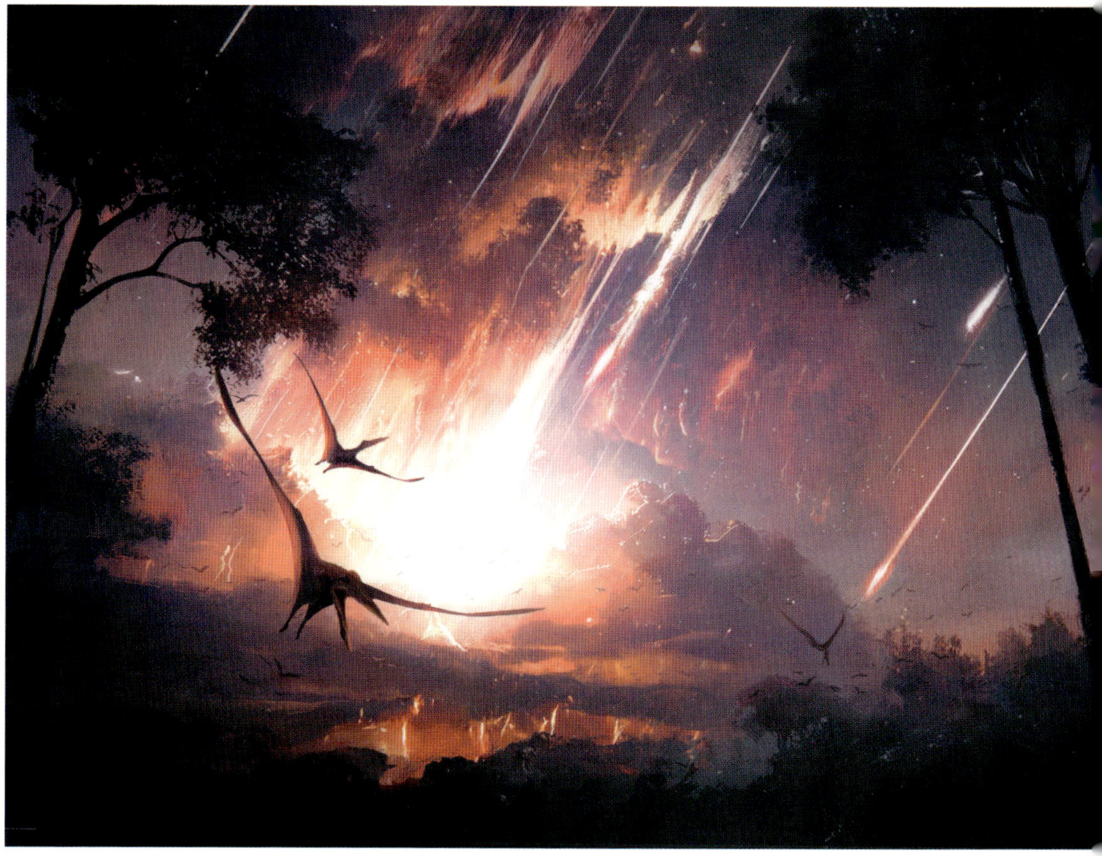

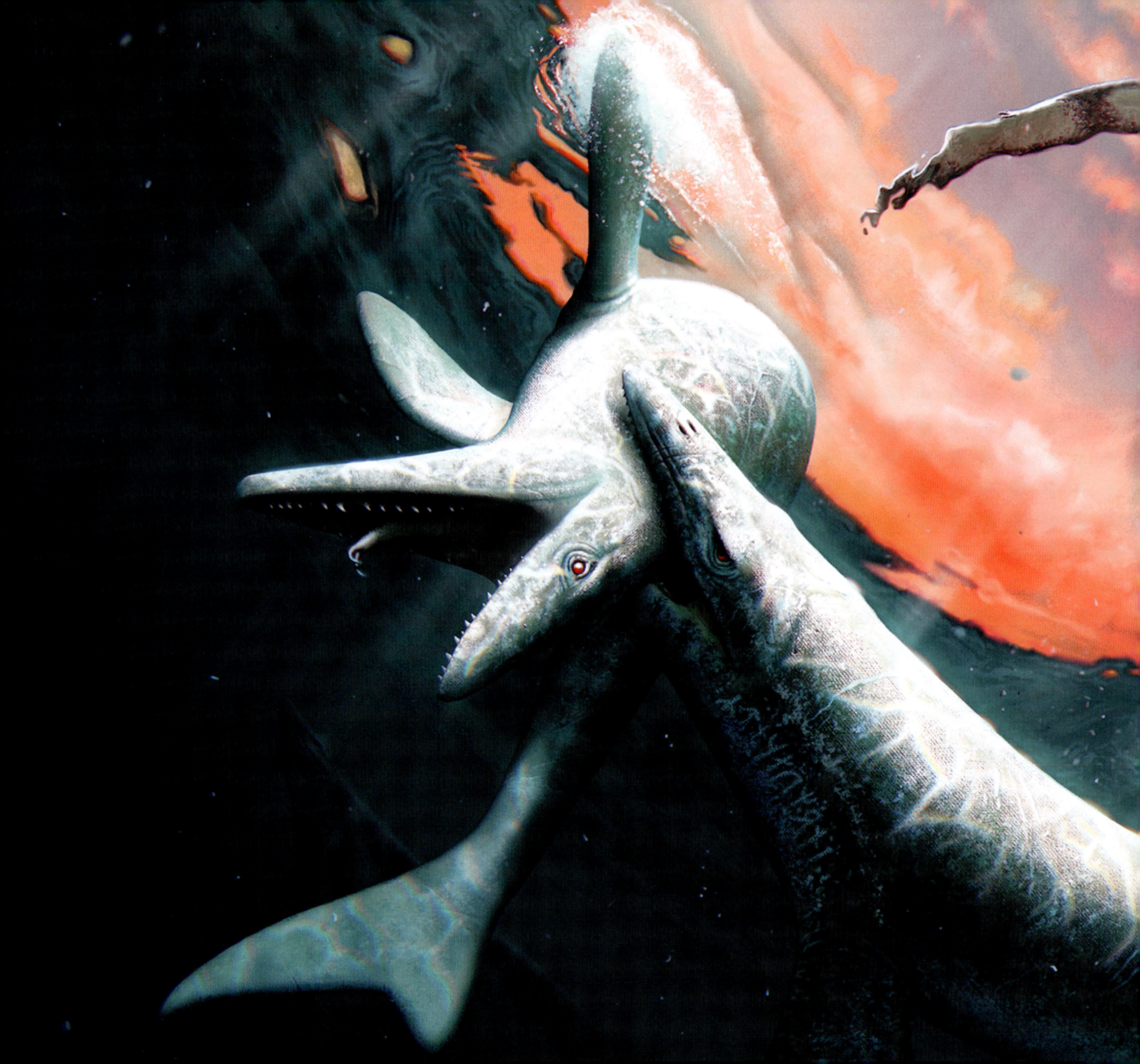

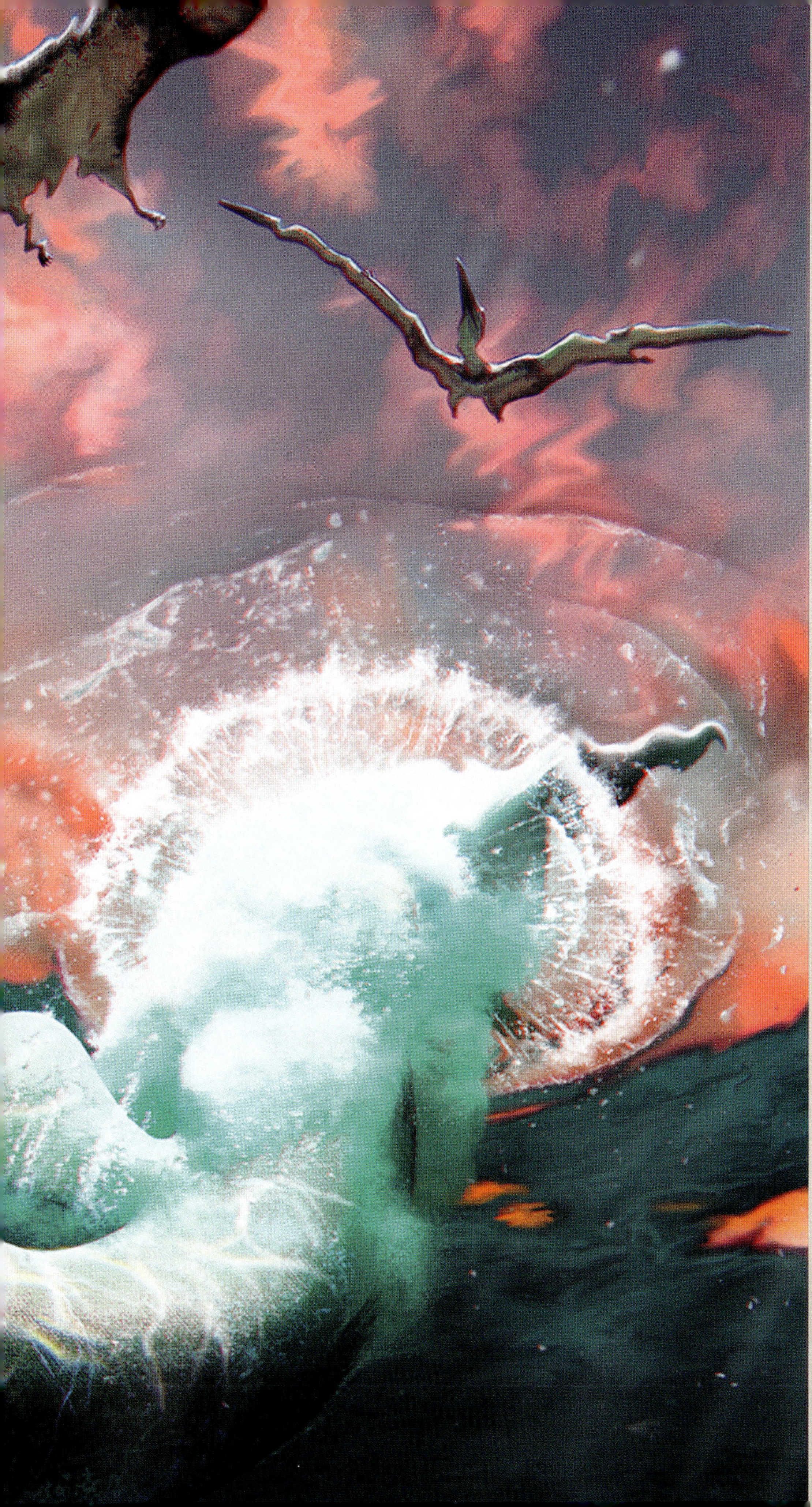

HENRY SHARPE

HENRY SHARPE is a palaeontology student and palaeoartist based in Edmonton, Canada. After *Jurassic Park* hooked him at age four, he began drawing and painting dinosaurs at every chance he could get. He spent the next two decades wiring his brain to produce crazy amounts of dopamine from making prehistoric creatures that look real. In addition to illustrating numerous scientific publications, Henry's art is now on display at the Royal Ontario Museum and Calgary Zoo and has been featured twice in the *New York Times*. In 2023, he was awarded the Society of Vertebrate Paleontology's Lanzendorf-National Geographic PaleoArt Prize for Scientific Illustration.

Henry's scientific work focuses on mosasaurs, hadrosaurs, soft tissue reconstructions, and macroevolutionary patterns. He finished his BSc at the University of Alberta, where he is now completing his MSc, studying European mosasaur anatomy and evolution. He spends his summers excavating dinosaurs for the University of Alberta in the badlands and river valleys of Alberta with his talented lab mates.

In his spare time, Henry plays the drums, chases the northern lights with his friends, and bakes.

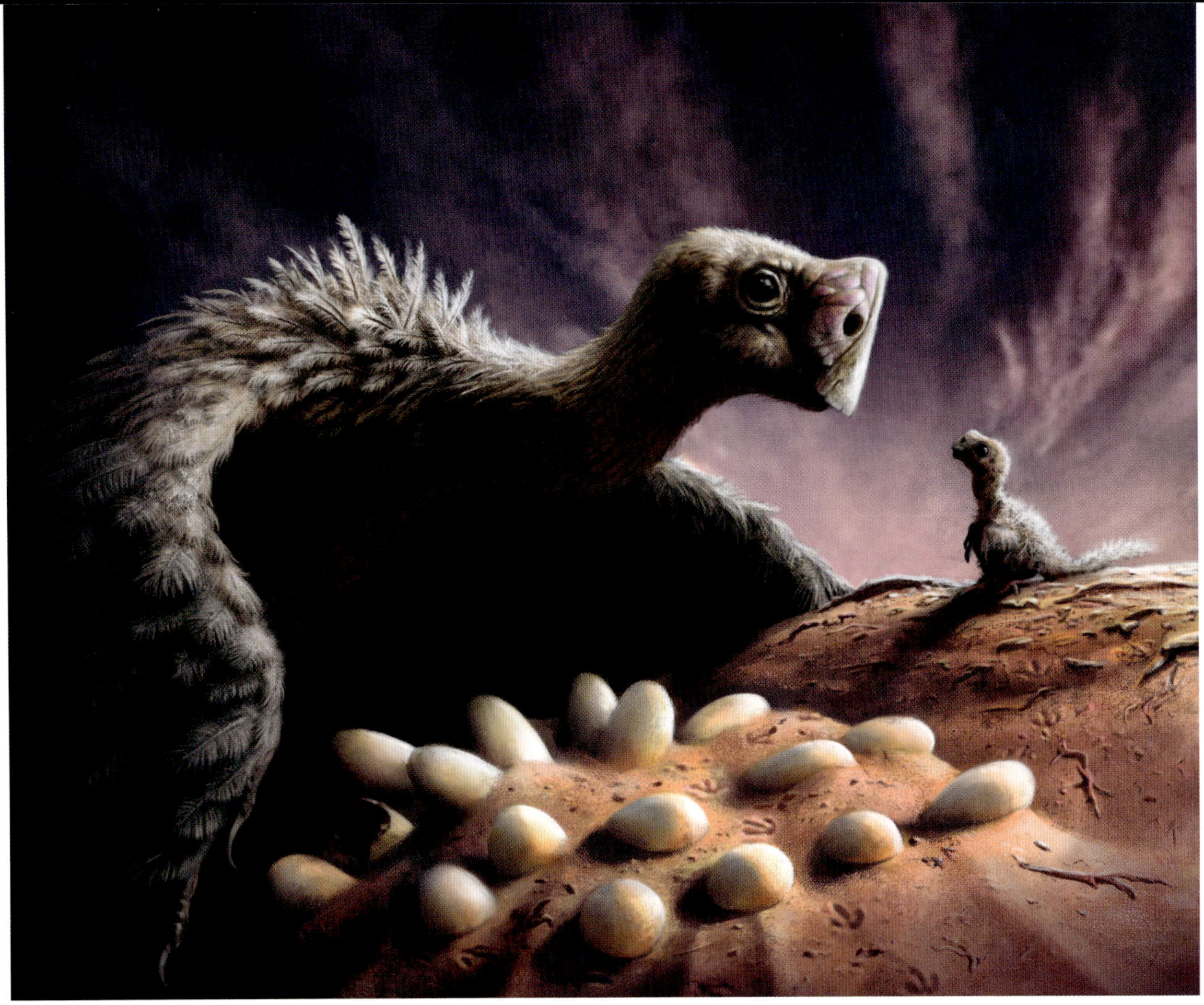

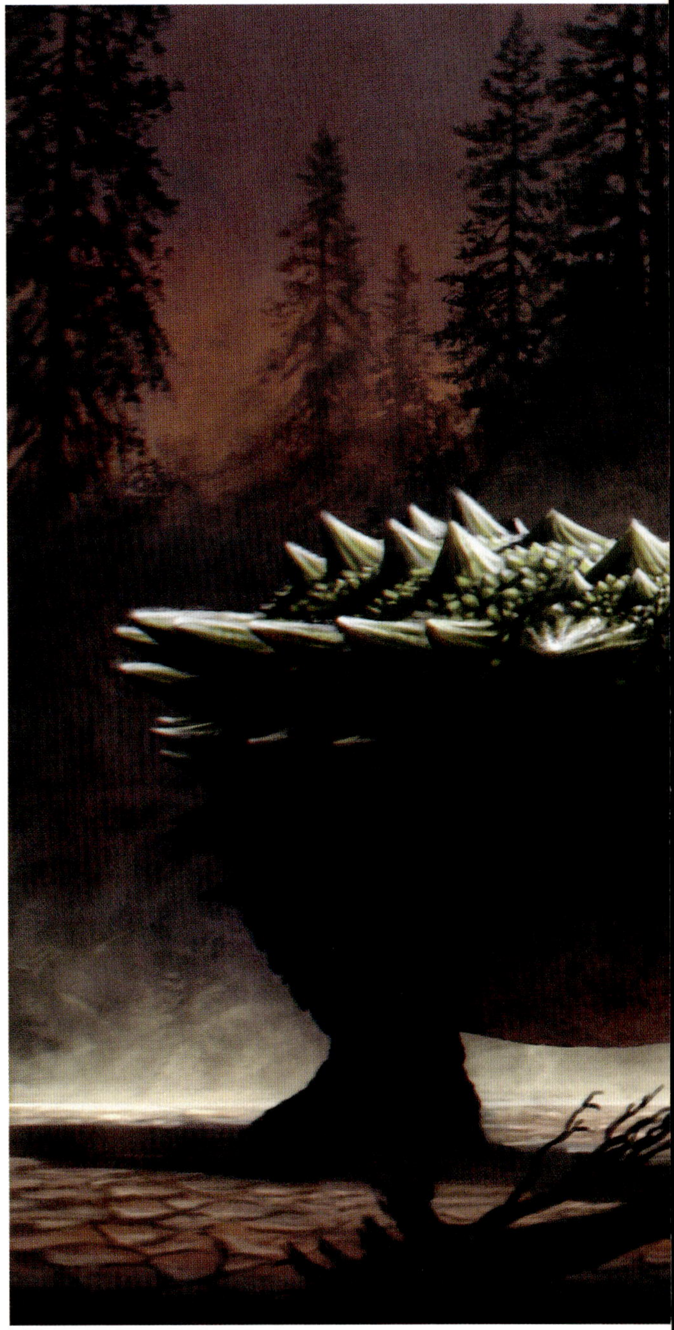

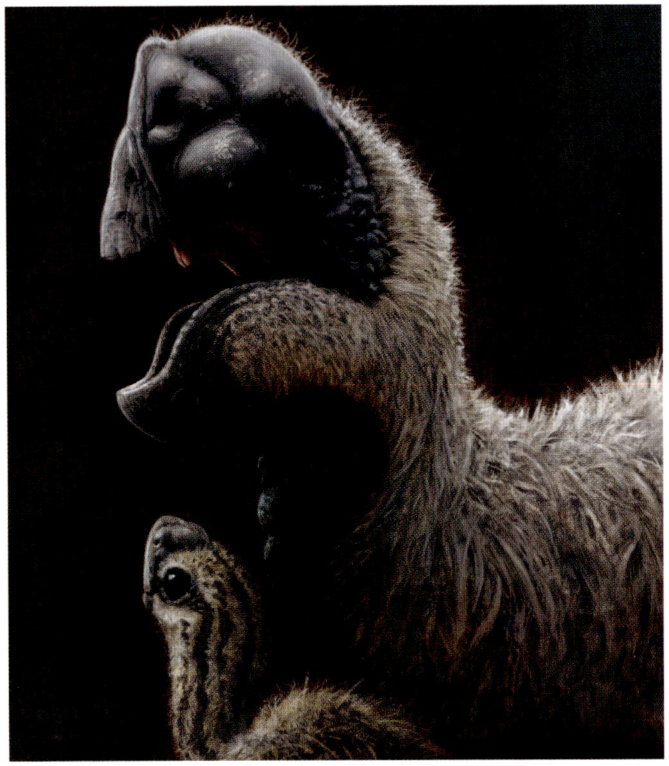

▲ *OVIRAPTOR*

Since its 1924 discovery, *Oviraptor* has been given several different looks. This is because it was long confused with other oviraptorosaurs now given different names and placed elsewhere on the family tree. *Oviraptor* in its true guise had a low crest over the snout and was small, at around 2m (6.6ft) in length. Preserved nests show that oviraptorosaurs laid their eggs in circles.
• Digital | 2022

[Previous pages]

JORMUNGANDR

Named after the mythical Norse serpent that encircled the world's oceans, *Jormungandr* is a Late Cretaceous mosasaur from North Dakota, USA, named in 2023. It's part of the specific mosasaur group that also includes the eponymous *Mosasaurus*. Unhealed bite marks on a bone from the spine indicate that the only known specimen received a fatal bite from another mosasaur. At around 7m (23ft) long, *Jormungandr* was large, but only about half the size of the largest mosasaurs known.
• Digital | 2023

◄ *CITIPATI*

Several oviraptorosaurs – this is *Citipati* from the Late Cretaceous of Mongolia – possess a bony crest over the snout and forehead. Partly due to its similarity to the casques of cassowaries, this has mostly been portrayed by artists as horn-covered and continuous with the beak. However, other coverings are possible, and some experts think it more likely that skin was the predominant external structure.
• Digital | 2021

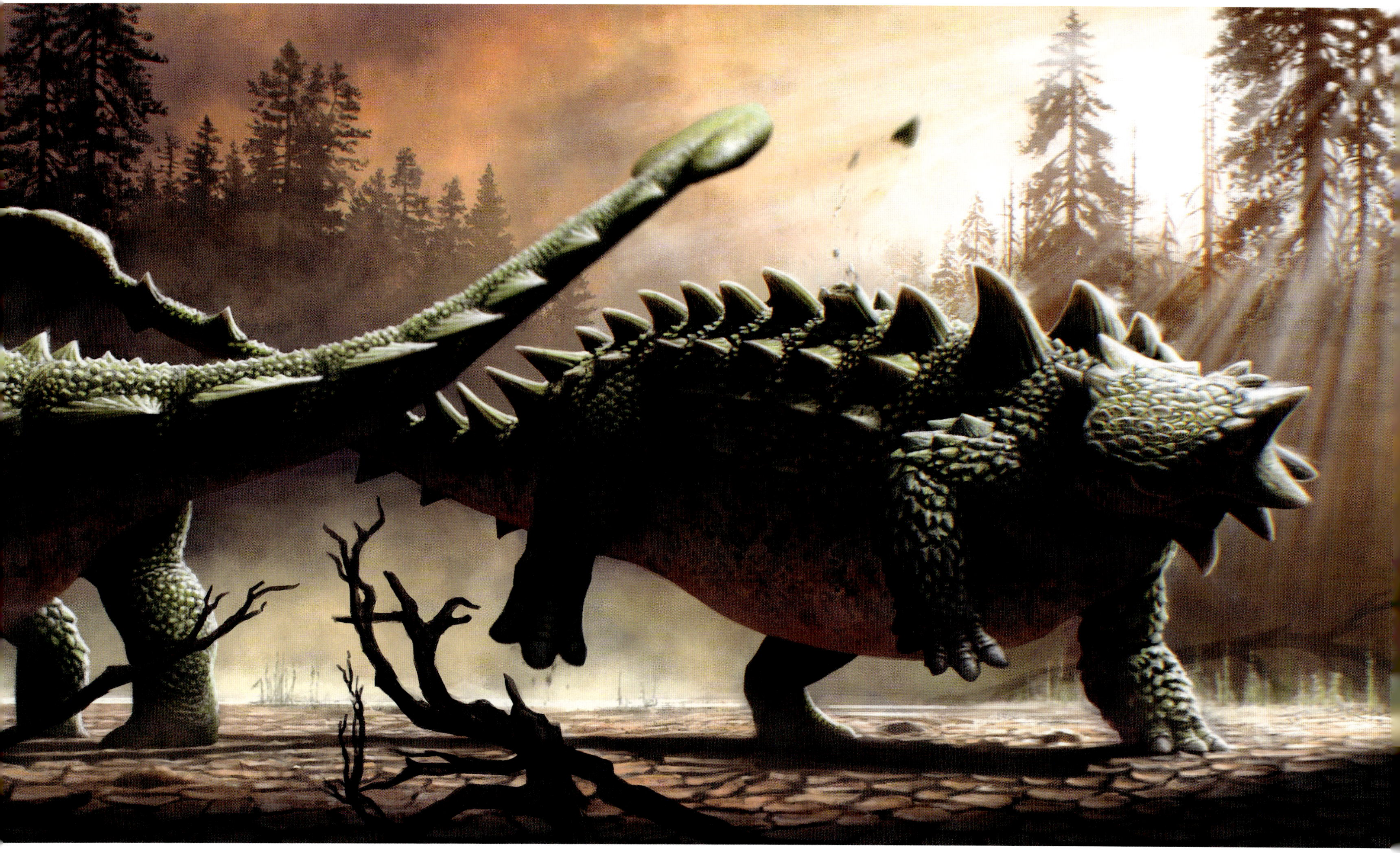

▲ *ZUUL*

Named from Canada in 2017 and noted for the prominent spikes on the sides of its tail, *Zuul* is one of several ankylosaurid specimens whose anatomy suggests that combat with rivals – not predators – was the main evolutionary driver for the armour plates, spikes and tail clubs in these animals. The low-angle view emphasises the low-slung, short-legged form of these dinosaurs.

• Digital | 2022

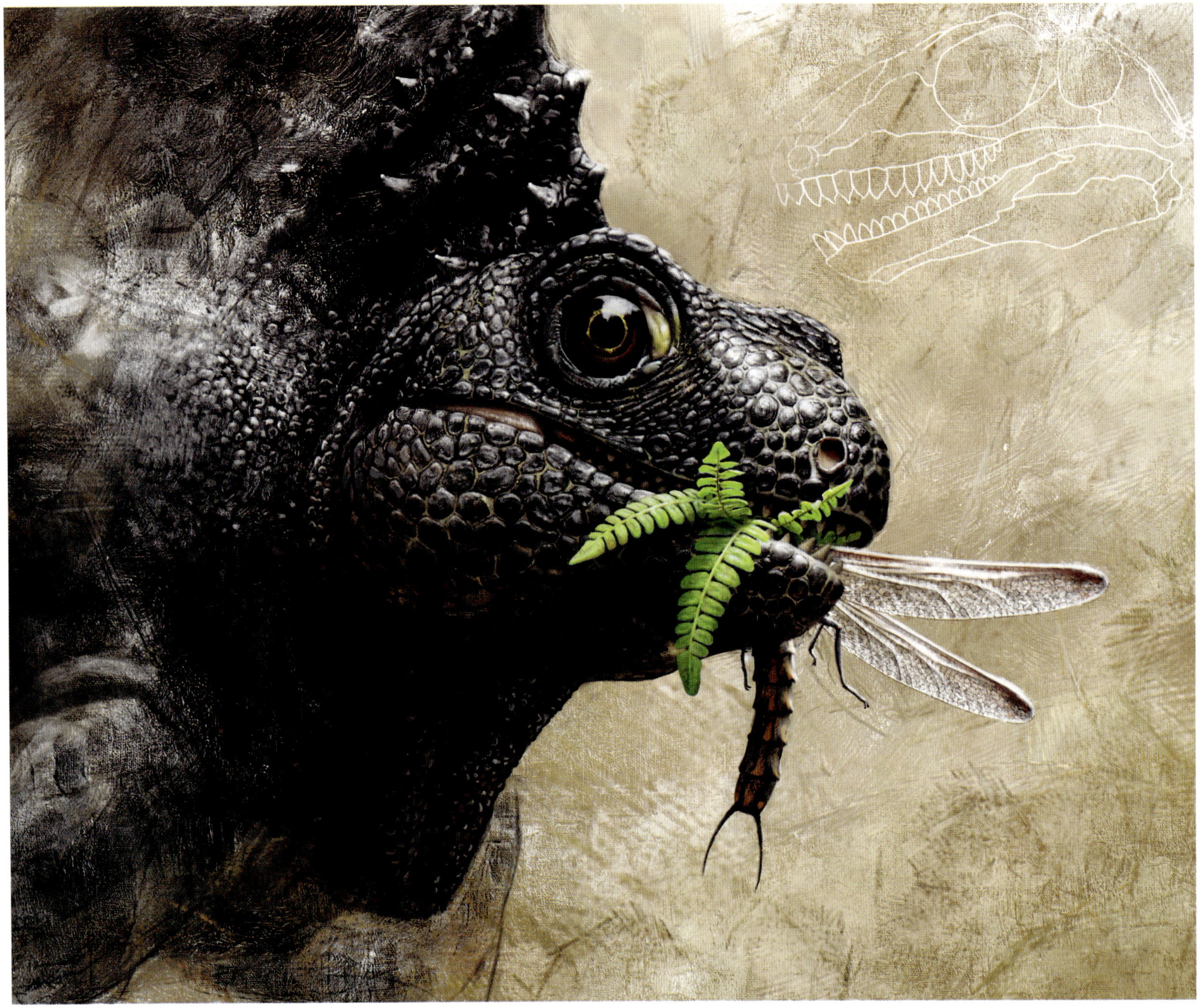

▲ *MELANEDAPHODON*

Melanedaphodon, named in 2023, is a North American edaphosaurid, a group of archaic mammal relatives that lived during the Carboniferous, about 310 million years ago. They were sail-backed, like their distant, predatory cousin *Dimetrodon*. The best known edaphosaurids were bulky-bodied herbivores, but early members of the group were likely omnivorous, an idea hinted at here.

• Digital | 2023

BOLOSAURUS ▶

Bolosaurids, like *Bolosaurus* major shown here, are lizard-like reptiles that lived across the Northern Hemisphere during the Carboniferous and Permian, between 300 and 260 million years ago. They are short-snouted, with robust jaws and broad, triangular teeth. Studies indicate a powerful bite. Notably, the group includes the oldest of all bipedal reptiles.

• Digital | 2024

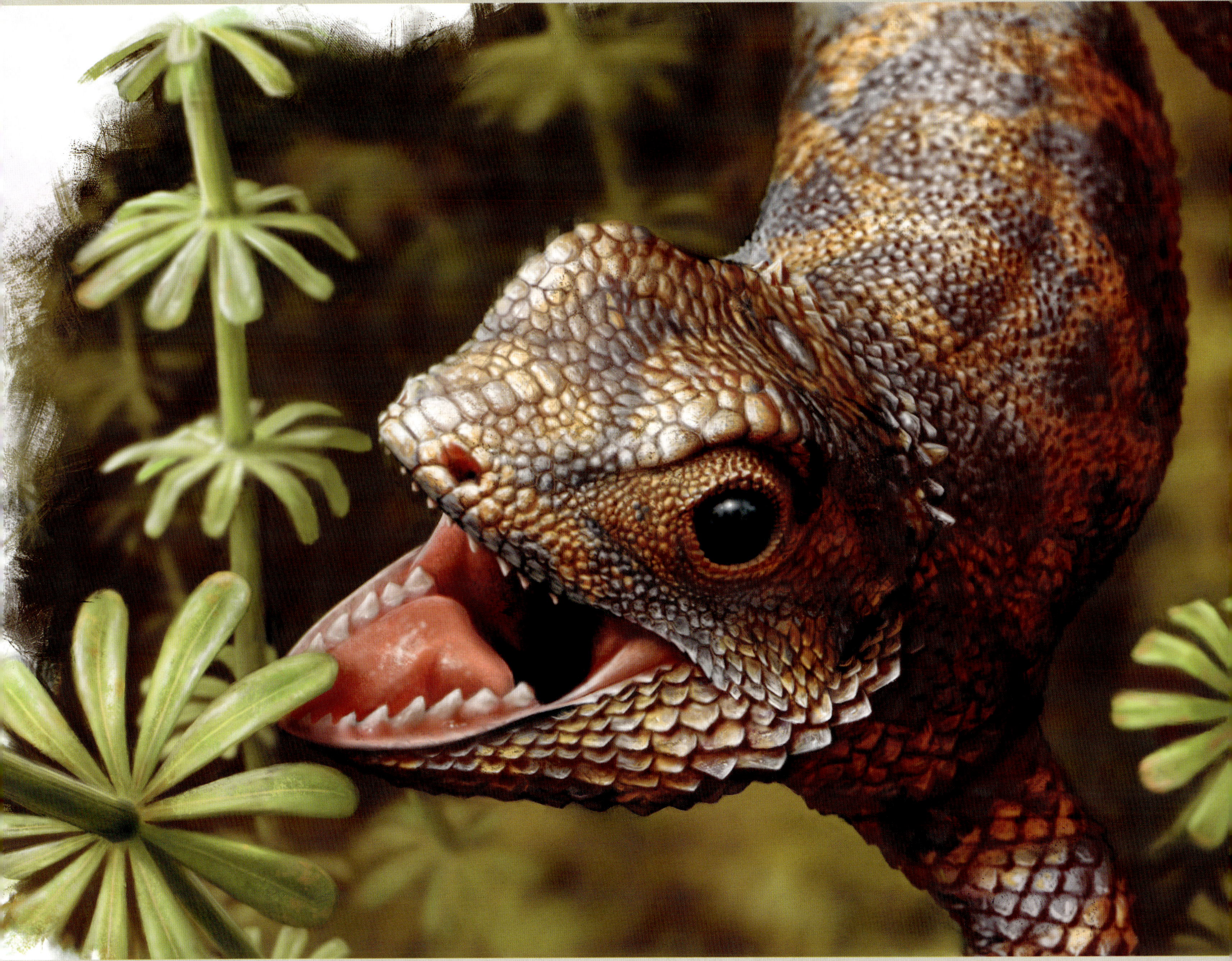

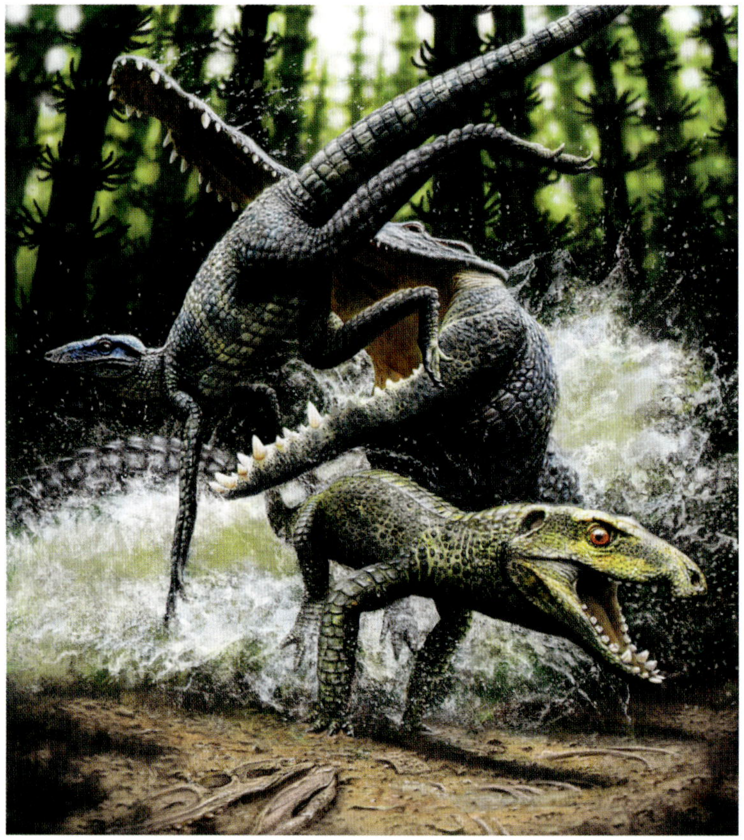

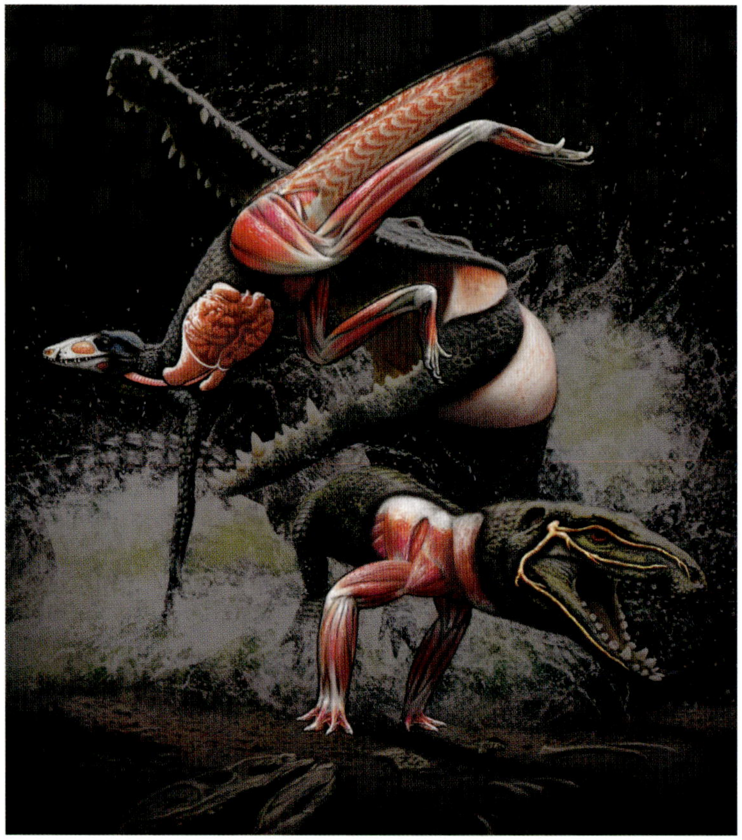

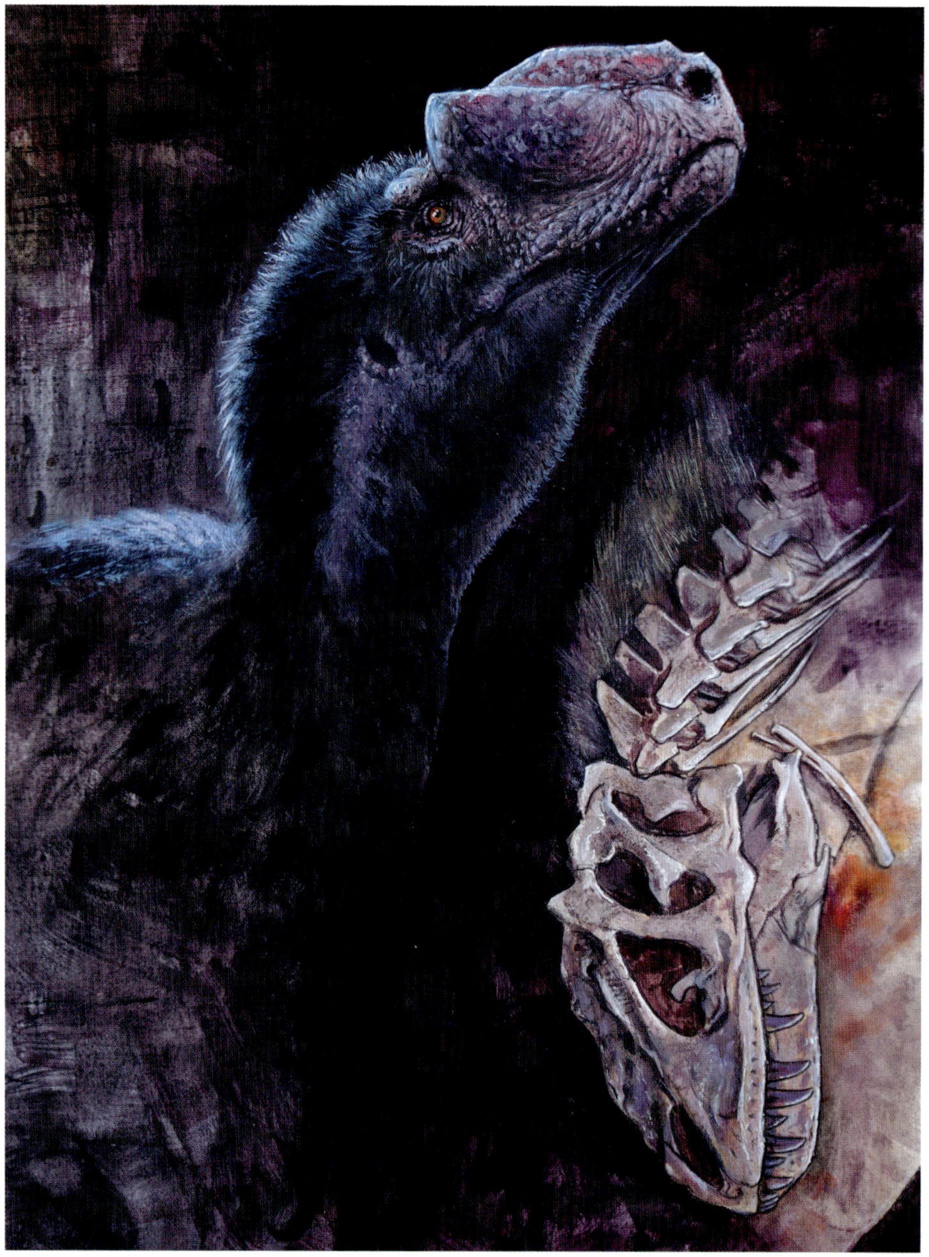

◂ KAYENTA CROCS

Crocodiles and kin – Crocodylomorpha – include leggy, terrestrial predators (like *Kayentasuchus*), long-jawed, amphibious predators (like *Calsoyasuchus*) and omnivores and herbivores with unusual teeth (like *Edentosuchus*). These illustrations showcase the crocodylomorph diversity of the Lower Jurassic Kayenta Formation of the Colorado Plateau, USA and highlight features of the musculoskeletal, respiratory and sensory system that make crocodylomorphs so special.

• Digital | 2022

▴ *YUTYRANNUS*

The three known specimens of *Yutyrannus*, from China's famous Liaoning fossil beds, seem to represent different growth stages of the animal. They indicate that the legs and arms shortened as it grew, while the head became deeper and more robust. These changes suggest that the prey of *Yutyrannus* changed as it aged – a feature seen in other large predatory theropods.

• Acrylic on illustration board | 2021

UNCINATE PROCESSES ▸

Bony prongs on the ribs of birds – the uncinate processes – form the attachment sites for muscles used in powering movement of the ribcage. These structures have long been known for some extinct bird-like dinosaurs, but a 2023 study argued that they were widespread across dinosaurs as a whole, and in other archosaurs too. This illustration – showing the location and muscular attachments of uncinate processes in a tyrannosaur – was produced to accompany this research.

• Digital | 2023

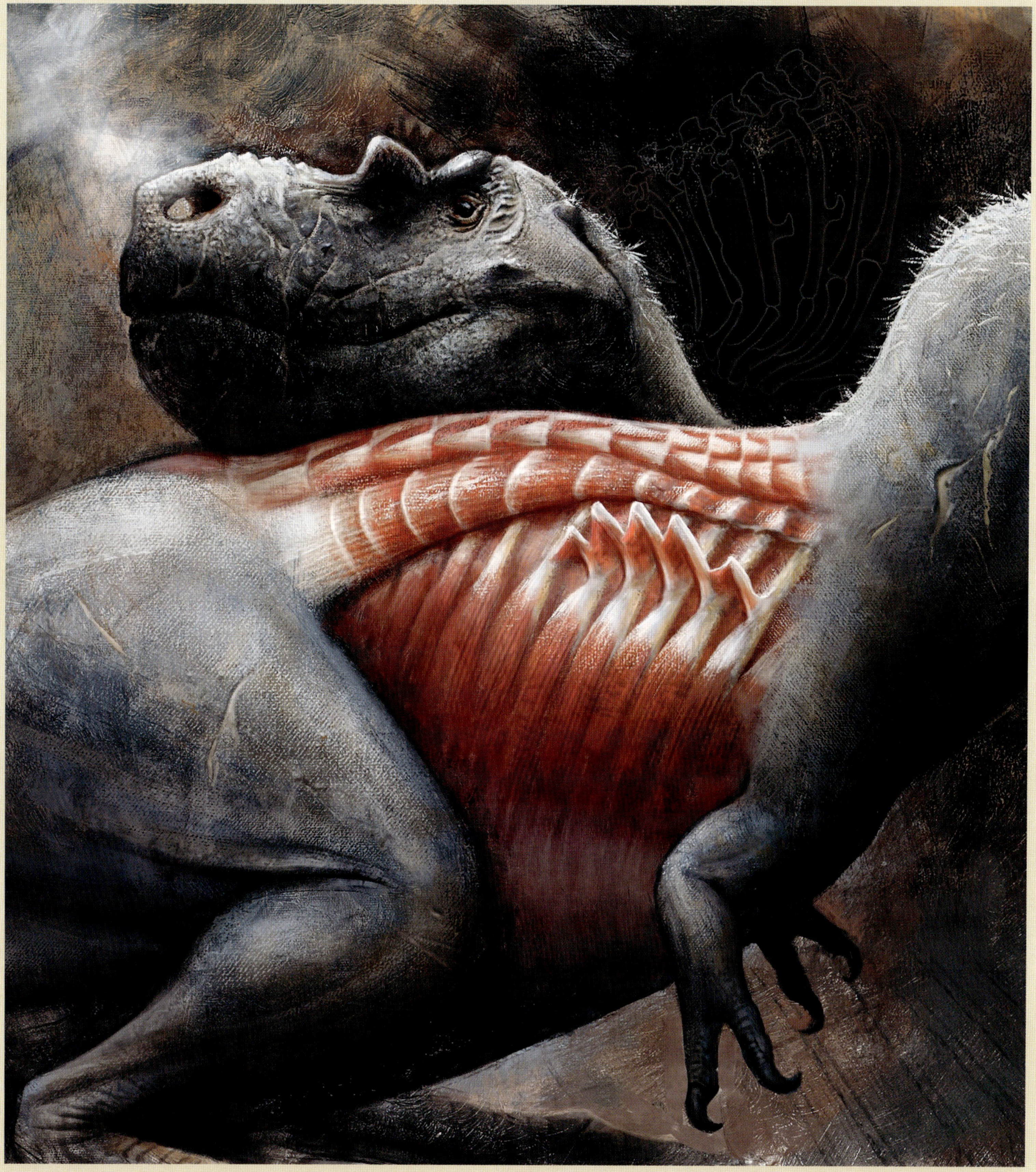

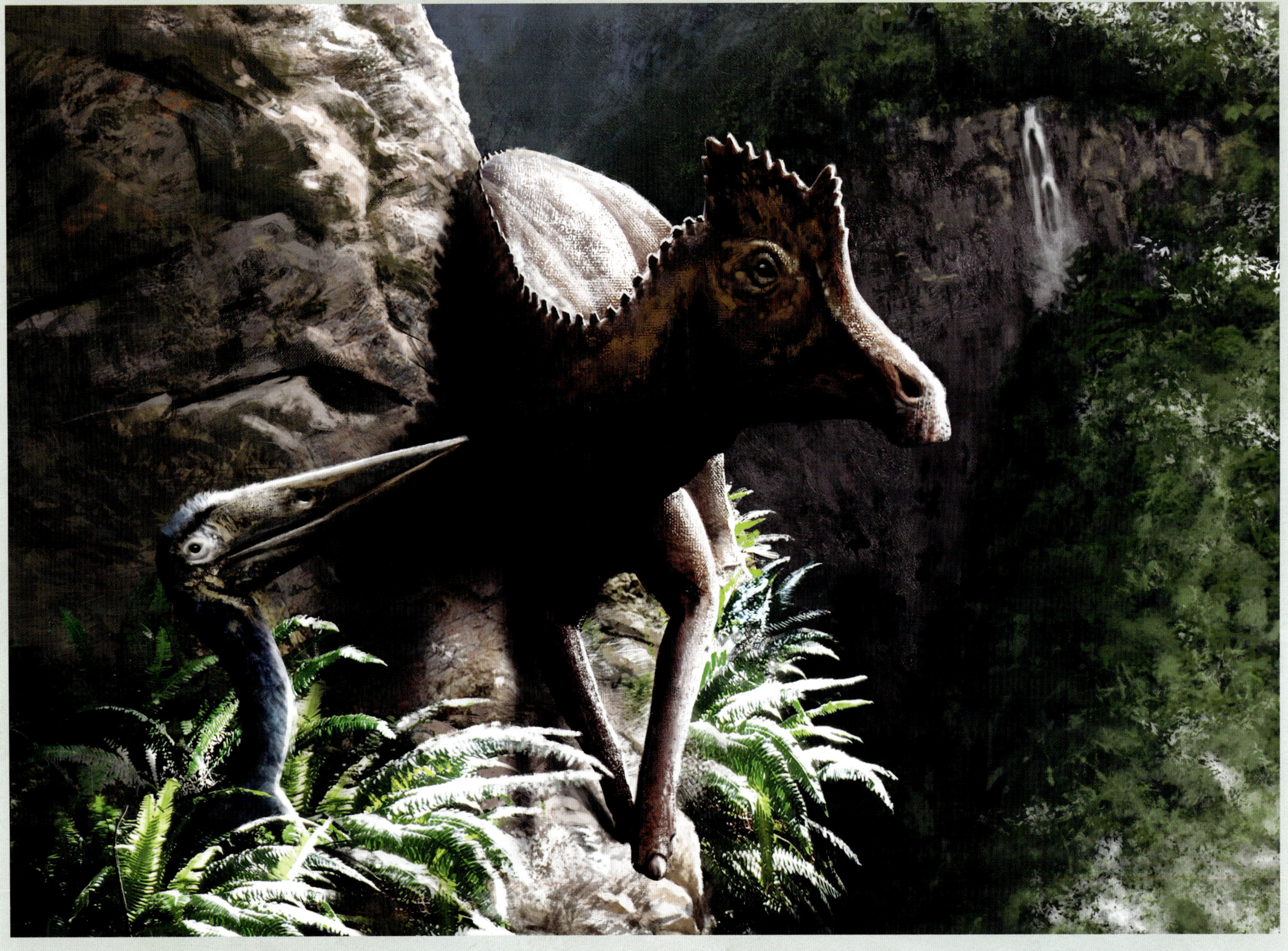

▲ SPECULATIVE MOUNTAIN HADROSAUR

On a narrow mountain pass, a mountain hadrosaur encounters a nesting azhdarchid pterosaur. A familiar complaint made by those interested in the animals of the past is that numerous species that once existed are unknown as fossils, and indeed might never be known. Might there once have existed small, slim-limbed hadrosaurs specialised for life on cliffs and rockfaces? We may never know, given the rarity of animals from mountainous environments, but this is plausible based on what we know.

• Digital | 2023

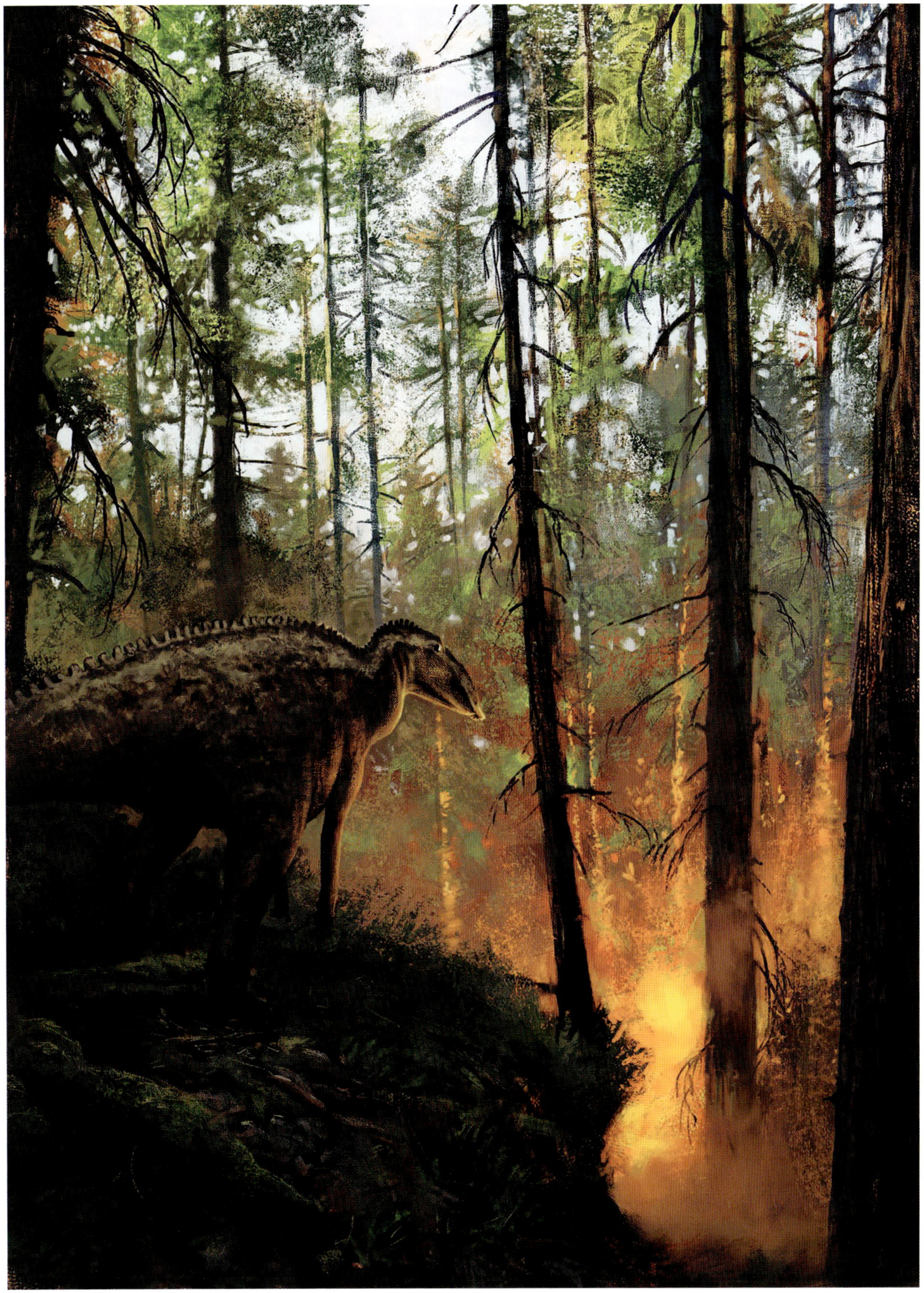

◀ *TETHYSHADROS* & FIRE

Tethyshadros is a Late Cretaceous Italian hadrosaur named in 2009, originally implied to be an island-dwelling dwarf. More recent finds show that its relatively small size (4.5m or 15ft) is probably due to its position on the family tree: all its relatives were small too. Raging forest fires – perhaps sparked by lightning – surely happened during the Cretaceous, but more common were 'slow burn' understorey fires that might have lasted months or even years.
• Digital | 2023

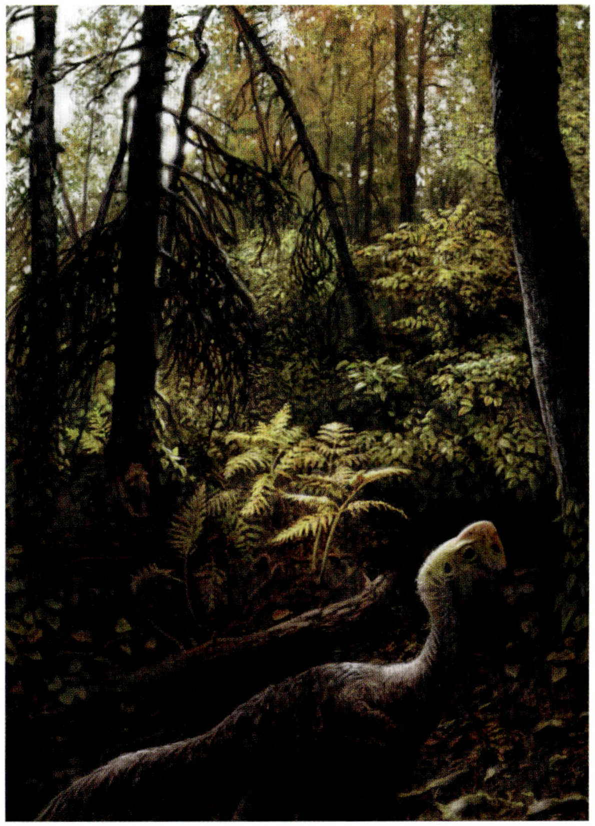

▲ CRETACEOUS CRYPSIS

Palaeoart has a long tradition of showing ancient animals as being part of their environment. While many animals – big, showy creatures like dinosaurs being high on the list – surely often stood out relative to their surroundings, others would have been cryptic – experts at disappearing. Here, in a Late Cretaceous forest, a small ceratopsian and an oviraptorosaur displays the latter ability.
• Digital | 2022

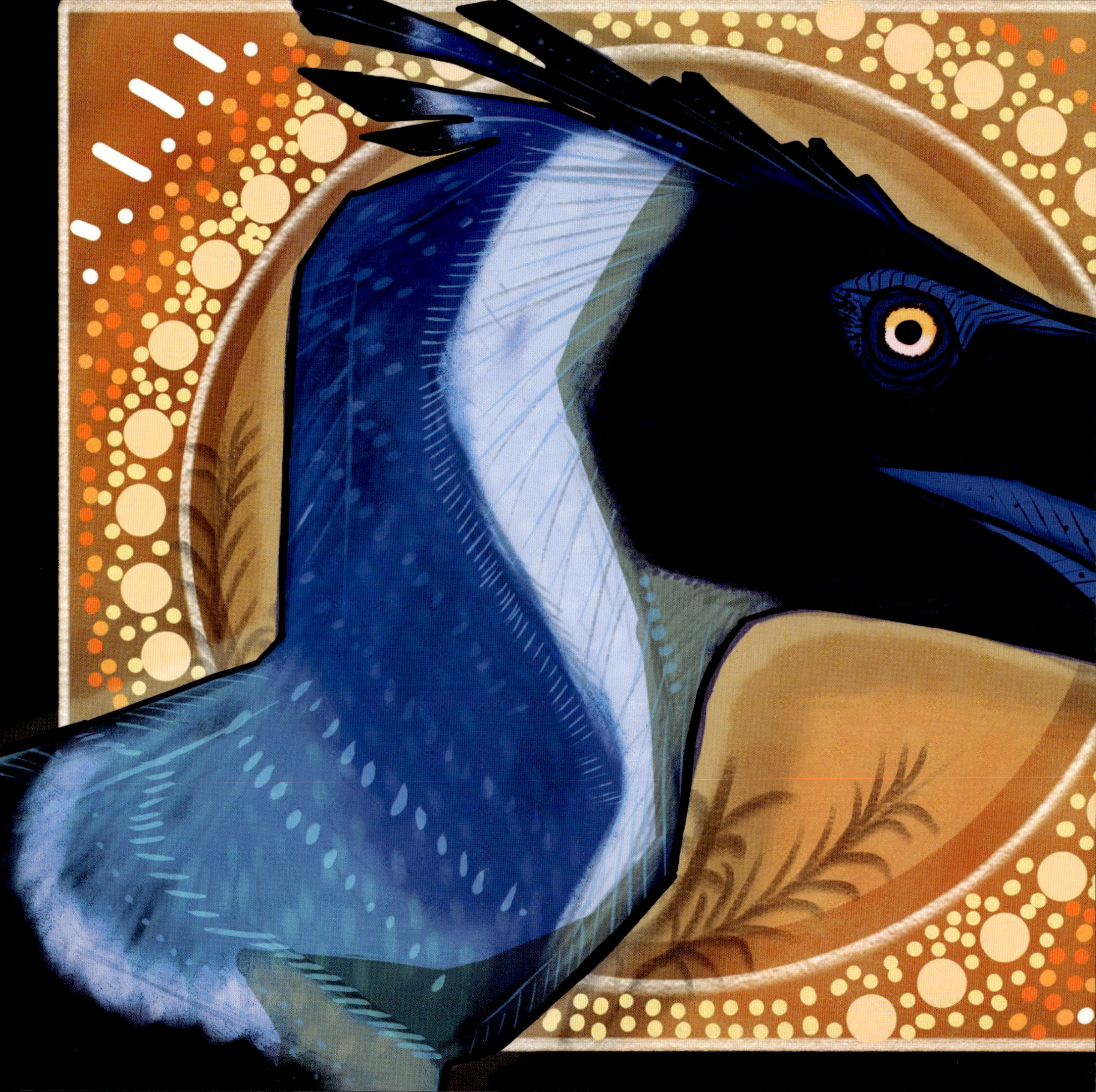

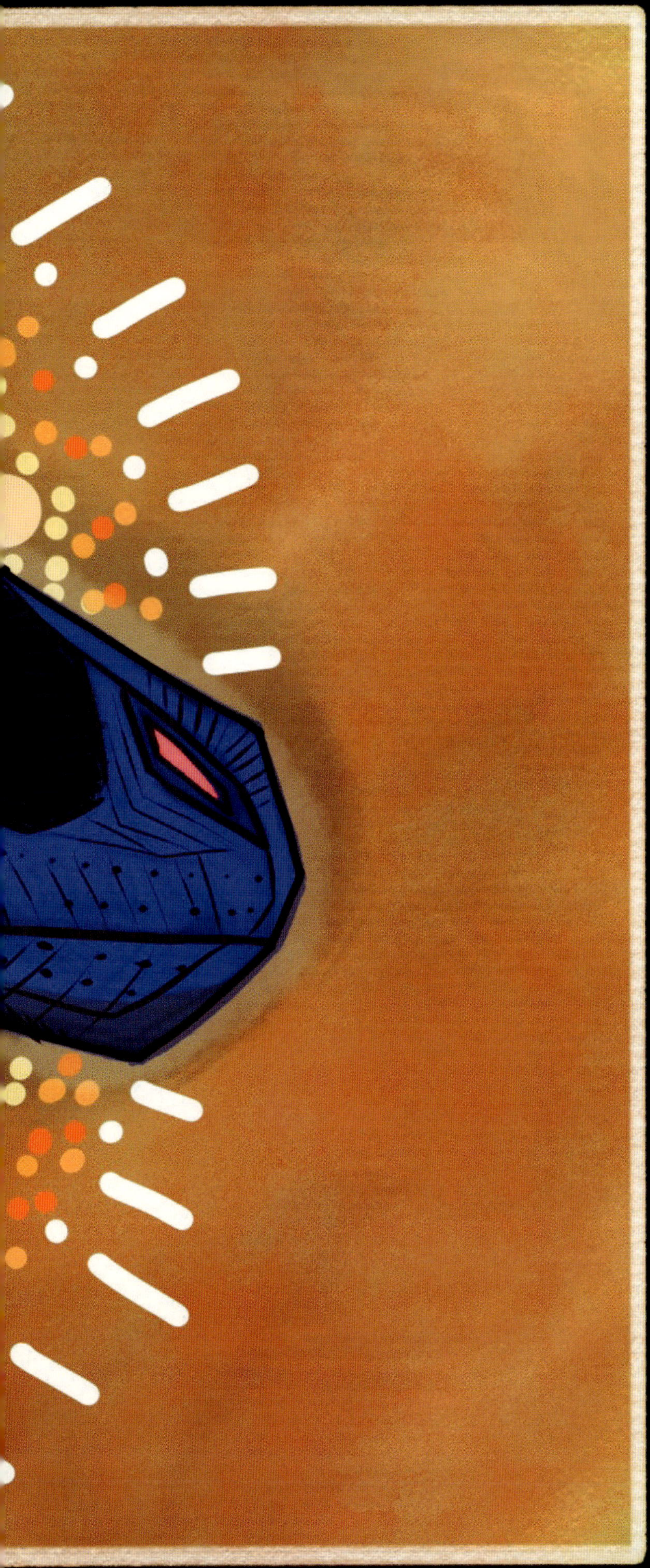

DJ WASHINGTON

DJ IS A SELF-TAUGHT digital artist from the USA and recent graduate from the University of Mary Washington where he majored in conservation biology. While mostly occupied in his professional capacity as an animal keeper specialising in extant dinosaurs, he owes the kindling of his love for the natural world to palaeontology.

From the age of three DJ was fascinated by dinosaurs. His most formative experiences were visits to his local natural history museum and an intense craving for palaeomedia, ranging from watching documentaries to amassing an extensive collection of books over the years. It was in seeing the various illustrations that adorned such books that he began to admire palaeoart's ability to bridge scientific research and artistic expression, a conduit he began practising in his own art. DJ's approach combines an appreciation for the way our picture of ancient animals can be shaped by new research with the fantastical feeling these animals have on the imagination.

[Previous pages]

DEINONYCHUS

This Art Nouveau-inspired rendition of *Deinonychus* – truly a pivotal animal in the history of palaeoart and science – imbues the animal with the intelligence and aesthetics of modern corvids. *Deinonychus* was named in 1969 from Montana and Wyoming but is now known from further east in the USA and may have ranged across much of Early Cretaceous North America.

• Digital | 2023

▼ BENEATH THE MOON, THE DEVIL STIRS

You can decide for yourself how literal this image is meant to be in terms of what *Carnotaurus*'s horns actually looked like. Here, DJ has played with the 'devilish' appearance of the famous 'meat-eating bull' of Late Cretaceous Argentina, enhanced by the light of a blood moon. The animal's horns have helped to enhance its pop culture reputation.

• Digital | 2024

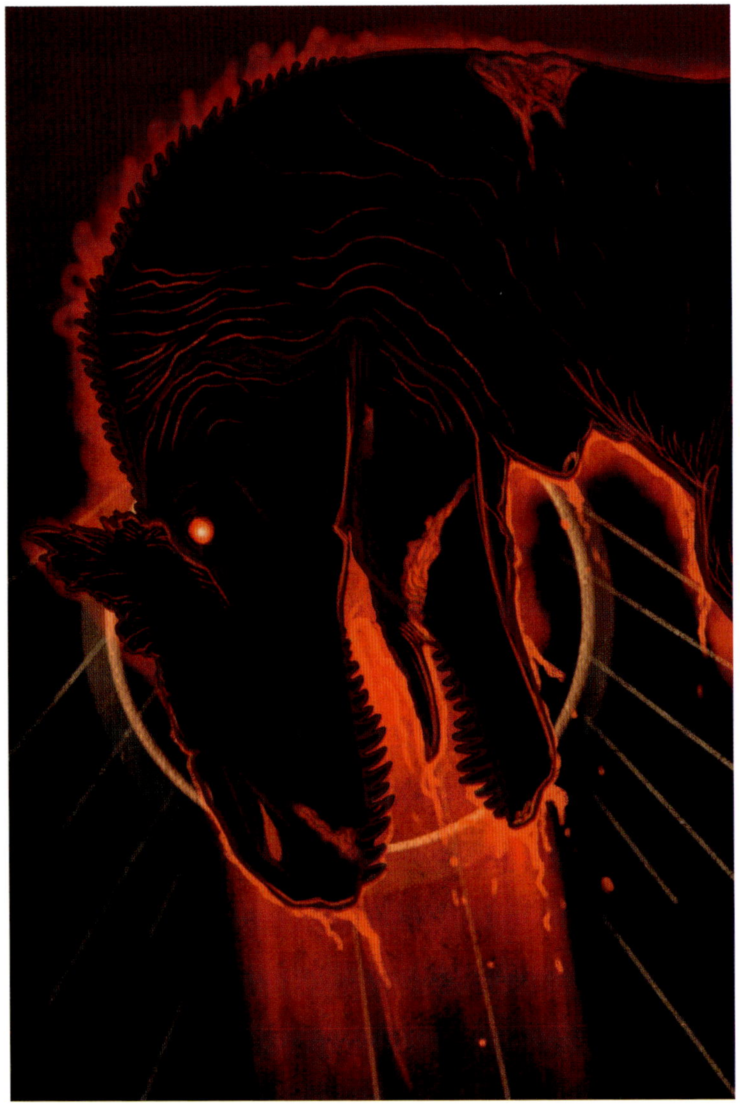

▲ A FALLEN DRAGON: *ALLOSAURUS*

This dramatic and powerful image of the Late Jurassic *Allosaurus* might be an allusion to the sometimes violent lives of these awesome animals. How often *Allosaurus* died a peaceful, natural death of old age will likely never be known, but we do know that some specimens died from disease, from combat with prey, or from injurious battle.

• Digital | 2024

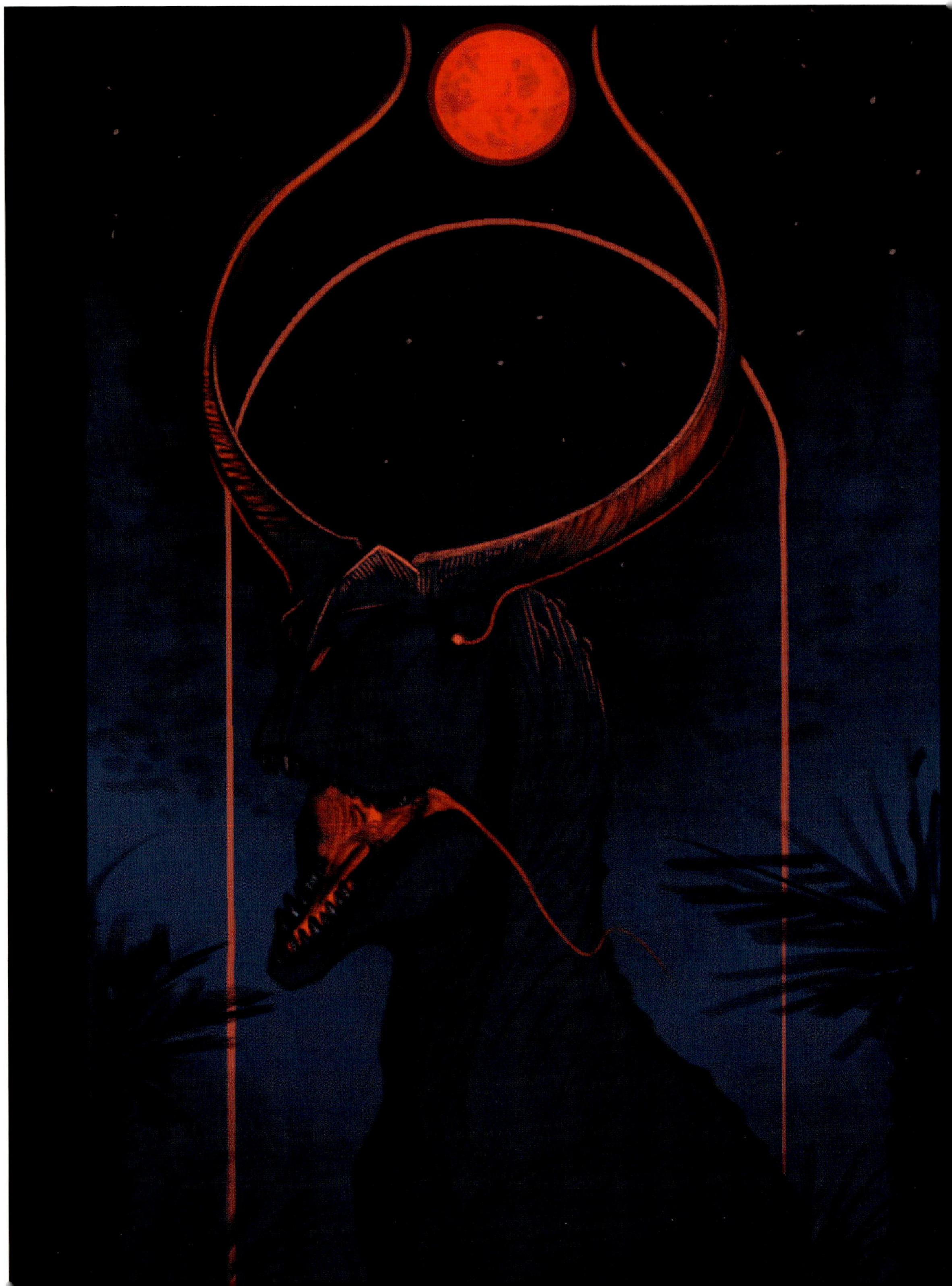

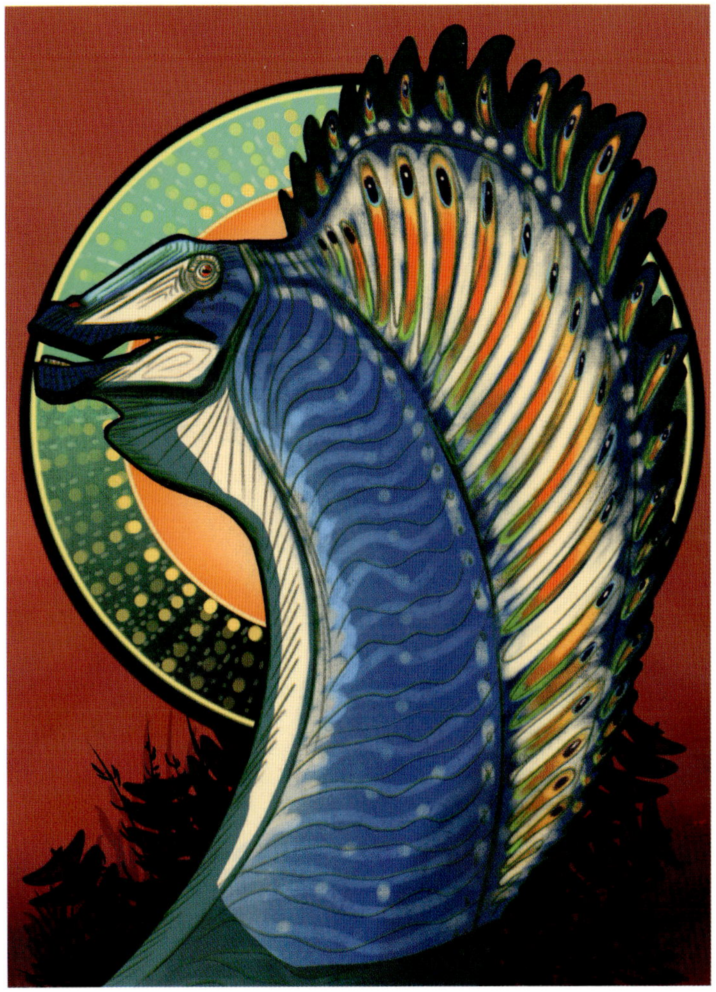

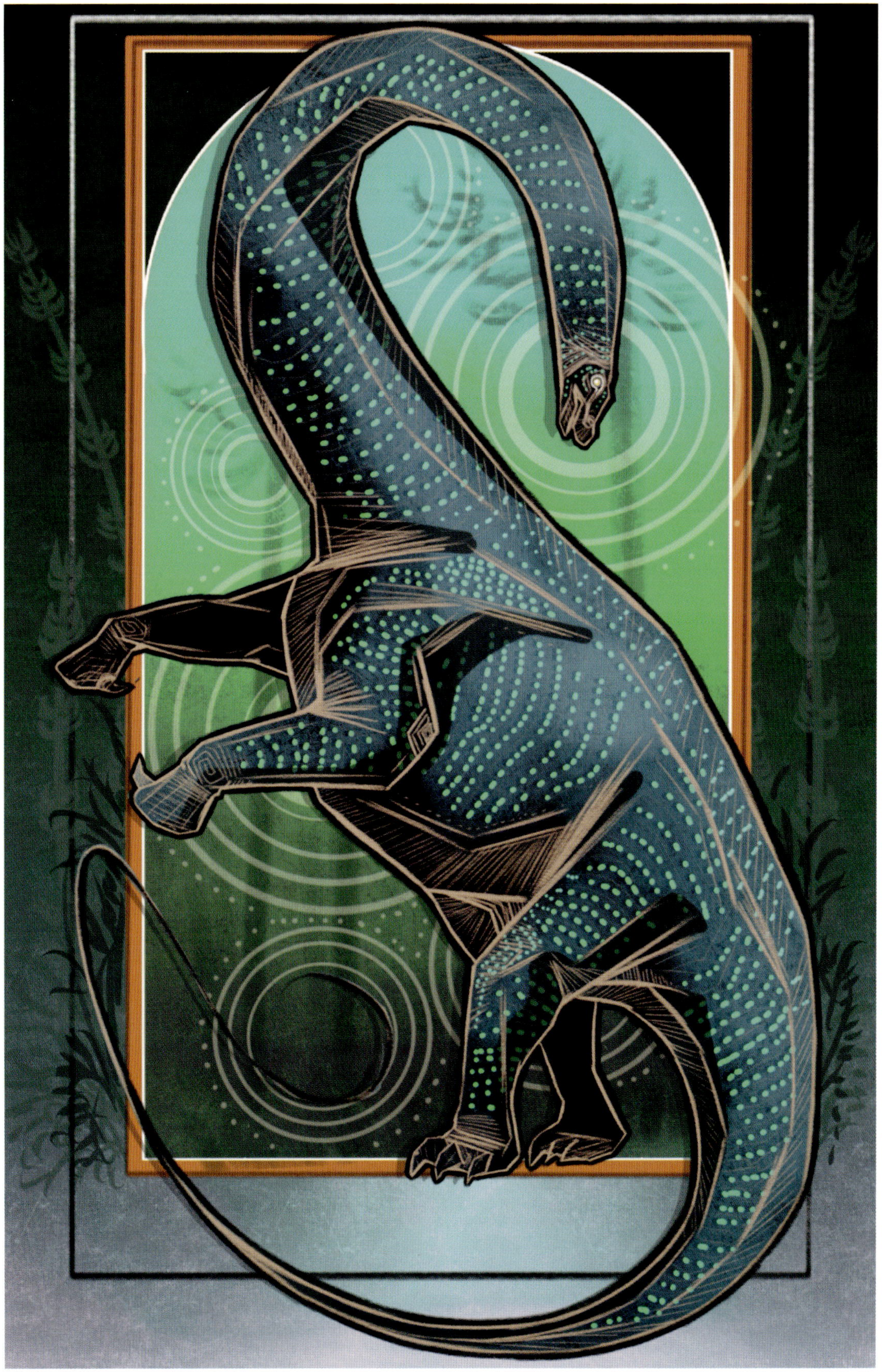

▲ SAILED ELEGANCE: *BAJADASAURUS*

Named from the Lower Cretaceous of Argentina in 2019, *Bajadasaurus* is a weird and still poorly known sauropod, a probable cousin of the better known spiny necked *Amargasaurus*. Long, forward-pointing, bony neck spines indicate the presence of a tall frill or deep, bilobed crest extending along the top of the animal's neck.

• Digital | 2023

EMERALD GIANT: *BAROSAURUS* ▶

A diplodocid cousin of the more familiar *Diplodocus*, *Barosaurus* was long-necked, long-tailed and – like probably all sauropods of this sort – capable of rearing up on its back legs. The repeated dotted patterns here imply a girth sometimes not conveyed in artistic images of diplodocids. The green hues link the animal to its verdant, forested environment.

• Digital | 2023

▼ THREE-HORNED GLORY

It seems surprising that certain artists feel compelled to portray dinosaurs like *Triceratops* as bland and monotone. It may be a matter of taste, but it's difficult to avoid the idea that the frill of *Triceratops* was likely patterned and colourful. The enormous horned snout and projecting cheek horns perhaps had advertising roles as well.
• Digital | 2023

A CROWN FIT FOR A KING ▶

Images throughout this book show how a relatively new idea on a specific dinosaur – namely a 2020 proposal about the Early Jurassic theropod *Dilophosaurus* – have widely, and quickly, spread throughout modern palaeoart. Dilophosaurs did not have thin, plate-like crests, but a robust, convex, and perhaps elaborately marked and coloured, casque.
• Digital | 2023

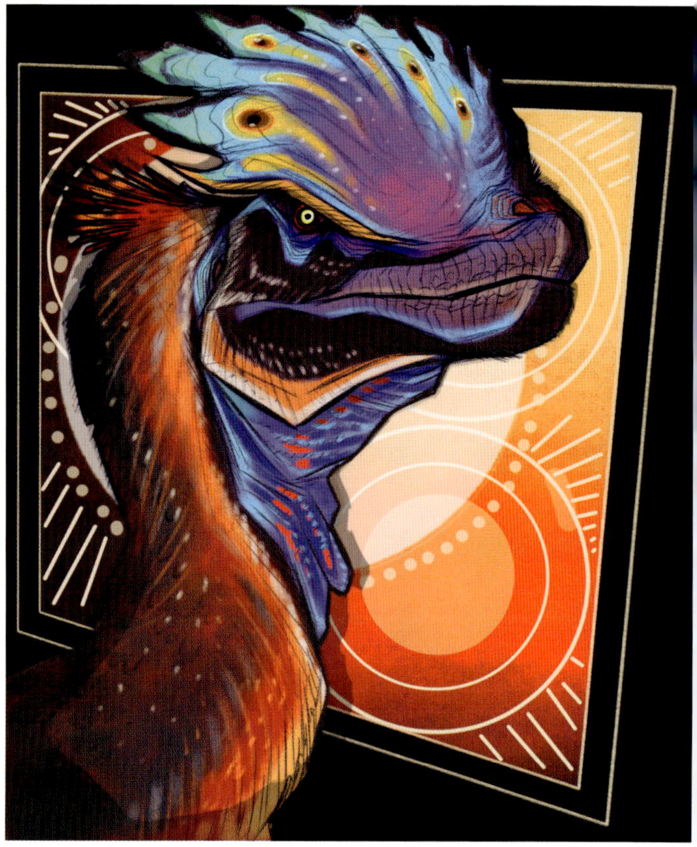

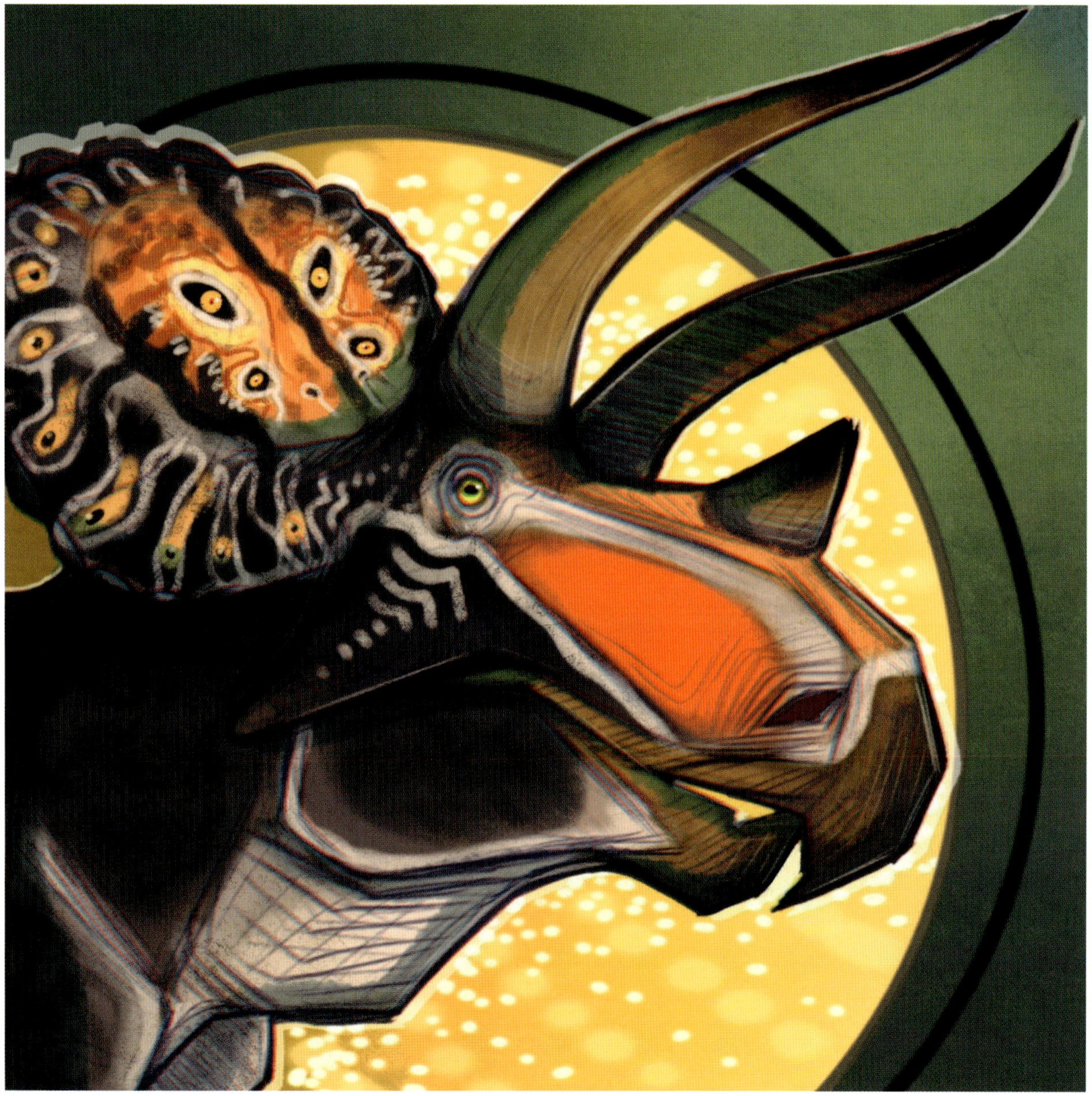

ANCIENT RIVER DRAGON ▶

The idea that the famously giant, famously controversial *Spinosaurus* might be another 'showy' dinosaur has mostly been limited to discussion about its sail (and, arguably, its paddle-like tail as well). New evidence presented in 2024 has revealed that some species or specimens of *Spinosaurus* had a prominent crest projecting from the base of the snout.
• Digital | 2023

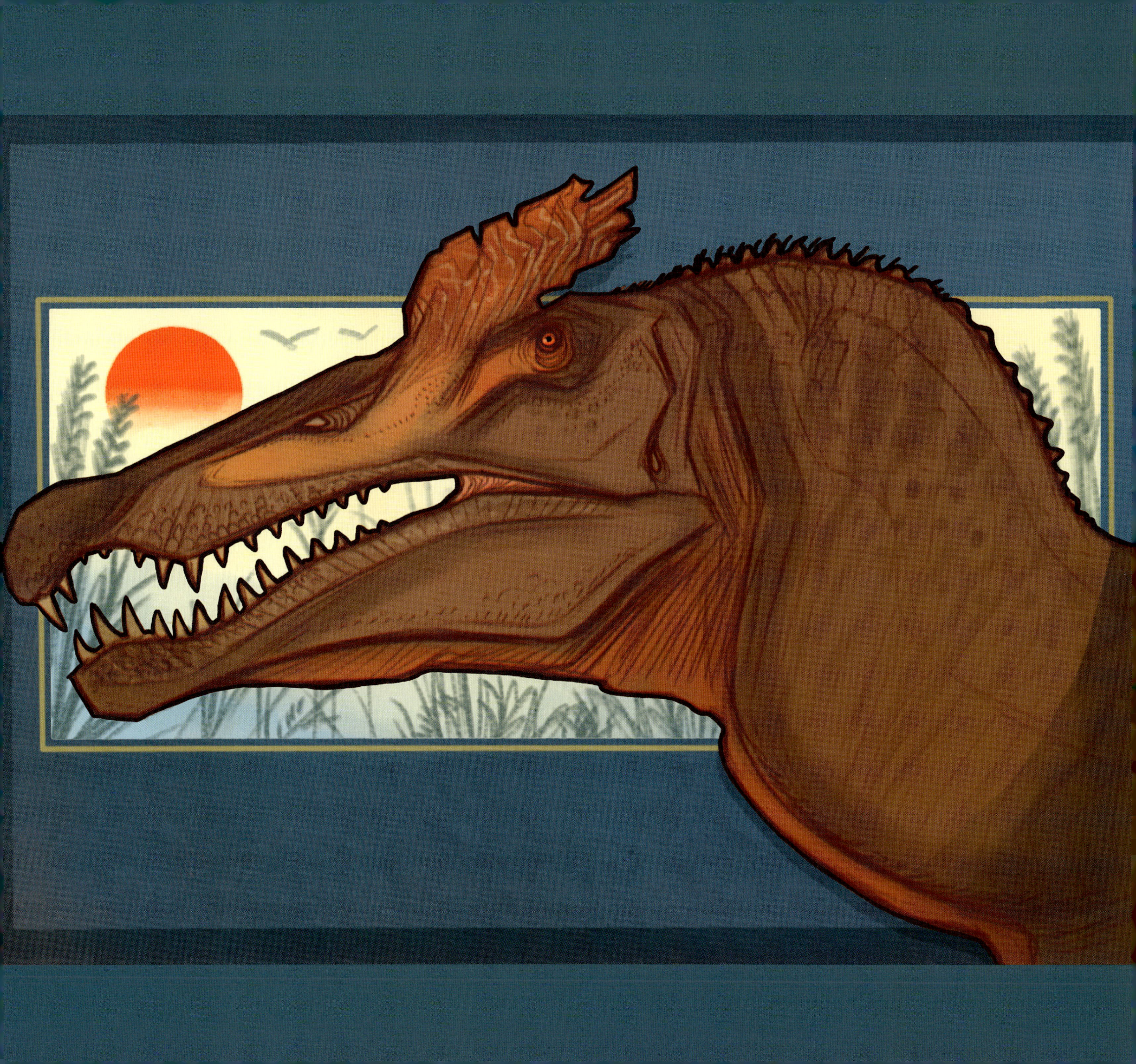

LUCID DREAM: *OLOROTITAN* ▶

While stylised, this image communicates the elegant, even graceful, proportions of hadrosaurs like eastern Russia's *Olorotitan*, a relative of North America's *Lambeosaurus* named in 2003. The crest of *Olorotitan* was narrow and hachet-shaped and its snout and jaws were stretched, giving it a long and slender face. The zebra-like striping is entirely speculative but very possible.
• Digital | 2023

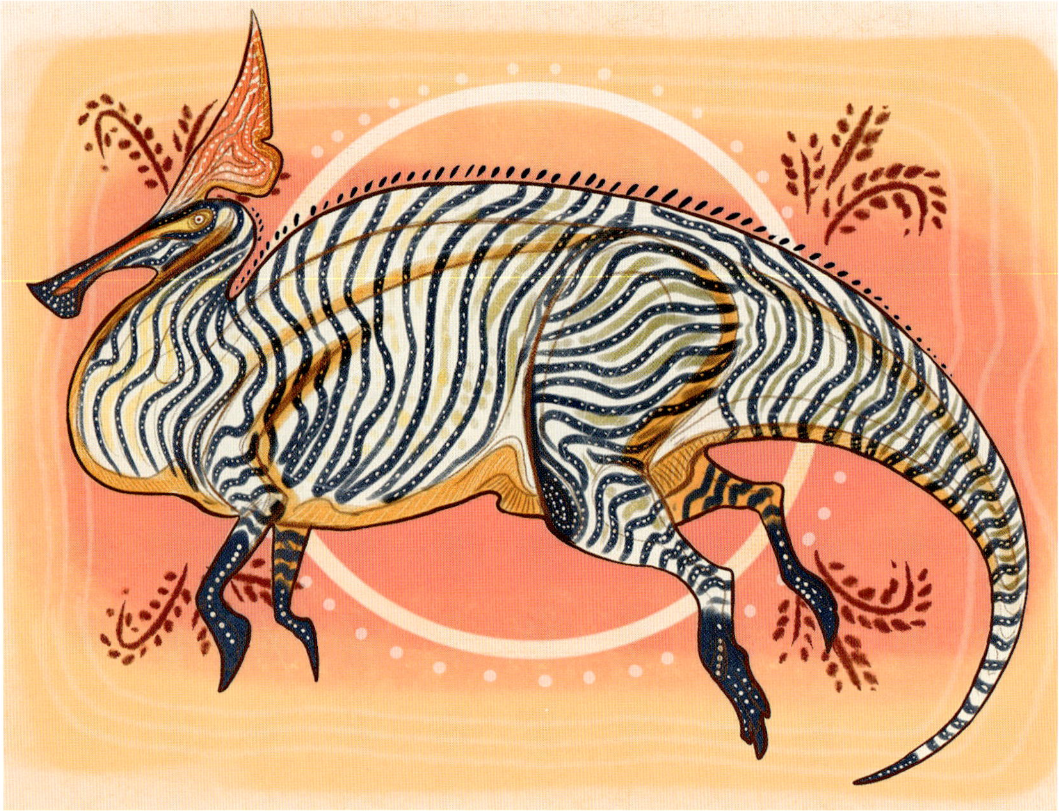

▼ GOLDEN AURA: *PARASAUROLOPHUS*

Parasaurolophus – named from Canada in 1920 but later discovered in Montana, Utah and New Mexico – is one of the most popular of hadrosaurs. While sometimes portrayed as svelte and leggy (as in *Jurassic Park*), it was actually heavyset and muscular. Its distinctive crest, differing in shape across several species, suggests a flamboyant look in life.
• Digital | 2023

▼ *CORYTHORAPTOR*

Named from southern China in 2017, *Corythoraptor* is a mid-sized oviraptorosaur and probably part of the same group as the also-crested *Citipati* and *Rinchenia*, both of Mongolia. Oviraptorosaurs were feathered. However, virtually all knowledge of what their plumage was like comes from turkey-sized species. Bright blue hues and massive feathers are present in some birds today, so may have been present in *Corythoraptor* too.
• Digital | 2024

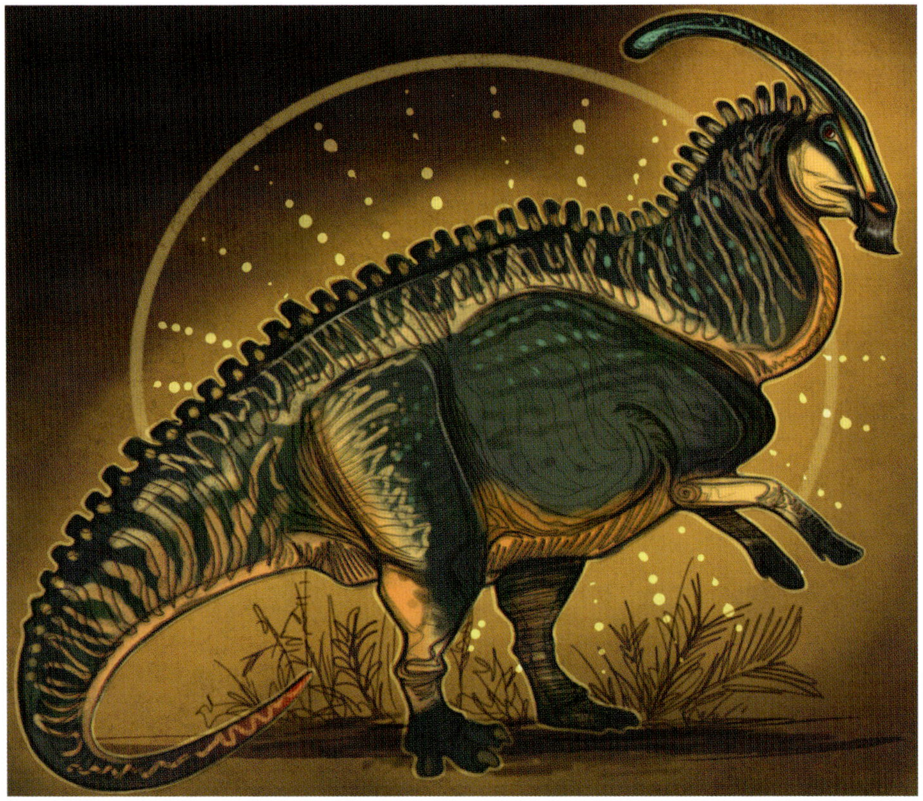

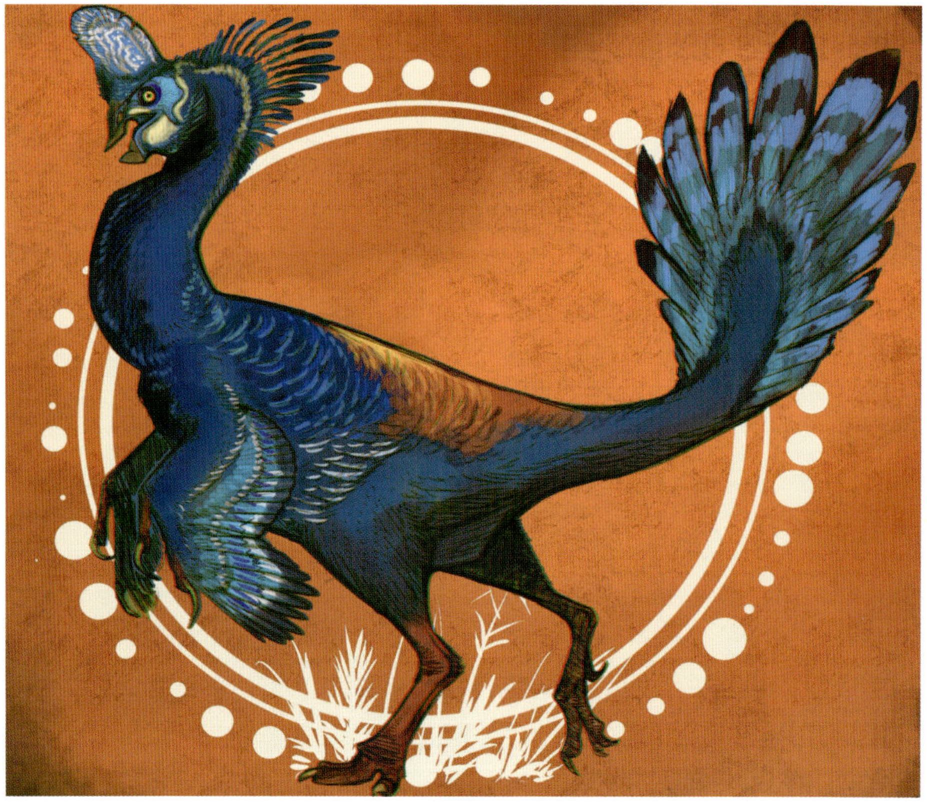

▼ *STEGOSAURUS*

Stegosaurus is yet another dinosaur that has to be included on the 'over-the-top' list in terms of how ornate it was. Suggestions that its staggered row of plates might have served in temperature control or self-defence remain less likely than that they mostly functioned in display. Attention-grabbing markings – eyespots, even – therefore seem quite likely.
• Digital | 2023

STYRACOSAURUS ▶

Originally described from the Upper Cretaceous of Canada in 1913, *Styracosaurus* – the 'spiked lizard' – is famous for the long spikes that project from the edges of its frill. In some specimens, the spikes are asymmetrical (a feature common in modern display structures like deer antlers). Keratin extensions would have made the horns longer in life than in the fossils, while the horn tips were possibly slimmer, pointier or more curved than fossils imply.
• Digital | 2023

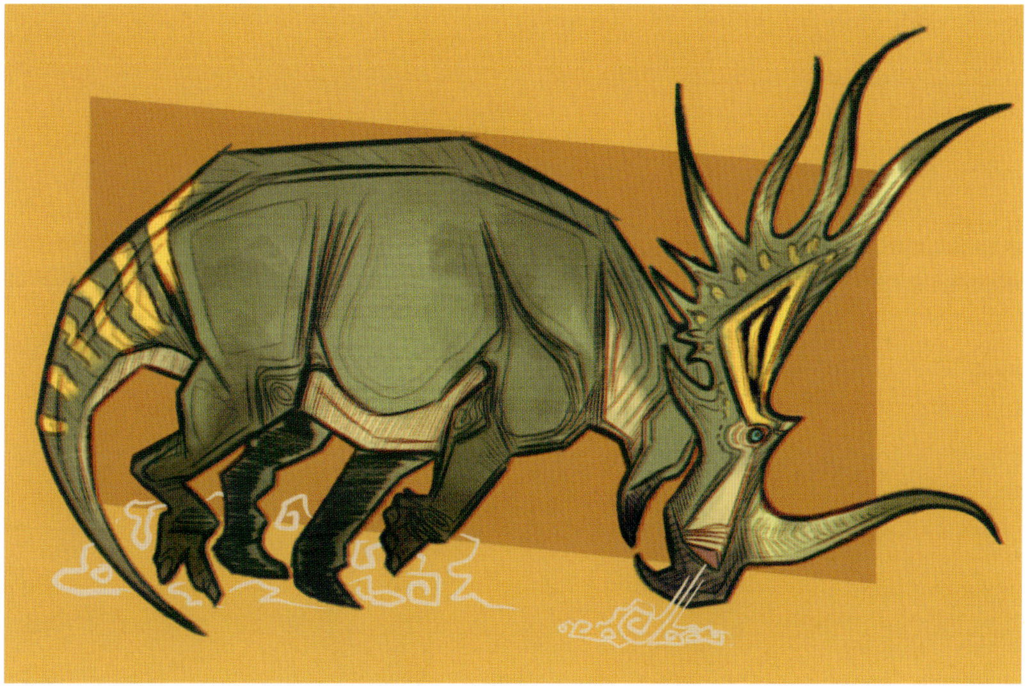

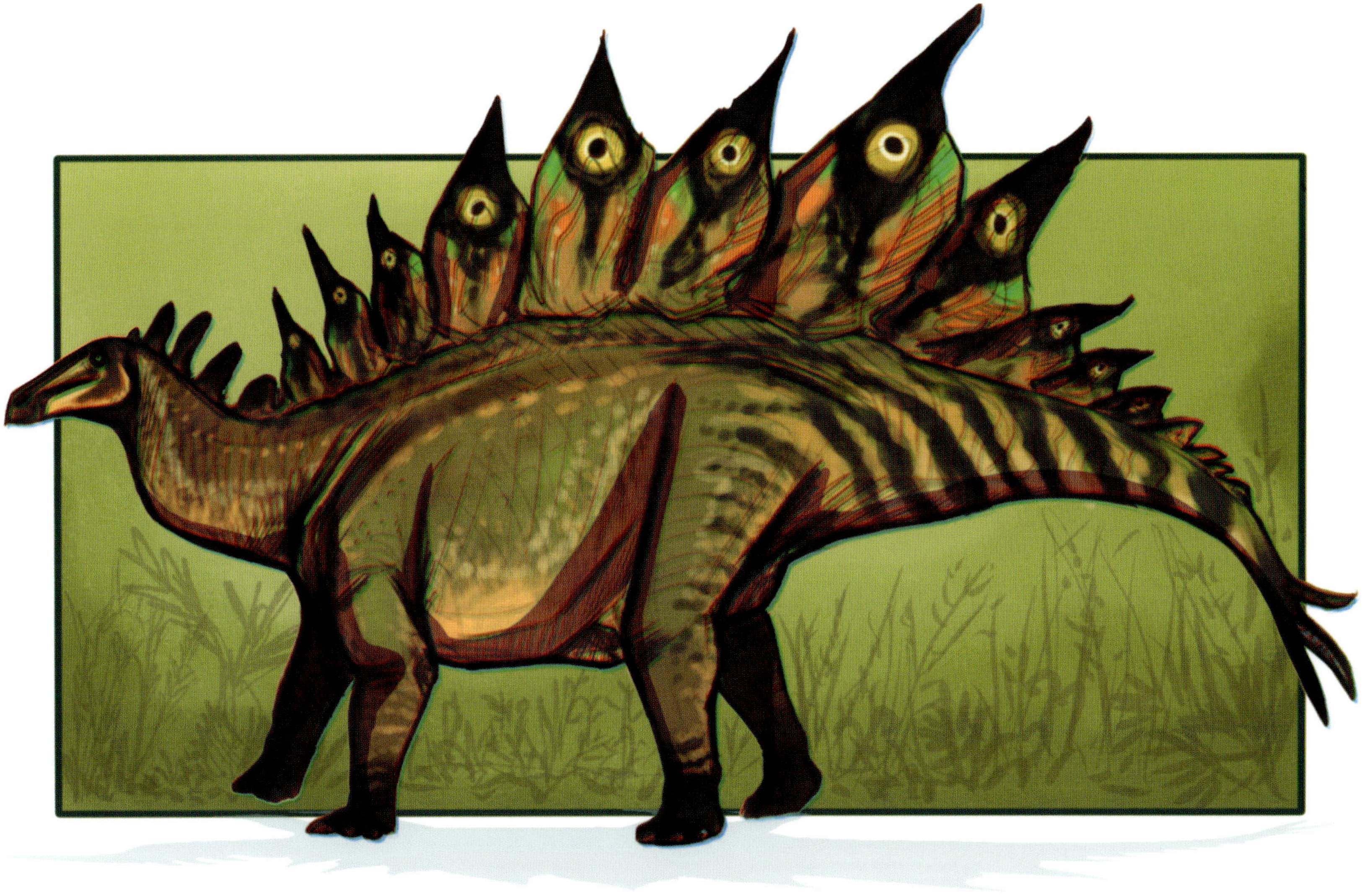

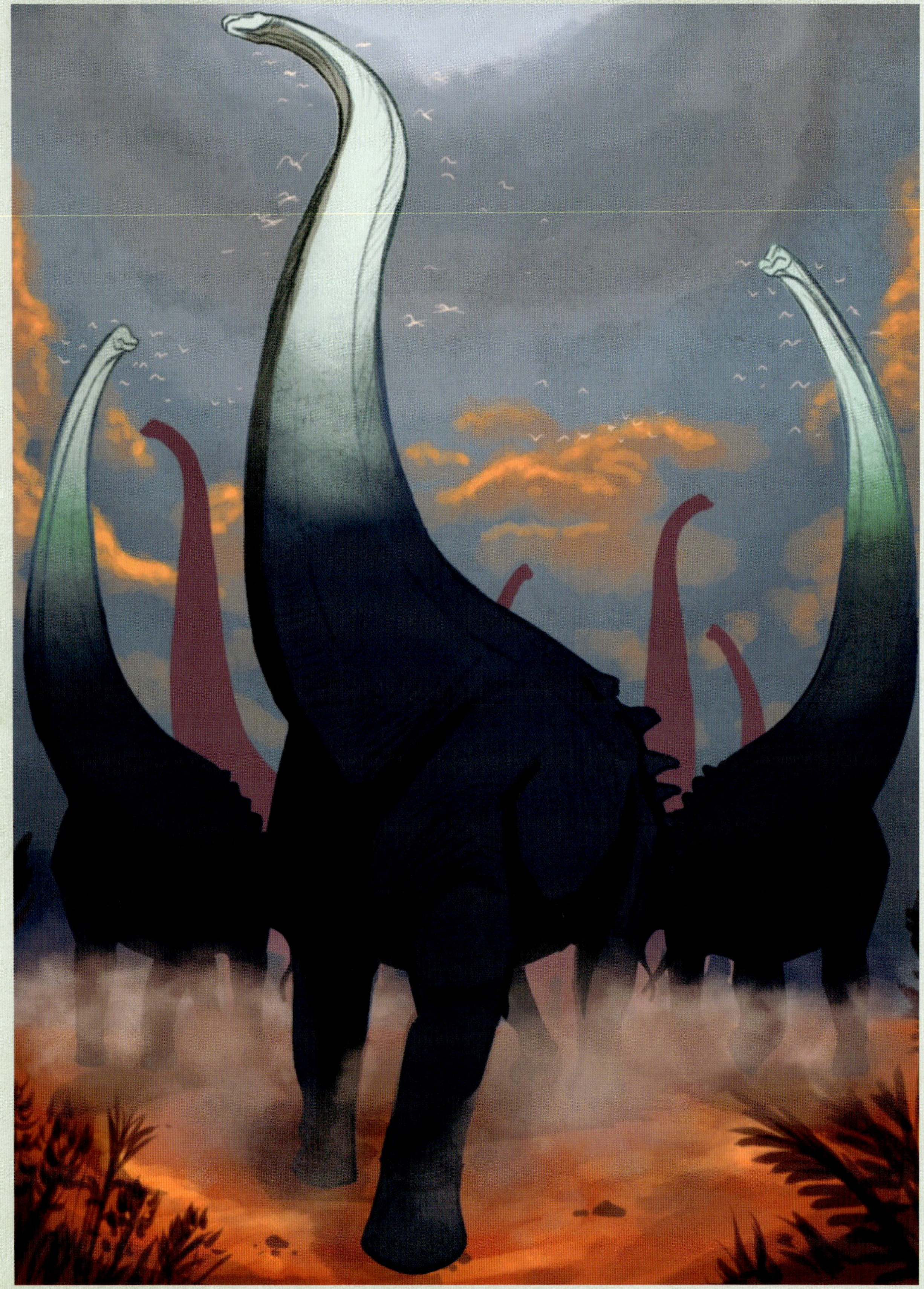

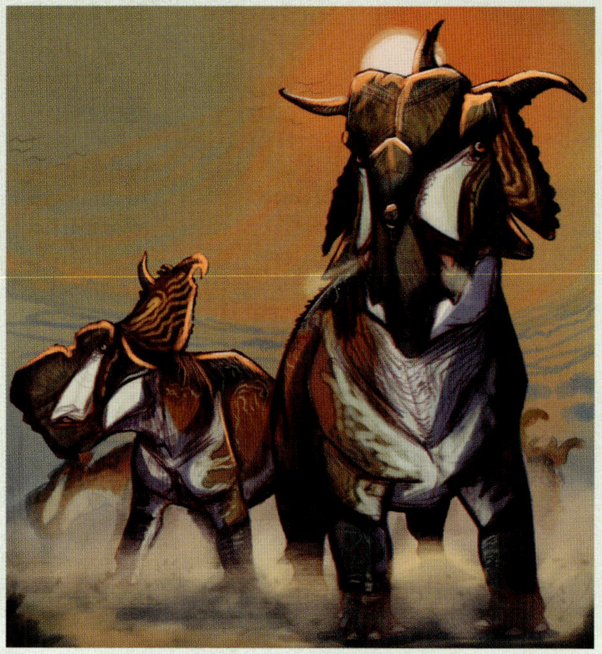

▲ BIG BOSSES

Pachyrhinosaurus was first named in 1950 and initially regarded as a real oddity within the history of its group (the ceratopsians). Today, we know of several pachyrhinosaur species: they were the last members of a successful, ornate lineage specialised for life in the north of North America. Their ornate nature invites speculation that bold 'dazzle' markings might have hidden or masked the animal's vulnerable features, such as the eyes.

• Digital | 2024

◄ MARCHING TITANS: *ARGENTINOSAURUS*

When announced in 1993, *Argentinosaurus*, a Late Cretaceous titanosaur, was hailed as the largest dinosaur of all. This idea was then challenged as additional super-sized titanosaurs, most of which were also from Argentina, saw print. Recent studies do, however, find *Argentinosaurus* to *potentially* be the biggest, at over 30m (98ft) long and perhaps 72 metric tonnes (80 tons). Picturing these animals alive is quite the feat of imagination – especially if they moved in herds!

• Digital | 2023

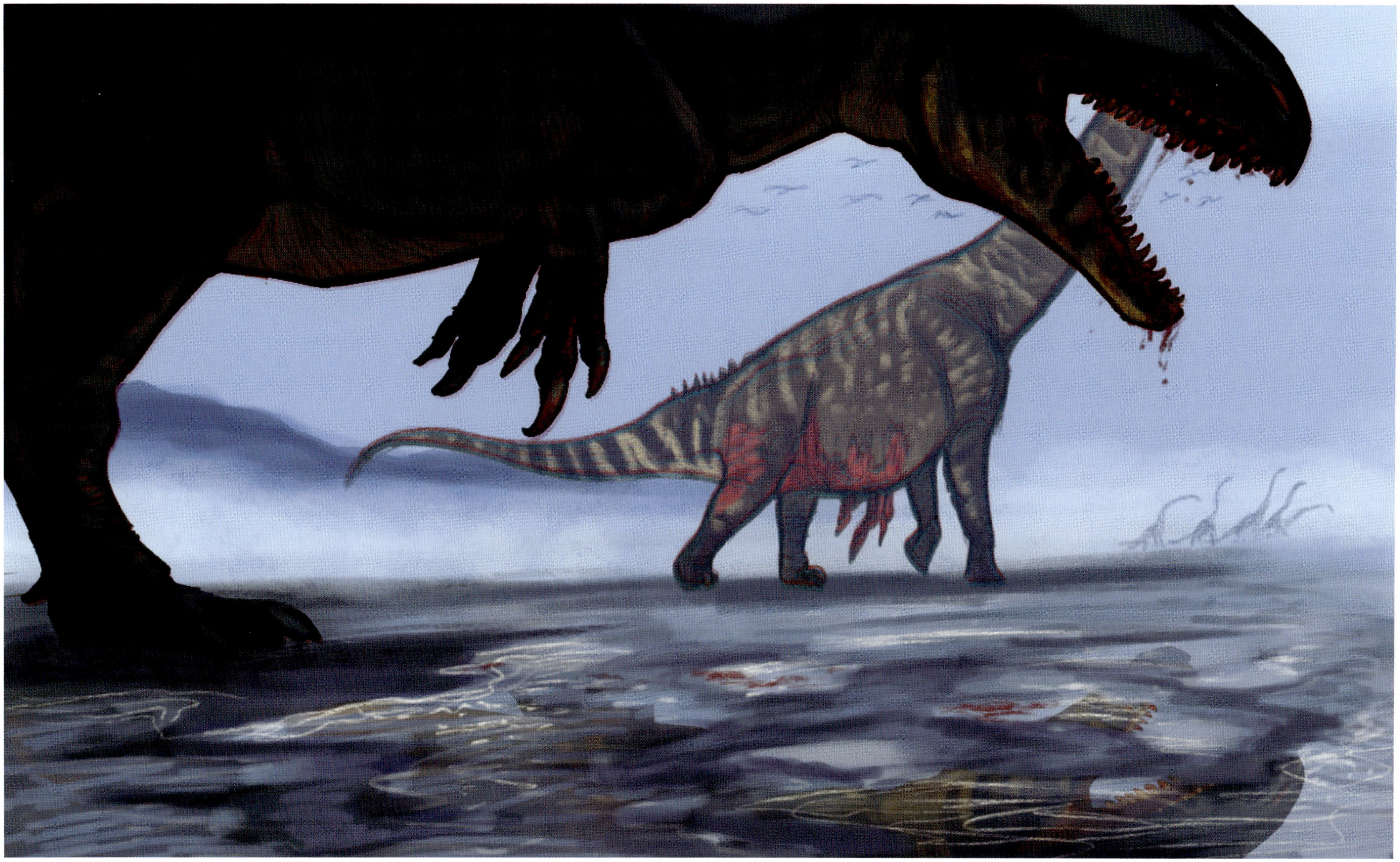

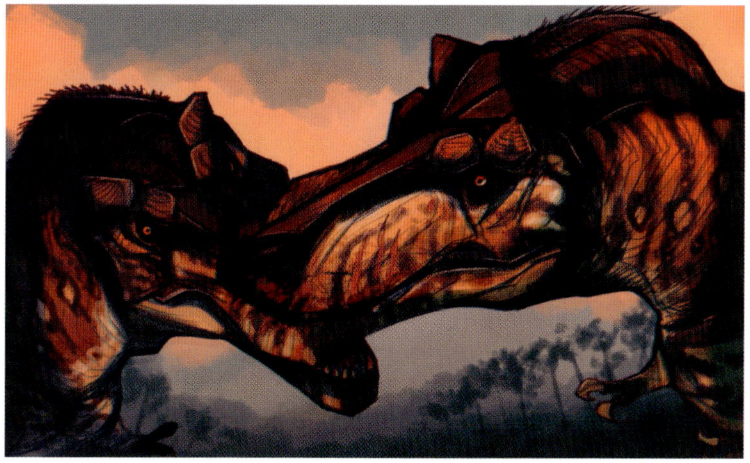

▲ CLOSING IN

Several experts have suggested that large allosauroid theropods, like the Early Cretaceous *Acrocanthosaurus* shown here, were 'giant killers', capable of killing even the biggest of sauropods. Maybe, as shown here, prey was disabled via massive wounds. The sauropod slows, weakened by its wounds, all while the predator shadows its victim, waiting to move in for the kill.
• Digital | 2023

T. REX WRESTLING ▶

Several fossils show that tyrannosaurs like *T. rex* (and other big theropods too) engaged in combat by raking, biting and grabbing one another by the snout and jaws. It might even be that the horn-covered bony bosses, hornlets and snout ridges of these dinosaurs evolved in step with this function, an idea implied here by the protective, plate-like look of the keratinised surfaces.
• Digital | 2023

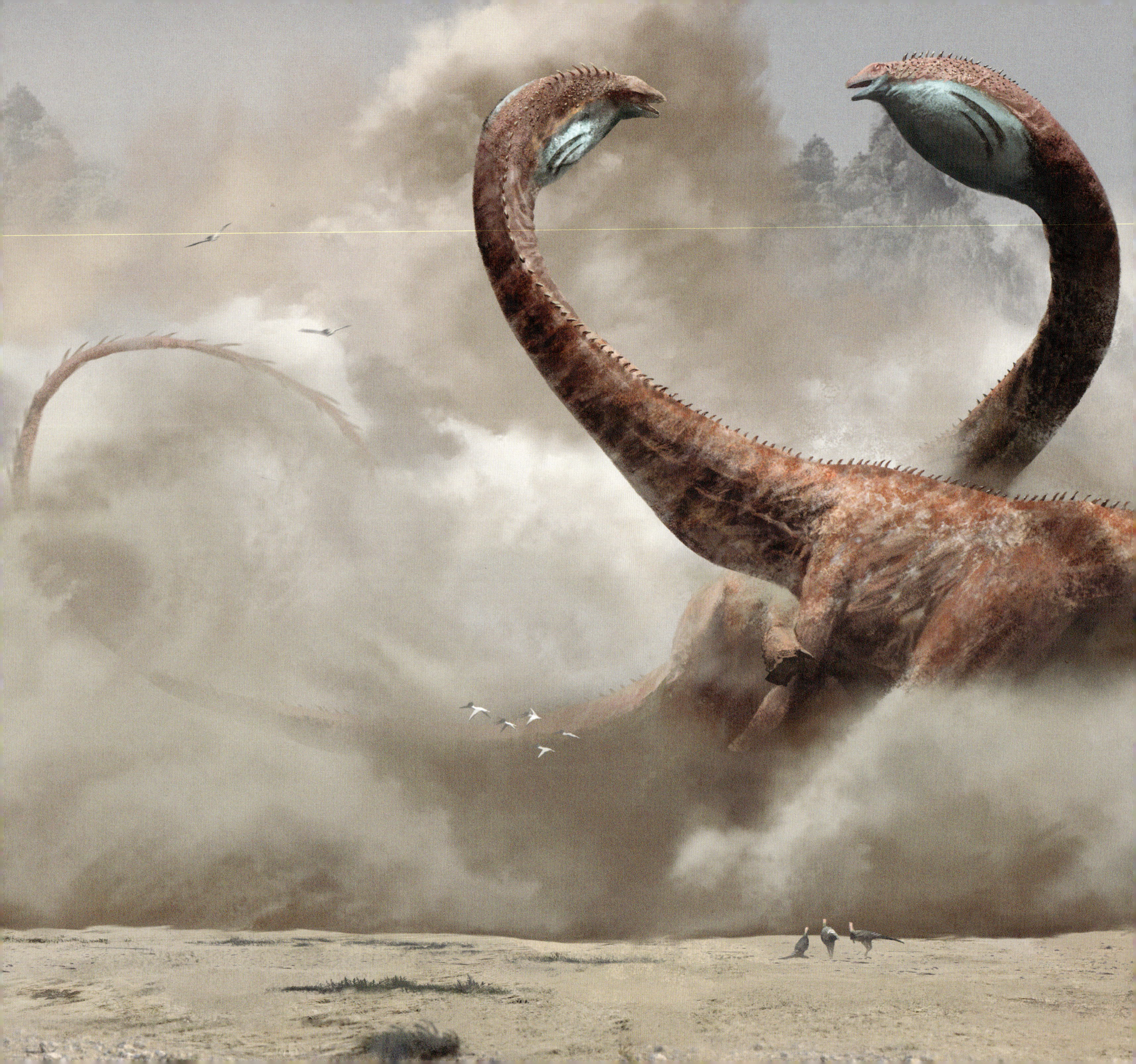

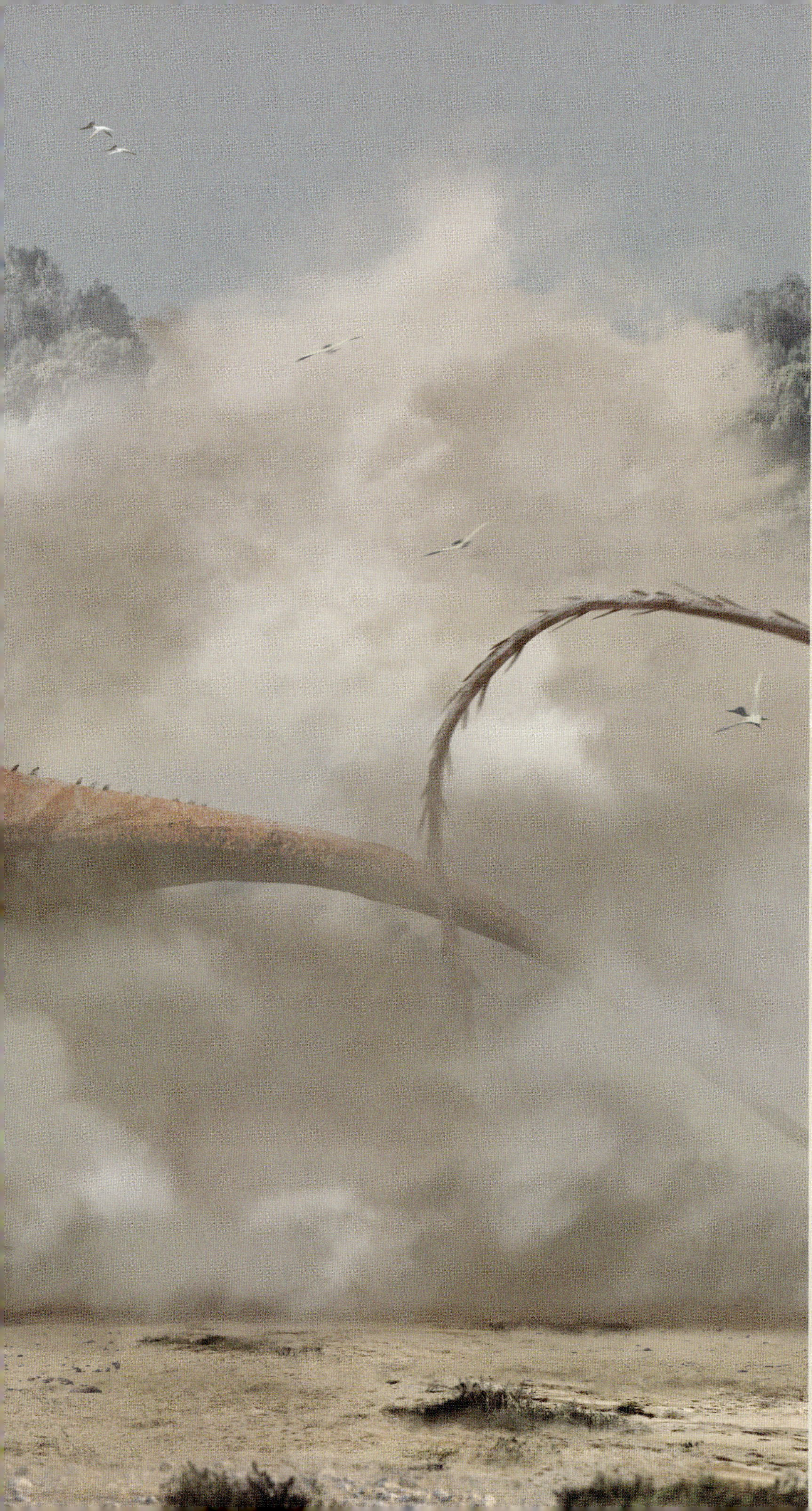

RUDOLF HIMA

THROUGH HIS PAINTINGS, Rudolf Hima portrays prehistoric animals not as monsters or props but as living, breathing animals that once walked the Earth. Rudolf grew up in Indonesia, a country where an interest in palaeontology and prehistoric life seemed somewhat out of place. Therefore, his influence came from overseas when the BBC's *Walking with Dinosaurs* aired on TV, portraying the dinosaurs as ordinary animals with their own share of fortune and tragedy, creatures with their own stories to tell. This fuelled him to draw and create his own little world, although initially drawing was pretty much just a hobby that came and went. It was not until his studies at University of Pelita Harapan that Rudolf began to take illustration more seriously. The Internet also allowed him to connect to people with shared interests, amateur and professional alike. These people were crucial to Rudolf's palaeoart journey, enabling his development through easier access to palaeontological information.

What followed were depictions of the prehistoric world through detailed, scientifically accurate renderings but with cinematic flair. These have been featured in several scientific publications. He uses colour and lighting to capture the mood of the scene, be it to evoke tranquillity or tension. Ultimately, however, his work is there to reflect an adoration of prehistoric life on Earth which words simply cannot.

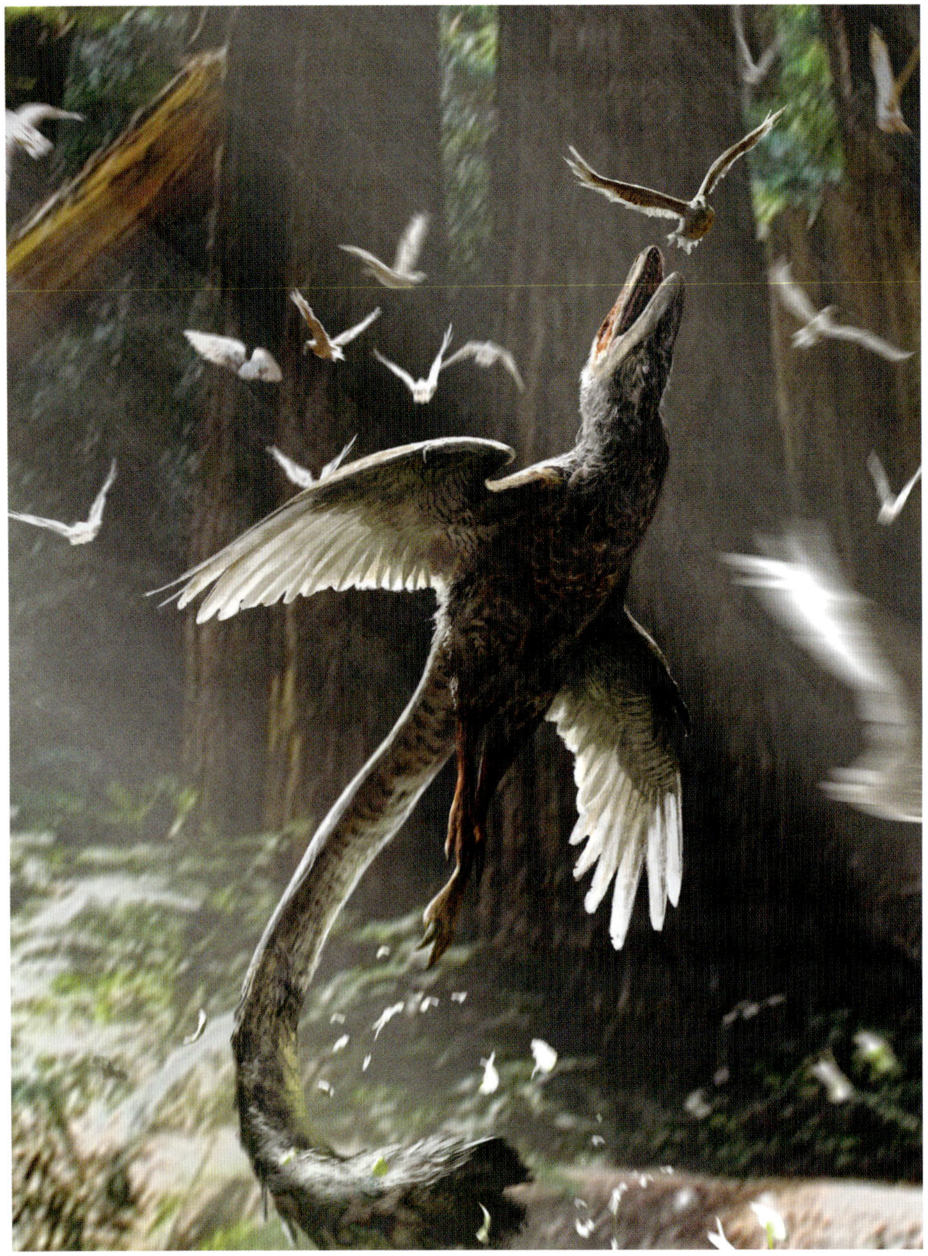

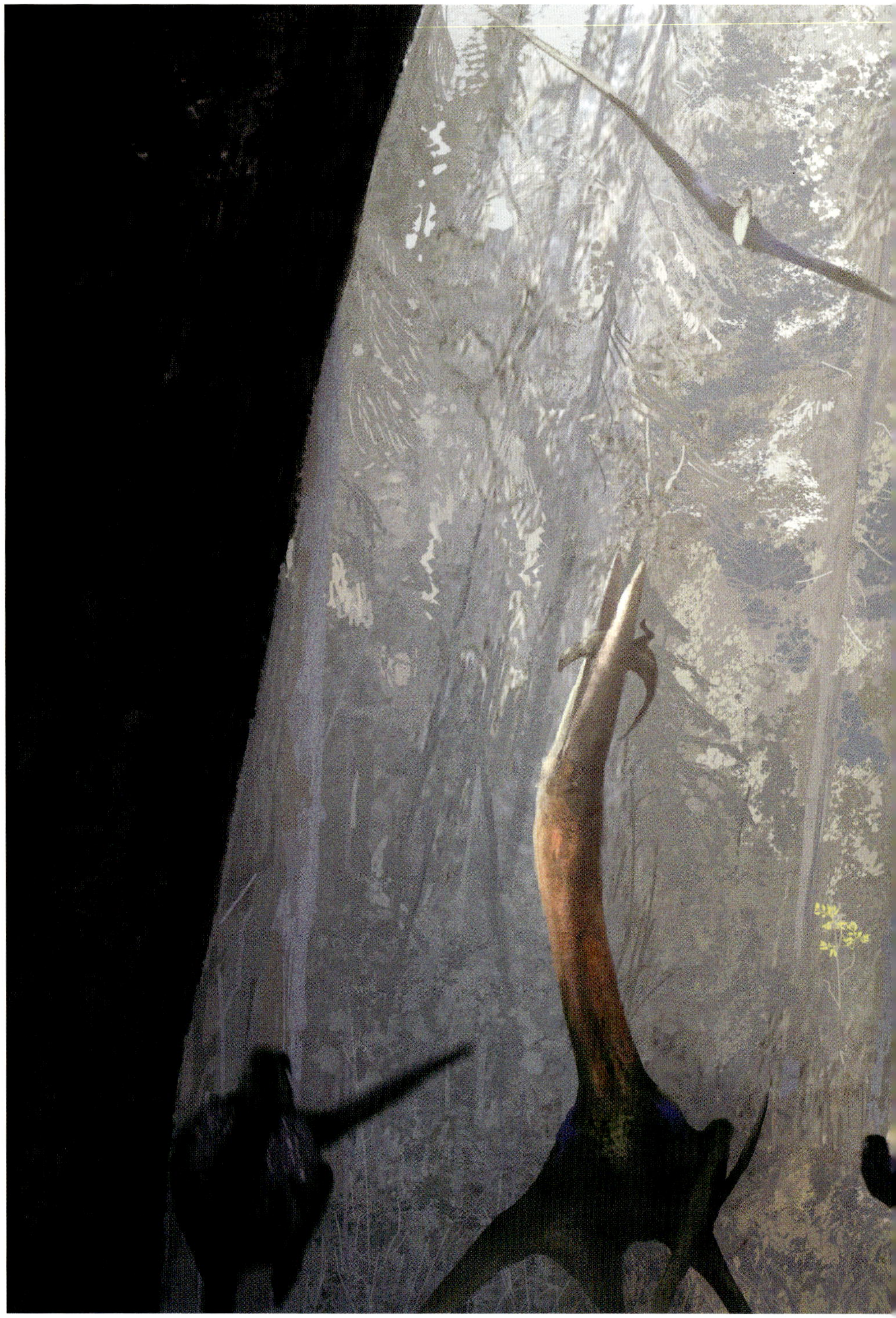

[Previous pages]

BAROSAURUS COMBAT

This spectacular image gives some idea of what combat might have been like among the biggest of the sauropods, in this case *Barosaurus*, a diplodocid from the Late Jurassic Morrison Formation of the USA. Also explored here is the idea that some sauropods might have been able to inflate sections of the body via their internal air sac system.

• Digital | 2022

▲ FLIGHT

In a sunlit forest glade, the Late Cretaceous dromaeosaurid *Acheroraptor* from Montana, USA, surprises some of its smaller cousins. The muscular hindlimbs, slender tails and lightweight frames of dromaeosaurids almost certainly made them highly adept at leaping and it's plausible that they could 'boost' a leap with their feathered forelimbs.

• Digital | 2022

▼ AMBUSHED

In their current guise as striding, predatory 'mega-storks', giant azhdarchids like this Romanian *Hatzegopteryx* – here shown ambushing a herd of *Zalmoxes* – surely rank as among the scariest of Cretaceous predators. The relatively thick neck and robustly built skull of this pterosaur is consistent with the idea that it was able to grab and consume house cat-sized dinosaurs.

• Digital | 2023

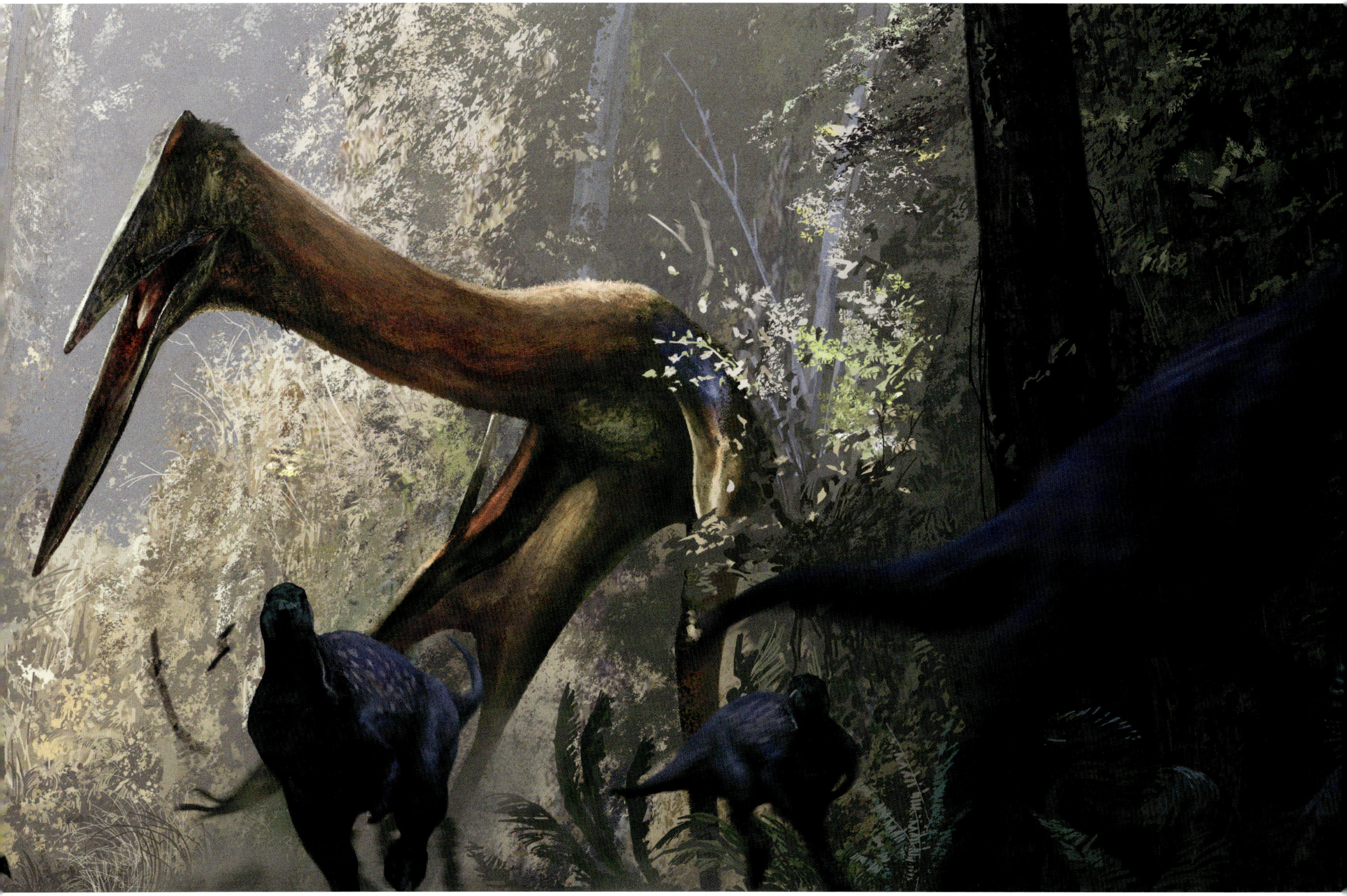

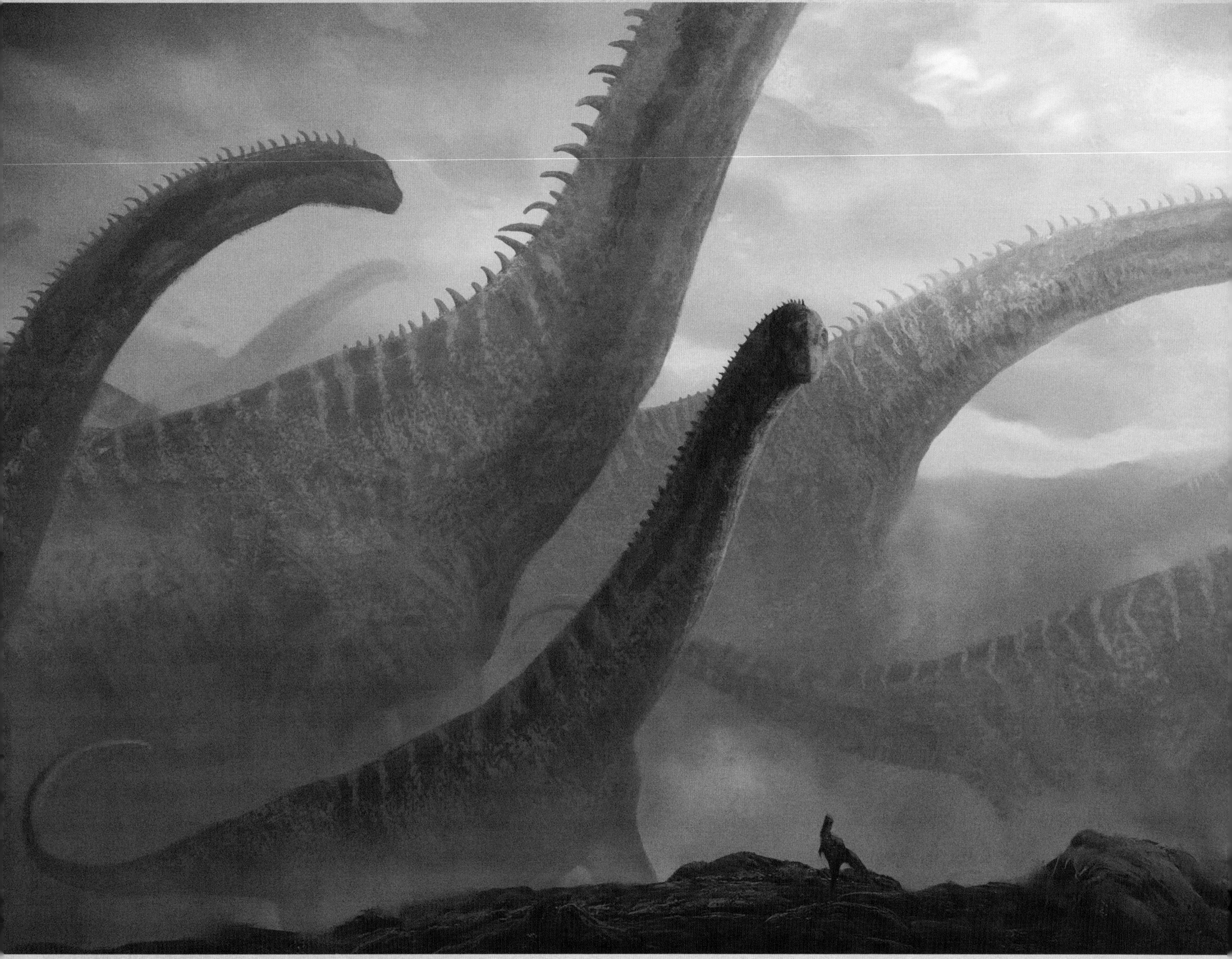

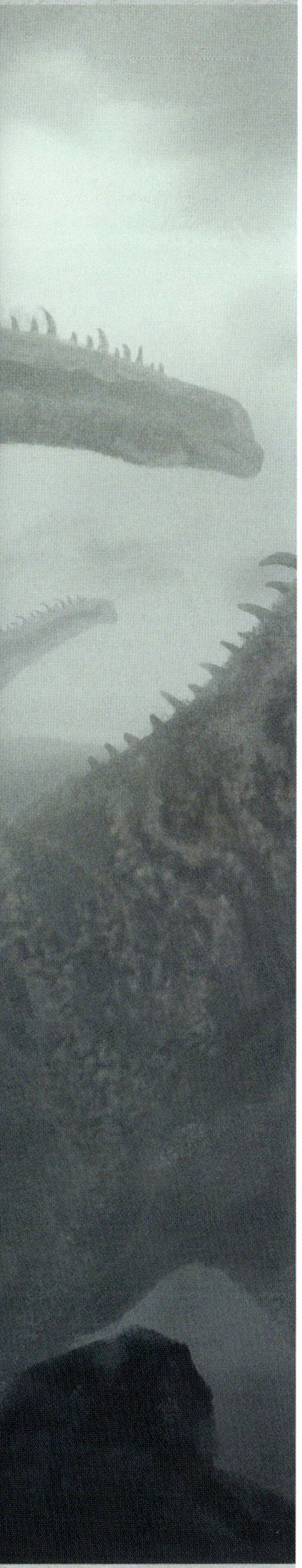

◂ FUTALOGNKOSAURUS

The gargantuan sizes achieved by many sauropods of the Jurassic and Cretaceous – this image shows the formidably named *Futalognkosaurus* from the Late Cretaceous of Argentina – requires that they had an almost *elemental* dimension, well portrayed here. As they were constantly exposed to wind, storms, hail and baking heat, we have to assume that they were built to withstand whatever was thrown at them.

• Digital | 2022

PTERANODON ▸

It is thought *Pteranodon* preyed on fishes and squids less than 60cm (2ft) long. However, large marine predatory birds today are opportunistic, and any prey is on the menu if it can be picked up and swallowed. Might *Pteranodon* have added juvenile mosasaurs to its diet? This is the long-crested species *P. longiceps*, of which a large male could have a wingspan of 5.6m (18ft).

• Digital | 2021

▾ SHONISAURUS

During the Triassic (between around 251 and 201 million years ago) the sea-going ichthyosaurs underwent a rapid burst of evolution and several gigantic lineages evolved. *Shonisaurus* from the Late Triassic of the USA (and perhaps Canada) was among the biggest, some specimens reaching 15m (50ft). However, they were seemingly predators of small prey such as squid.

• Digital | 2021

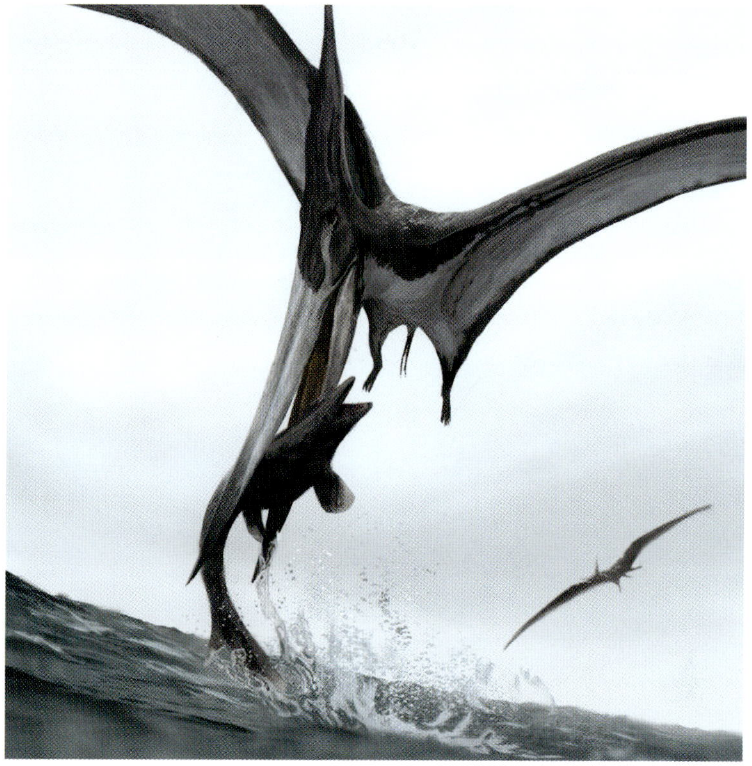

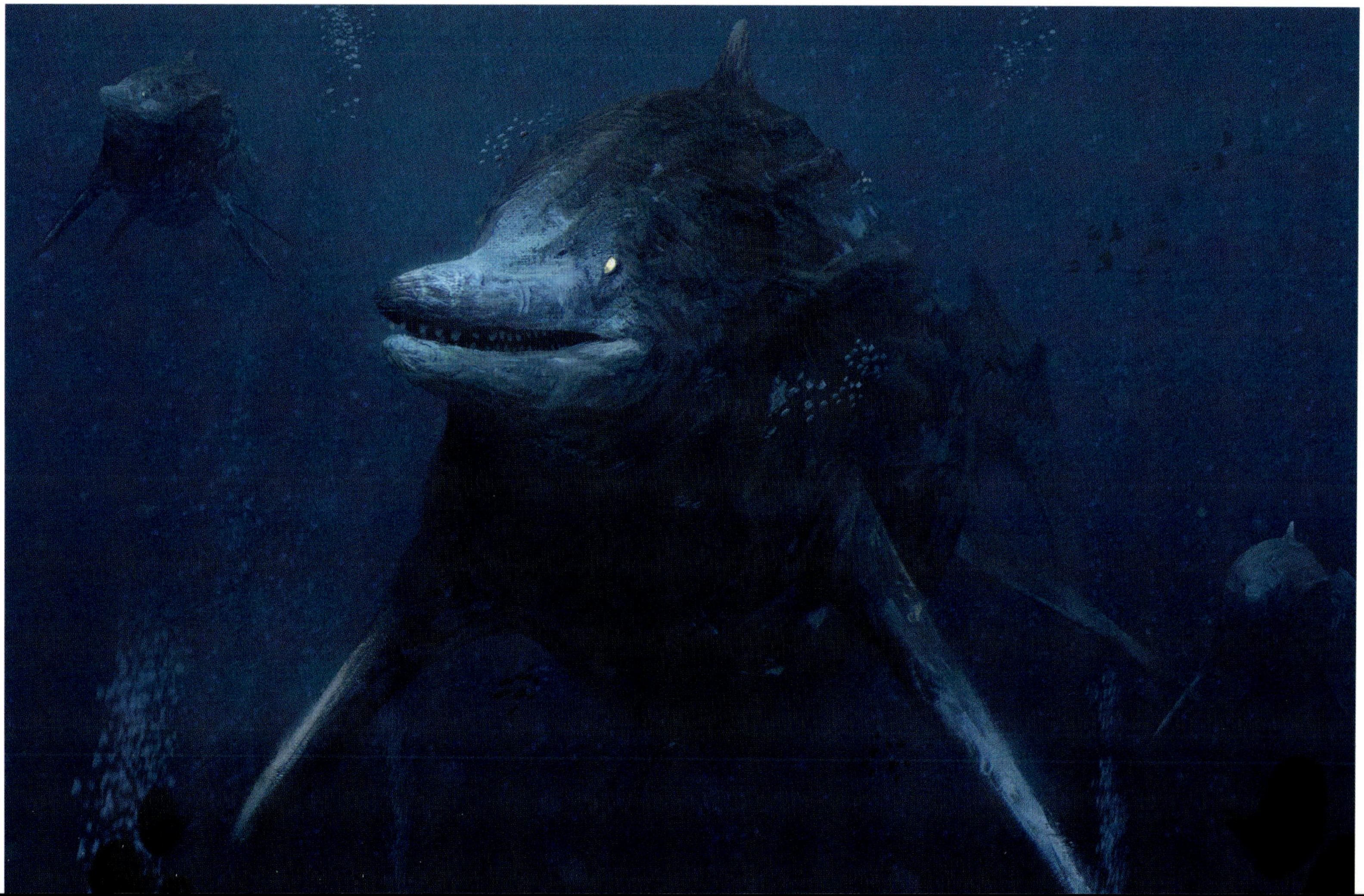

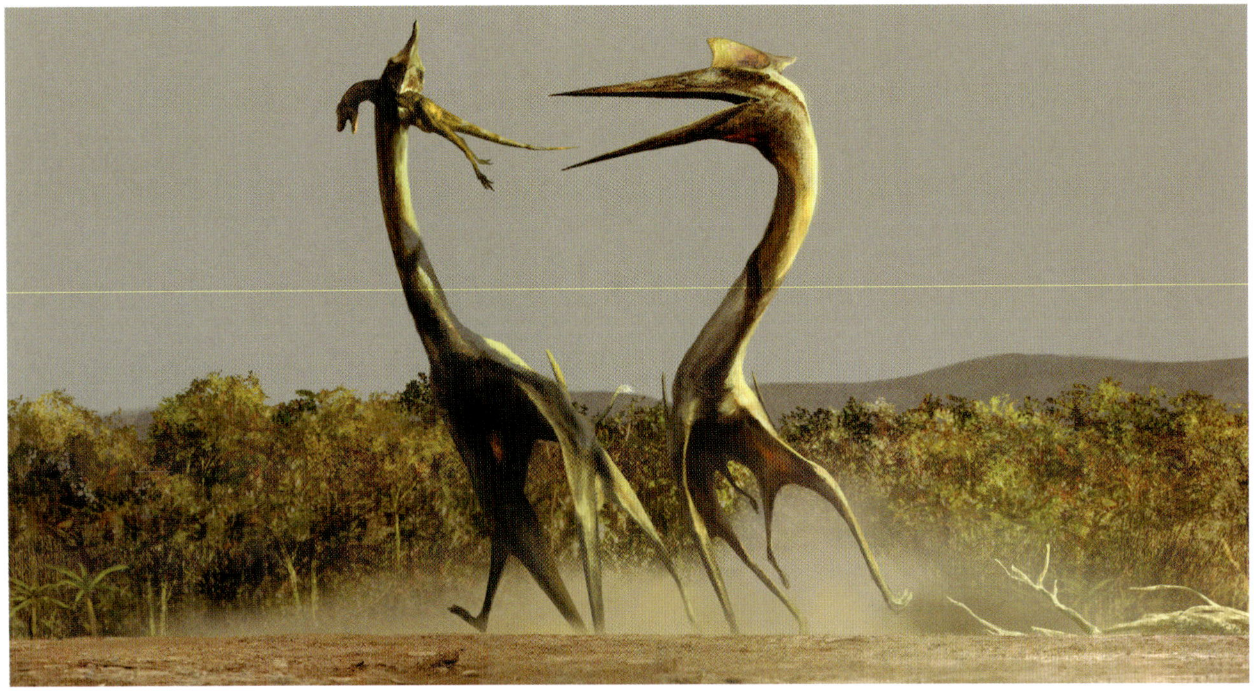

◄ QUETZ QUARREL

It's looking increasingly likely that azhdarchid pterosaurs – once visualised as 'ultra-light air beings' – lived a more grounded existence. Giraffe-sized *Quetzalcoatlus* of Late Cretaceous North America probably had a spindly, leggy appearance, but strong bones and large hip and upper arm muscles still gave it a turn of speed and an ability to pivot and turn when running.

• Digital | 2023

▼ *VARANUS*

The monitor lizard family – the Varanidae – dates back to the Mesozoic, but the biggest was *Varanus priscus*, from Pleistocene Australia. It was once thought to grow up to 7m (23ft) in length, but modern estimates are more conservative. Most experts now think that a total length of around 5.5m (18ft) was perhaps more typical. *V. priscus* did not survive Australia's megafaunal extinction event 50,000 years ago.

• Digital | 2022

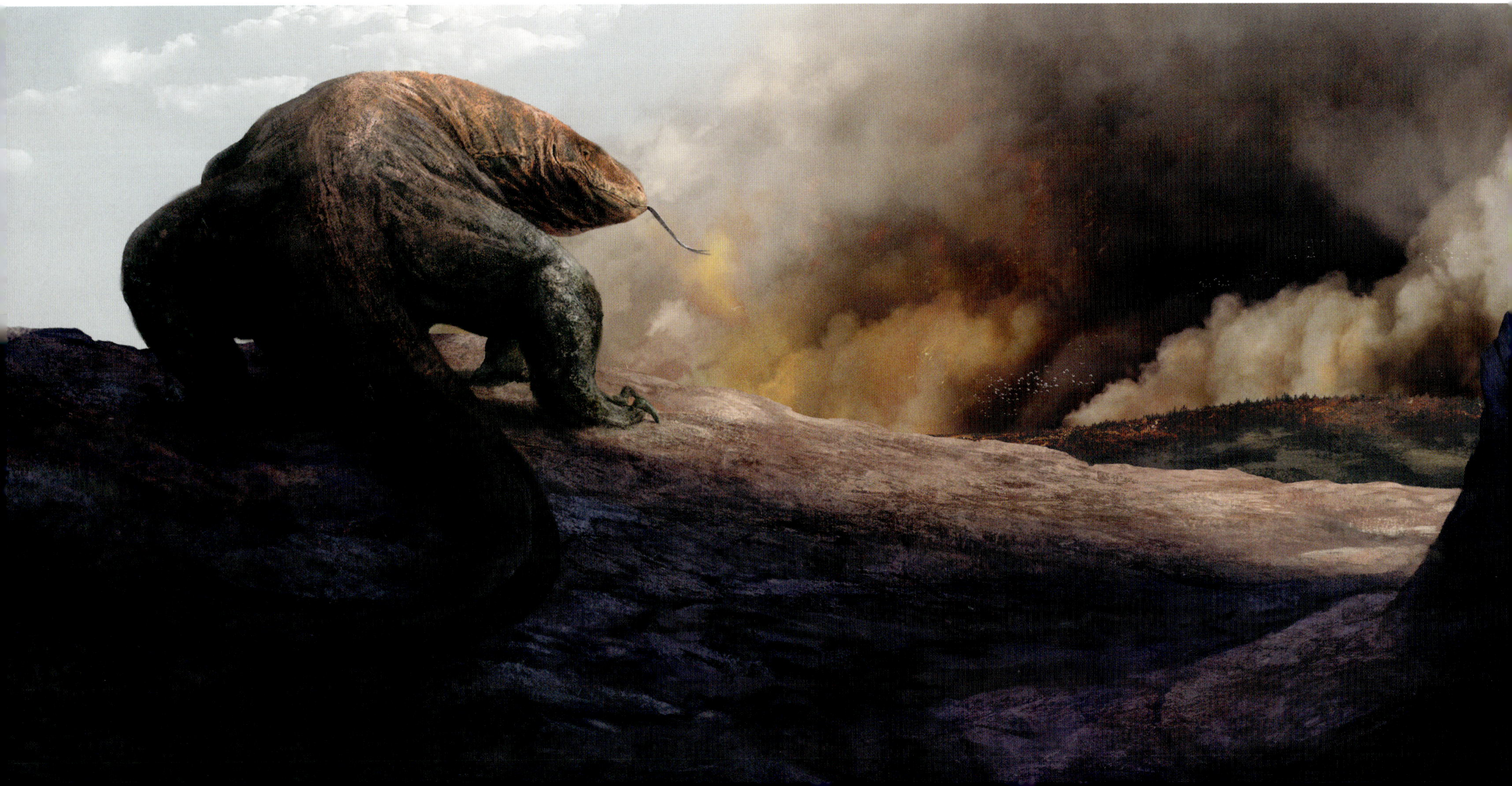

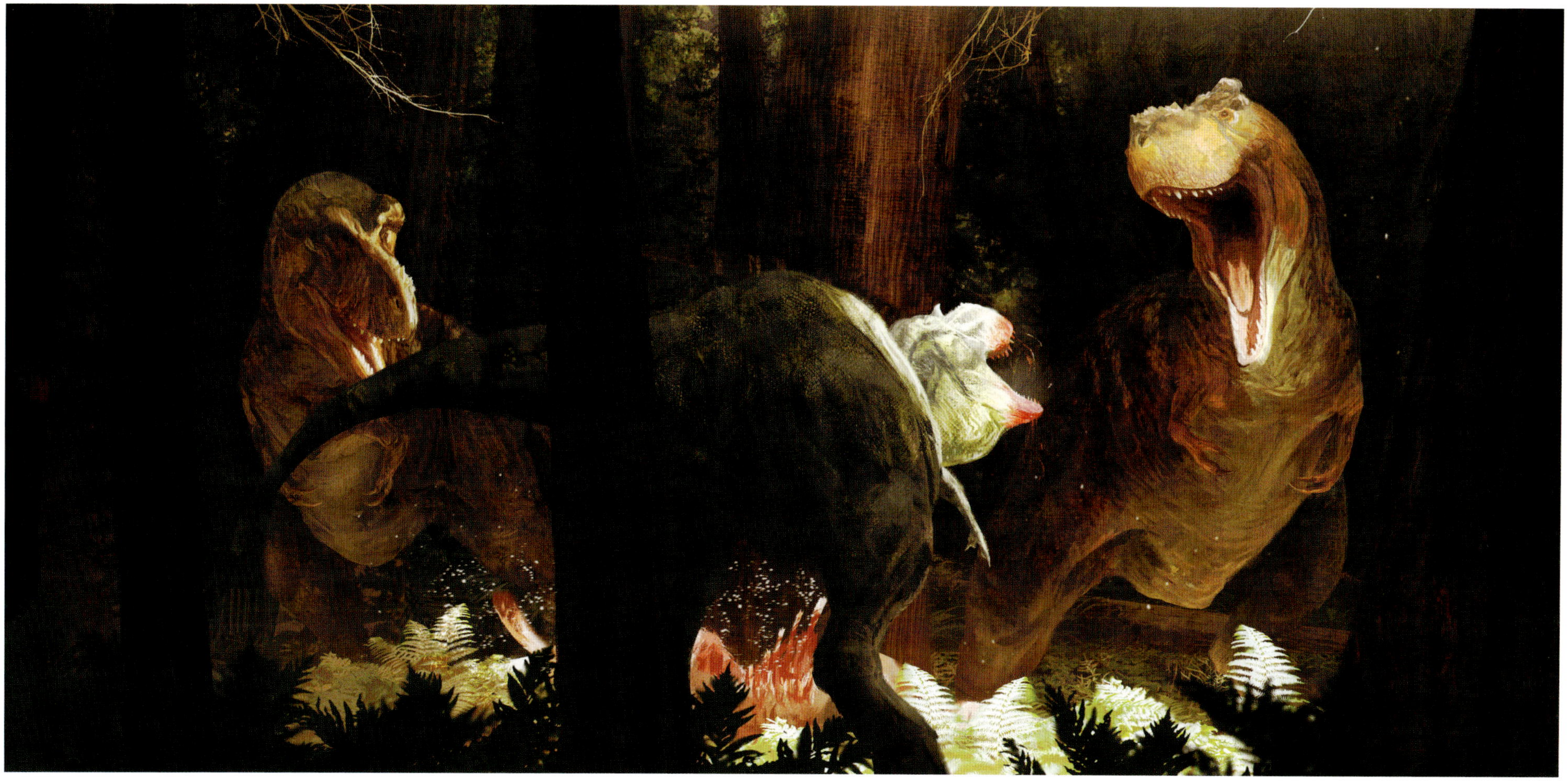

▲ USURPERS

This piece depicts an interaction between differing clades of the tyrannosaurid family: hypothetical Asian tyrannosaurines and a North American albertosaurine, *Albertosaurus*. The former type has crossed the Beringian land bridge from Cretaceous Asia into North America, where they will supplant the latter and evolve into *T. rex*. While studies have suggested such a scenario, it currently remains just a hypothesis.
• Digital | 2021

YUTYRANNUS ▶

Described in 2012 from the Early Cretaceous of northeastern China, *Yutyrannus* caused a sensation, covered as it was in a coat of filamentous feathers. It is, at 9m (29.5ft) long, the largest known dinosaur with a definite feathery covering. It lived in a cool, upland environment where the flora and weather must sometimes have recalled that of modern northern Europe.
• Digital | 2021

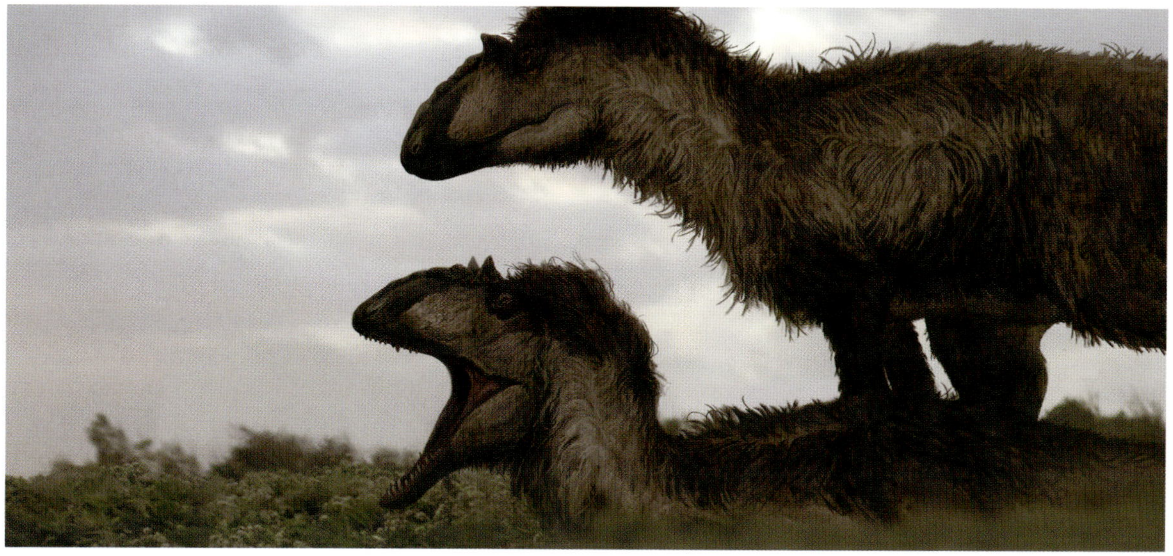

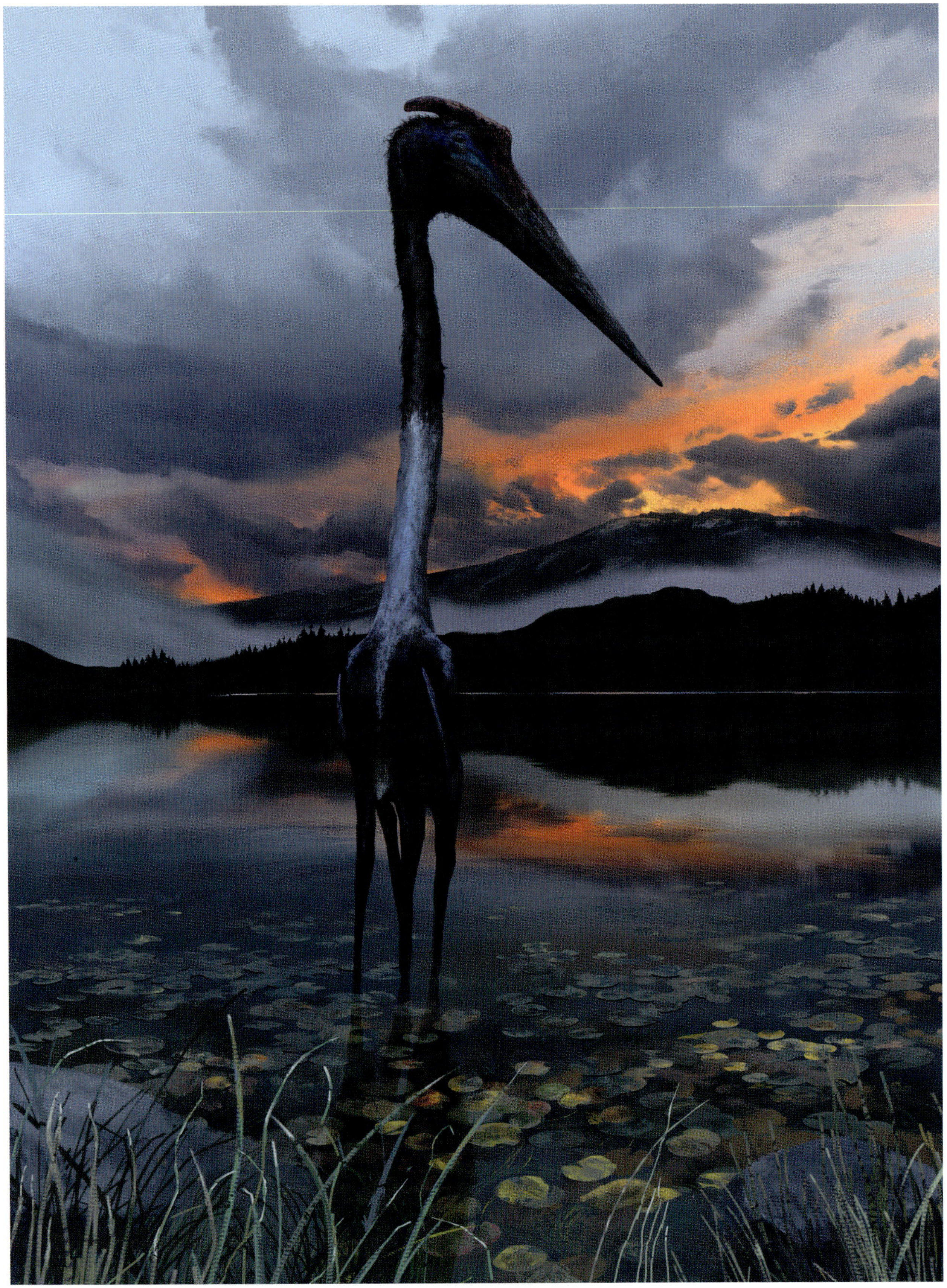

◂ *ARAMBOURGIANIA*

Long ignored and known only to specialists, the Jordanian giant azhdarchid *Arambourgiania* – discovered in the 1940s and named in 1959 – is an extreme member of its group, possessing an almost ridiculously long, slender neck. It presumably looked skinny and impossibly slight relative to robust members of its group, like *Hatzegopteryx* of Romania. The small, narrow feet of azhdarchids meant that they were not specialised for habitual wading. But anatomy, they say, is not destiny.

• Digital | 2023

FIRE HUNTER ▸

In recent years, it has become increasingly appreciated how fire-prone some Mesozoic environments were; common enough that animals may even have adapted to their regularity. Azhdarchids – able to arrive and depart from the skies, equipped with excellent vision and good at grabbing the deceased or injured from the edges of a burn – look well suited to exploit fire-prone landscapes.

• Digital | 2021

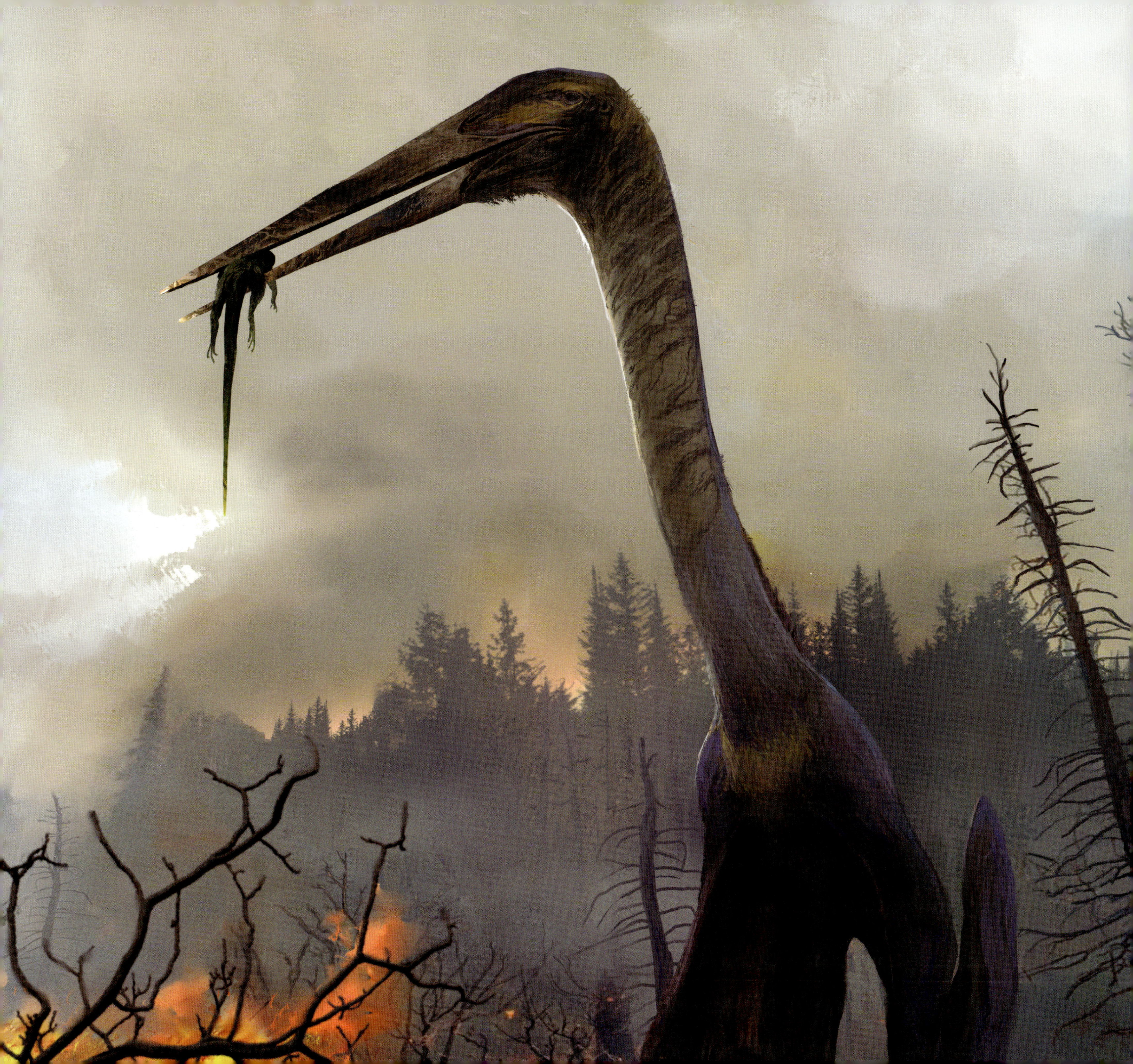

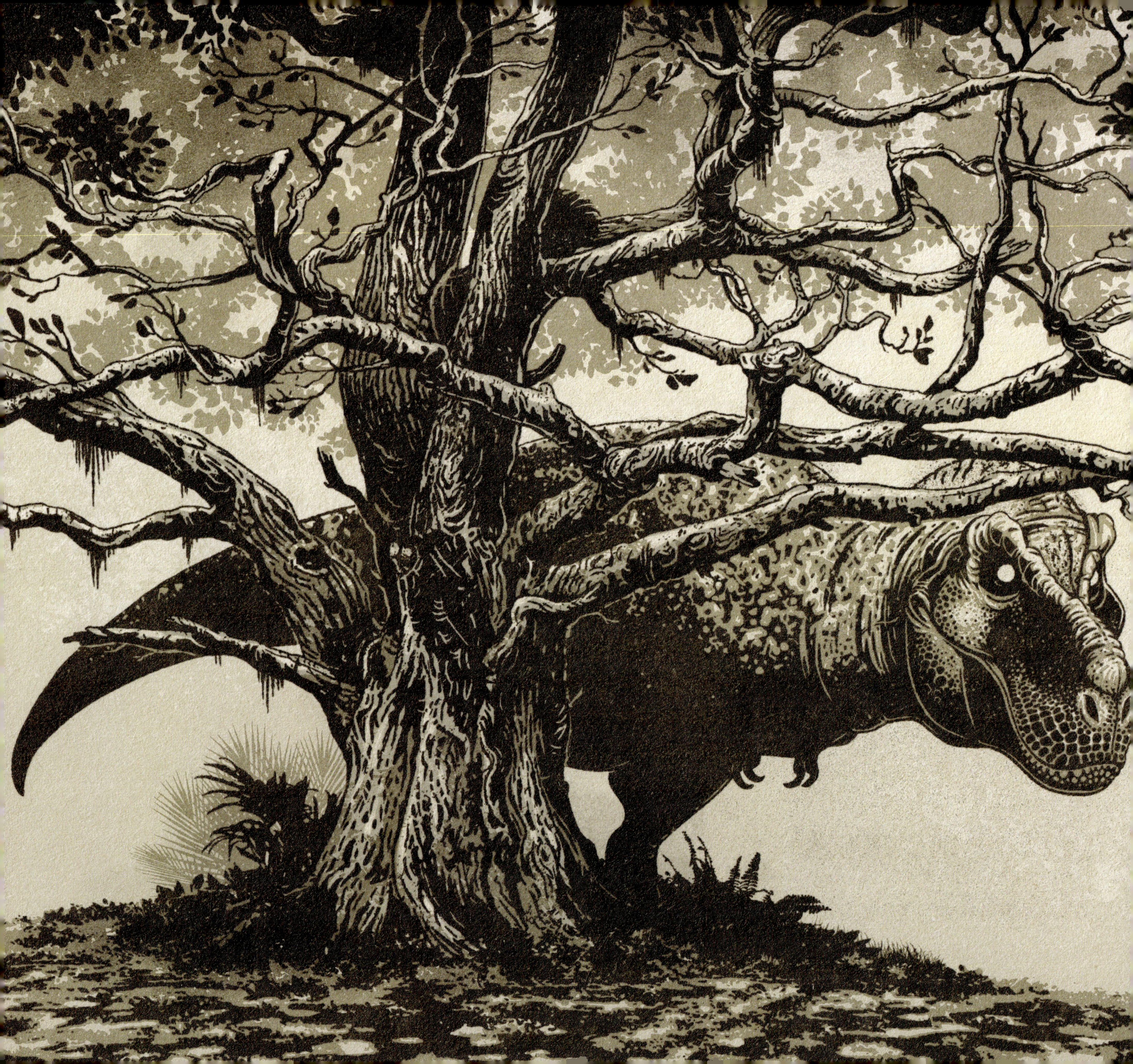

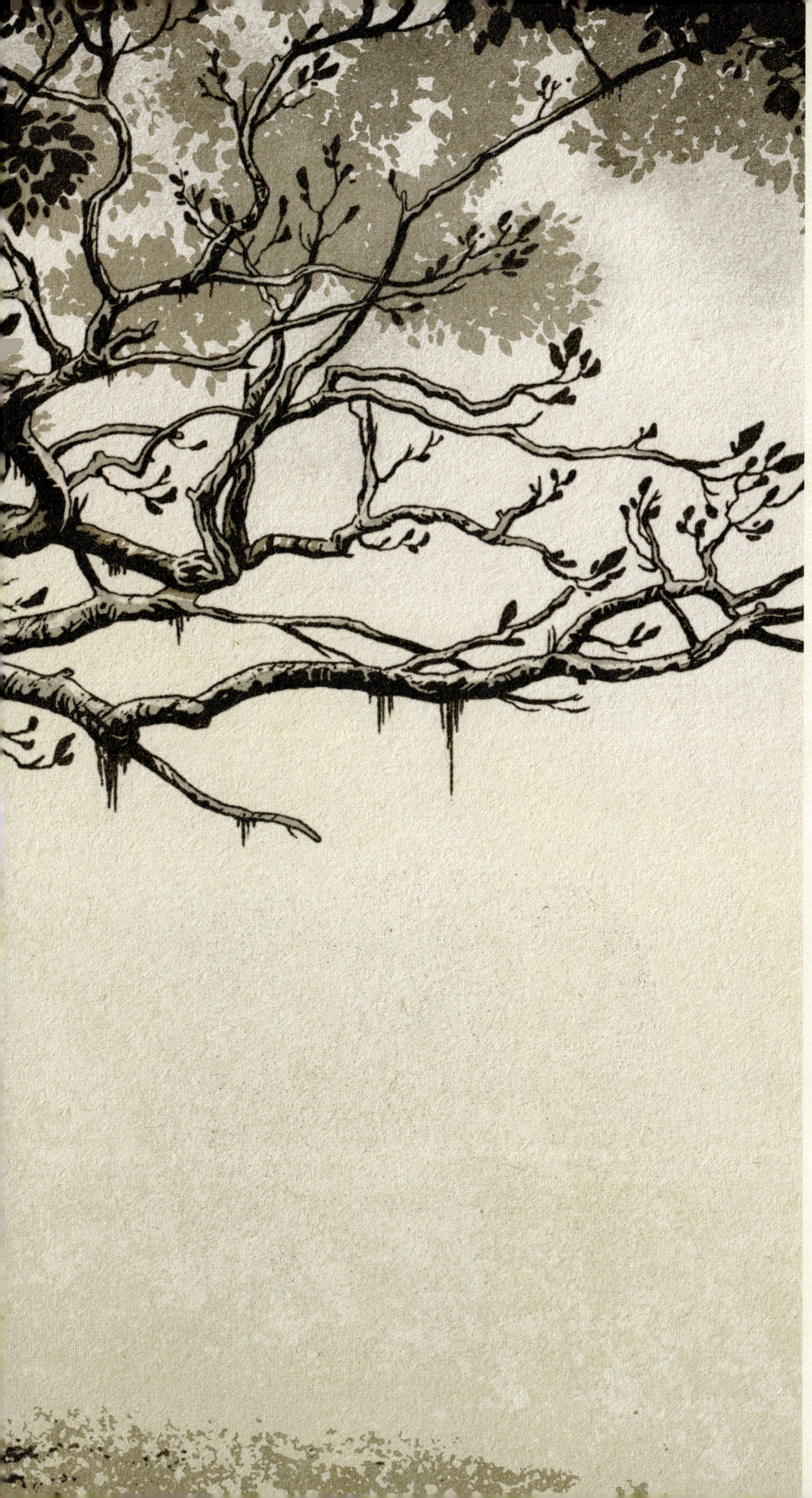

STIEVEN VAN DER POORTEN

STIEVEN IS A freelance illustrator and graphic designer from the Flemish part of Belgium. A few years ago, he felt an urge to get back to his roots and draw what he really loved: dinosaurs and other extinct creatures (which took some studying and re-educating). Stieven is mostly known for his stylised but up-to-date depictions of prehistoric animals, which often place them in 'slice of prehistoric life' dioramas and cutaway-style scenes that focus on a specific geological formation. These reflect Stieven's fondness for animal books, informative posters, infographics and other educational art. He is particularly inspired by vintage palaeoart and educational pieces, sometimes experimenting with a certain look or style directly inspired by these older artworks. He also practises the use of art techniques from around the world.

He recently illustrated a children's book, *The Tyrannosaur's Feathers*, which won *The Week Junior* Book Awards: Animals and Nature category.

Besides drawing prehistoric animals, Stieven has a serious interest in food, travelling, science and all creatures great and small.

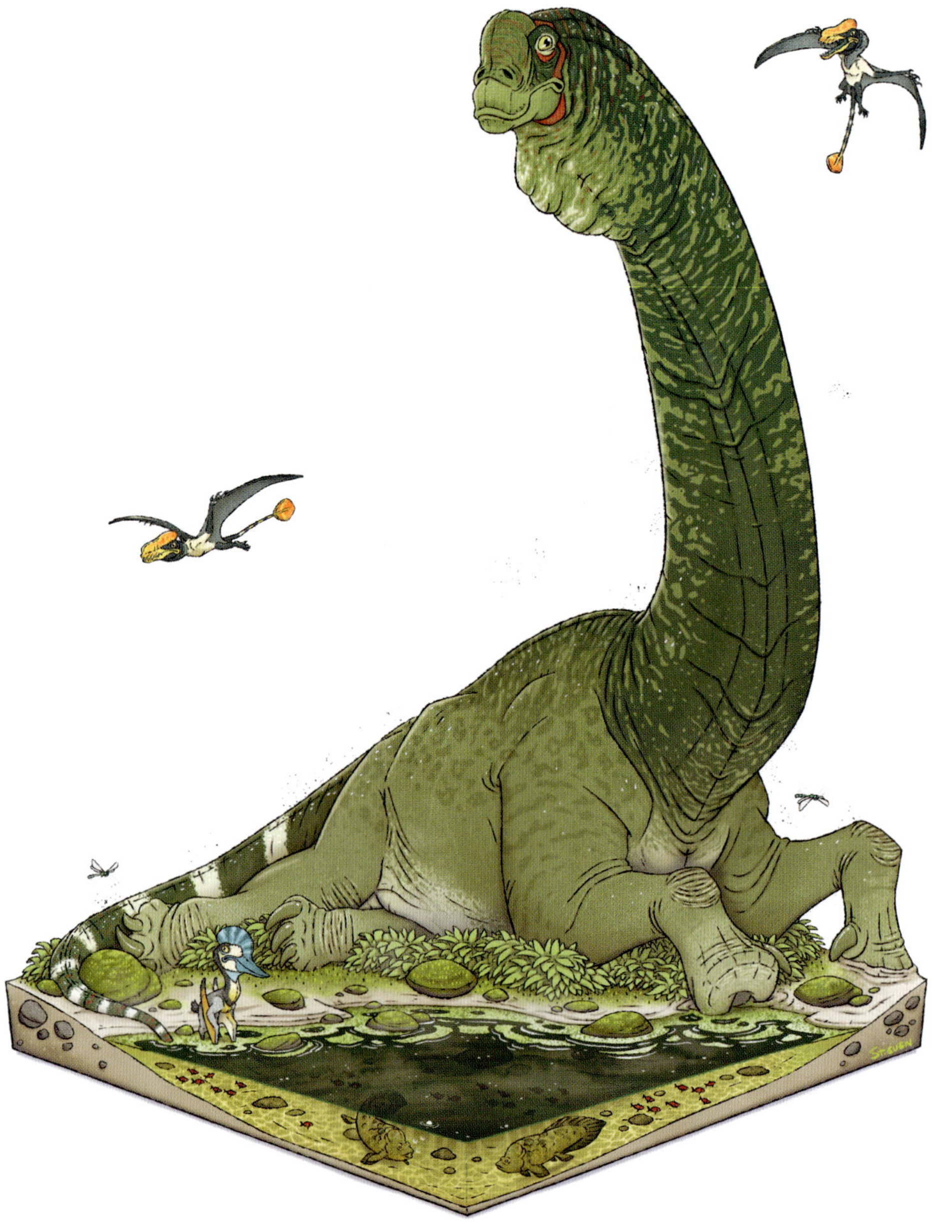

THE MORRISON

This image shows Late Jurassic animals from the famous Morrison Formation of the USA. Among the Morrison's many sauropods is the famous and gigantic *Brachiosaurus*, while Morrison pterosaurs include the long-tailed *Harpactognathus* and short-tailed *Kepodactylus*. Also visible are large ceratodontid lungfishes of a sort known from numerous Morrison fossil sites.
• Digital | 2022

[Previous pages]

T. REX

This stark, moody illustration shows the great and mighty *T. rex* in a monotone that might evoke the mist or fogginess typical of the forested, often humid, world this predator inhabited. Textural details of the facial scales and ridges covering the dinosaur's gigantic head have a rough-hewn quality not unlike the bark of the tree.
• Digital | 2023

JURASSIC ANTARCTICA

The rocks of Antarctica may hold a rich assortment of fossils, but they're hard to get to *and* mostly covered by ice! Partial confirmation came from the 1994 description of *Cryolophosaurus*, a theropod with a forward-curving crest. Smaller animals include insects, a dimorphodontid pterosaur and a cynodont. All are known from the Lower Jurassic Hanson Formation of the Transantarctic Mountains.
• Digital | 2023

NIOBRARA SCENE

At the edge of the Niobrara Sea – a warm, shallow marine incursion that covered Kansas and nearby during the Late Cretaceous – pterosaurs and seabirds gather. The massive size difference between *Pteranodon* and *Nyctosaurus* is obvious. The bird in flight is the gull-sized, tooth-jawed *Ichthyornis*; also here is the less well-known bird *Iaceornis* and the squid-like *Glyphiteuthis*.
• Digital | 2021

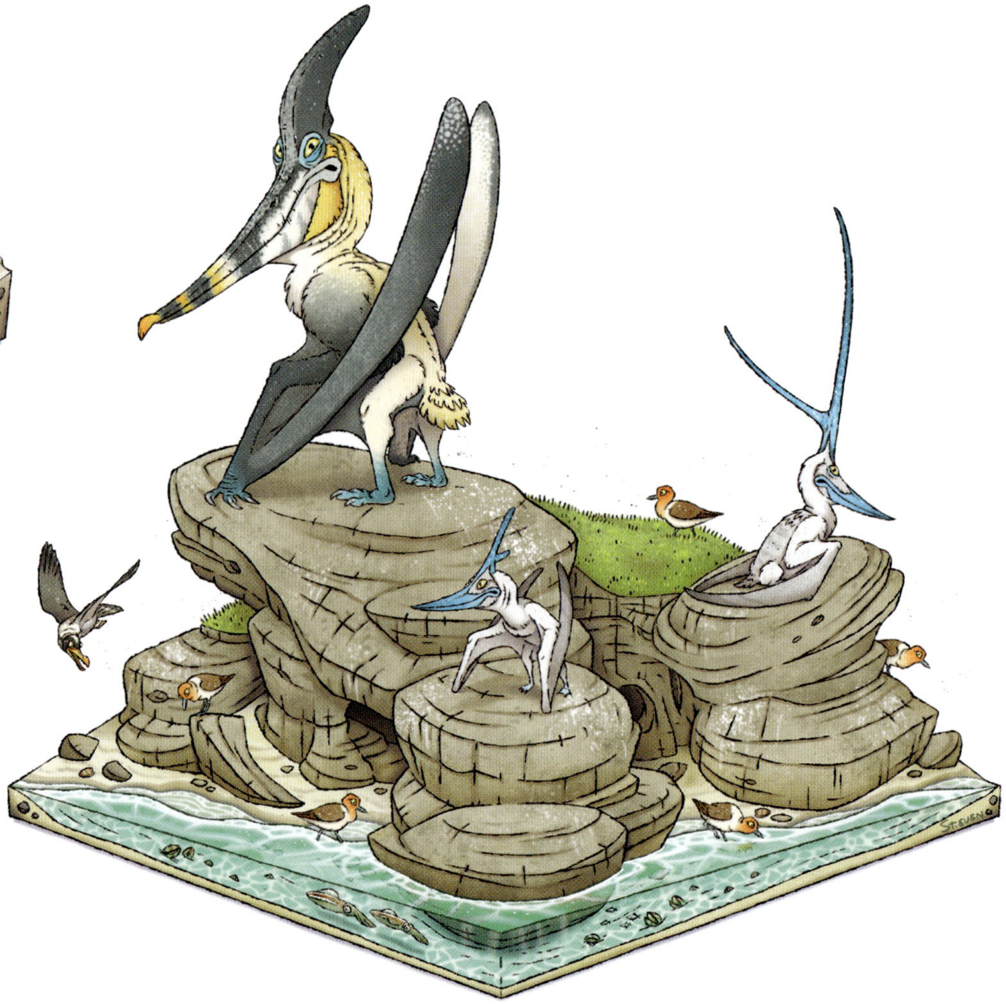

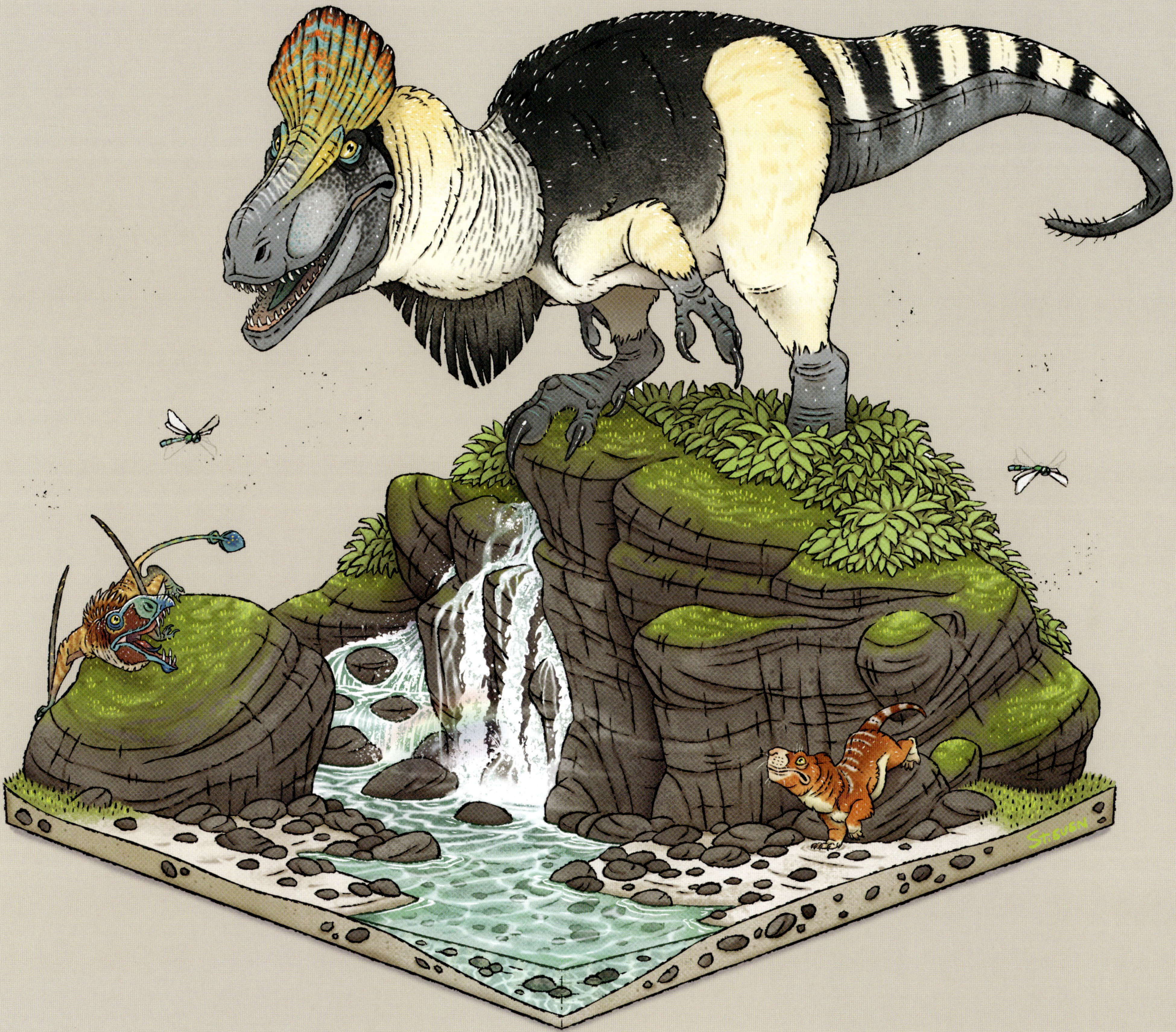
ST.EVEN

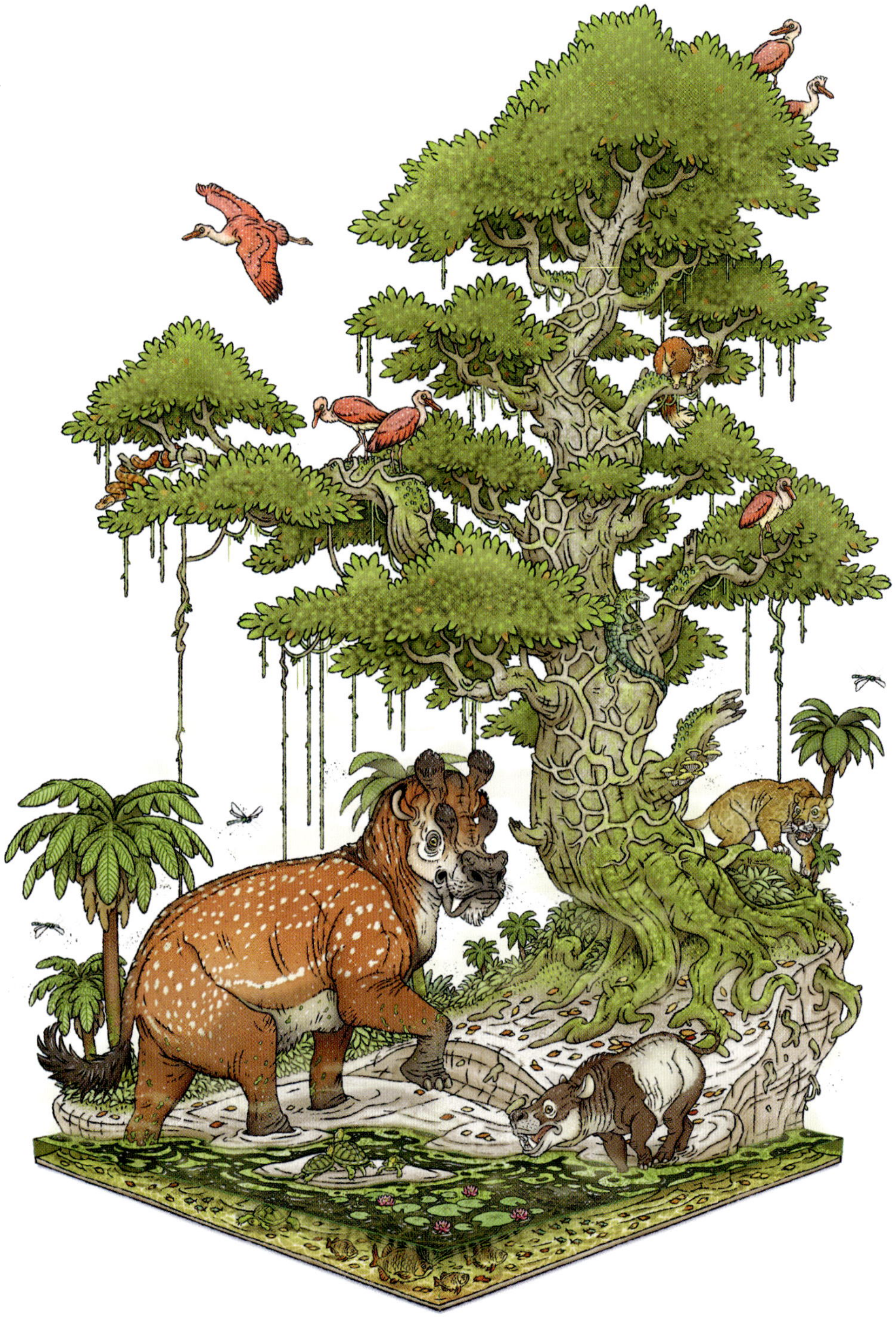

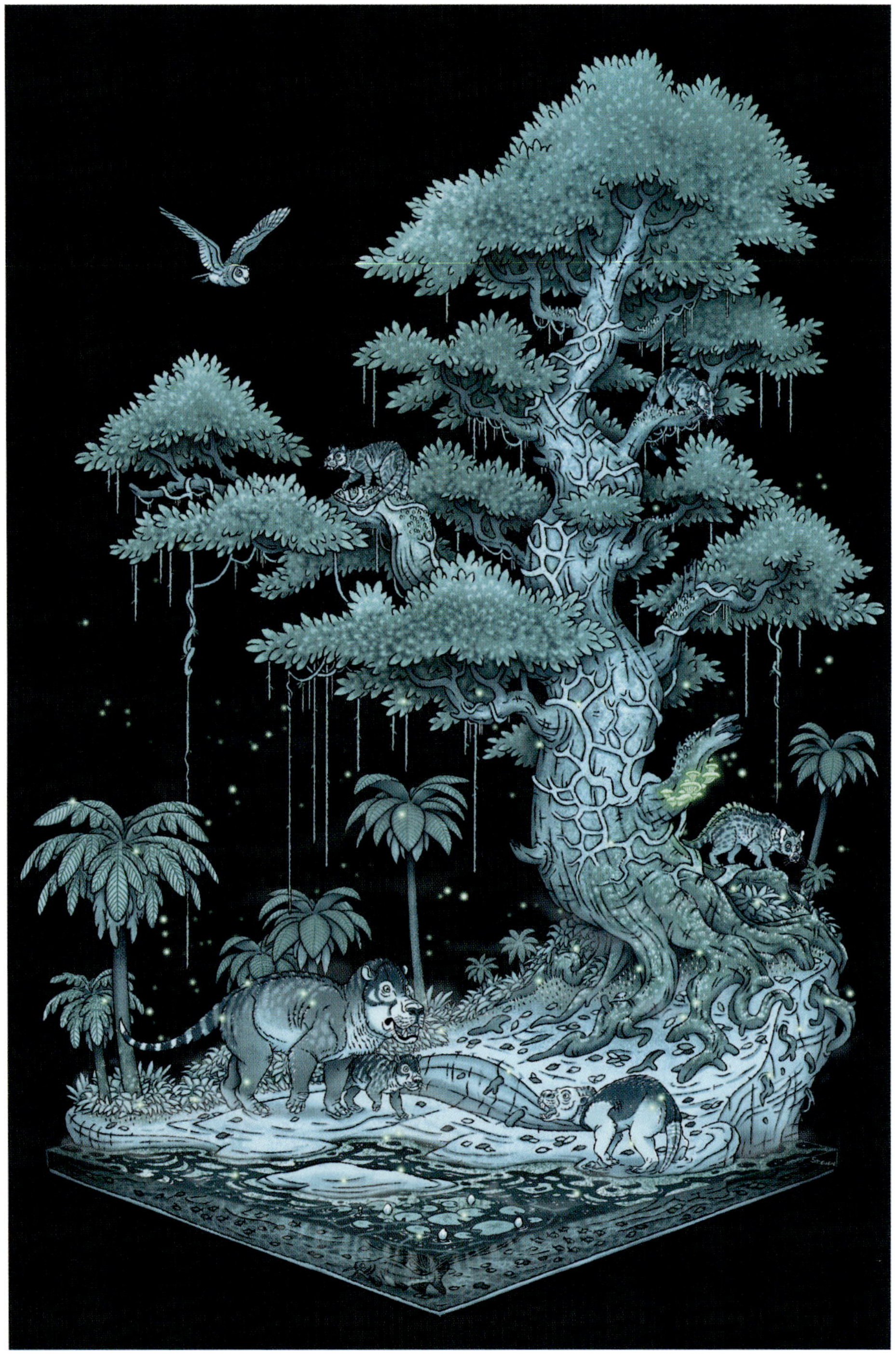

▲ BRIDGER FORMATION BY DAY

In a subtropical forest of the Bridger Formation of Wyoming, USA (and thus about 50 million years ago), a rhino-sized *Uintatherium* passes *Achaenodon*, an ancient relative of hippos, camels and deer. *Patriofelis*, a cat-like predator, lurks behind a tree where the ancient rodent *Paramys* hides high up on a branch. The bird is *Juncitarsus* – a cousin of flamingos.

• Digital | 2024

▲ BRIDGER FORMATION AT NIGHT

This night-time version of the Bridger Formation scene opposite shows a menagerie of strange mammals: the mesonychian *Harpagolestes* (at left, with juvenile) and the tillodont *Tillodon*. In the branches, vaguely cat-like *Machaeroides* is at left, civet-like *Tritemnodon* at right, and the superficially fox-like *Sinopa* is behind the tree. All belong to archaic, wholly extinct groups.

• Digital | 2024

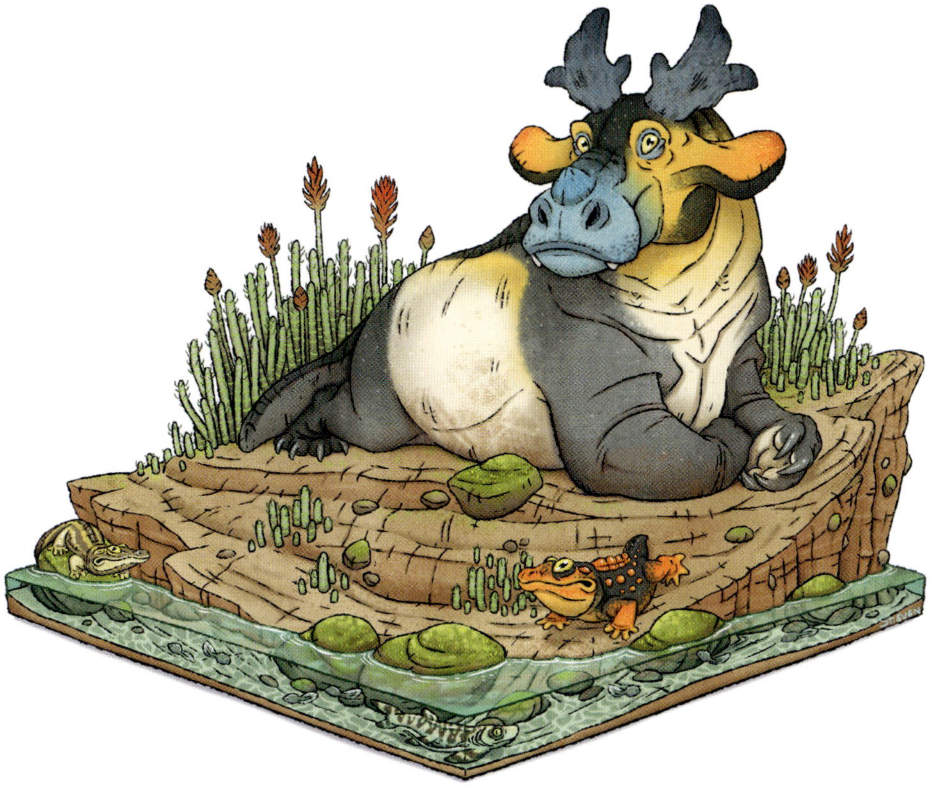

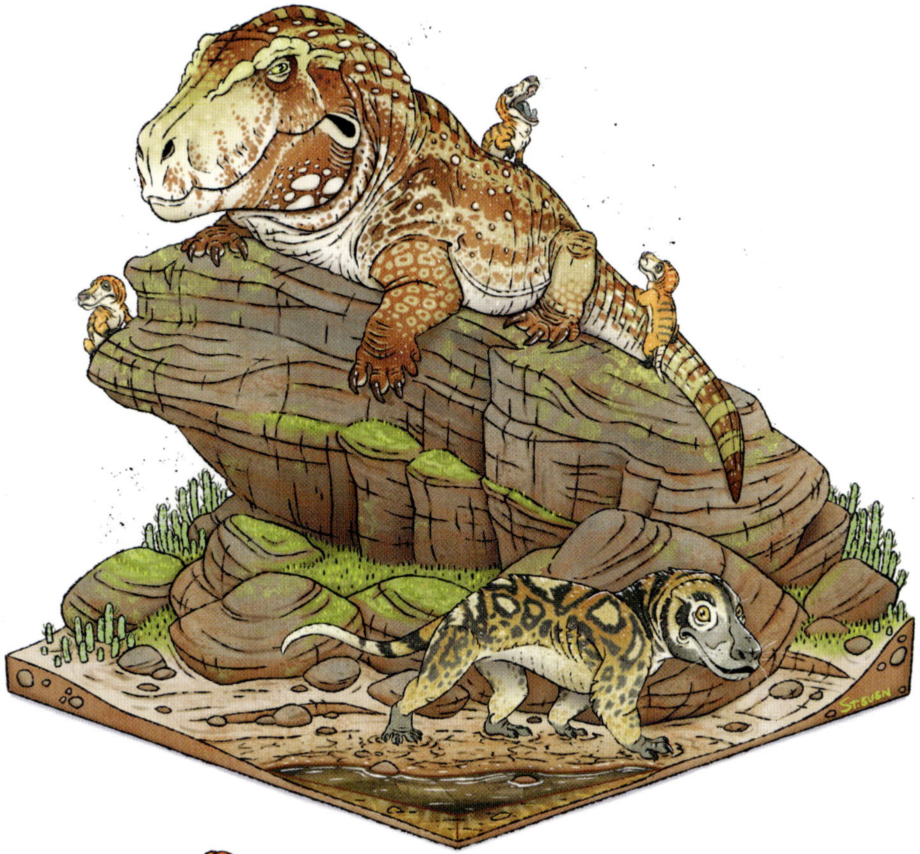

▲ BELEBEI FORMATION SCENE

In Permian Russia about 265 million years ago, the cow-sized synapsid *Estemmenosuchus* reclines on a riverbank. Its antler-like crests likely functioned like those of deer and antelopes, and we think that males had bigger, more elaborate structures than females. Other animals discovered in the Belebei Formation of Russia's Ural Mountains include *Kamacops*, a temnospondyl amphibian.
• Digital | 2021

CHINLE SCENE ▶

This Late Triassic environment shows animals and plants known from the Chinle Formation, an important geological unit associated with the Colorado Plateau and its adjacent landscape features. The big predator is *Postosuchus*, a rauisuchian, and hence part of the same great archosaurian group as crocodylians, not dinosaurs. A *Coelophysis* and a temnospondyl amphibian keep a safe distance.
• Digital | 2020

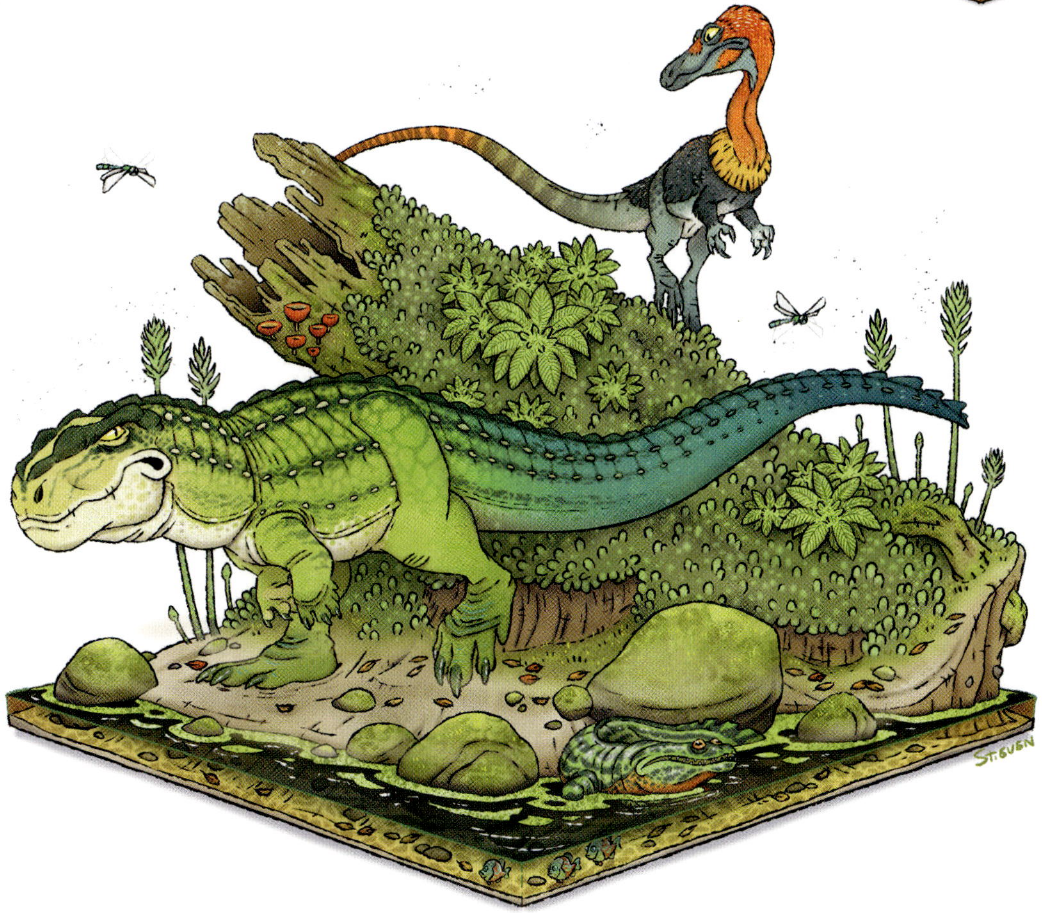

▲ TRIASSIC SOUTH AFRICA

During the Early and Middle Triassic, members of the archosaur lineage shared ecosystems with synapsids, cousins of the earliest mammals. The grumpy-looking reptile is *Erythrosuchus*, a South African relative of archosaurs, 5m (16.4ft) in length with an especially large, long skull. Its neighbourhood synapsids included the small, probably omnivorous *Bauria* and *Cynognathus*, a larger, dog-sized predator.
• Digital | 2024

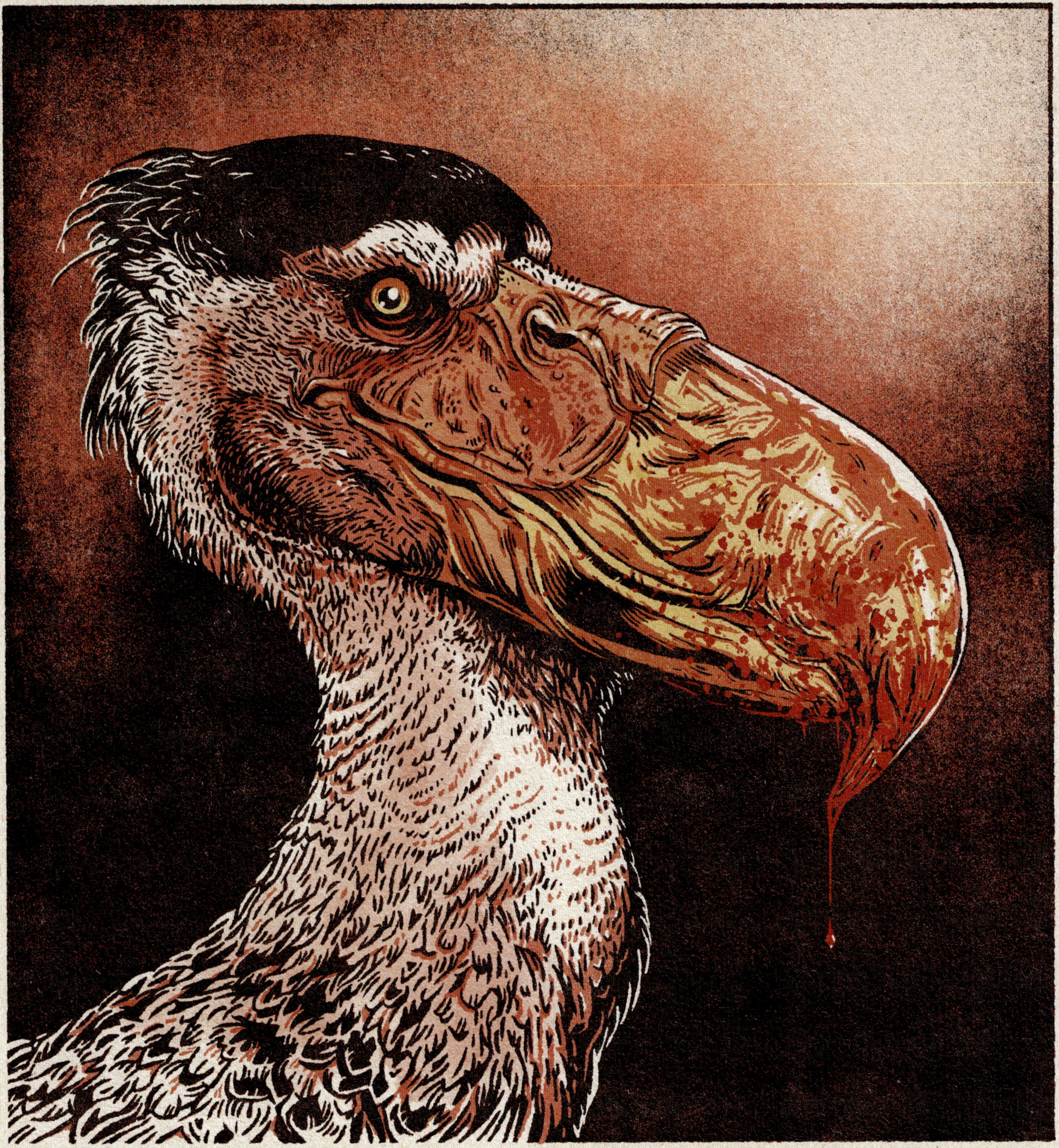

◂ *KELENKEN*

Named in honour of a mythical spirit visualised as a great bird of prey, Argentina's *Kelenken* – named in 2007 – sported a great hooked bill, borne on a skull 71cm (2.3ft) long, allowing it to kill prey as big as an adult human. *Kelenken* is a phorusrhacid or terror bird, a group that inhabited South, Central and southern North America.

• Digital | 2023

▾ *MICRORAPTOR*

This dynamic portrait of *Microraptor* highlights the agility and complex anatomy of these feathered, crow-sized dinosaurs. Views on the aerial abilities of *Microraptor* and its kin vary greatly. However, most evidence indicates that they were gliders of cluttered forested environments. As such, they were probably expert at making dynamic, sweeping moves and plunging dives.

• Digital | 2022

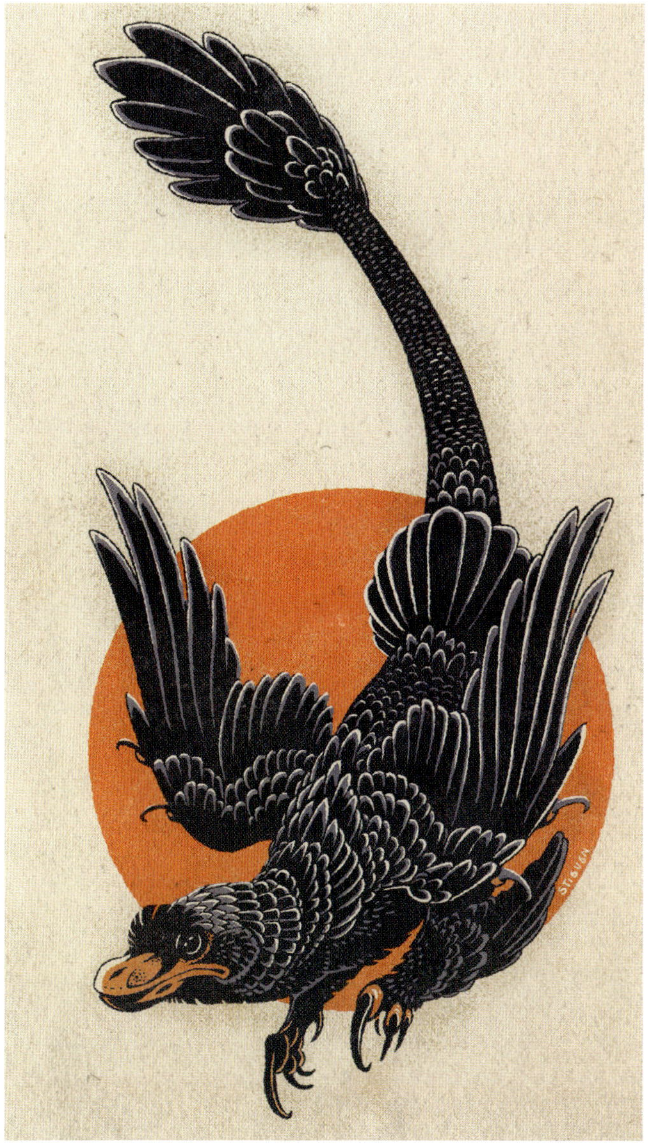

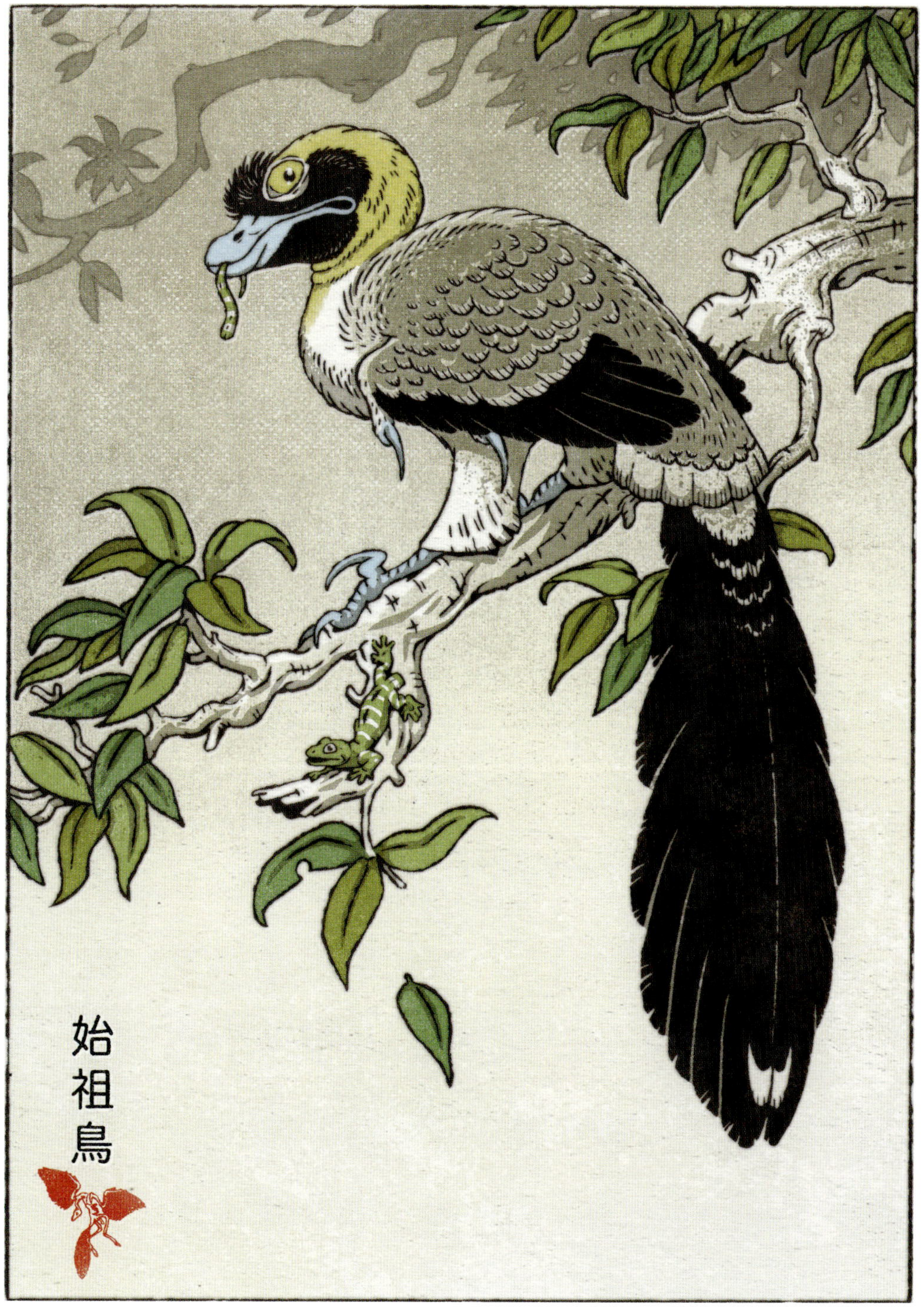

▴ UKIYO-E *ARCHAEOPTERYX*

Depicted in the ukiyo-e Japanese woodcut style, this arboreal *Archaeopteryx* has the colour scheme of a forest bird and is foraging for small prey among the branches. Decades of argument surround the tree-climbing abilities of this animal and most evidence suggests it was predominantly a ground-dweller. This does not, however, preclude an occasional climb amongst trees and shrubs.

• Digital | 2021

◂ *SKORPIOVENATOR* NEST

Skorpiovenator is a Late Cretaceous abelisaurid theropod from Argentina, a cousin of *Carnotaurus*. In this infographic, produced as part of Stieven's merchandise range, we see the speculative internal arrangement of its nest mound. A fun detail – inspired by behaviour in modern reptiles and fossil finds too – is that lizard-like rhynchocephalians are shown taking advantage of the site for their own nesting purposes.

• Digital | 2023

▾ *CYNTHIACETUS*

Certain whales of the Eocene – those living between 56 and 34 million years ago – would have resembled modern whales in some respects but be very unfamiliar in others. This is *Cynthiacetus*, a relative of the giant *Basilosaurus* originally described from Mississippi, USA, and later reported from Peru. Its archaic features include visible hindlimbs, the possible presence of facial whiskers and only partially retracted nostrils.

• Digital | 2023

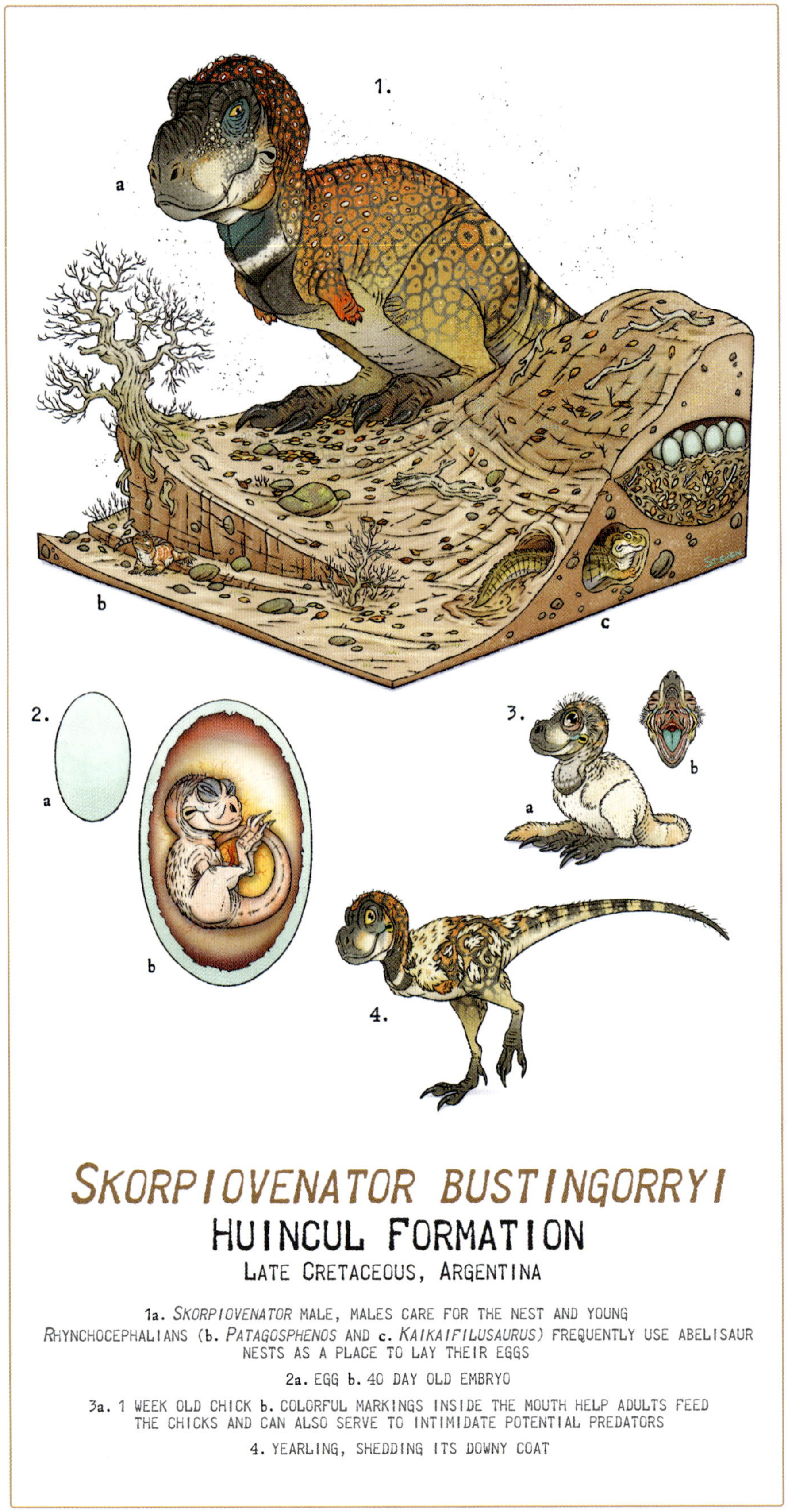

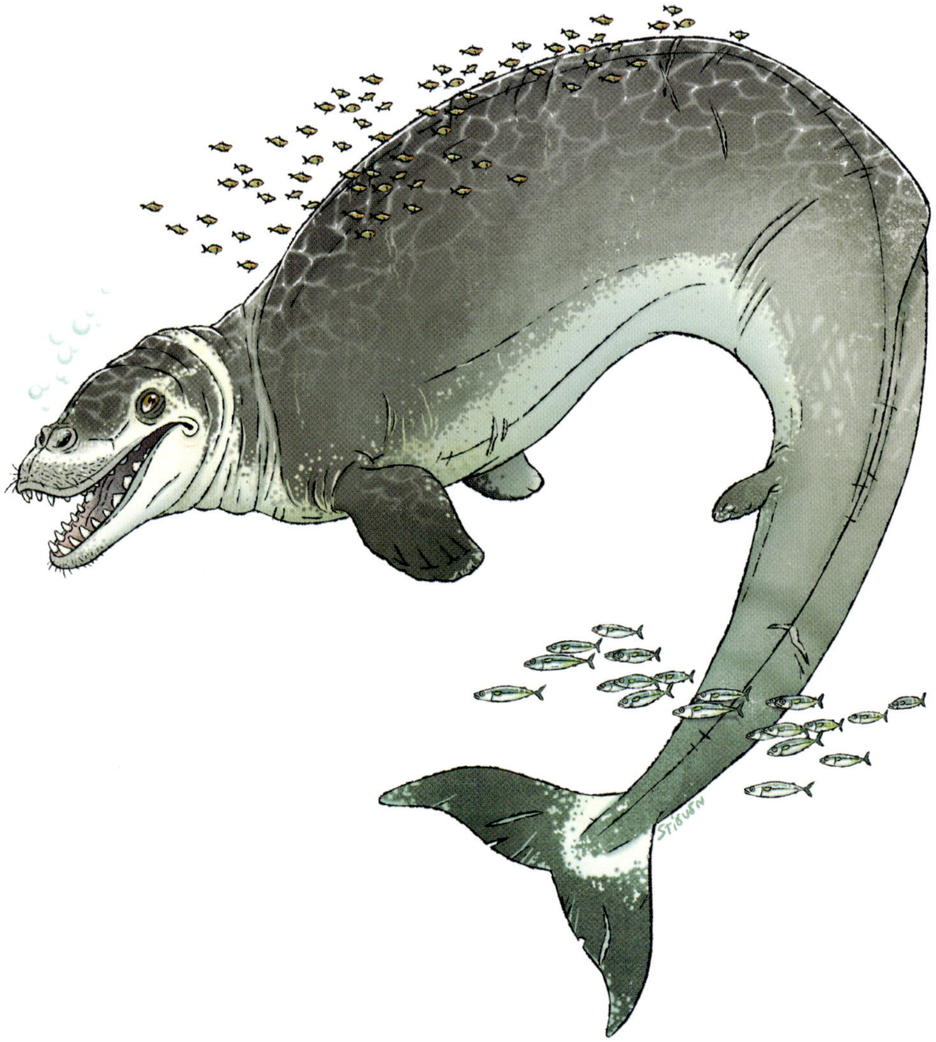

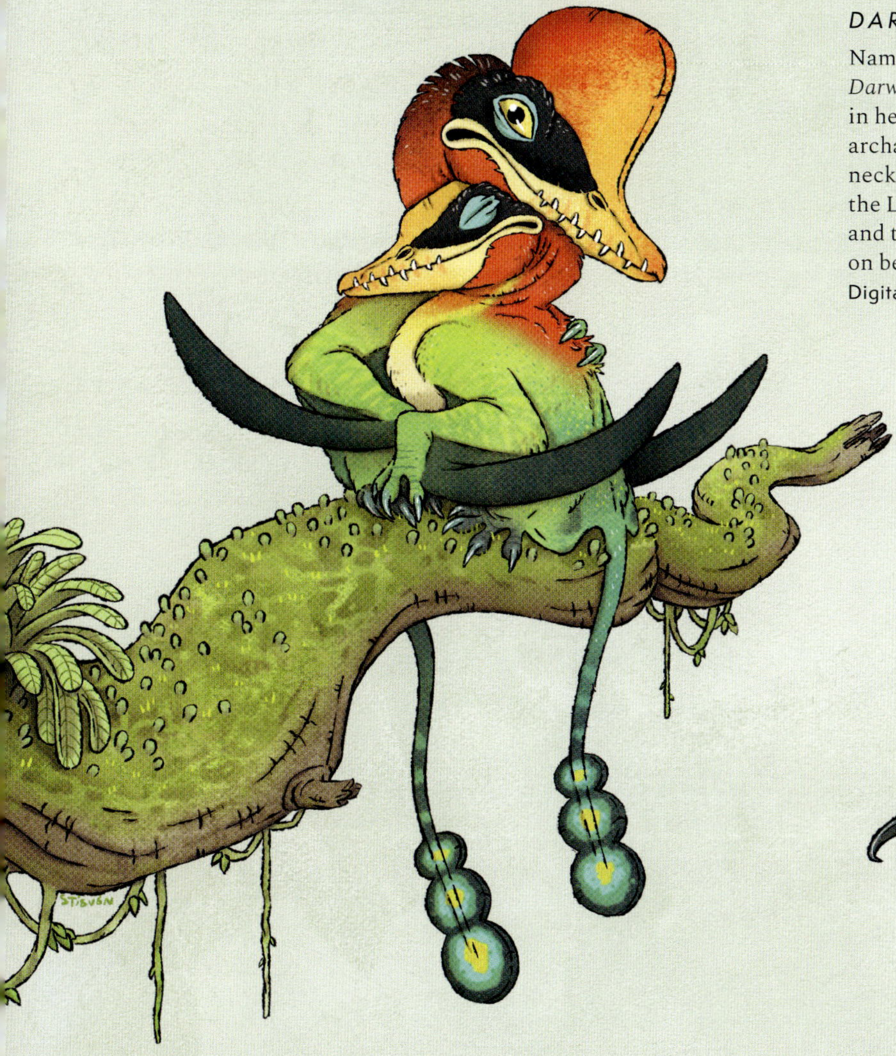

DARWINOPTERUS ▸

Named from the Middle Jurassic of China in 2010, *Darwinopterus* ('Darwin's wing') proved sensational in helping to bridge the evolutionary gap between archaic, long-tailed pterosaurs and the longer-necked, short-tailed group mostly associated with the Late Jurassic and Cretaceous. It was crested and the shape of its teeth suggest that it preyed on beetles and similar insects.
Digital | 2020

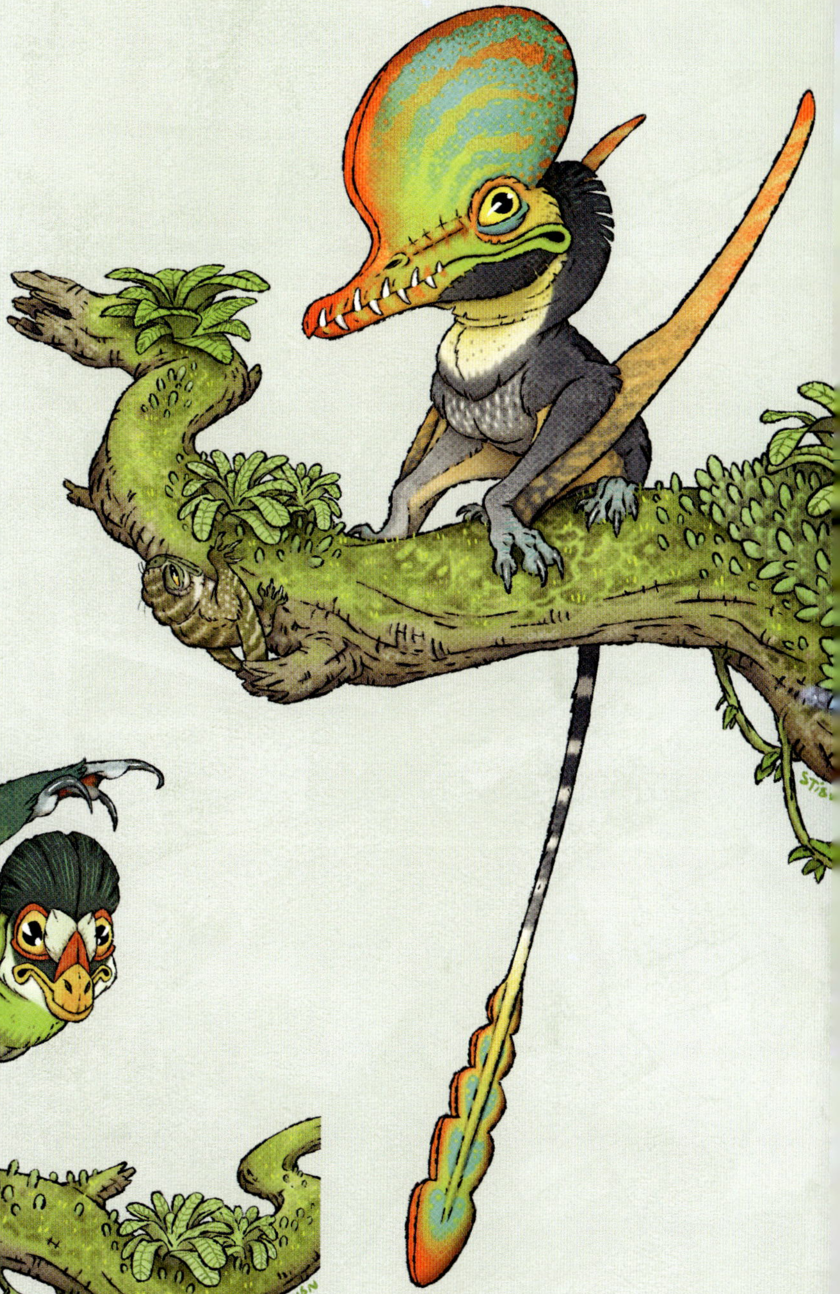

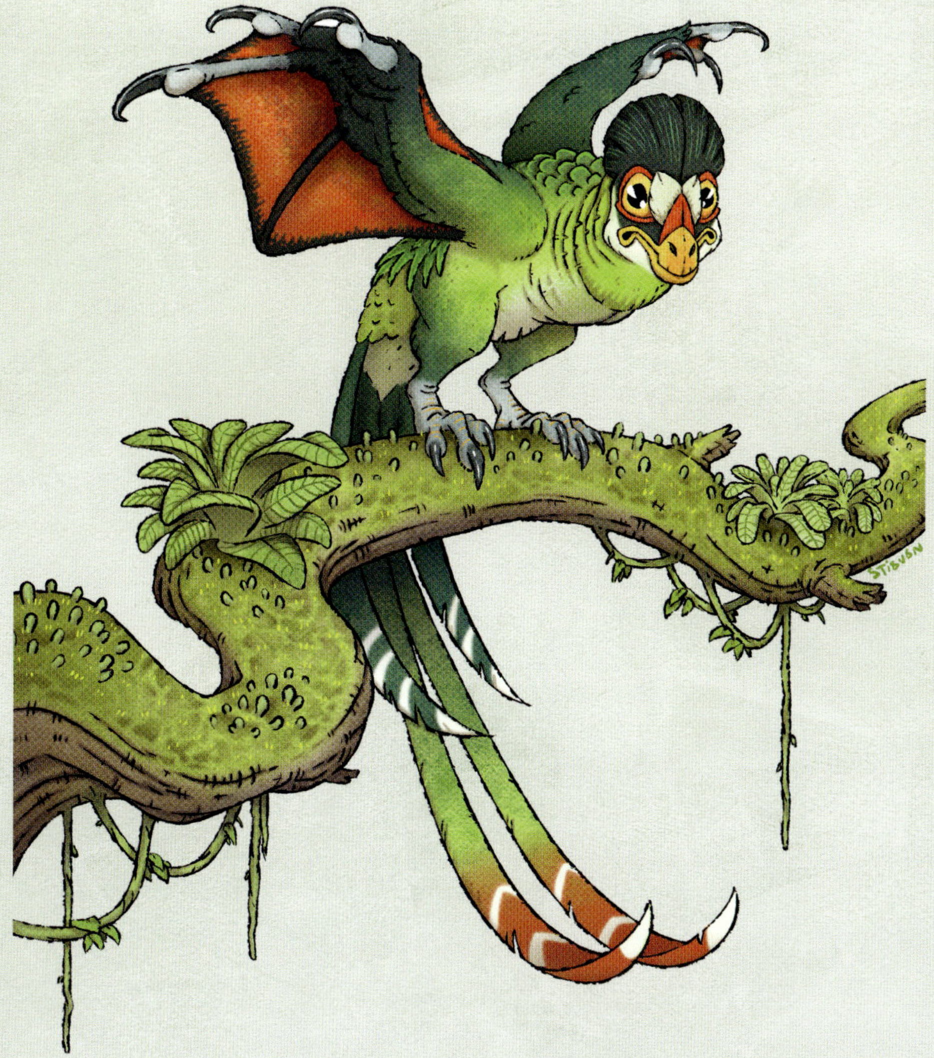

▴ *KUNPENGOPTERUS* LOVEBIRDS

The idea of paired individuals of pterosaurs like these *Kunpengopterus* snuggling together lovebird-style might seem fanciful but is not – despite preconceptions – out of keeping with behaviour seen in modern lizards and crocodylians. Pterosaurs of this sort appear to have been sexually dimorphic, with only males possessing the big, rounded crests.
• Digital | 2021

YI ▸

Deliberately named to have the shortest name in all of dinosaurdom, *Yi* (pronounced *ee*), from the Middle or Late Jurassic of China, was a short-snouted, membranous-winged little theropod. Its hindlimbs possess features suggestive of a climbing ability. Several animals of this sort – known as scansoriopterygids – are known from the Chinese Jurassic record.
• Digital | 2021

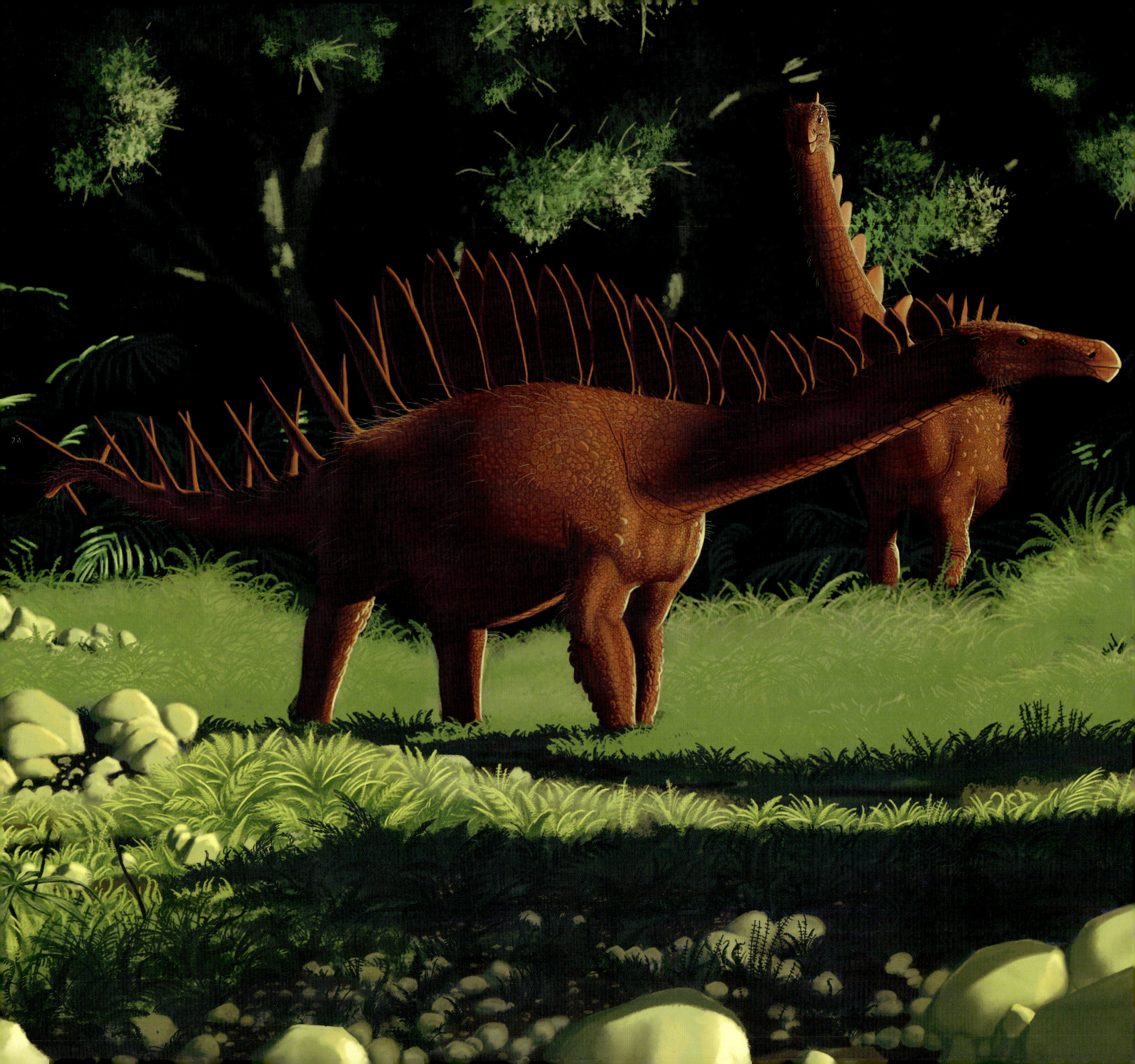

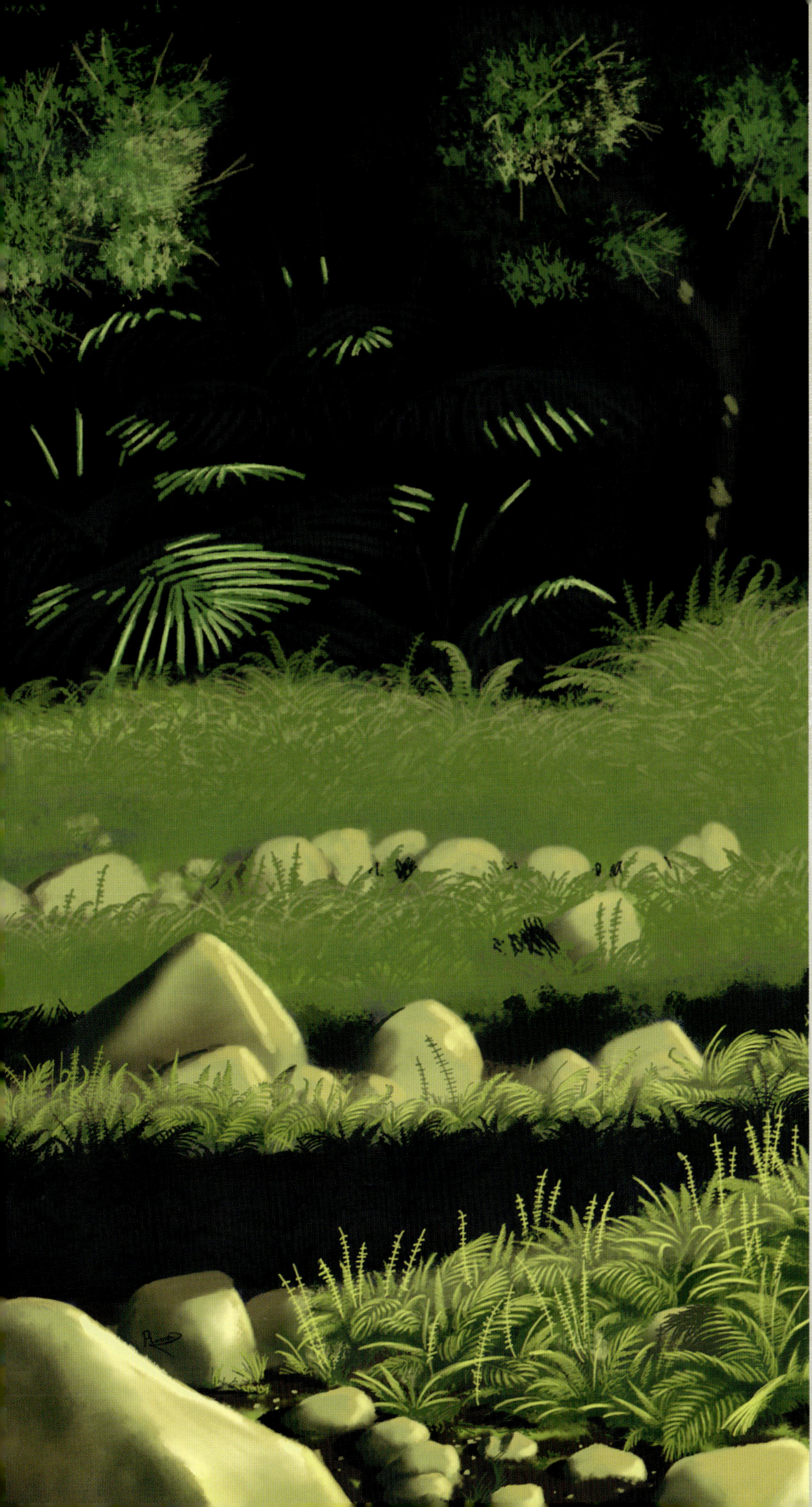

RAMÓN M GONZÁLEZ

THOUGH CURRENTLY AN agricultural engineering student at Universidad de La Rioja, Ramón was born and raised in Madrid, amongst 10-storey buildings that his young mind often transformed into sauropod necks. Being a typical nerdy kid who enjoyed scribbling, he found dinosaurs an ideal muse after watching his first dinosaur documentary. His growing fascination meant he never really grew out of his 'drawing phase', even managing to depict some dinosaurs on the cover of his high school science magazine.

Ramón believes that one hobby supported the other. With more practice came the desire to know more. Childhood magazines turned into books, which were later joined by the Internet's vast wellsprings of papers, blogs, forums – and art, lots of it. He was quickly starstruck by the most incredible pieces of work by palaeoartists, people who he began to revere. This online revolution let him meet many people who he could chat to and even paint with. They helped ease his transition to digital art while encouraging him to publish his pieces online.

His future plans include studying for a master's degree, perhaps as a researcher in pest control, soil microbiomes – or something else entirely! Meanwhile, he hopes his artistic career develops into something he can be proud of, and that he keeps learning and having fun as he continues along his creative journey.

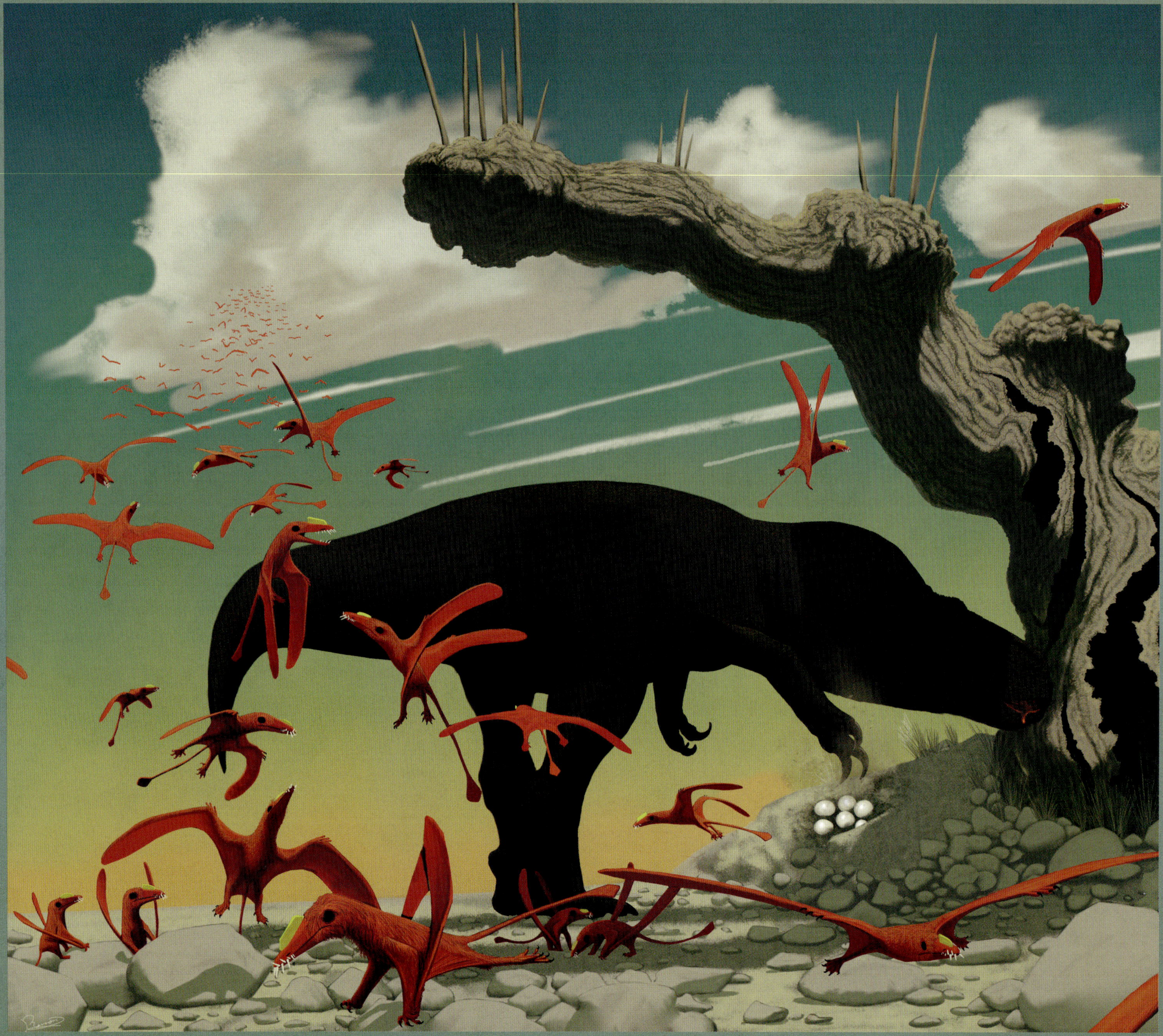

◂ *TORVOSAURUS & HARPACTOGNATHUS*

Better known as an apex predator of the Late Jurassic Morrison Formation, *Torvosaurus*, a large megalosaurid, may not have been above helping itself to the contents of an unguarded sauropod nest. In doing so, it's attracted the attention of a *Harpactognathus* flock, also out for an easy meal.

• Digital | 2023

▾ *CENTROSAURUS & VITIS*

High on a hill is not a lonely goatherd but the ceratopsian *Centrosaurus*, pausing its grazing of a *Vitis* grapevine to study approaching storm clouds. This Late Cretaceous horned dinosaur – originally described from Canada's Dinosaur Park Formation – is well known because of the hundreds of individuals found in bonebeds caused by perhaps flooding or drought.

• Digital | 2024

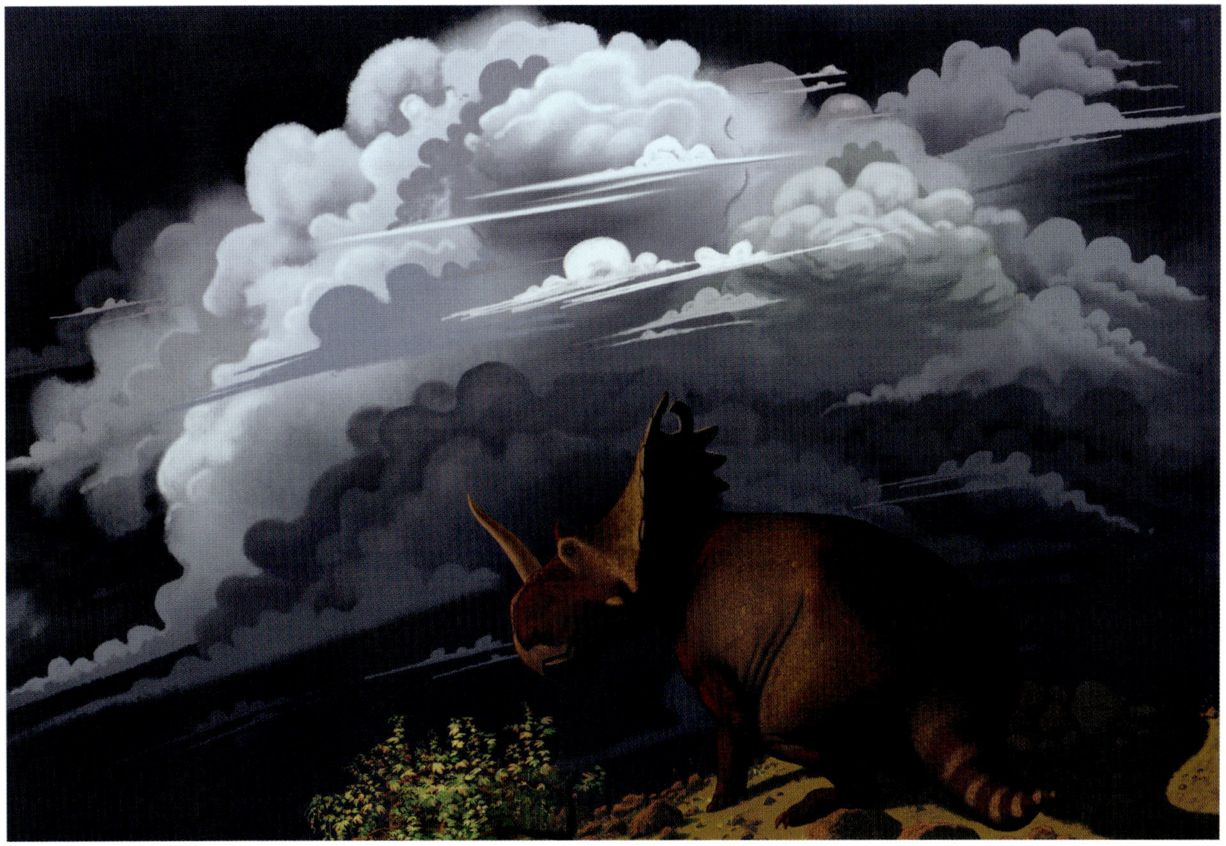

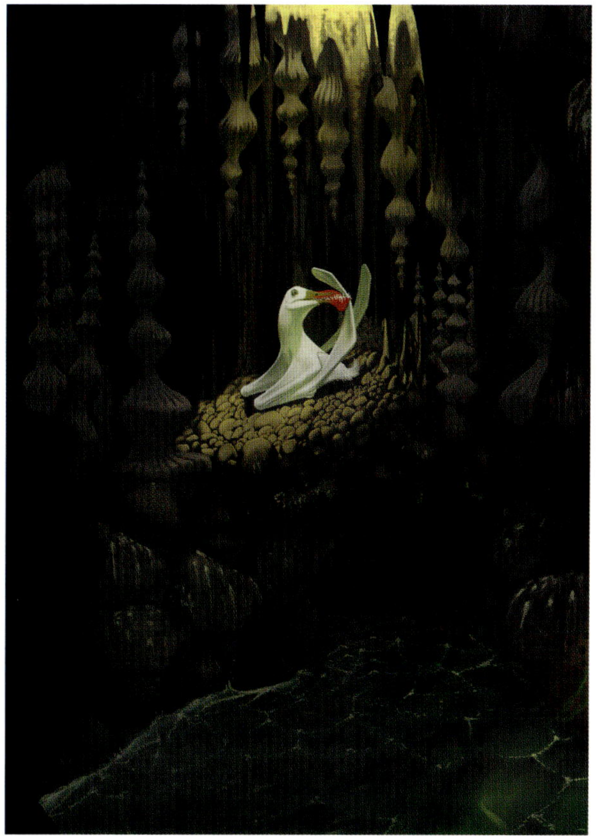

▴ *ANHANGUERA*

A piscivorous pterosaur with a 4.6m (15ft) wingspan, several species of *Anhanguera* are known from Early Cretaceous Brazil and Late Cretaceous Morocco. Ramón's portrayal shows an *Anhanguera* roosting in a cave system connected to the seashore. Like many of its relatives in the anhanguerid family, it sports semi-circular crests at the tip of both jaws.

• Digital | 2023

[Previous pages]

MIRAGAIA

Two individuals of *Miragaia*, a long-necked, Late Jurassic stegosaur from Portugal, pause in a clearing. This image is notable for giving the animals a detailed, complex scalation but combining it with the presence of hair-like filaments. Some experts think that *Miragaia* is one and the same as *Dacentrurus*, a stegosaur initially named from England.

• Digital | 2022

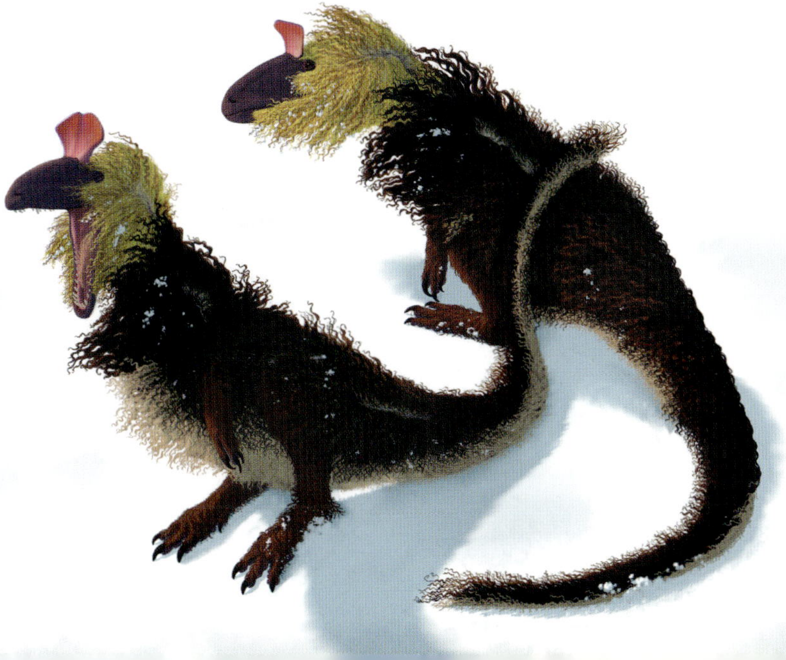

◂ *CRYOLOPHOSAURUS*

The Early Jurassic theropod *Cryolophosaurus* is regularly portrayed amidst ice and snow, thereby living up to its name – the 'frozen crested lizard'. However, during the Early Jurassic, the global climate was warmer than today and Antarctica, still part of the supercontinent Gondwana, was further north. Still, the environment could cool enough for wintery weather, especially in more hostile inland areas.

• Digital | 2024

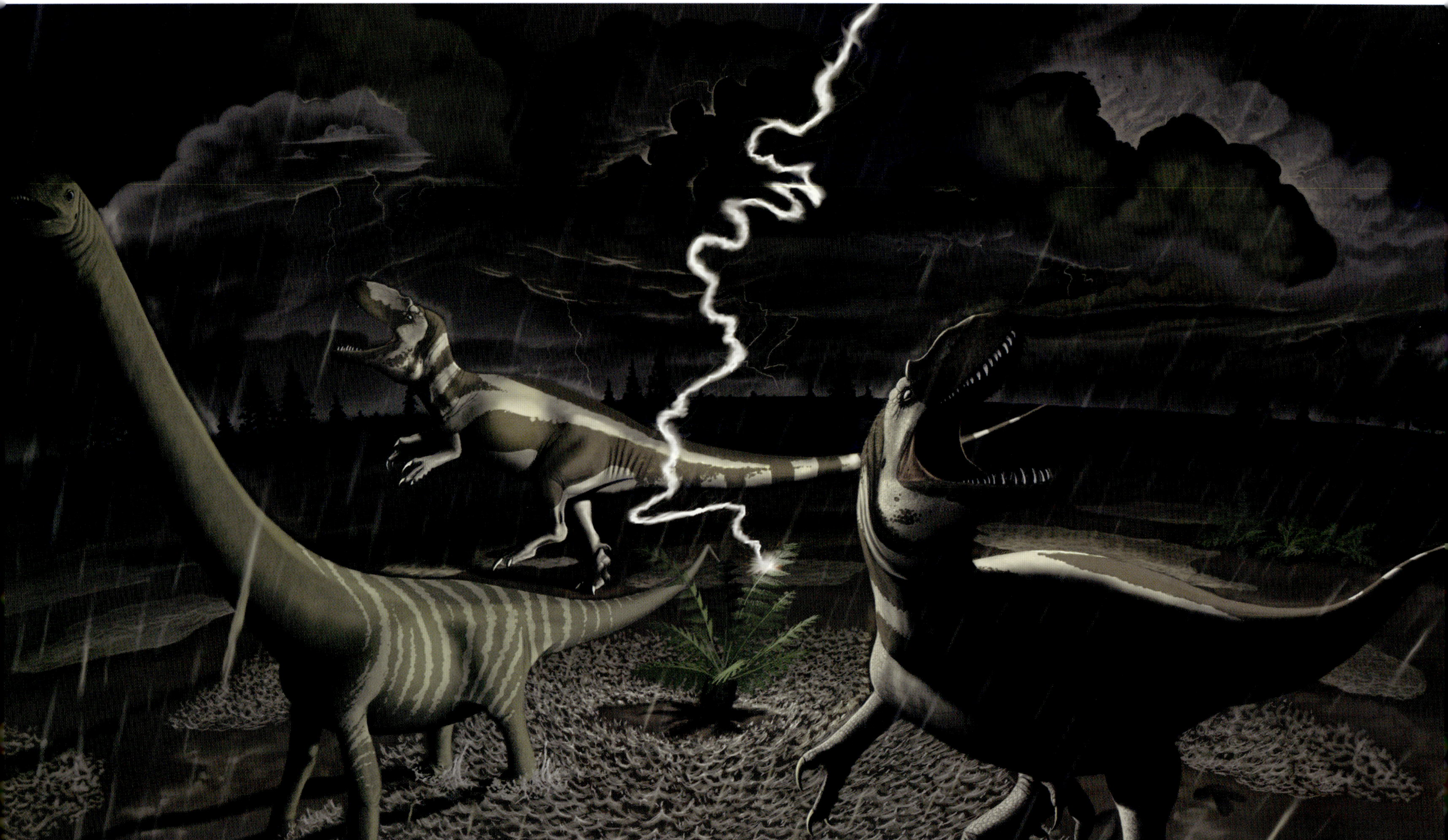

◀ ALMOST!

A pair of the Late Jurassic theropod *Sinraptor* find their attack on a juvenile *Mamenchisaurus* rudely interrupted by an uncomfortably close lightning strike. Lightning doesn't have to strike animals directly to kill them: currents discharged through the ground can also be fatal, and this is more of an issue for quadrupeds than bipeds.
• Digital | 2023

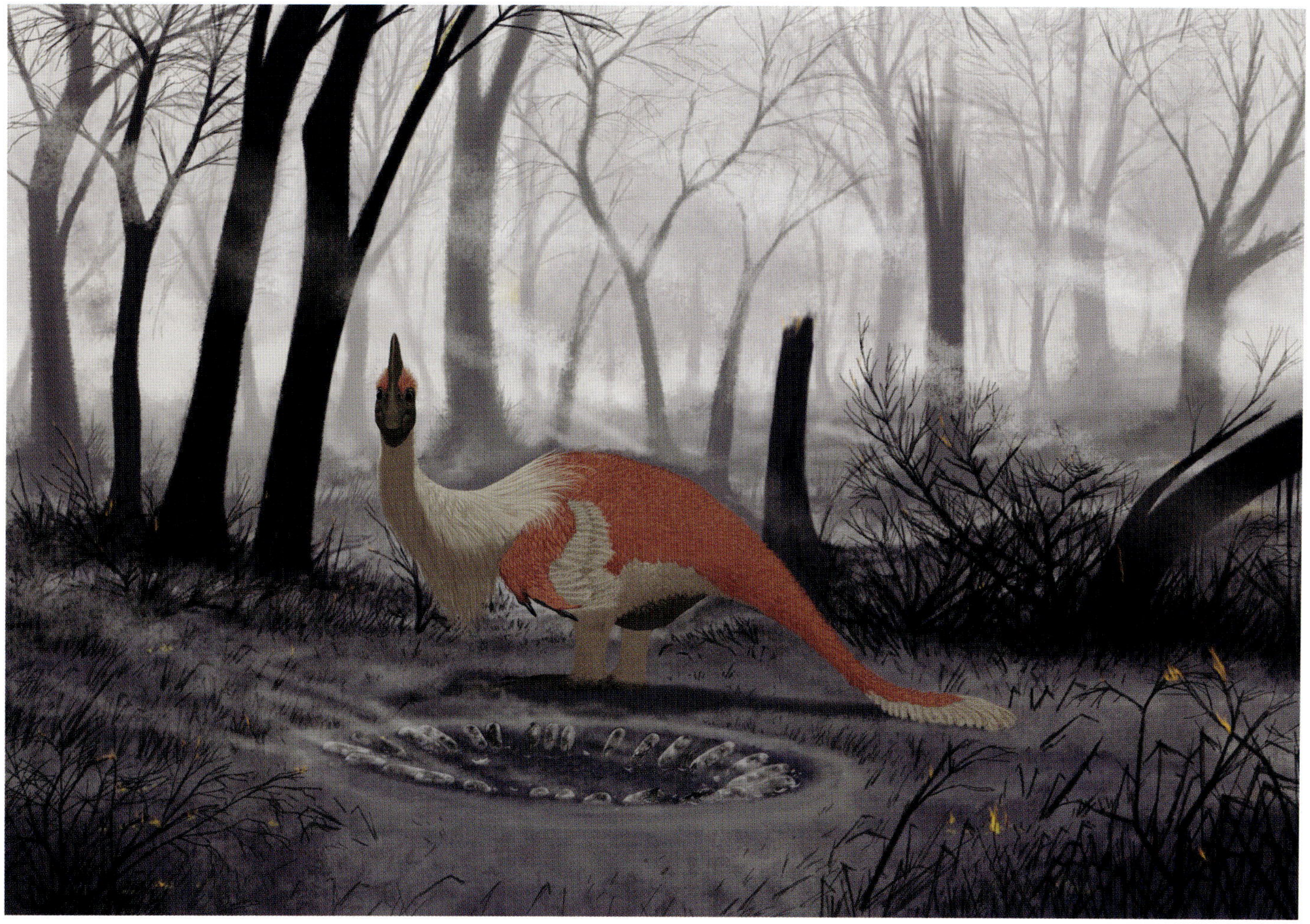

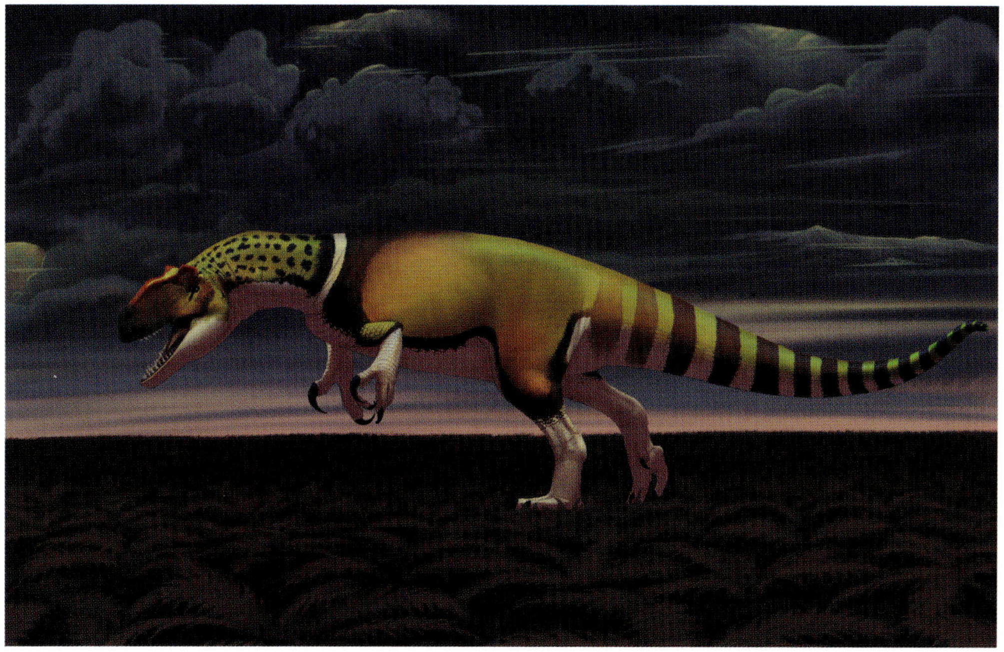

▲ *CORTHYORAPTOR*

The mid-sized, crested oviraptorosaur *Corythoraptor* of Late Cretaceous southern China was named in 2017. Like several other members of its group, it appears to have been a forest-dwelling omnivore or herbivore that probably lived in small groups. The individual shown here has returned to its nest after a quick-burn forest fire. All is not well.
• Digital | 2023

◀ *ALLOSAURUS*

One of the most intensively studied of non-bird theropods, *Allosaurus* of Late Jurassic USA (plus Portugal and maybe elsewhere) has been given many looks over time. The excellent colour vision of its prey makes it possible that even giants like this were boldly marked and coloured for camouflage, perhaps with colour schemes recalling those of modern lizards and birds rather than those of large mammals.
• Digital | 2023

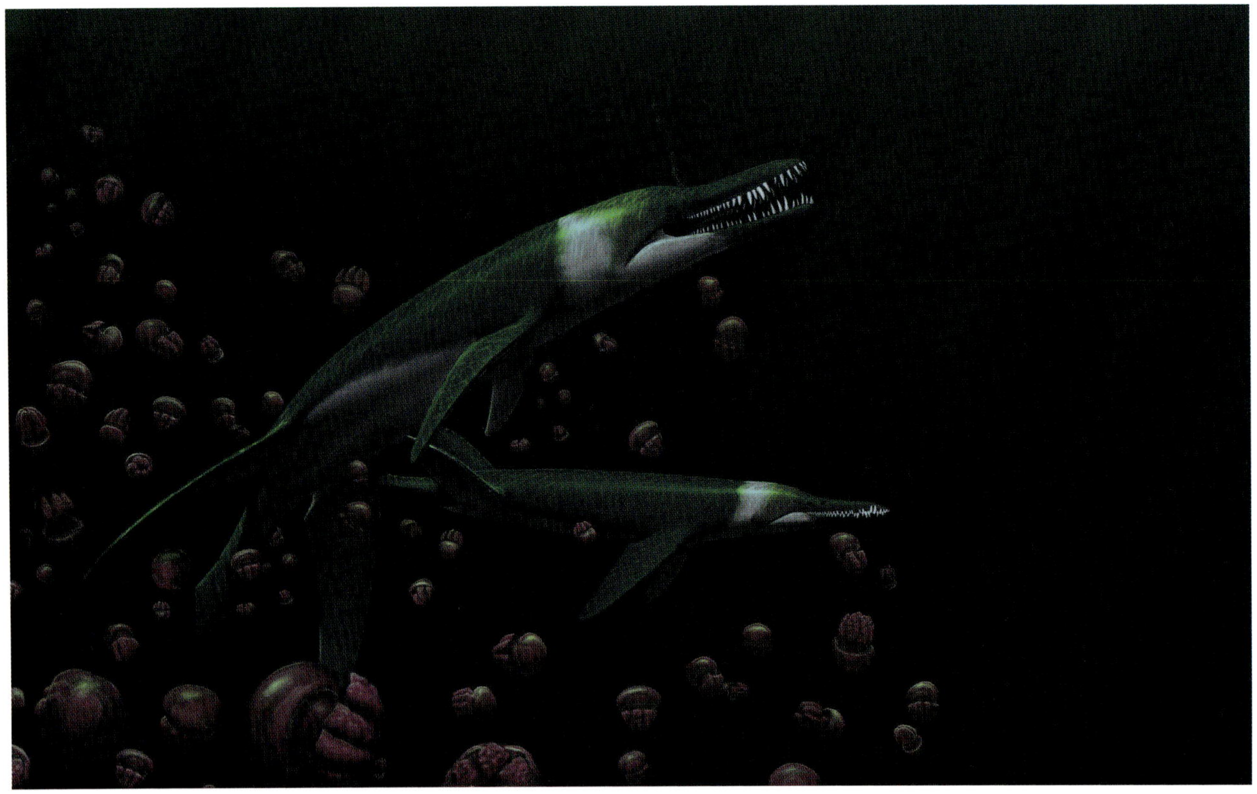

◂ *PLIOSAURUS*

Pliosaurus – a denizen of the Late Jurassic seas that covered Europe, South America and elsewhere – was as likely to have encountered occasional jellyfish swarms as any living marine animal. Dark conditions – caused either by water depth or the rotation of the Earth – would have been as familiar to these animals as sunlit ones.

• Digital | 2023

▾ *PLESIOPLATECARPUS*

Mosasaurs like these North American *Plesioplatecarpus* lived alongside modern-looking sharks. We know that sharks scavenged on mosasaurs, and that other confrontations between the two occurred as well. The possibility also exists that mosasaurs mobbed them (a sort of territorial bullying) or even used them as scratching posts, both behaviours seen in marine mammals today.

• Digital | 2024

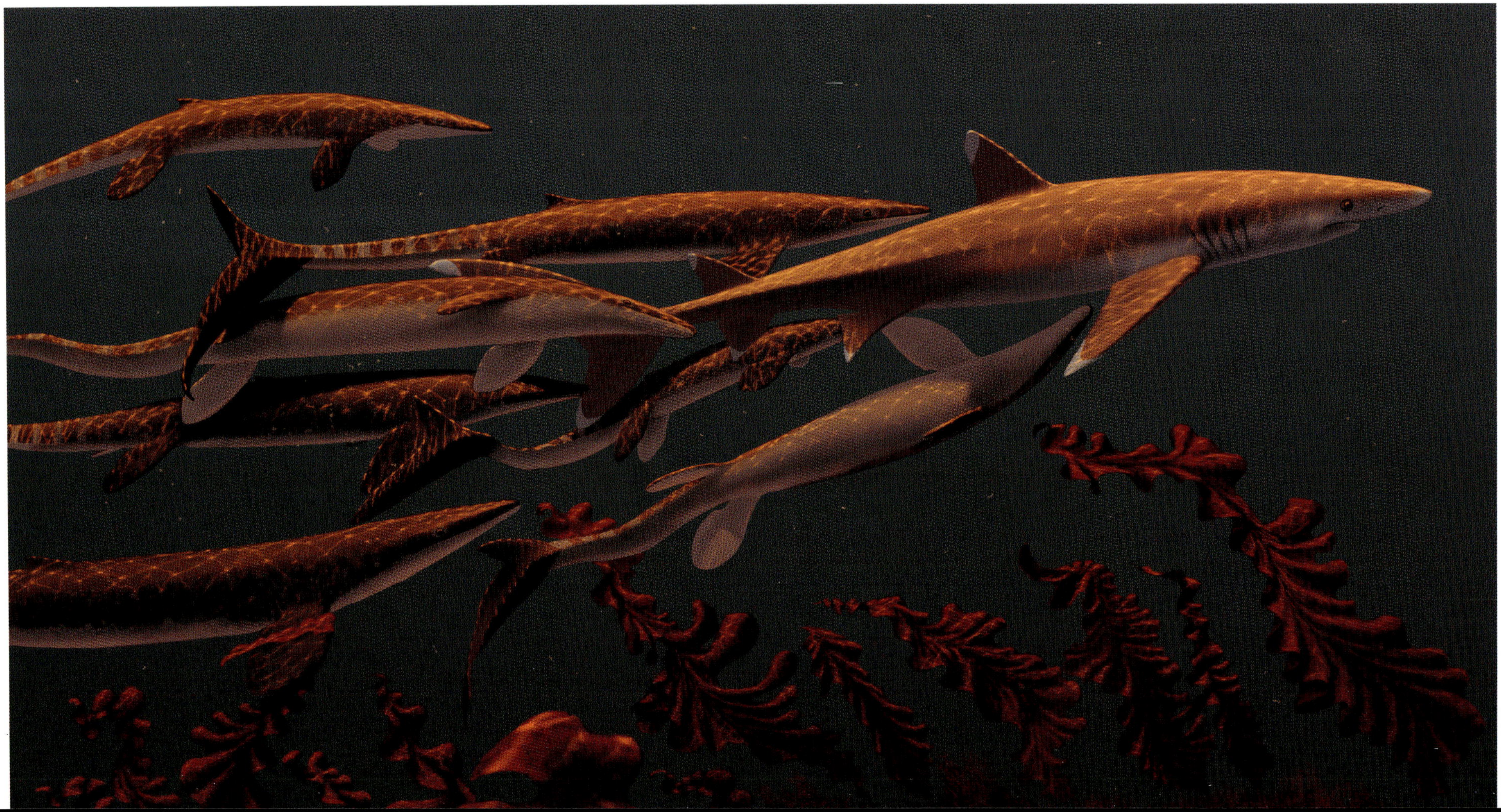

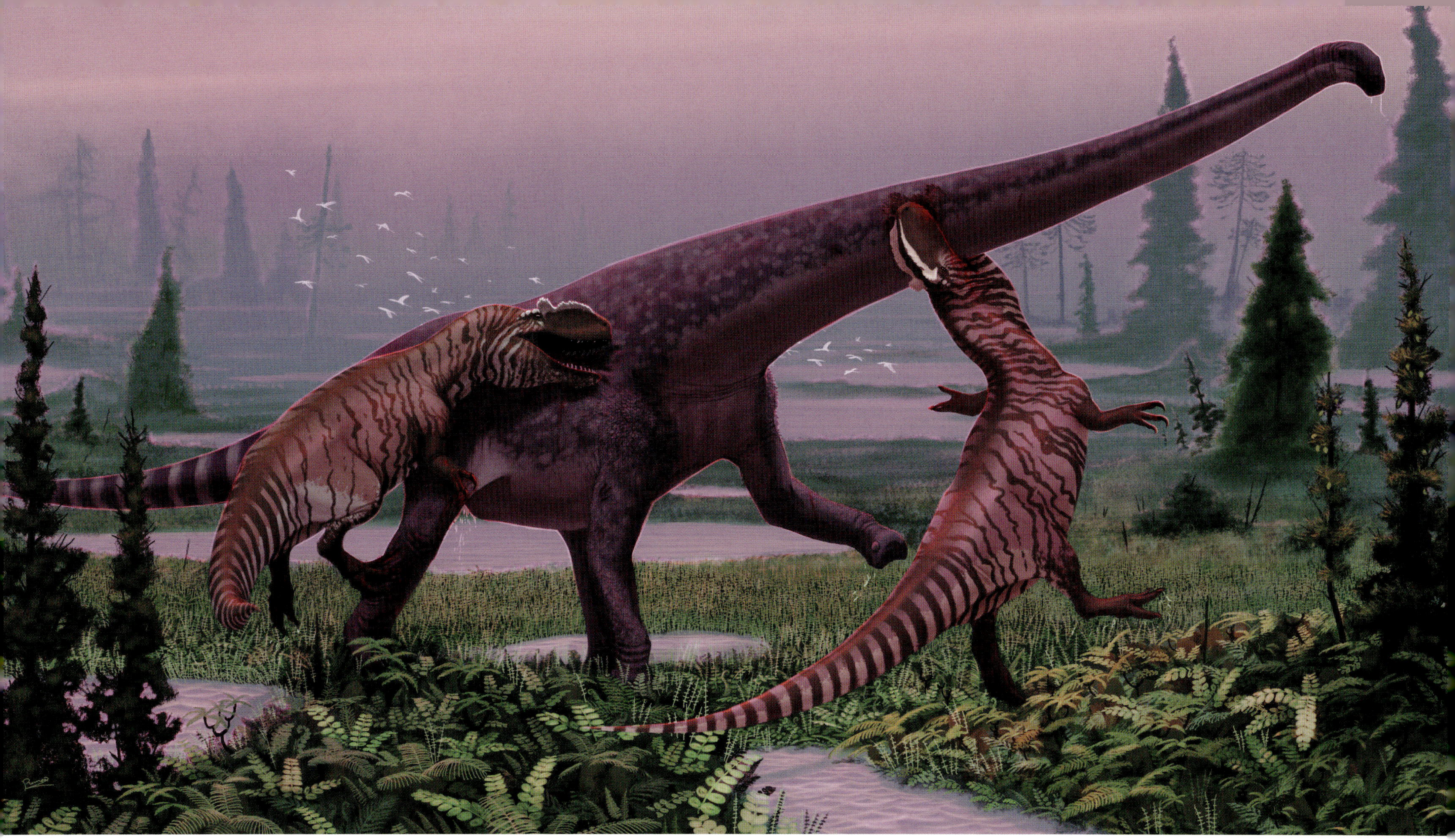

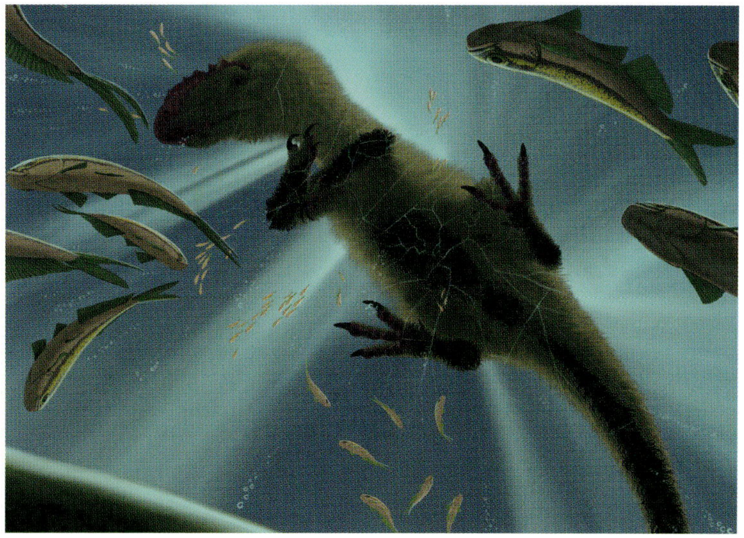

▲ *YANGCHUANOSAURUS*

A *Yangchaunosaurus shangyouensis* pair attempt to fell a *Mamenchisaurus* in what will be the Shaximiao Formation of Late Jurassic China. Even with their vulnerable and insanely long necks, hunting these huge sauropods must have needed a joint effort by even the biggest theropods. This is one of Ramón's personal favourites.

• Digital | 2024

YUTYRANNUS ON THIN ICE ▶

The Early Cretaceous tyrannosaur *Yutyrannus* inhabited an environment where ice may well have formed over lakes, and both curiosity and hunger might then result in a scene like this fish's-eye-view. But walking on ice can be dangerous, as much so in the Cretaceous as it is today.

• Digital | 2023

REX OR MCRAEENSIS ▸

A *Tyrannosaurus* freshens up after a meal. But rather than depicting *T. rex*, this image shows *T. mcraeensis*, a species from New Mexico named in 2024. The specimen was found in 1983 and initially identified as a *T. rex*. Similar in size to *T. rex* but apparently 6–7 million years older, *T. mcraeensis* is claimed to be unique in numerous details and has smaller bony crests than *T. rex*. However, tyrannosaur experts are not in agreement over its validity, and debate is underway.
• Digital | 2024

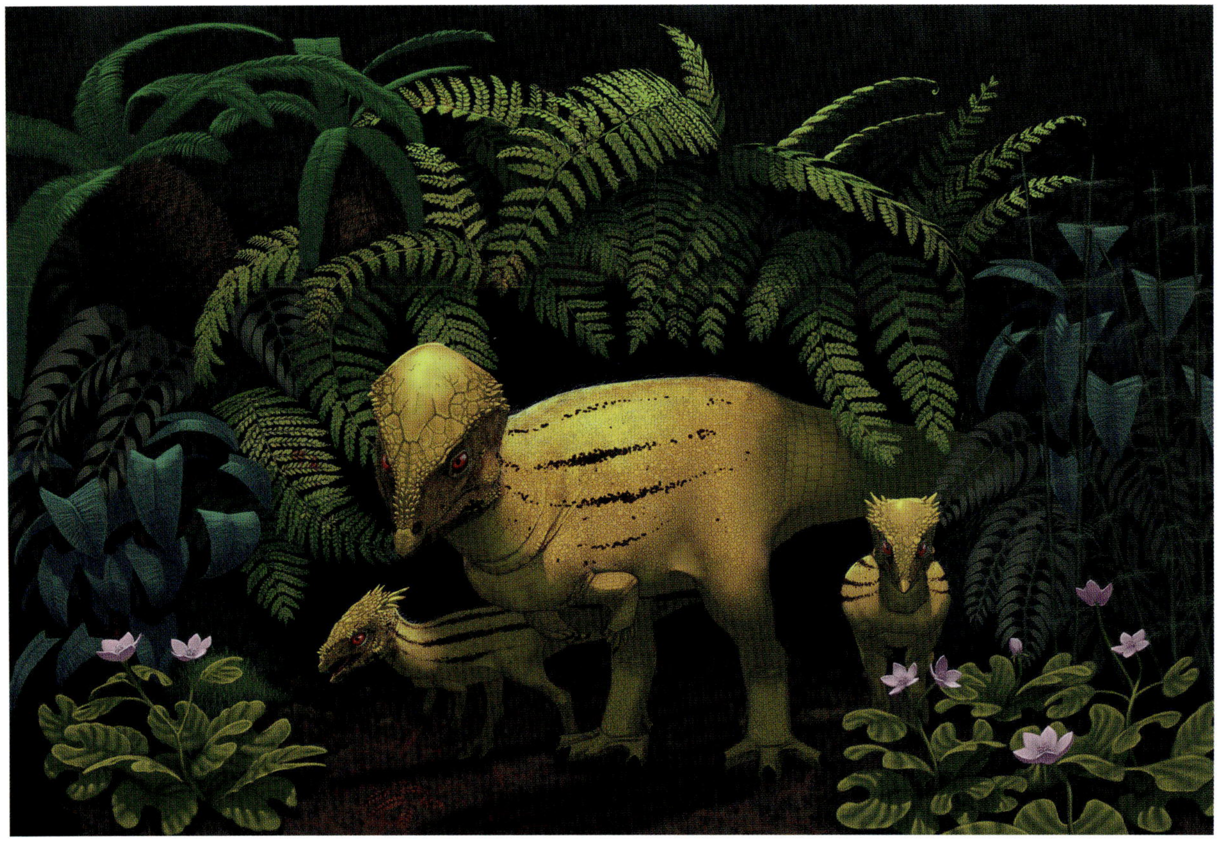

▾ *ANUROGNATHUS*

The comical proportions of anurognathid pterosaurs make them look amusing when depicted from certain angles. This image highlights a theme important to many palaeoartists: getting the plants right! This is virtually always harder than getting the animals right, since the relevant information is so hard to find. This *Anurognathus* is picking its way through gingko nuts.
• Digital | 2023

▴ *PRENOCEPHALE*

Pachycephalosaurs – the bone-headed, sometimes dome-skulled, ornithischians of Late Cretaceous North America and Asia – were mostly quite small, as emphasised in this depiction of Mongolia's *Prenocephale*. This image explores the view that juvenile dome-skulled pachycephalosaurs had a very different look from the adults, a topic very relevant to scientific debates about this group.
• Digital | 2023

SINORNITHOSAURUS NEST ▸

It's possible that dromaeosaurids like China's *Sinornithosaurus* – an animal similar in mass to a Bald Eagle – could climb and even nest in trees. This would have been advantageous in environments where other theropods, plus mammals, were a danger to nests and their contents, but we may never know the answer, since nests made of vegetation are vanishingly rare as fossils.
• Digital | 2023

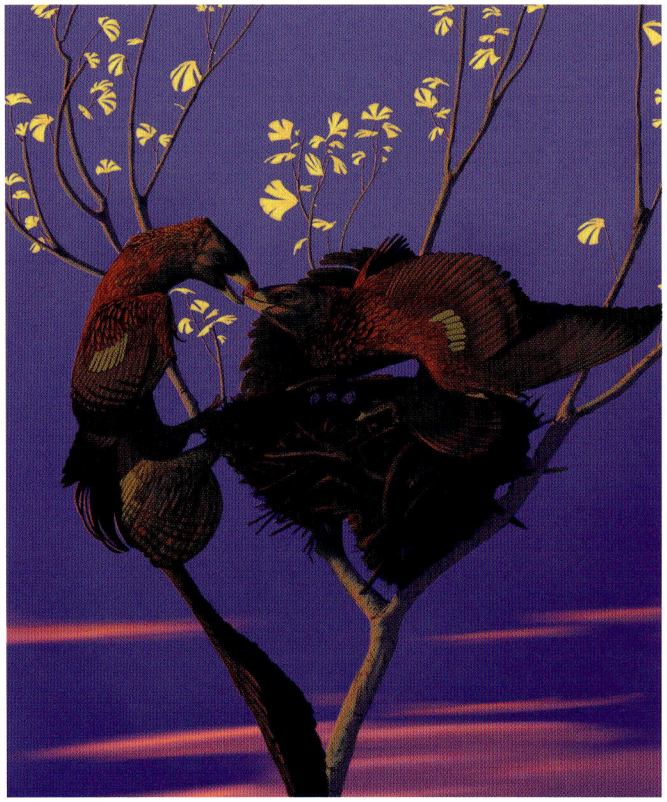

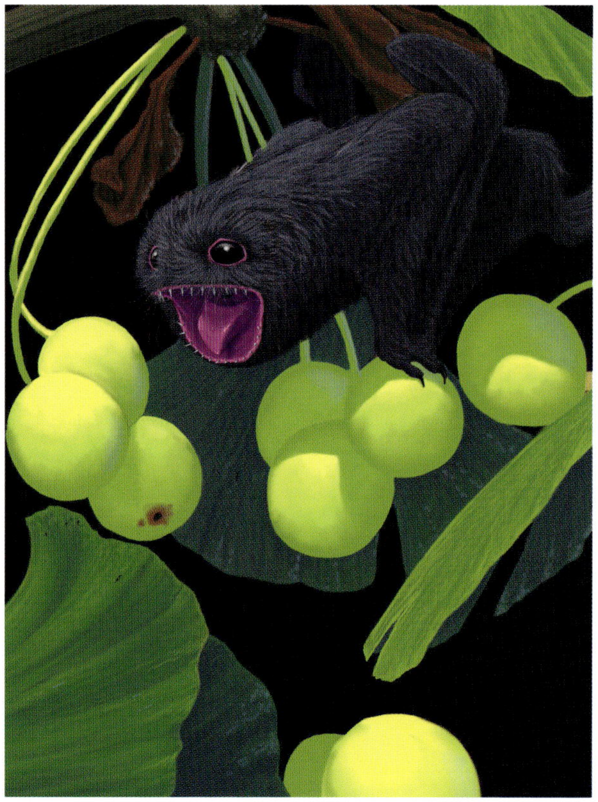

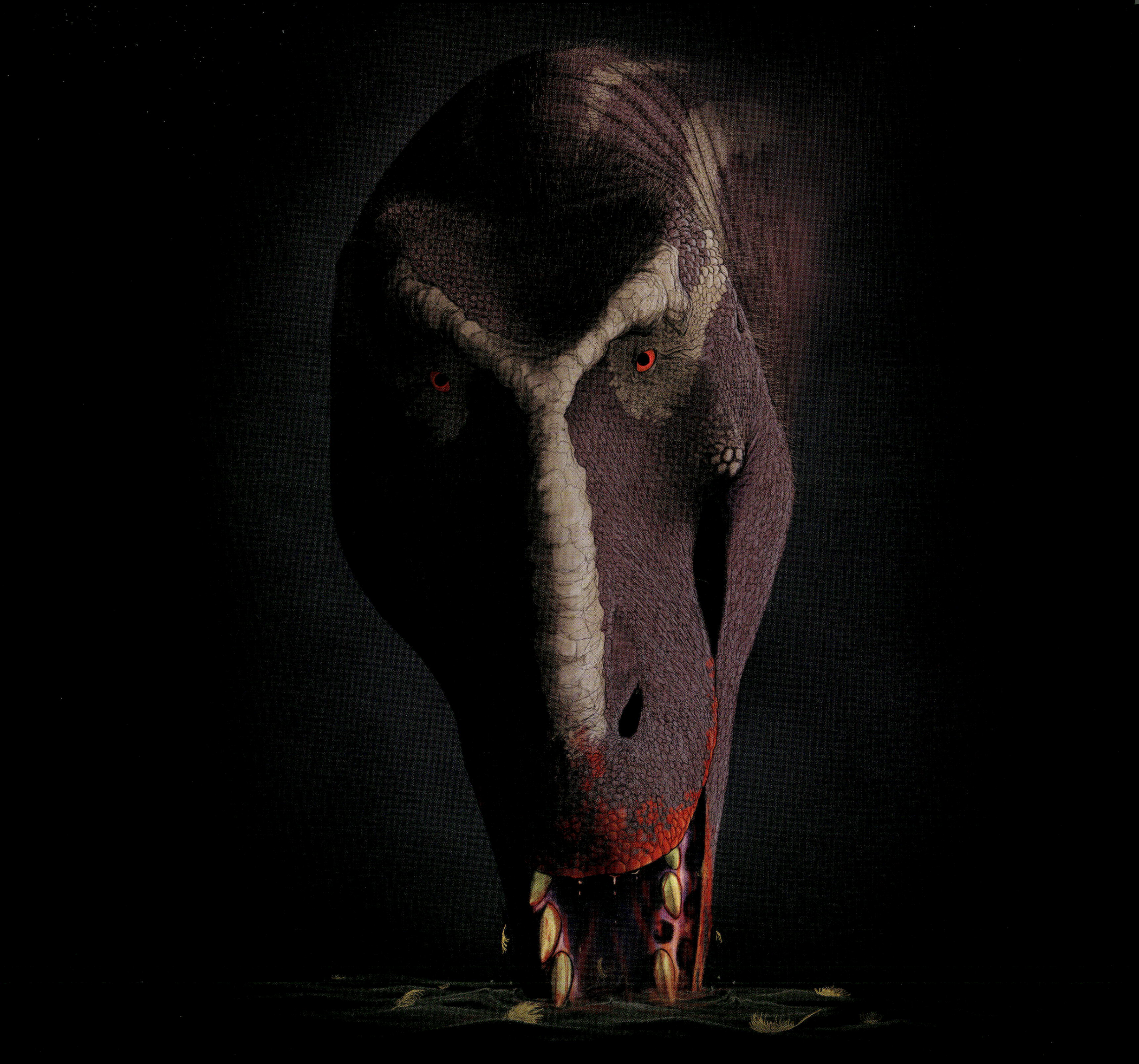

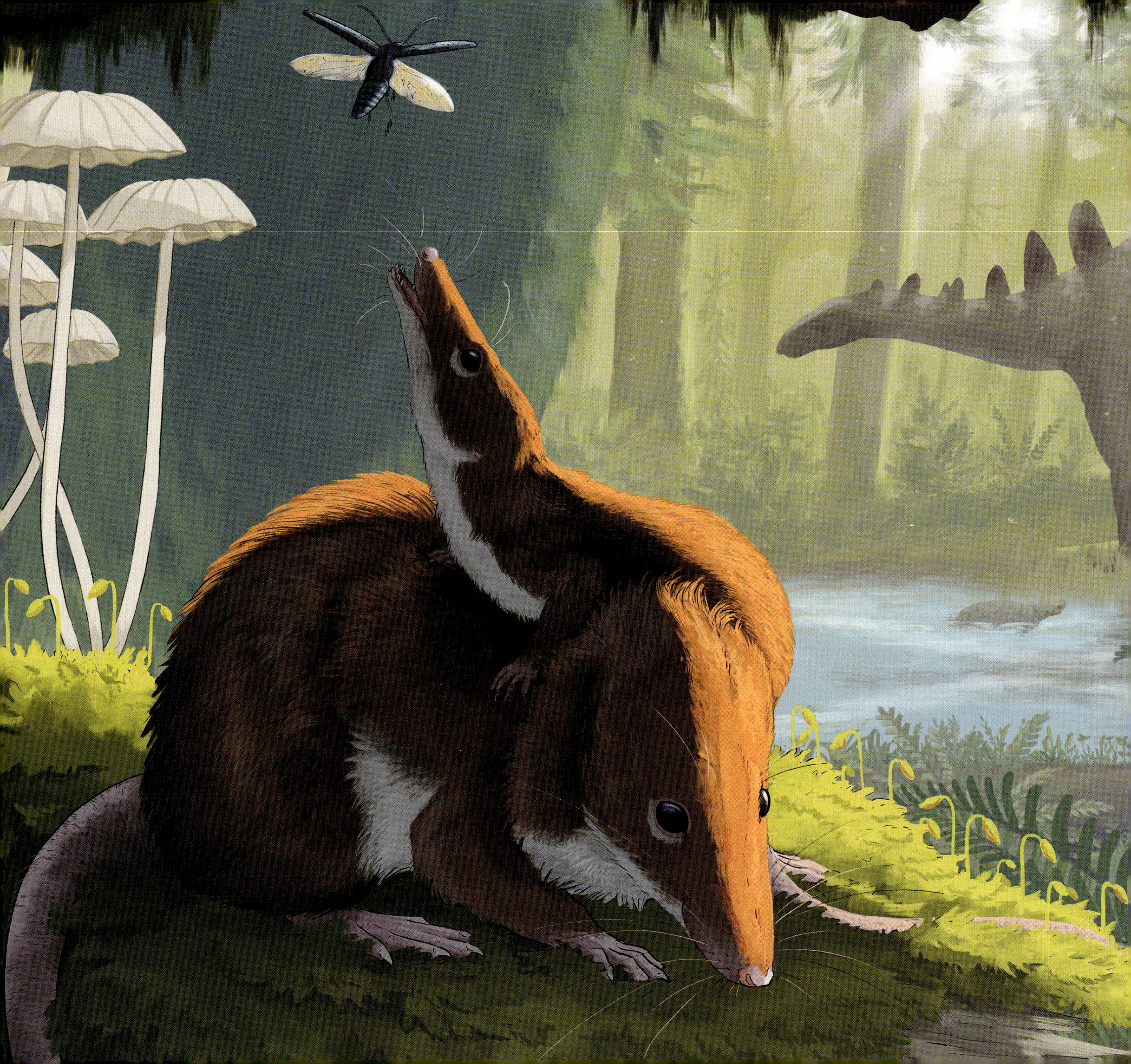

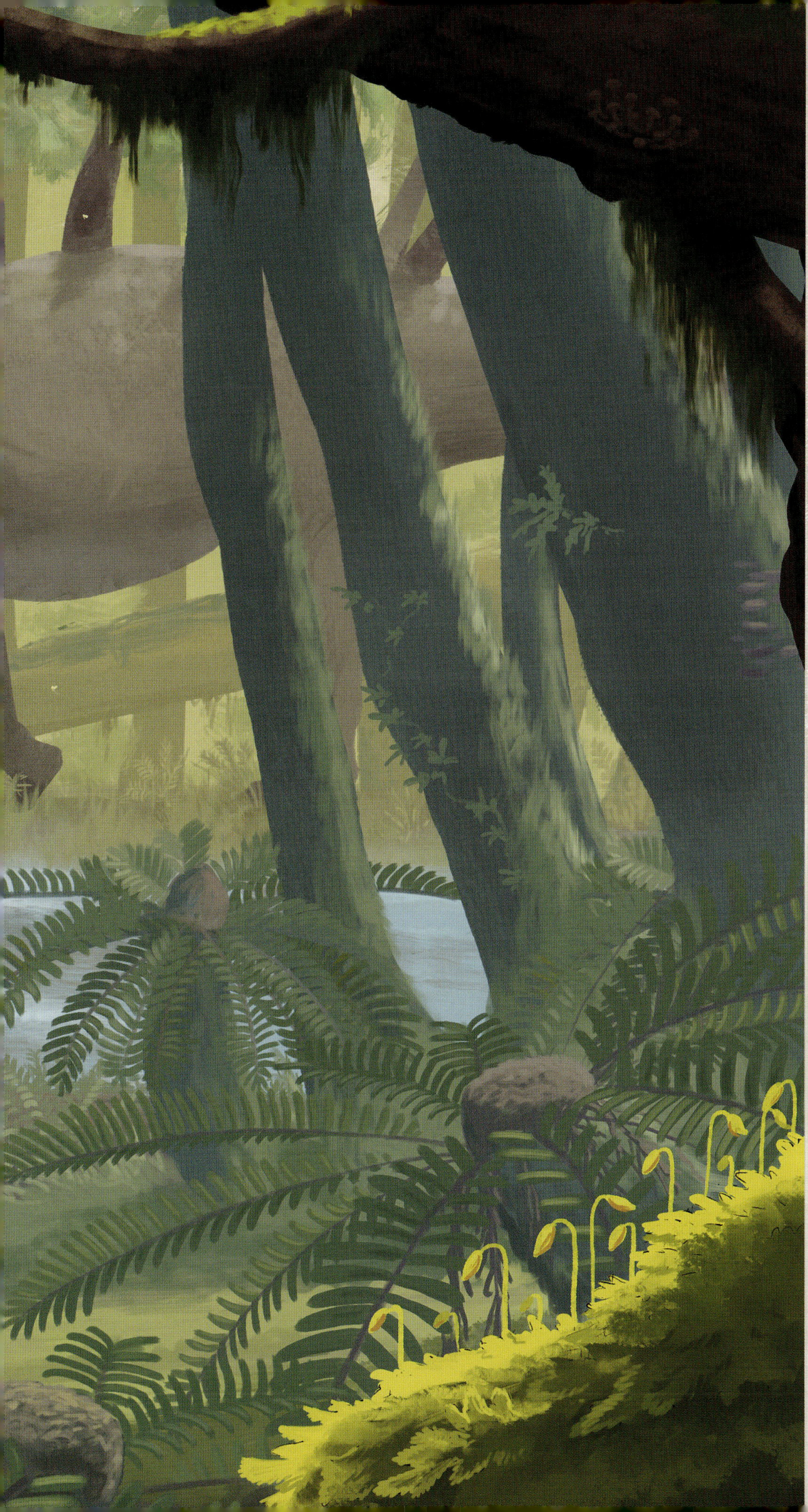

MAIJA KARALA

MAIJA IS A FREELANCE ILLUSTRATOR and science writer from Finland, with a master's degree in Ecology and Evolutionary Biology from the University of Jyväskylä. She writes popular articles and consults and fact-checks works of palaeontology and evolution across a variety of mediums. Over the years, illustration – which was supposed to have been just a hobby – has slowly been taking over more and more of her time. She has illustrated biology textbooks, a comic book, magazine articles, museum exhibits and has written and illustrated her own book, *Dinosaurusten aika selkokielellä* ('Time of the Dinosaurs in Easy-Read Finnish'), published in 2021.

Her main interests are palaeontology, and animal behaviour and cognition, which she feels are connected to her broader understanding of evolution, ecology and climate change. With her palaeoart, Maija tries to depict believable ecosystems and animals that have individual life stories and personalities, using it to share ideas and information about ancient ecosystems.

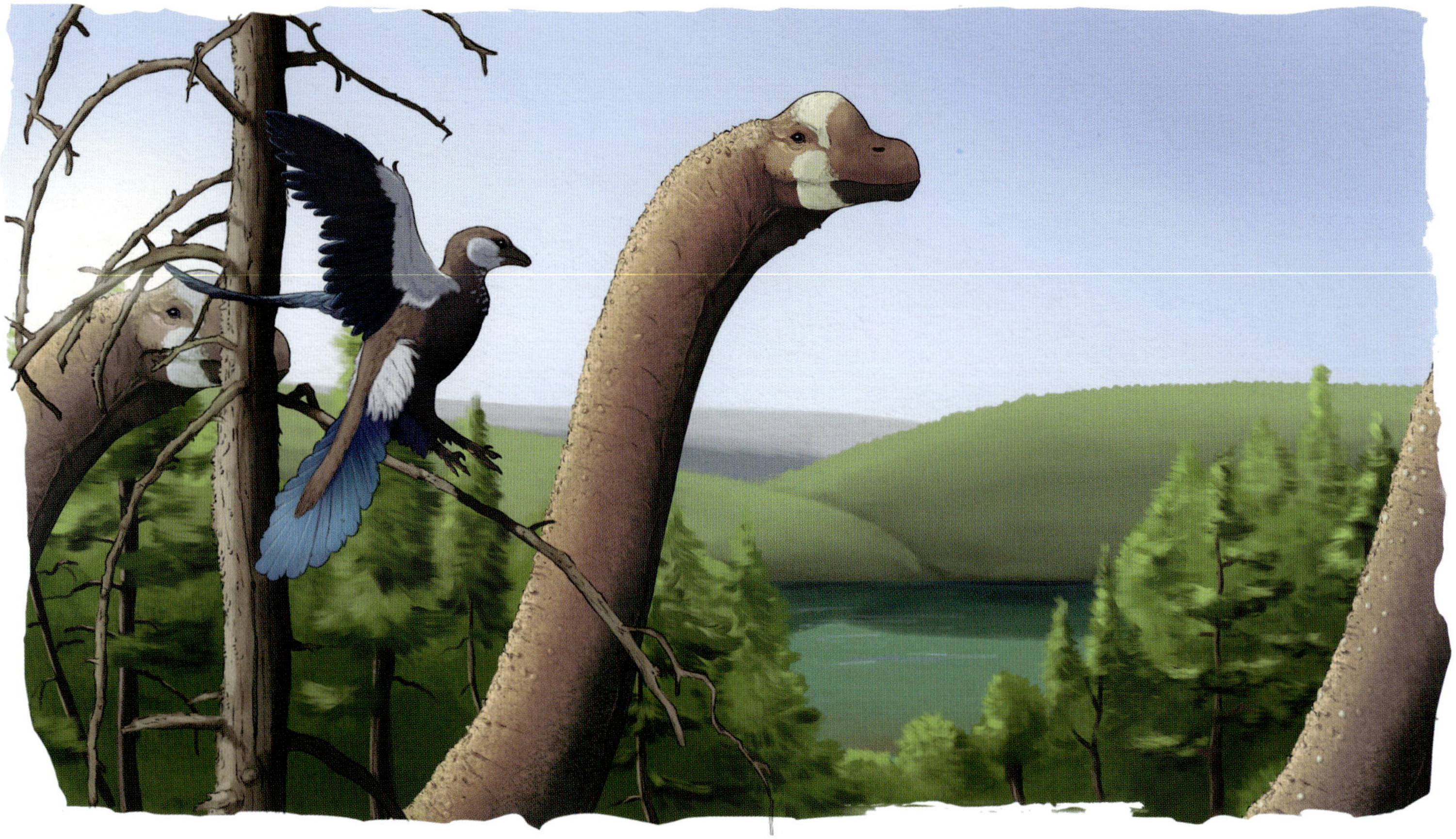

LONGISQUAMA ▸

Famously weird, and – still! – famously enigmatic, little *Longisquama* is a Late Triassic reptile from Kyrgyzstan in Central Asia, best known for the long, curved, ribbon-like structures projecting from its back. *Longisquama* is so unusual that it is associated with several surprising ideas. One is that the dorsal structures were paired and used in gliding; another is that it was close to the ancestry of birds. Both ideas contradict other lines of evidence.

• Digital | 2022

▴ BEFORE THE IMPACT

This is one of a series of speculative landscapes set around the rim of Lake Lappajärvi, an impact crater in western Finland dated to the Late Cretaceous. This scene illustrates the Late Jurassic, some 85 million years before the impact. Shown are brachiosaurs, known from Europe, Africa, North America, and *Archaeopteryx*-like birds.

• Digital | 2020–2023

THE DINOSAUR EATER ▸

Many discoveries – mostly from China – have shown that the mammals of the Jurassic and Cretaceous were diverse in size, shape and lifestyle. The badger-sized *Repenomamus* was a danger to small animals, including small or young dinosaurs. It may have been able to kill animals larger than itself.

• Digital | 2020

[Previous pages]

KRUSATODON

In a lush Jurassic forest, a young *Krusatodon* studies an ommatid beetle. Fossils of an adult and young individual were discovered on Scotland's Isle of Skye in 2024. Studies indicate that these ancient mammals grew more slowly and matured later than similar-sized living mammals. The young may also have stayed with their parents far longer. The juvenile was aged at seven months but had only just developed adult teeth.

• Digital | 2024

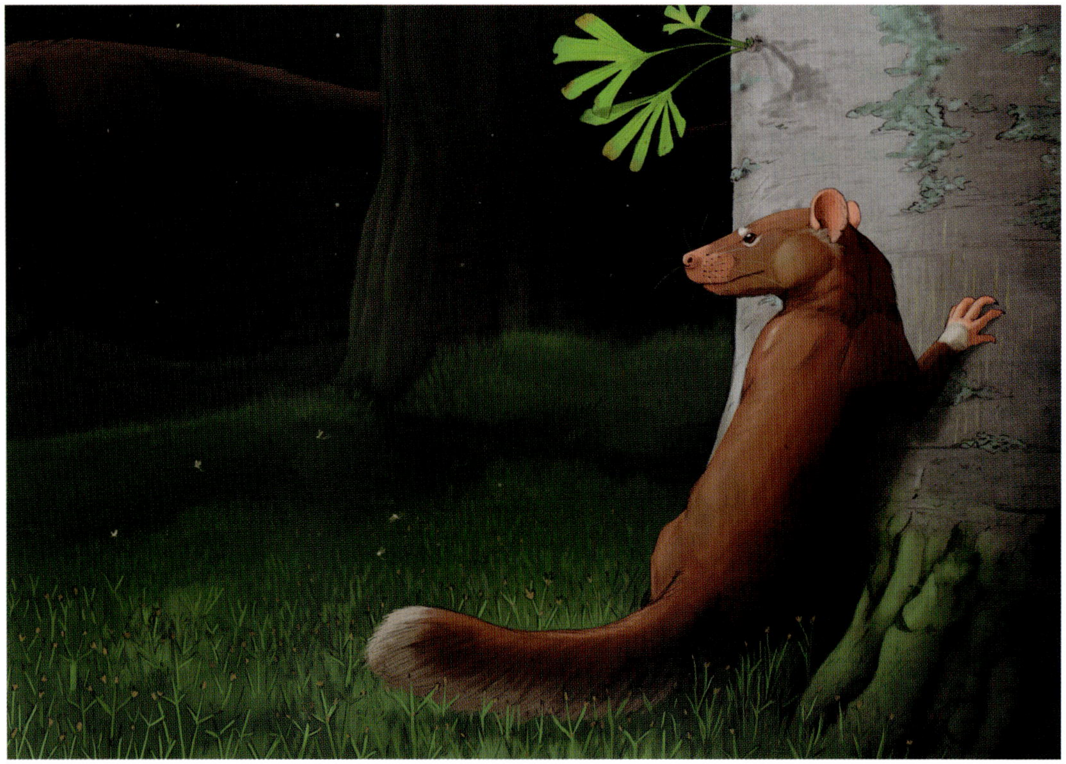

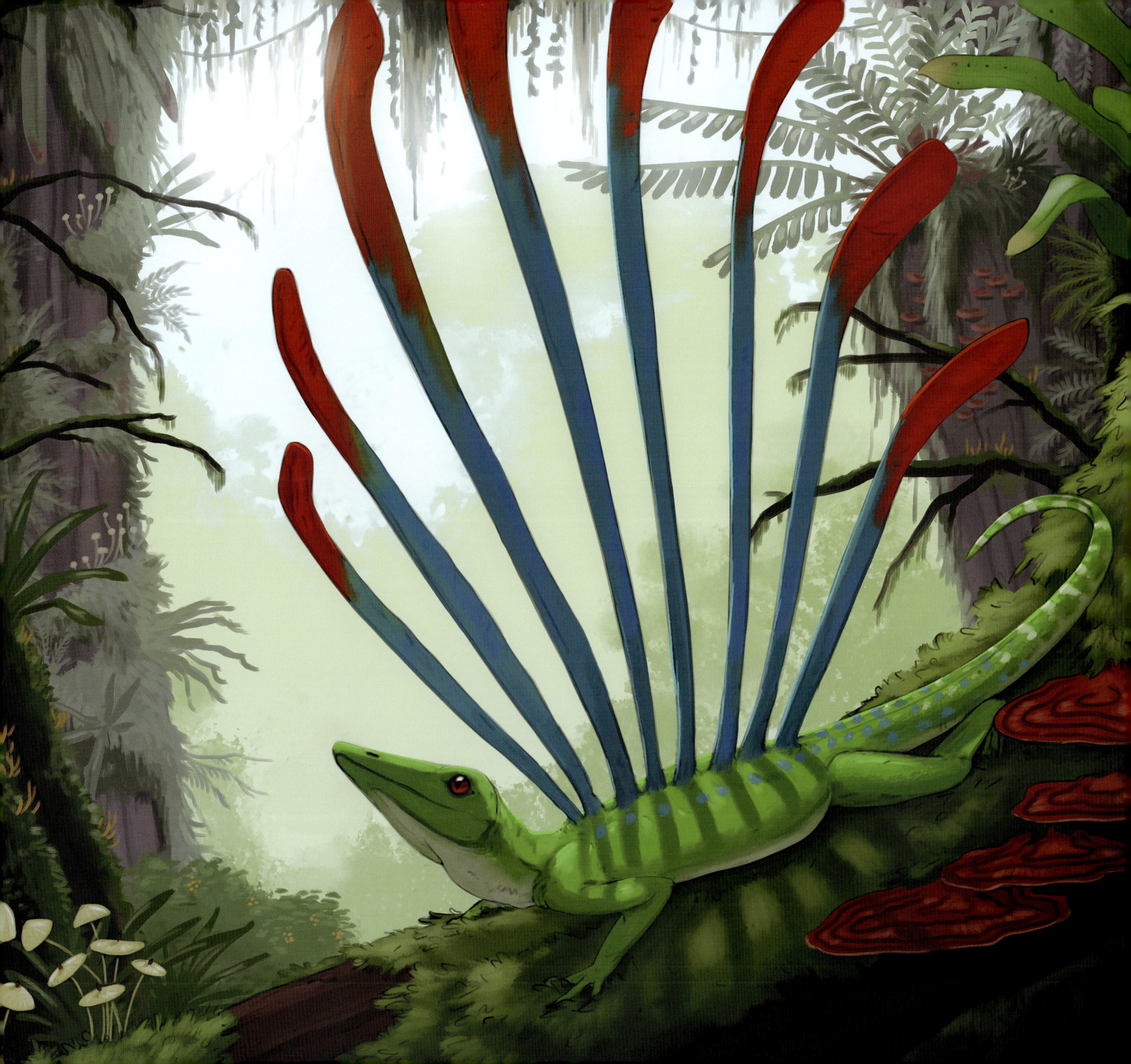

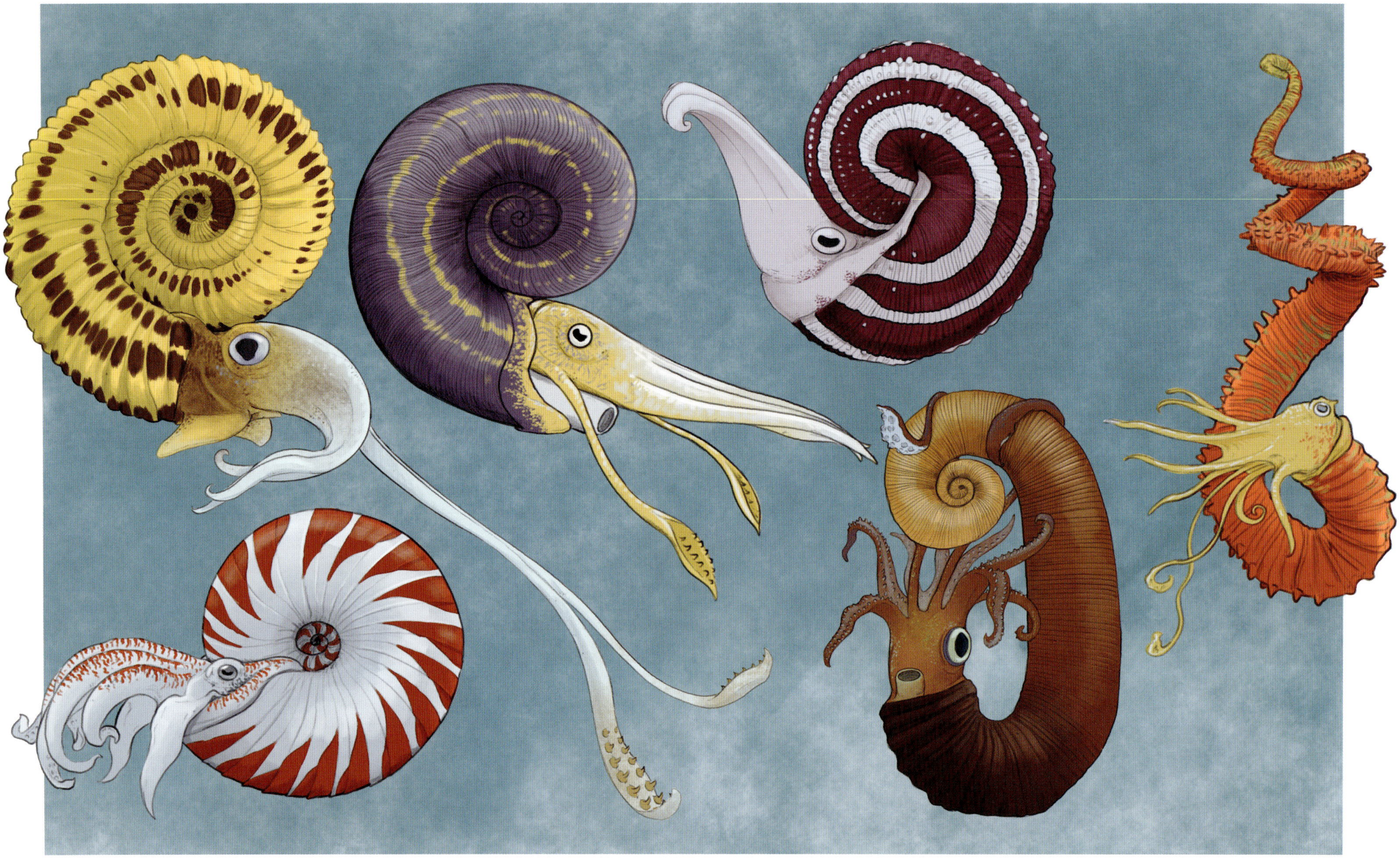

▲ AMMONITE DIVERSITY

Ammonites – shelled swimming molluscs related to squids and octopuses – existed in a bewildering variety of sizes and shapes. Argument remains about their soft parts and experts disagree on how similar they were to modern squid. What's clear from their many shell shapes is that they were occupying a huge number of ecological roles in the world's ancient oceans.

• Digital | 2020

FORELIMB-HEAVY *IGUANODON* ▶

The name *Iguanodon* is specifically associated with *I. bernissartensis*, an Early Cretaceous species named from Belgium in 1881. This was a massive, heavily built animal, its notable features including its comically huge hands up to 30cm (11.8in) wide with a spike-shaped thumb 20cm (7.8in) long. That's without the keratin sheath that would have added to its length.

• Digital | 2020

SINOSAUROPTERYX **IN DEATH ▸**

Named from north-east China's Liaoning Province in 1996, *Sinosauropteryx* is the little dinosaur that just keeps on giving. It was among the first non-birds to preserve a coat of filamentous feathers, and more recent studies have revealed pigments that indicate a red-brown banding along the length of its tail, a pale belly, and a dark 'bandit mask' on the face.

• Digital | 2023

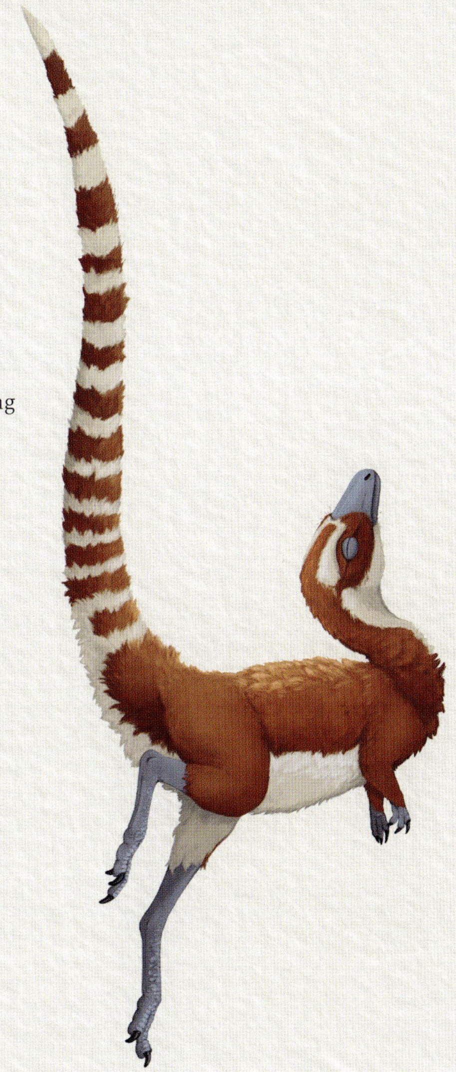

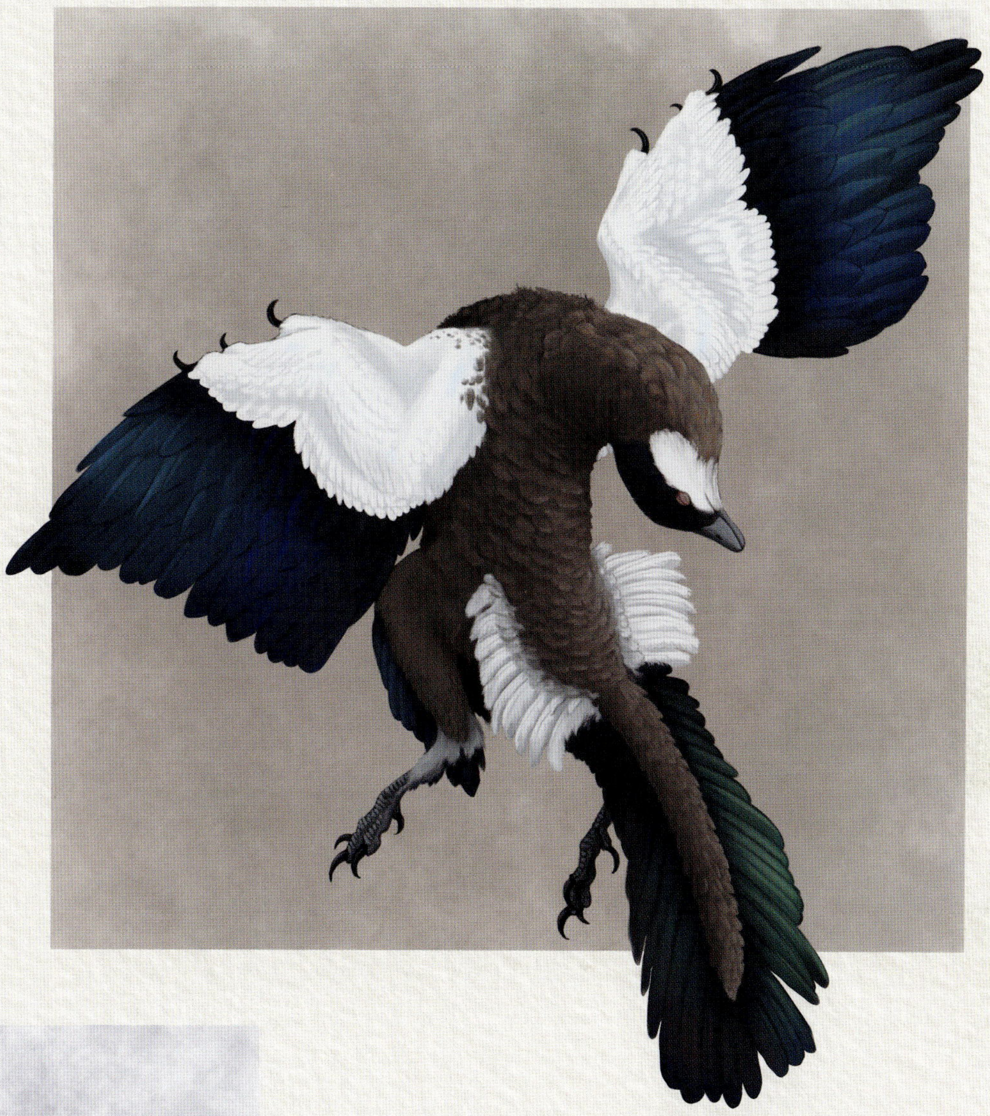

▴ *ARCHAEOPTERYX* DEATH POSE

It's almost ironic that modern visions of the famous Jurassic bird *Archaeopteryx* have gone full-circle relative to reconstructions of about 100 years ago. Once again, Archaeopteryx looks superficially magpie-like. But it was distinct and anatomically archaic too, not just in having prominent wing claws and a 'Mesozoic-style' tail, but also in having prominent tail fans at the tail's base. This image depicts the famous Berlin specimen, discovered in 1874 or 1875.

• Digital | 2020

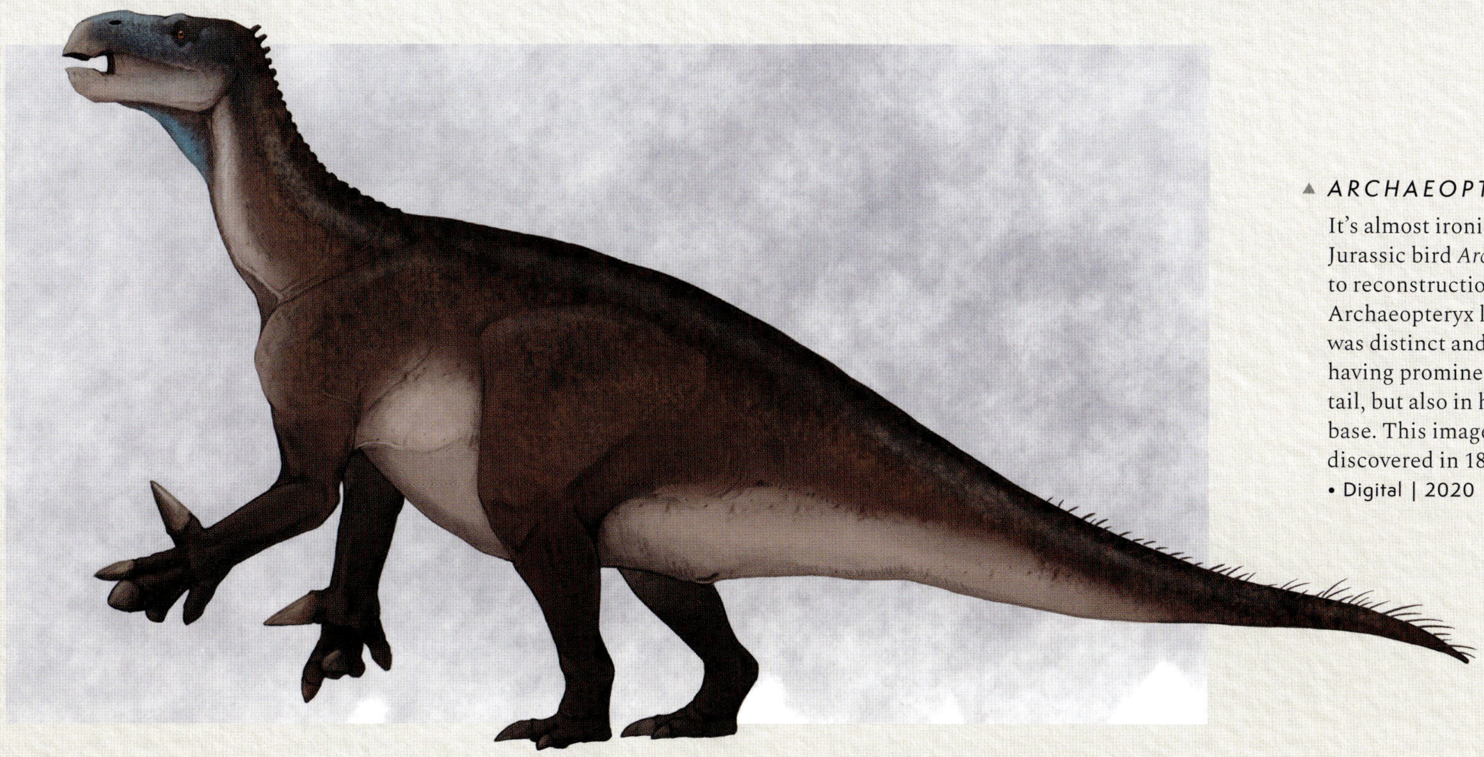

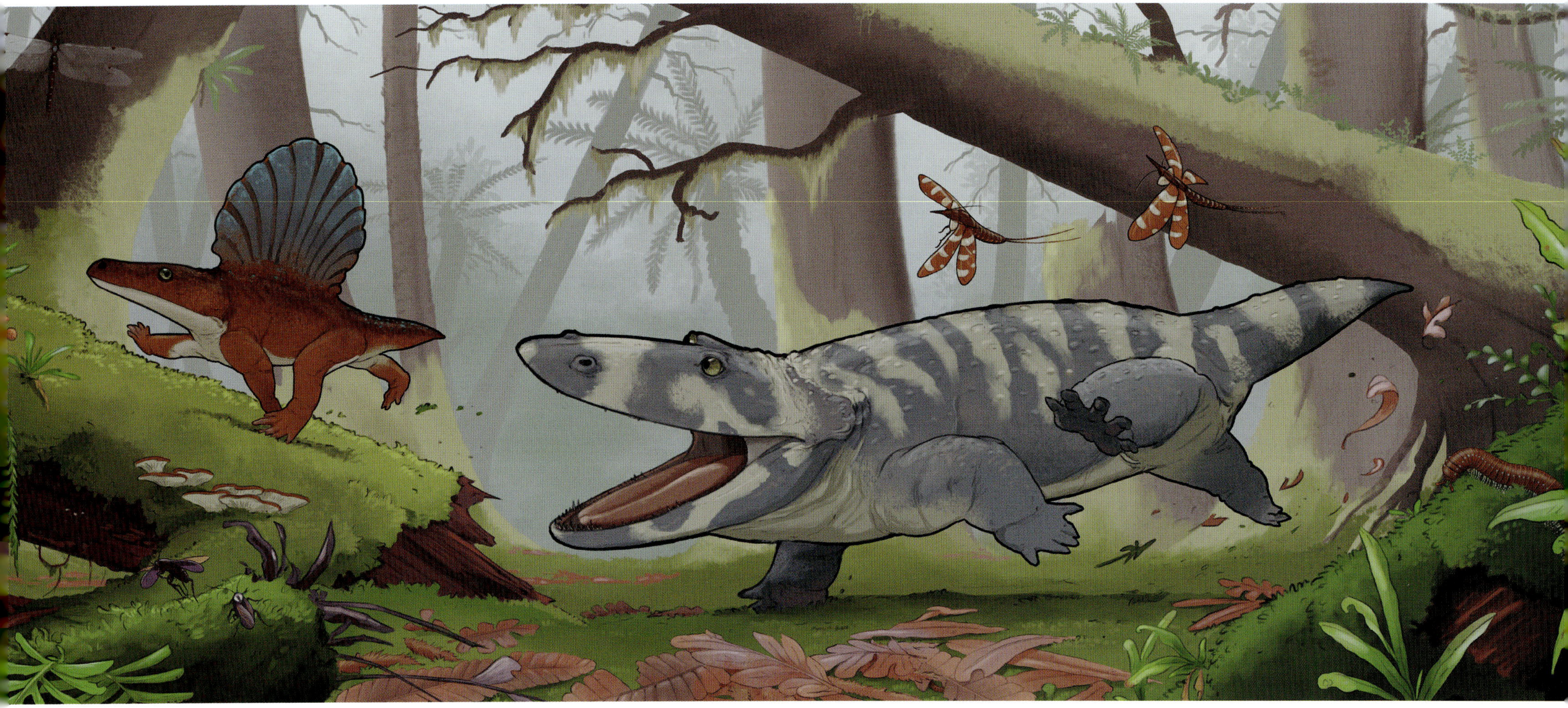

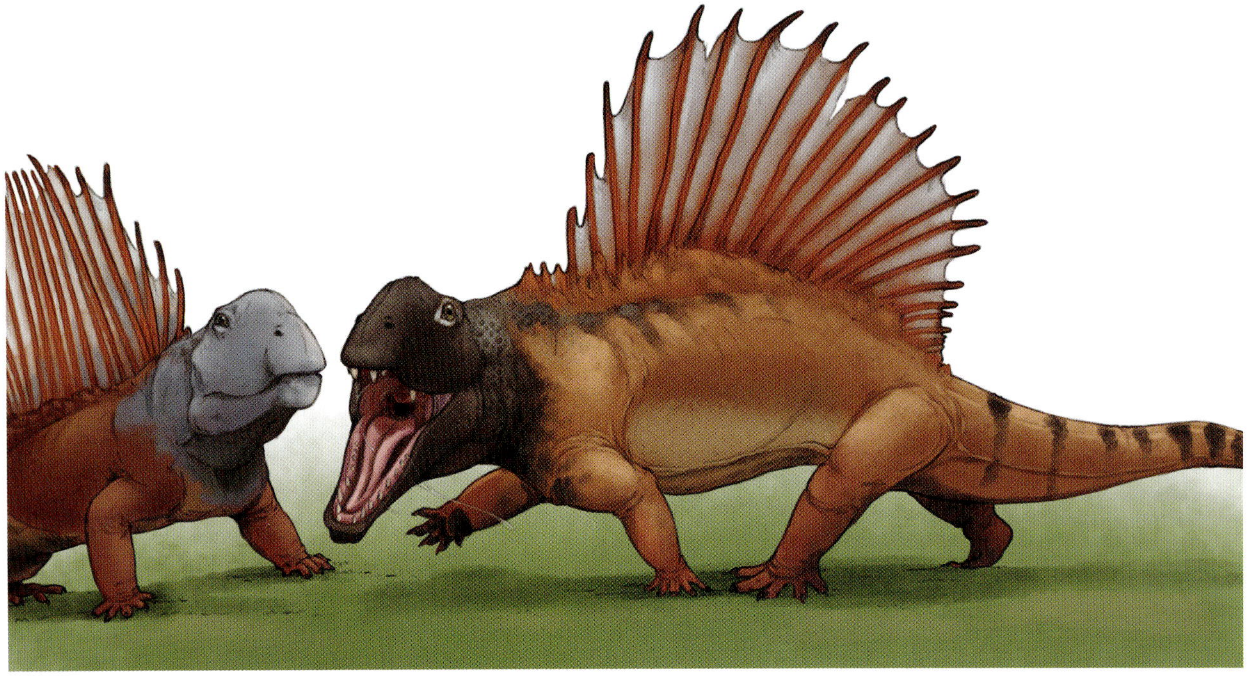

▲ THE CHASE

This Early Permian scene (set about 280 million years ago in the Texas Red Beds) hints at the diversity within the temnospondyls, a huge group of amphibians that includes the ancestors of today's frogs, salamanders and caecilians. *Platyhystrix* – notable for the tall bony spines on its back – belongs to a group well suited for terrestrial life. The much bigger *Eryops* was more an animal of pools and lakes, but forays on land are at least plausible.

• Digital | 2023

◀ PUSHY

The Permian stem-mammal (or, if you prefer, non-mammalian synapsid) *Dimetrodon* is a highly familiar animal. Numerous species are known, spanning a range of body sizes. Views on what the sail was like have varied. Recent studies suggest that the spine tips projected beyond the webbed skin. The sail's function is also disputed. This image follows the view that its primary role was display.

• Digital | 2022

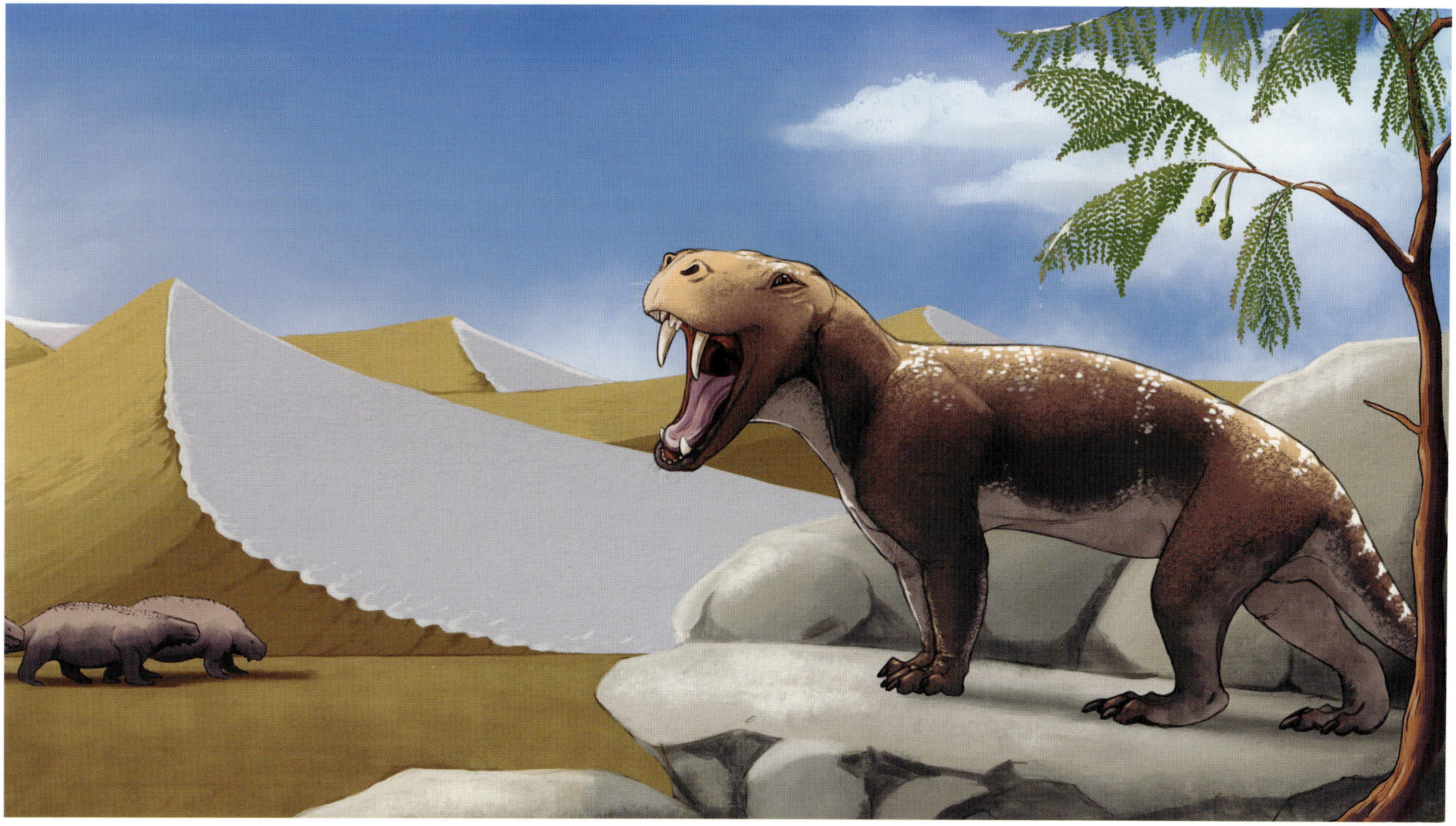

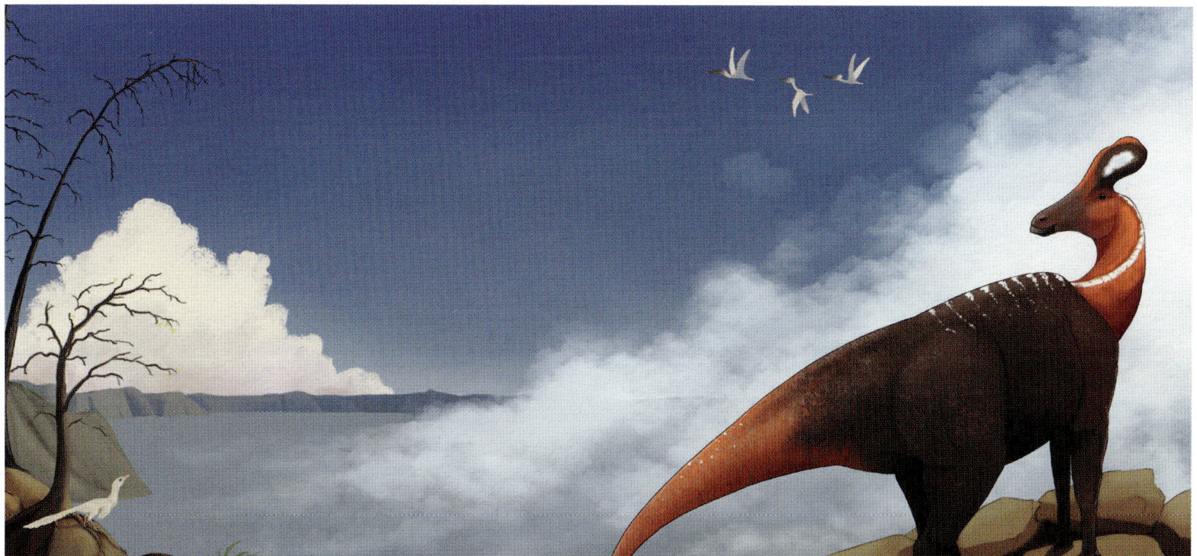

▲ POSTCARD FROM THE PERMIAN

The most important large land animals of the Permian Period (between 299 and 252 million years ago) were the synapsids (like *Inostrancevia*, the sabre-toothed gorgonopsian shown here), but large reptiles like giant pareiasaurs such as *Scutosaurus* were important too. These animals are known from a cold desert environment in the Arkhangelsk region of northwestern Russia.

• Digital | 2020–2023

◀ CRETACEOUS CRATER

One of Maija's 'postcards' shows Lake Lappajärvi 77 million years ago, some years after the impact that created it. The crater is still almost boiling hot, but life is returning. The dinosaur resembles the hadrosaurid *Tlatolophus* – a cousin of the trombone-crested *Parasaurolophus* recently named from Mexico.

• Digital | 2020–2023

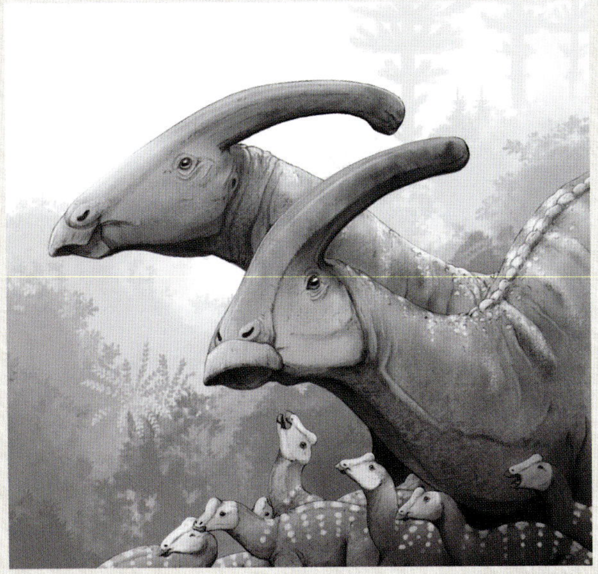

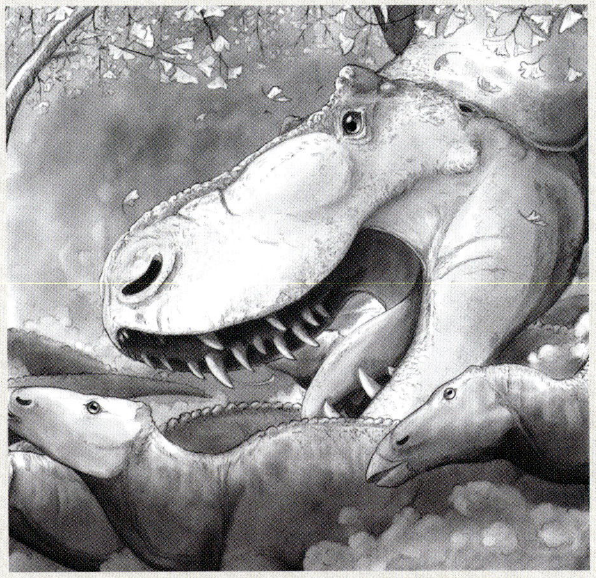

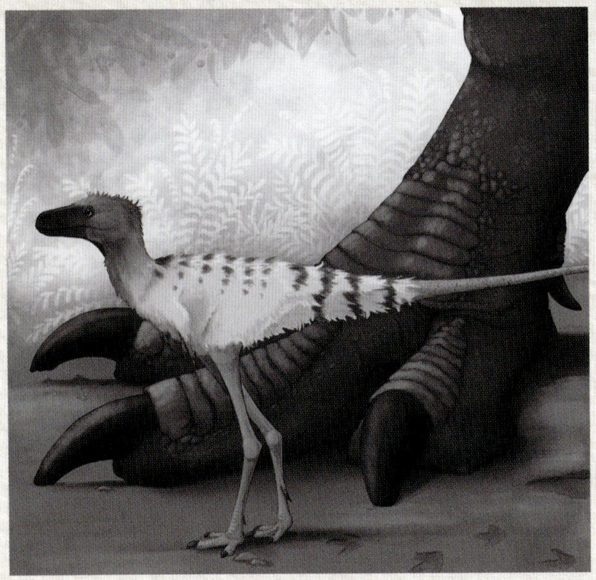

▲ LARAMIDIA IN BLACK & WHITE

This black-and-white vignette, arranged like a storyboard, depicts life among Late Cretaceous dinosaurs of Laramidia, the narrow continent corresponding to modern far western North America. We see parental behaviour among a long-crested species of *Parasaurolophus*, predation on *Edmontosaurus* youngsters by a tyrannosaur, and a fluffy tyrannosaur juvenile with its parent.
• Digital | 2023

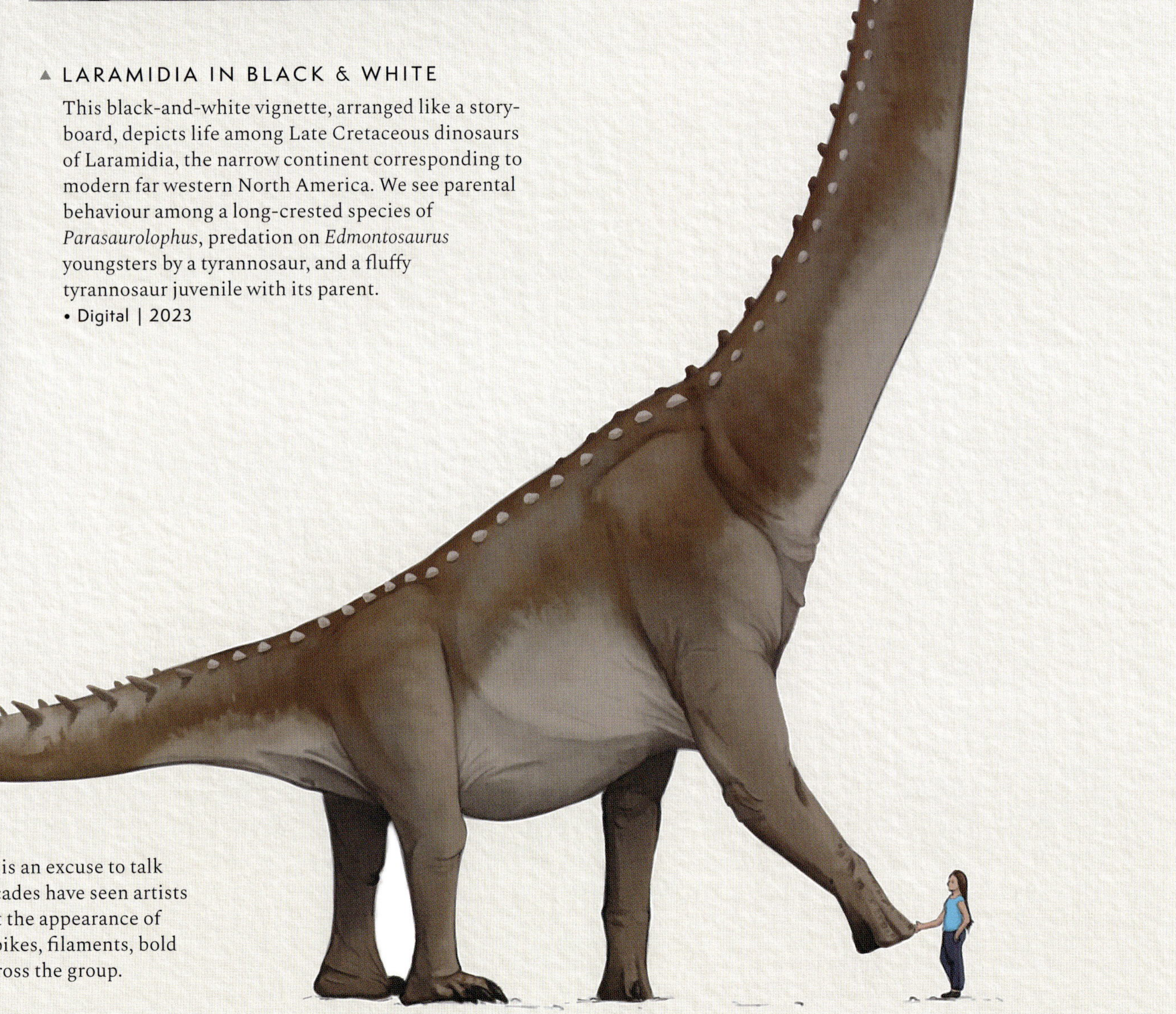

GIRAFFATITAN ▸

It might seem every artistic impression of a giant sauropod is an excuse to talk about the giant size of these animals, but...wow...Recent decades have seen artists and at least some scientists move away from the notion that the appearance of these animals must have been dull and bland. Decorative spikes, filaments, bold markings and flashes of colour might have been present across the group.
• Digital | 2021

TRIASSIC DEAD ZONE ▶

At the end of the Permian, 252 million years ago, our planet suffered its greatest mass extinction...so far. Entire suites of animals died out entirely, and for millions of years, well into the Early Triassic, tropical and sub-tropical regions remained barren. Yet there were survivors. In a vast and largely empty landscape, a small archosaur – an ancestor of dinosaurs and their kin – lurks among the rocks...

• Digital | 2022

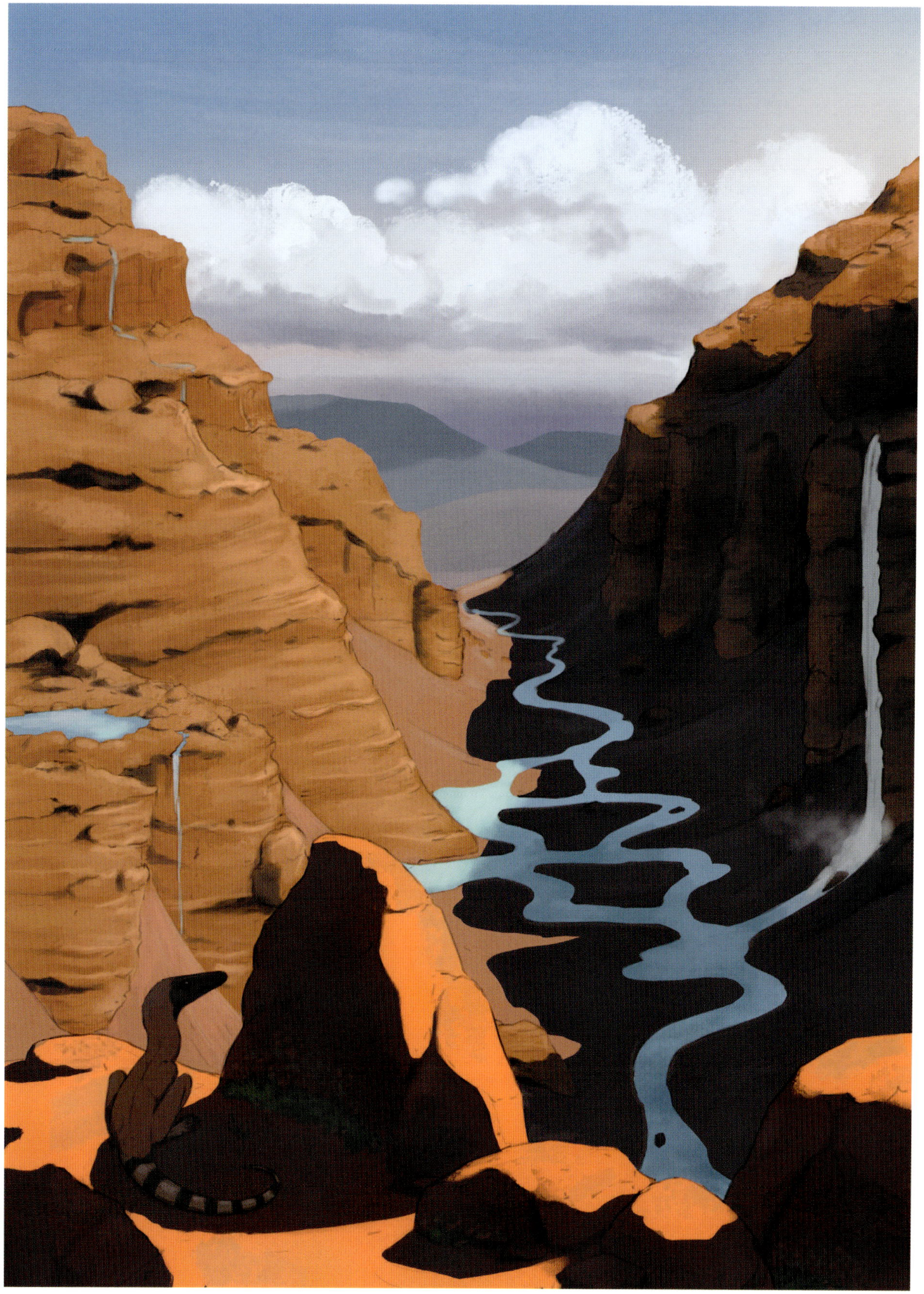

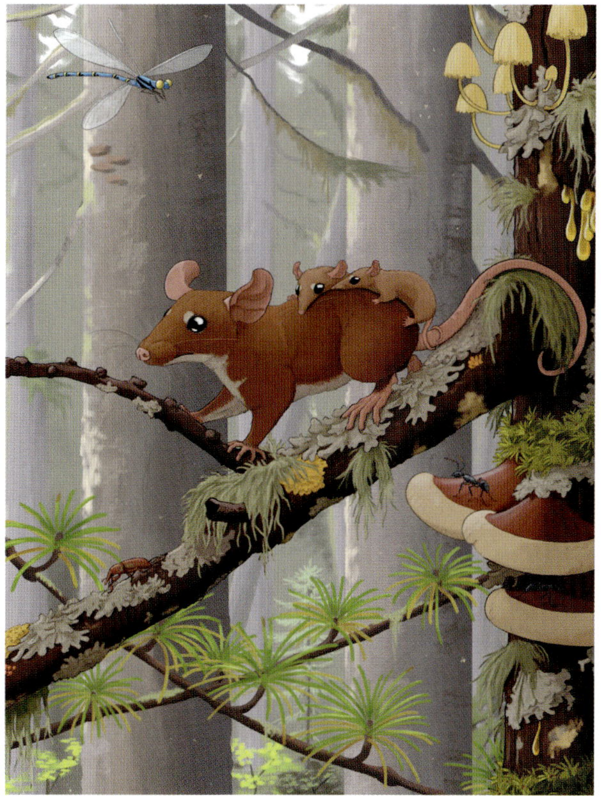

▲ EOCENE BALTIC AMBER FOREST

A speculative reconstruction of the warm coniferous forests located in Europe's Baltic region during the Eocene, which created the area's massive deposits of hardened tree sap – amber. The remains of numerous plants and animals were preserved in amber, including fungi, mosses, insects (like the ants, weevils and damselflies shown here) and even strands of mammal hair, perhaps belonging to the *Peradectes* shown here.

• Digital | 2022

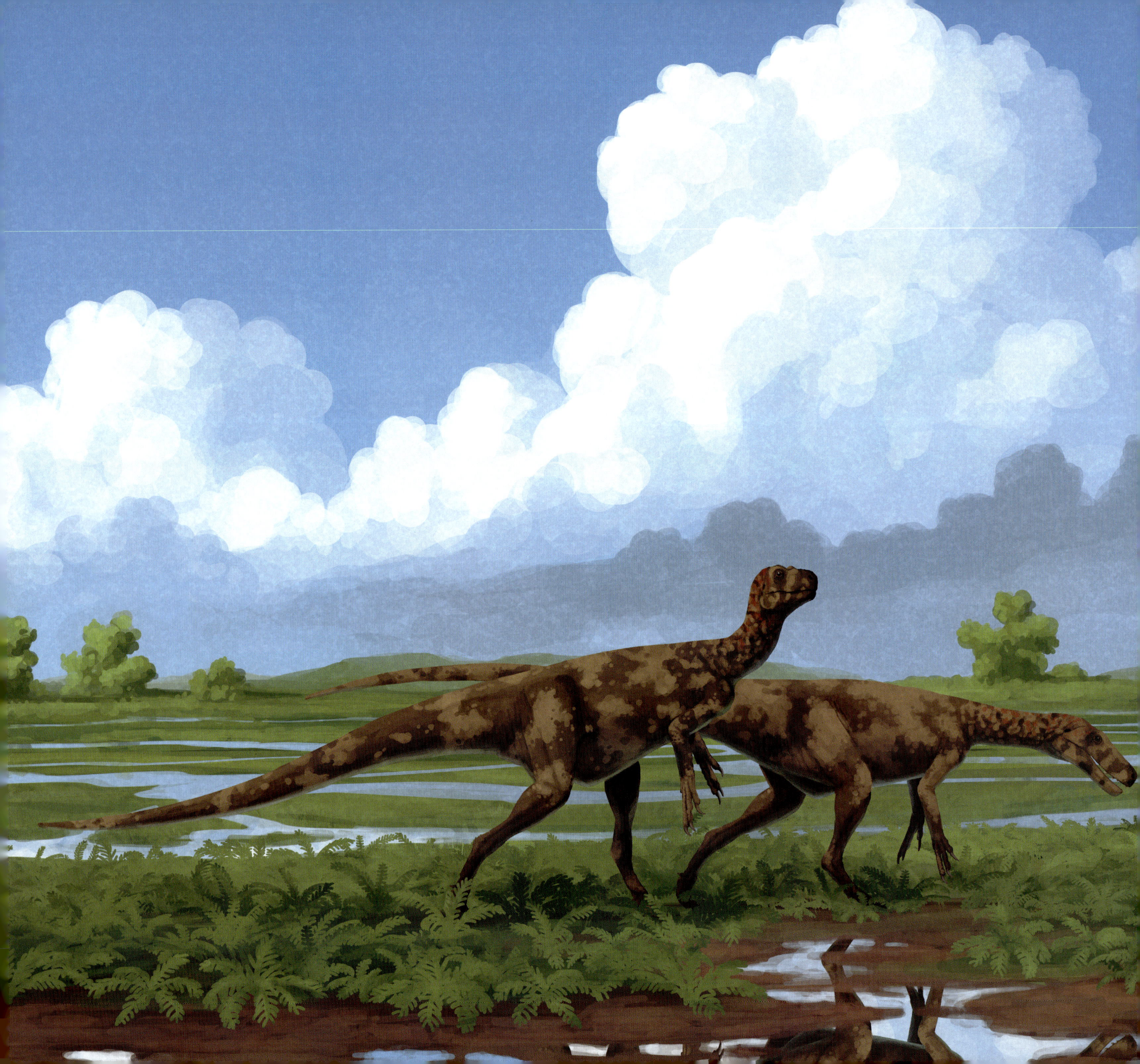

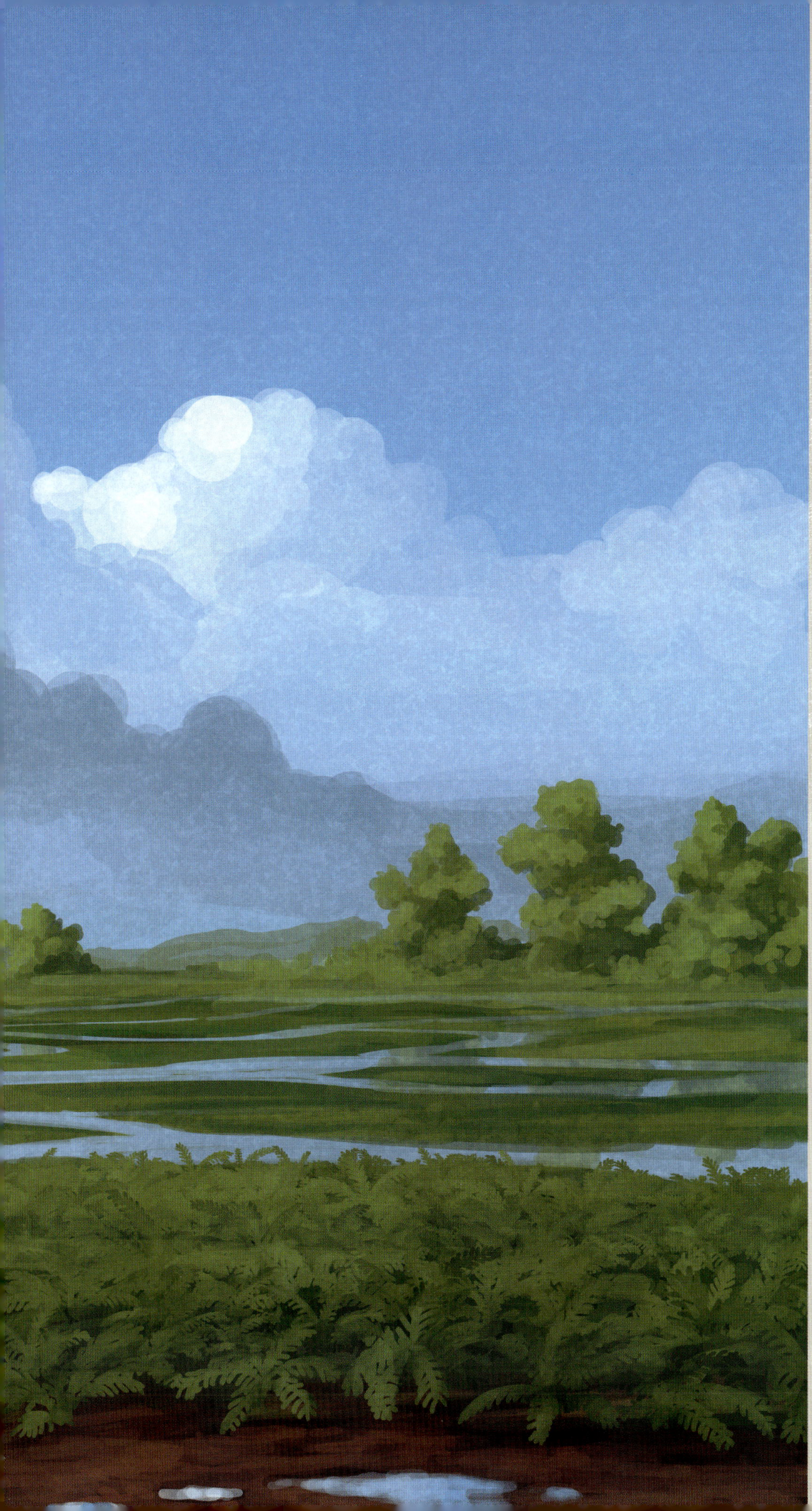

ASHLEY PATCH

ASHLEY WAS BORN AND RAISED in Australia. They grew up with one foot in the country and one in the city, regularly hiking and camping in the Australian outback and bushlands. These wildernesses have been a persistent inspiration throughout their life.

They were introduced to palaeontology through their mother, who, when Ashley was young, was at university majoring in geology and minoring in botany. This inspired them to begin drawing dinosaurs from the moment they could hold a pencil, taking a specific interest in their anatomy and how they are reconstructed. All the skeletons in their childhood books have scribbles over them as they tried to figure out how to accurately place flesh over the bones.

In 2013 Ashley joined DeviantArt and became known for their prolific output, posting nearly daily during their first few years on the platform. After finding artists like James Gurney, John Conway and Wayne Barlowe, they learnt that, in palaeoart, the artwork itself is as important as the scientific rigour. As such, they began applying themselves seriously into learning how to make an effective painting that just so happens to include dinosaurs.

Recently, however, Ashley has preferred to use art as a form of relaxation and is currently working on becoming an audio engineer.

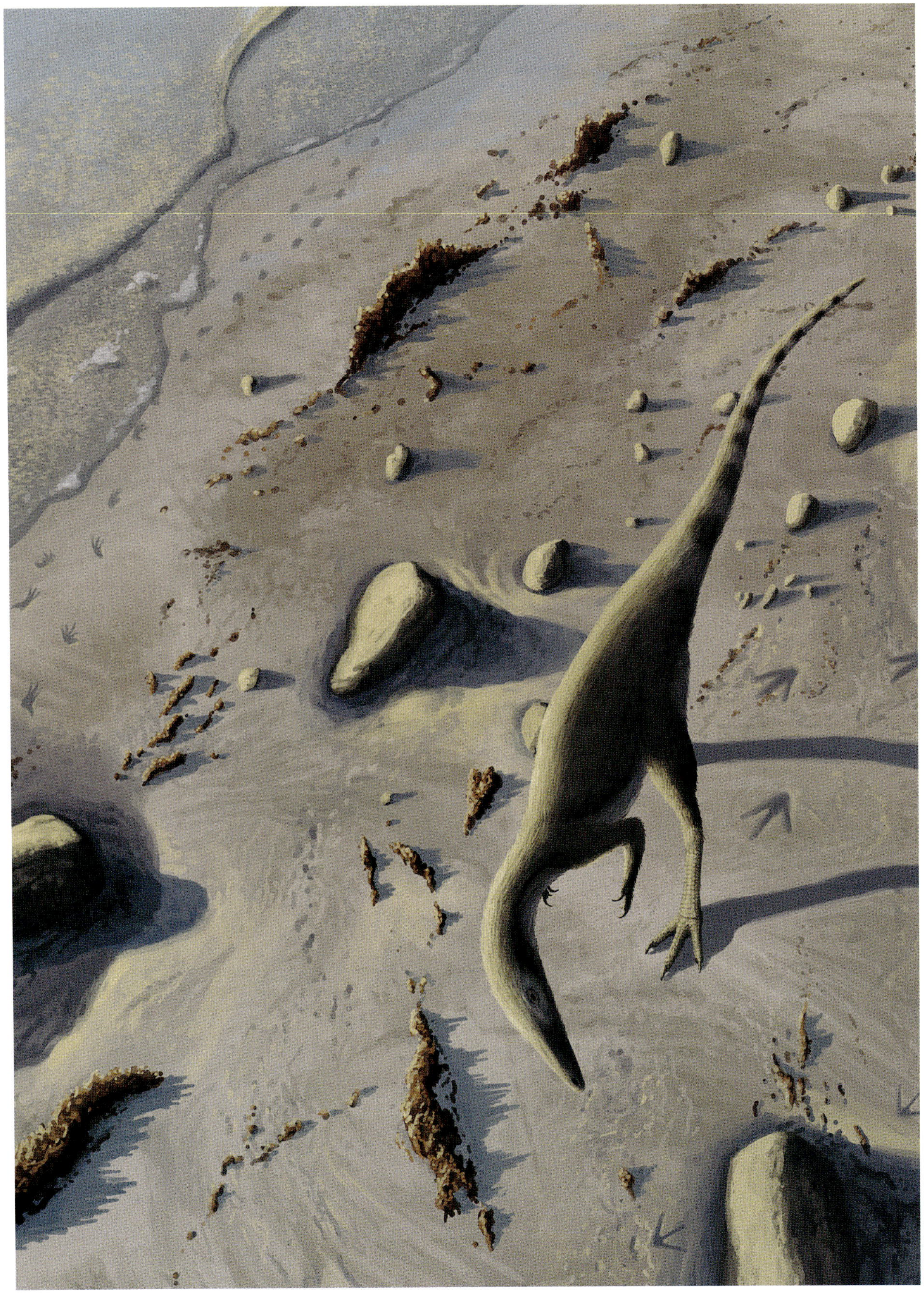

◂ **BEACHCOMBER *COMPSOGNATHUS***

Compsognathus is a Late Jurassic coelurosaur from western Europe, often stated in older texts to be among the smallest of non-bird dinosaurs. Debate continues over whether *Compsognathus* and similar theropods are the juveniles of larger forms. Its remains are associated with coastal and wetland locations, implying that *Compsognathus* foraged on shorelines.

• Digital | 2020

[Previous pages]

HERRERASAURUS

A pair of the archaic dinosaur *Herrerasaurus* stalk a Triassic wetland, somewhere in what is today Argentina, 230 million years ago. Little is known about the ecology or behaviour of Triassic dinosaurs like this, beyond that they were predatory. This reconstruction breaks with tradition in associating the animal with green, well-watered conditions rather than with desert. The former is more accurate based on geological evidence.

• Digital | 2021

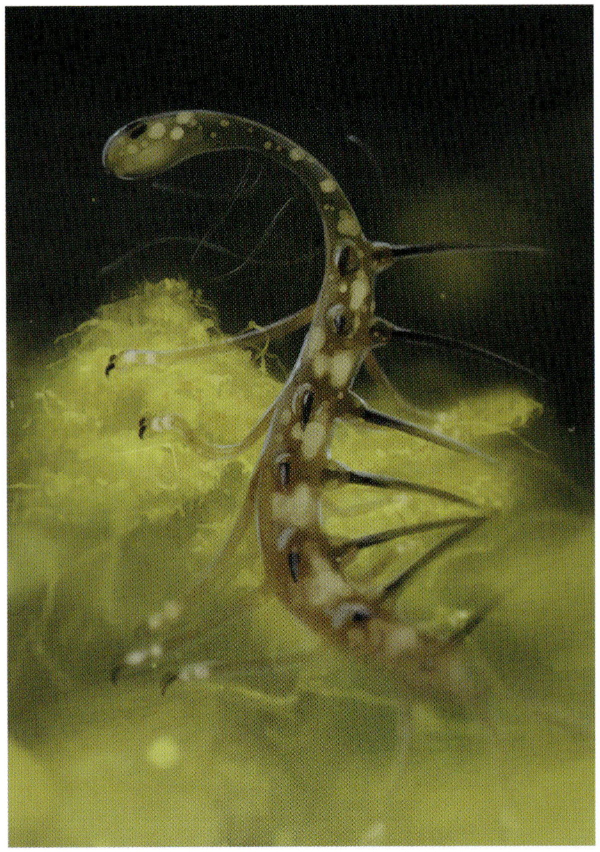

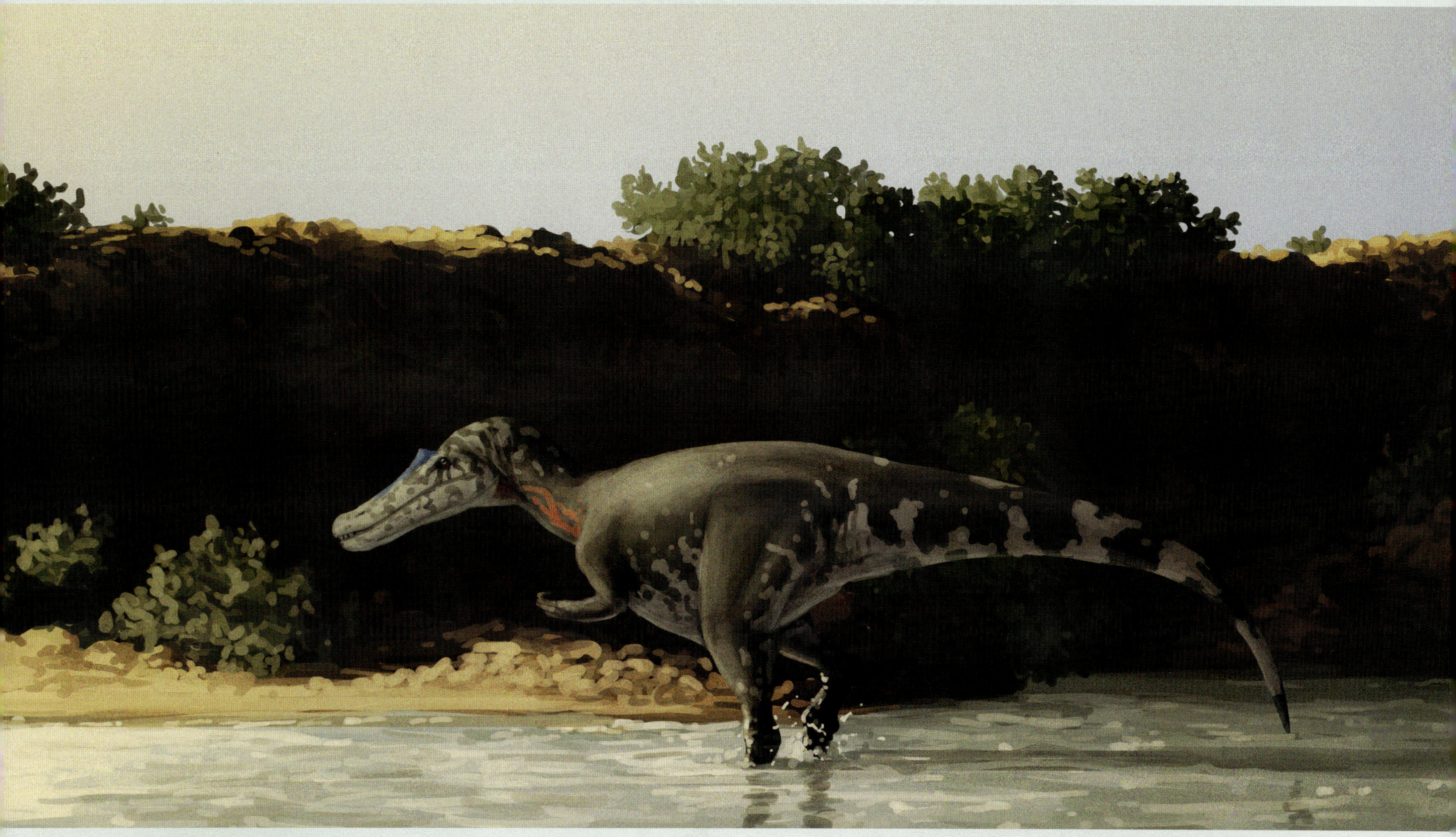

◂ ***HALLUCIGENIA***

The bizarre lobopods (or lobopodians) of the Cambrian existed in seas worldwide between around 530 and 430 million years ago, and *Hallucigenia* is the best known of them. It was small, at less than 5cm (2in) long. It's assumed that these were animals of reef environments and they've sometimes been depicted clambering on sponges or algae. However, their specific habitat preferences are unknown.

• Digital | 2024

▴ WADING *BARYONYX*

The spinosaurid *Baryonyx* – initially described from England – wades in an Early Cretaceous river, 135 million years ago. Jaw and tooth shape implies that these dinosaurs were generalist predators of diverse prey, but an ability to prey on fish was probably important in part of their range. Raised midline skull crests are a typical feature of spinosaurids, and these perhaps had eye-catching markings or colours, as shown here.

• Digital | 2021

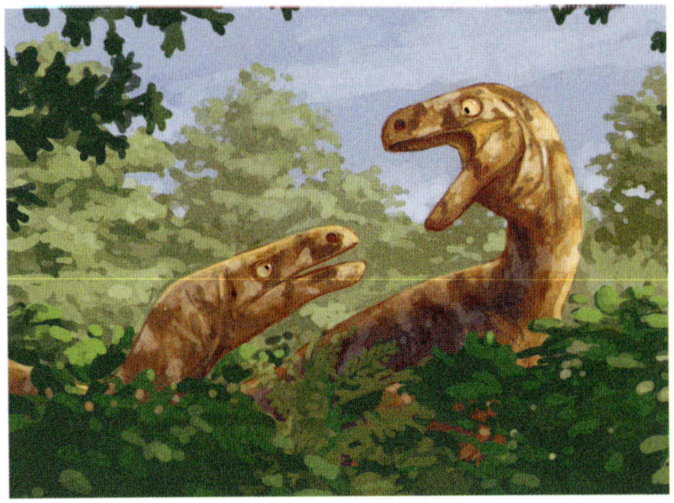

◄ MASIAKASAURUS

The Late Cretaceous Madagascan theropod *Masiakasaurus* has typically been portrayed so that the radiating, protruding look of its lower-jaw teeth is emphasised. In reality, soft tissue and a corrected orientation of the jaw bones means that they were likely obscured in the live animal. This was a smallish theropod (2m [6ft] long) that probably ate small prey such as arthropods, lizards and amphibians.

• Digital | 2022

▼ GALEAMOPUS

Galeamopus, known in 2015, is one of the many diplodocid sauropods named from the USA's famous Upper Jurassic Morrison Formation. Some *Galeamopus* specimens had previously been mistaken for remains of its cousin *Diplodocus*. Neck flexibility and tooth shape implies that these dinosaurs could feed both from the ground and high up in trees as well. The tree here is a ginkgo.

• Digital | 2021

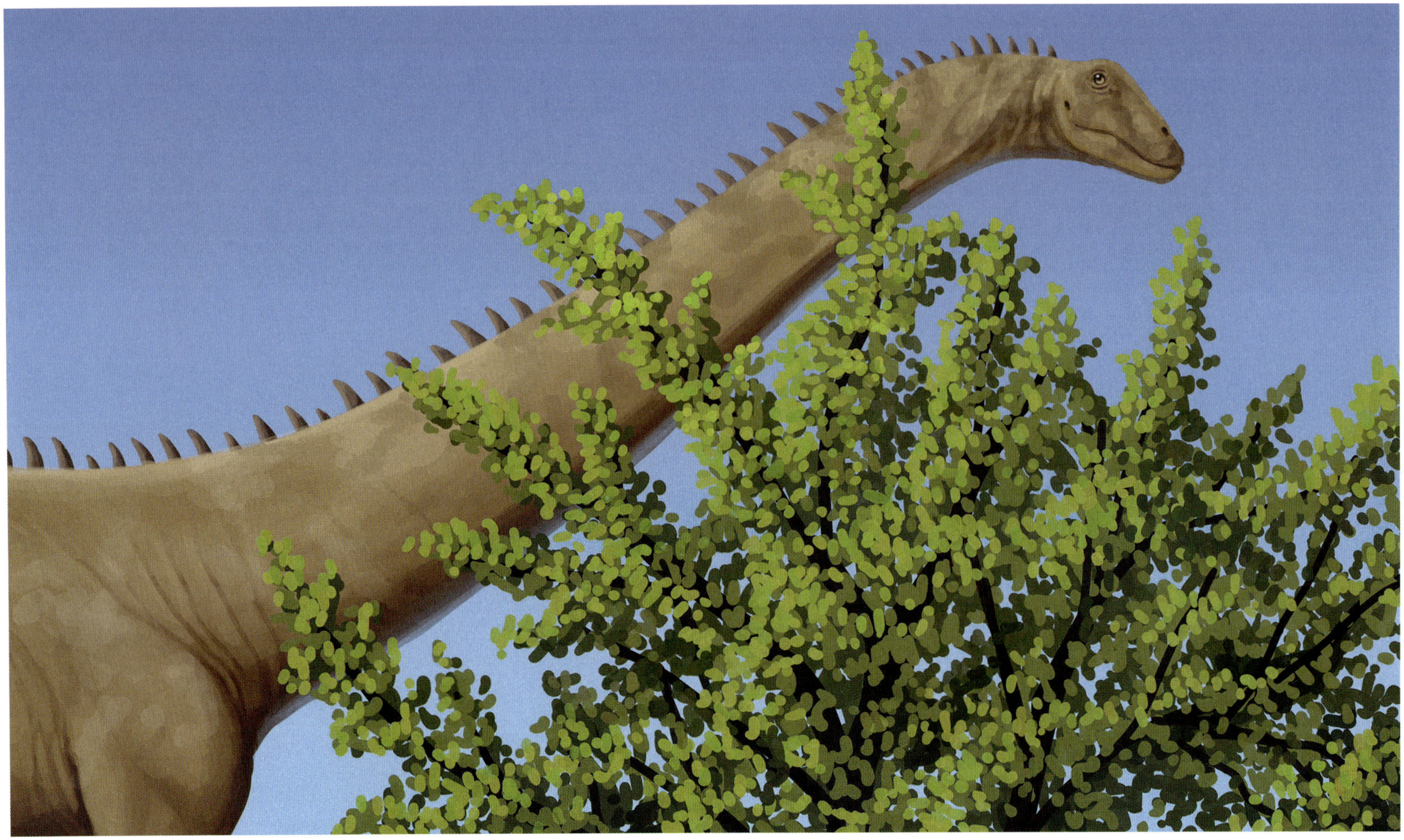

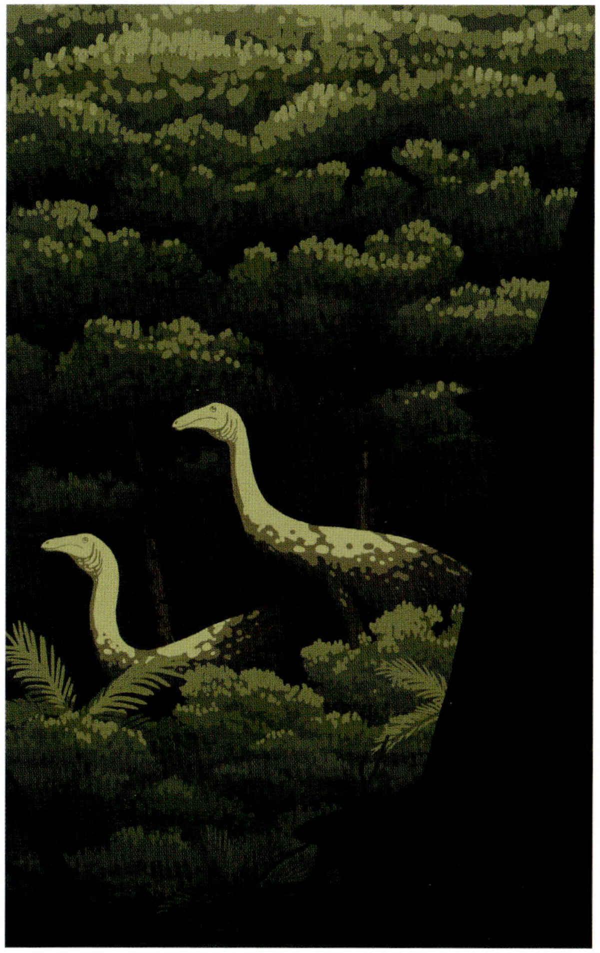

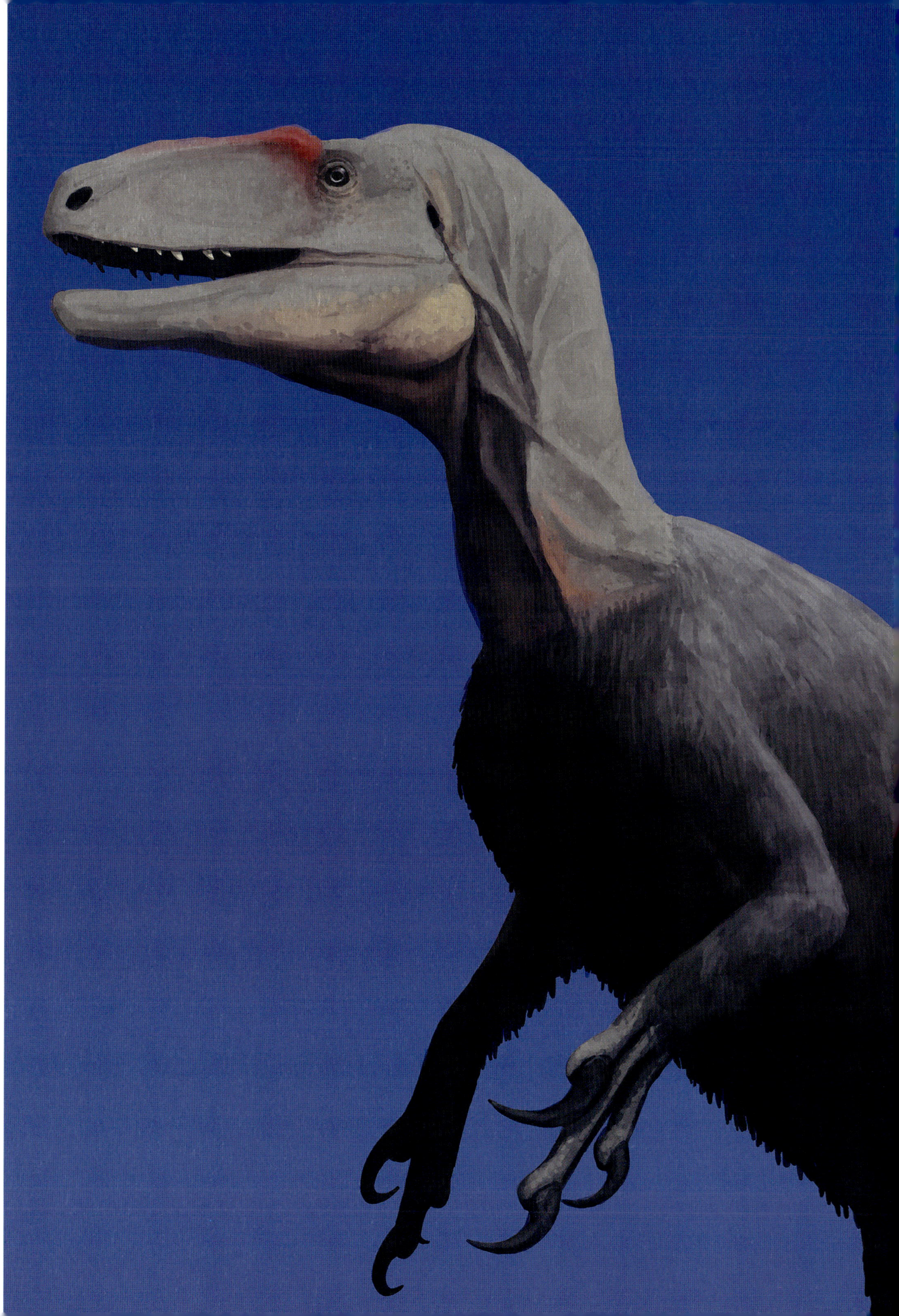

▲ *GALLIMIMUS*

Among the largest of the long-necked, long-legged ornithomimids or 'ostrich mimics', *Gallimimus* is an East Asian Late Cretaceous dinosaur, typically associated with desert and semi-desert environments. But lush forest existed within its range too, and the dappled colouration shown here suggests adaptation to wooded habitat.
• Digital | 2020

EOTYRANNUS ▶

One of several important Early Cretaceous theropods from the Wealden sediments of southern England, the tyrannosaur *Eotyrannus* was a mid-sized predator with a rectangular skull and long, slender hands. Thick, gnarly bones formed the midline of its snout. The size and striking curvature of theropod hand claws has sometimes been understated in reconstructions, but not here.
• Digital | 2021

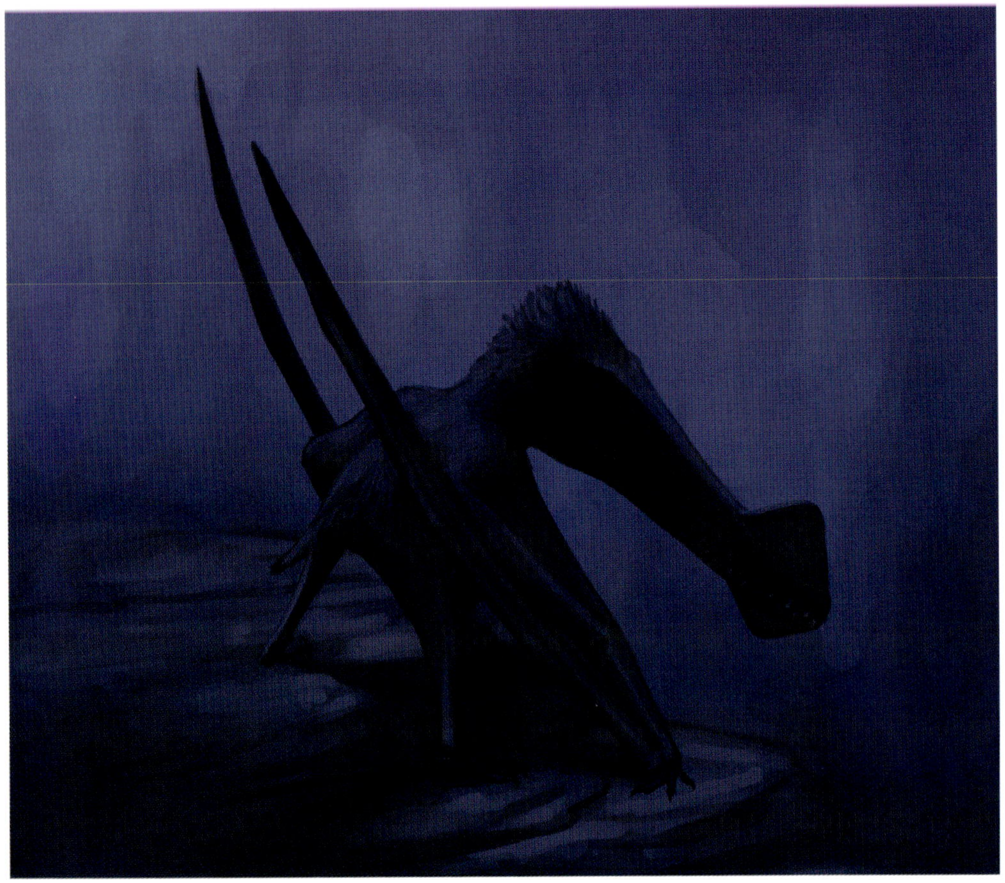

CERATOSAURUS ▸

The unique profile of the Late Jurassic theropod *Ceratosaurus* – the quintessential 'horned lizard' of the western USA, Portugal and elsewhere – has made it an artist's favourite. This moody reconstruction implies pair-bonding in these animals and hints at a habit of hunting either at dawn or in the evening. The unfortunate prey animal is a mid-sized sauropod.
• Digital | 2020

▴ *FERRODRACO*

Ornithocheiroid pterosaurs like this Australian *Ferrodraco* were extraordinary animals, with extremely long, slender wings, an absurdly large head, and tiny hindlimbs. That last trait suggests that they were slow, cautious walkers, as implied in this low-light scene. Several members of the group, including *Ferrodraco*, possessed prominent crests and keels at the jaw tips.
• Digital | 2021

PLATYPTERYGIUS IN LOW LIGHT ▸

The large eyes of ichthyosaurs – the so-called 'fish lizards' of the Mesozoic – have often been linked to a deep diving ability. They would also, however, allow hunting at night, as well as during dawn and dusk. Originally described from Germany, *Platypterygius* is an Early Cretaceous ichthyosaur that reached up to 5m (16ft) in size and possessed a long, low skull profile.
• Digital | 2021

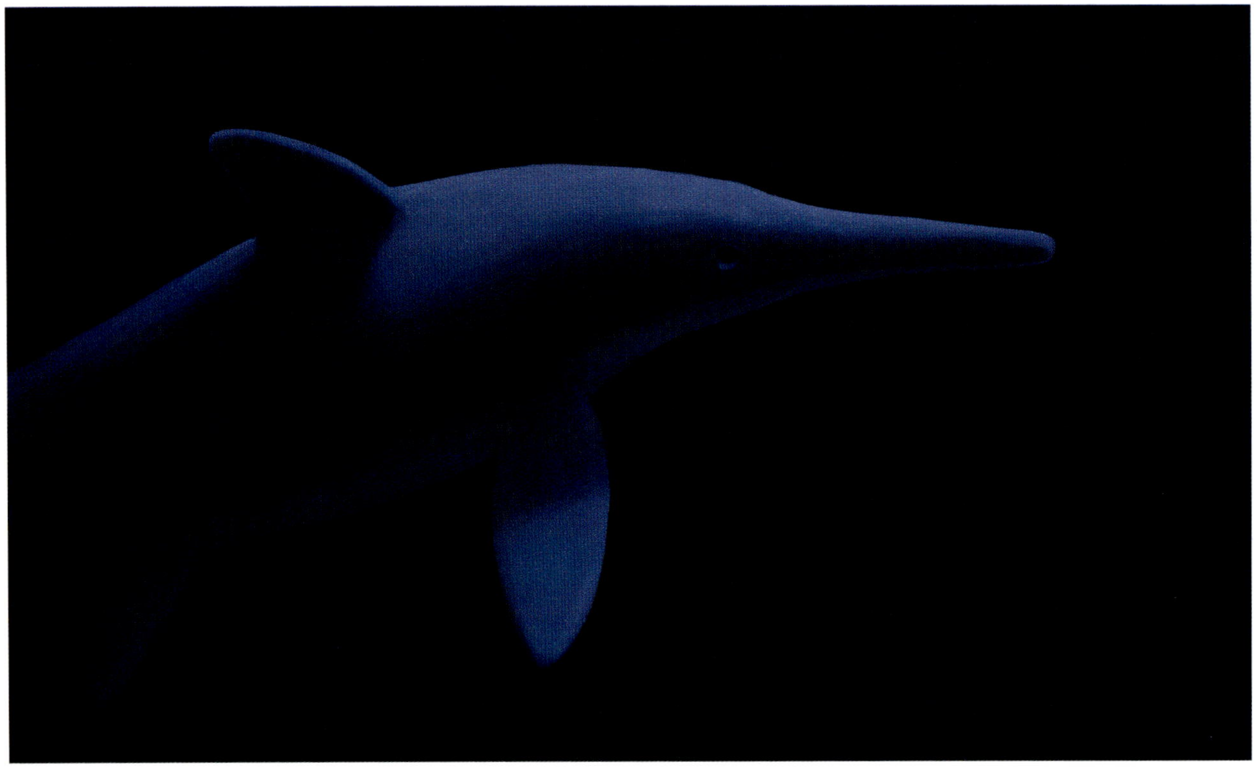

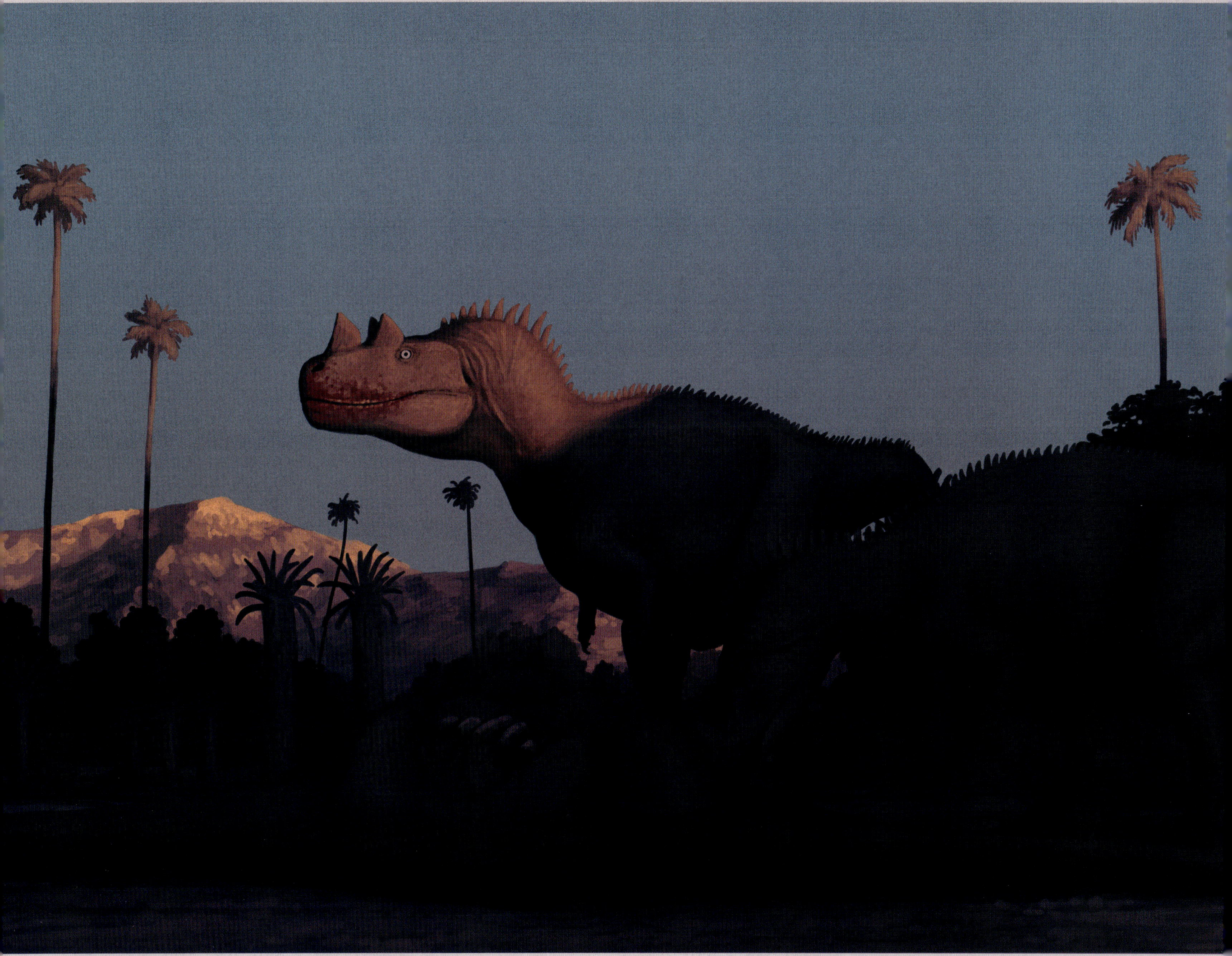

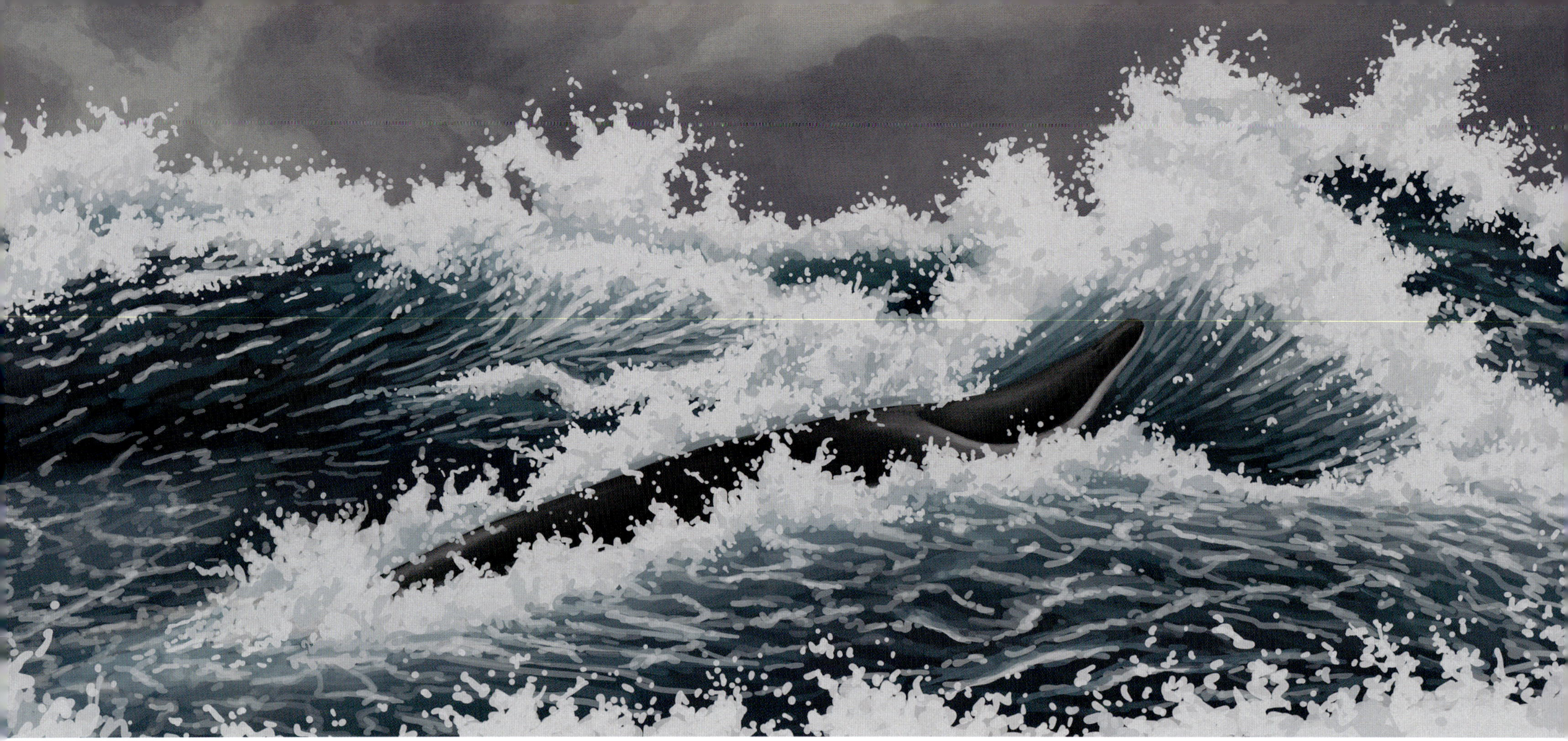

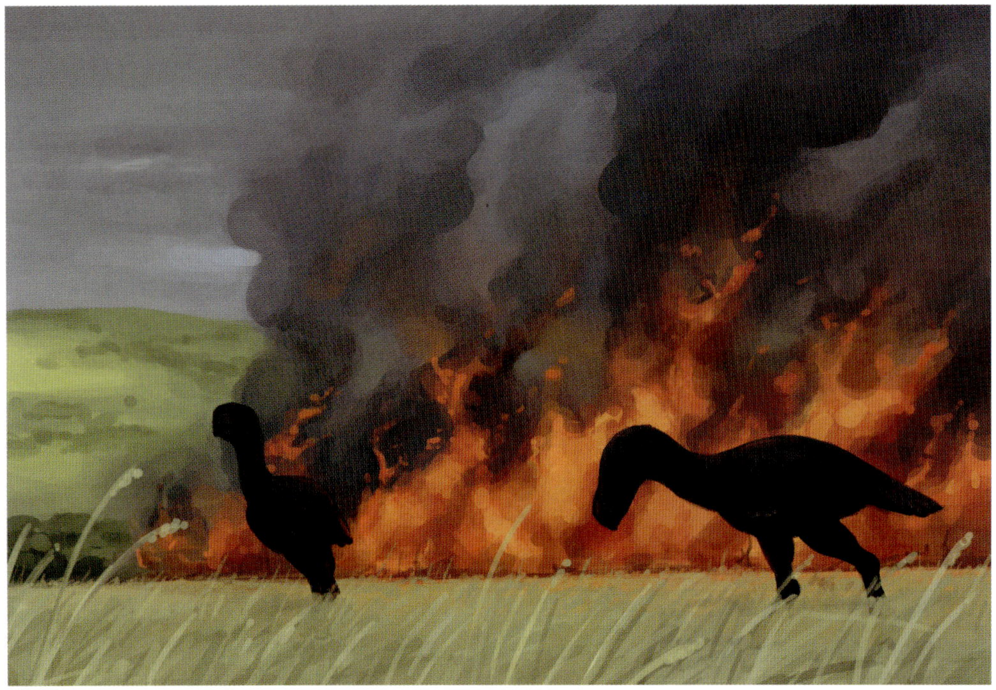

▲ *BASILOSAURUS*

The stem-whales of the Eocene – *Basilosaurus*, shown here, is the largest of them (it reached 20m or 66ft) – are virtually always depicted in calm seas. But they must have contended with storms and surges too, and as they were fully aquatic animals, we have to assume that they exhibited behaviours that allowed survival in these conditions.

• Digital | 2020

◀ GRASS FIRE

Phorusrhacids stalk the edge of a brushfire, looking for prey flushed out by the flames. This behaviour is seen amongst many extant predatory and omnivorous birds out for an easy meal. Phorusrhacids – colloquially known as 'terror birds' – were a family of carnivorous flightless birds, mostly associated with South America, that lived from the Middle Eocene, 43 million years ago, until the end of the Pleistocene.

• Digital | 2024

TAPEJARA

The strikingly narrow, often crested skulls of certain pterosaurs – like Brazil's *Tapejara*, shown here – surely made them bizarre and perhaps comical when seen from the front. Both the size of the eyes and anatomy of the brain shows that eyesight was excellent in these animals. The eye's position also implies a near wrap-around field of vision.

• Digital | 2022

DIMETRODON

The famed 'sailback' of the Permian, *Dimetrodon* existed across North America and Europe in numerous habitats. Though classically depicted in desert environments, it was more often associated with wetlands. If the sail really was a display structure, as some experts think, it's plausible that it was spectacularly colourful and highly visible from distance.

• Digital | 2021

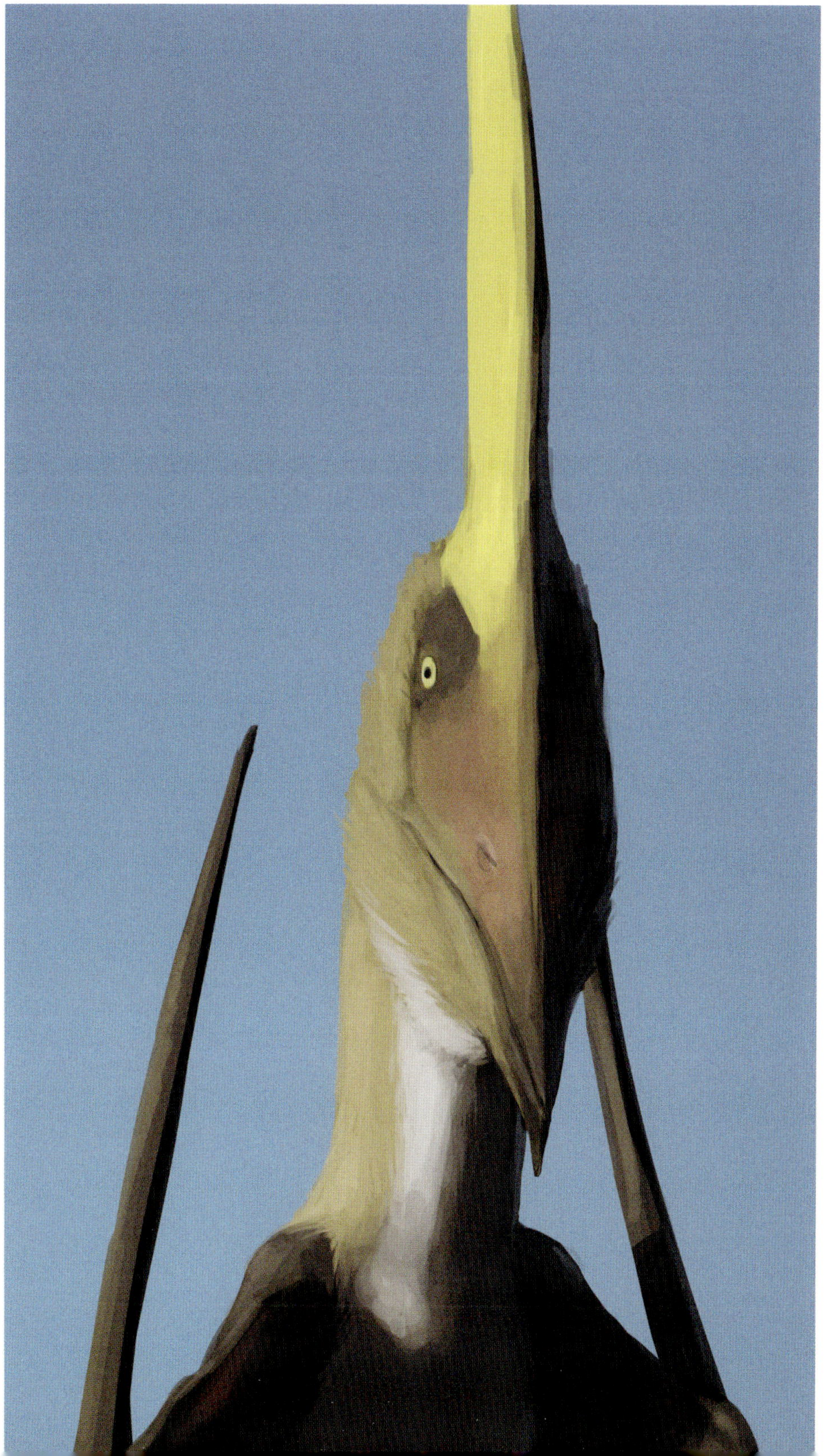

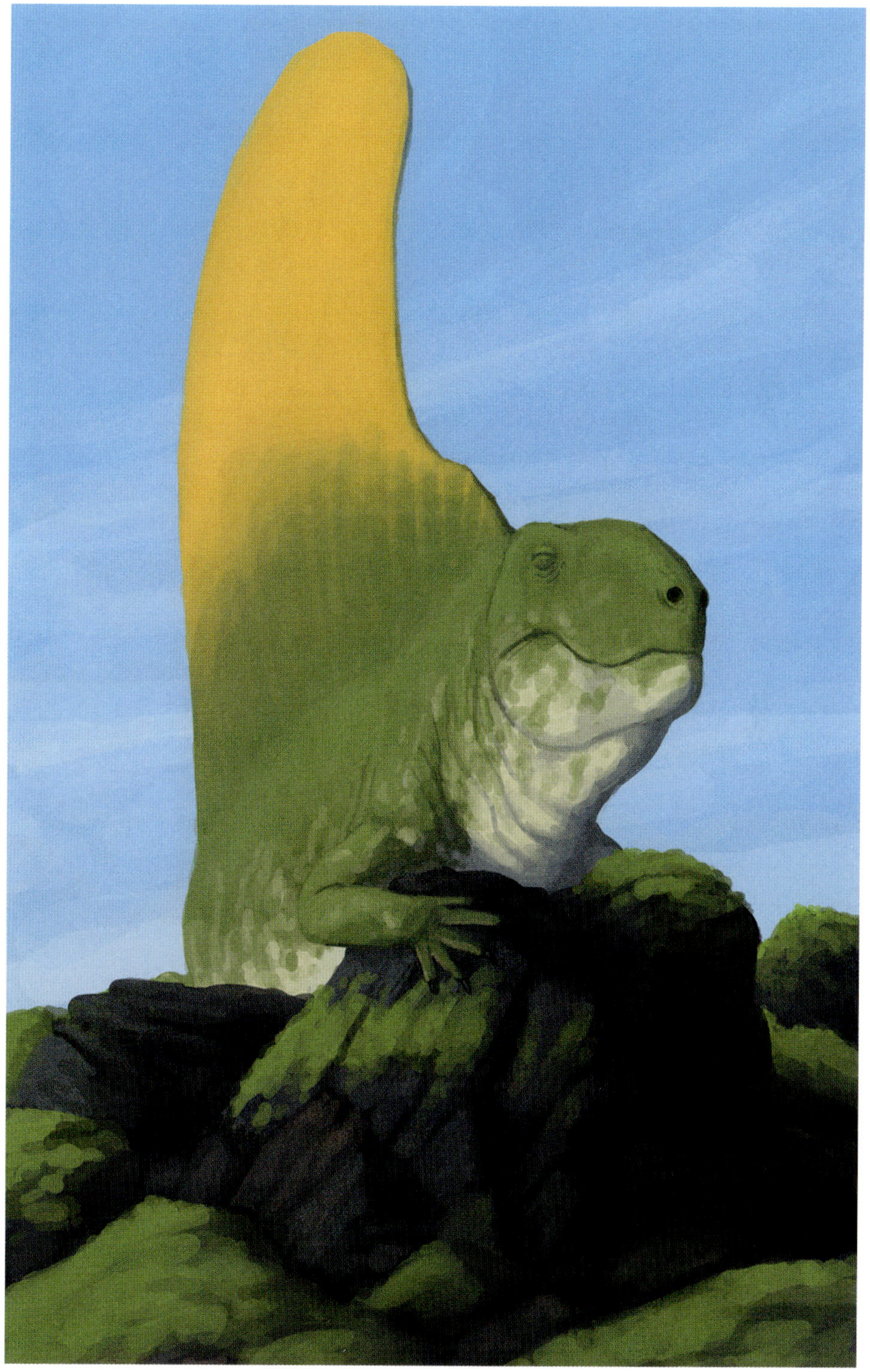

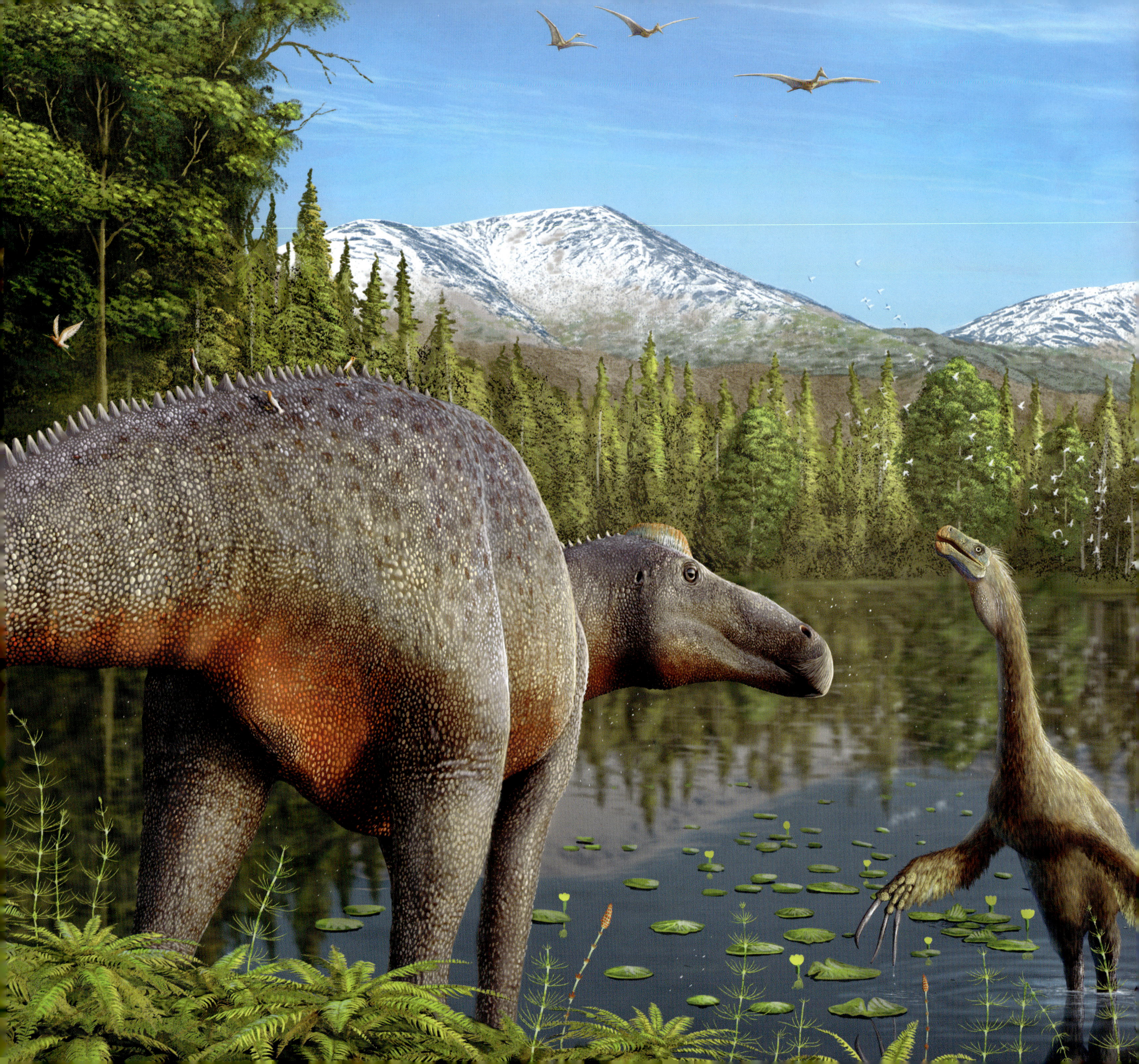

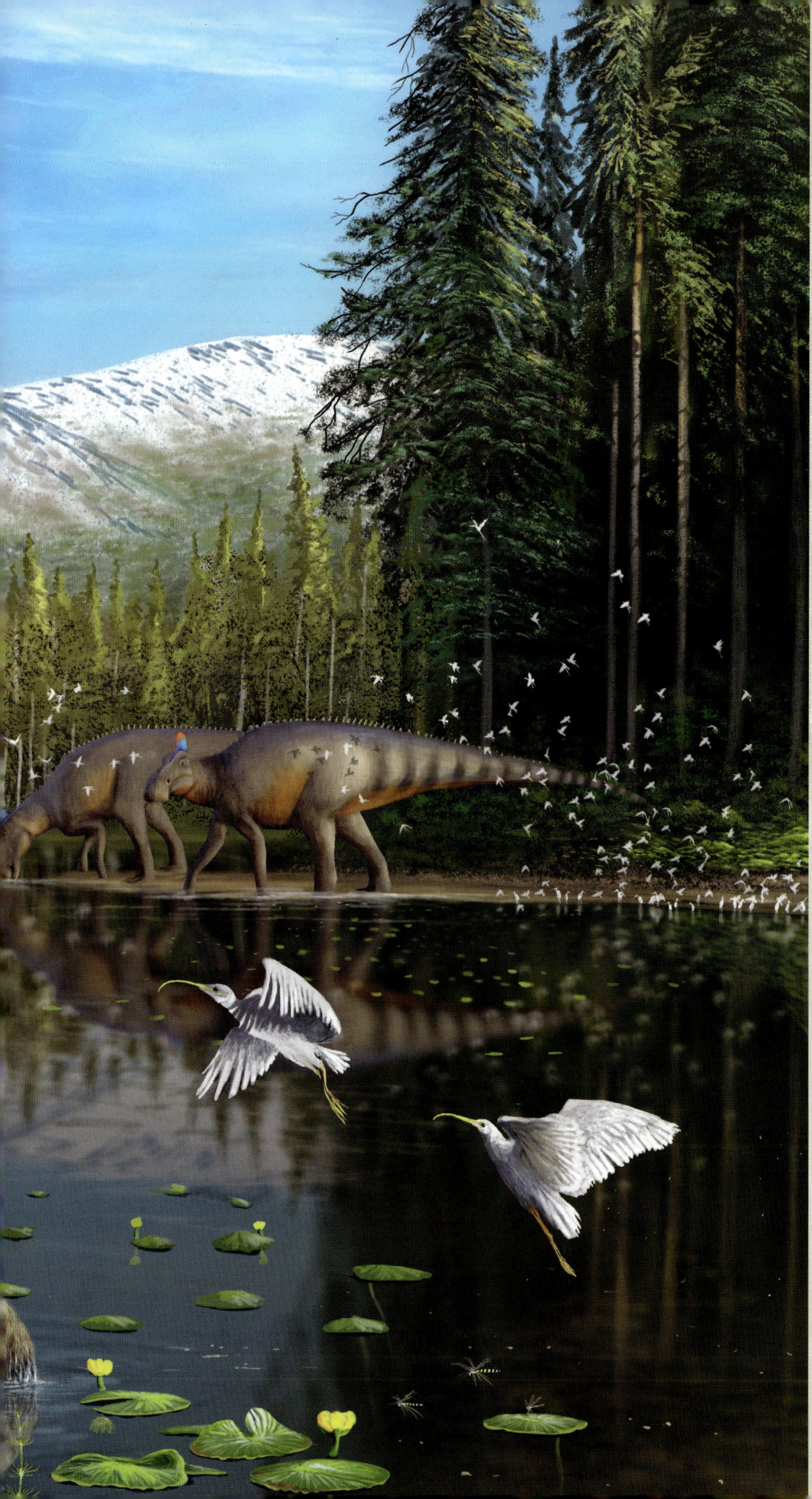

MATTIA YURI MESSINA

MATTIA YURI MESSINA, from Sardinia, Italy, graduated in Natural Sciences from the University of Cagliari and has a master's degree in Palaeontology from La Sapienza, University of Rome. While working on his thesis, he participated in several palaeontological excavations in Italy and Patagonia. At the time of writing, he is studying for a PhD in Palaeontology at the University of Cagliari.

Throughout his life, his passion for palaeontology has led him to use art as a means of representing the environments and life forms of the past. His principal medium is 2D digital drawing and 3D art, although he continues to regularly draw with pencils as practice. Some of his illustrations are currently exhibited in the virtual gallery of the Società Paleontologica Italiana, with one of these images featured on the cover of the gallery's printed catalogue. In 2022, one of his illustrations was used as the cover image for the scientific journal of the Società Paleontologica Italiana (Volume 61. N.3). He has also worked on palaeoart projects for various exhibitions featured at conferences and museums. In 2023, he won first prize at the Italian Award for Paleoart with the illustration 'Meetings Between Continents'.

His approaches to studying palaeontology and producing palaeoart have the same goal: to learn more about lost worlds and shed light on the living beings that inhabited them.

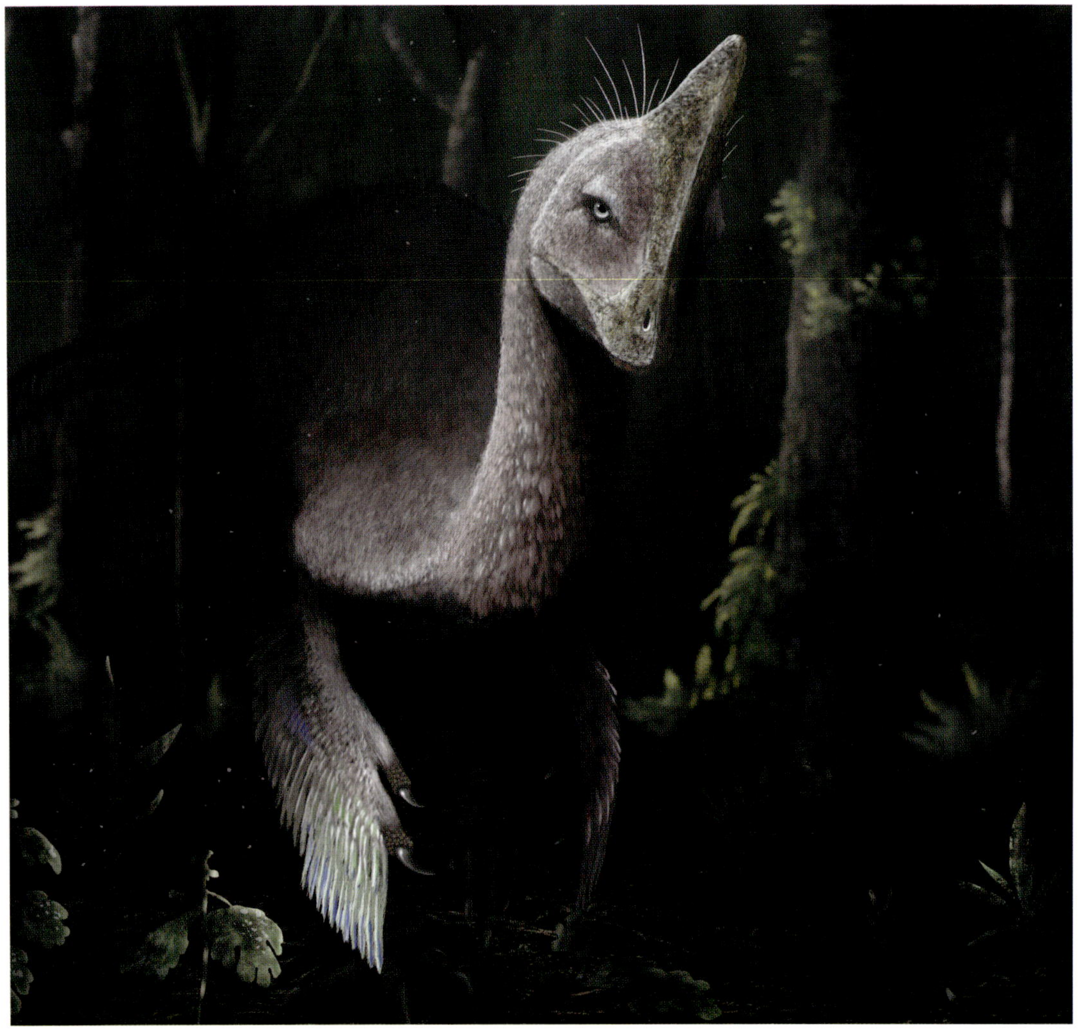

[Previous pages]

MEETINGS BETWEEN CONTINENTS

On a Late Cretaceous day in Alaska, USA, a bathing therizinosaur warns an approaching *Edmontosaurus* not to come any closer. This scene represents the Cantwell Formation, where no bones of dinosaurs and pterosaurs have been found but footprints attest to their presence. The long-billed birds are inspired by the discovery of bird tracks dubbed *Magnoavipes*.
• Digital | 2023

COELOPHYSIS ▶

Coelophysis, named in 1889, is the quintessential predatory dinosaur of North America's Early Jurassic. While it was conventionally portrayed as scaly, our knowledge of dinosaurian skin coverings and of the environmental temperatures that affected these animals suggest the possibility of a fuzzy covering. There's little doubt that these dinosaurs were predators of small, lizard-like reptiles, as shown here.
• Digital | 2024

▲ *ANZU*

The large oviraptorosaur *Anzu* – named from the USA in 2014 – possessed a tall, rounded crest and toothless, beaked jaws. The plumage of *Anzu* remains speculative. It might have been as luxuriantly feathered as modern pheasants, or pehaps have a sparser look with a shorter, tighter plumage. Unlike many related forms from Asia, *Anzu* was a forest animal.
• Digital | 2022

DREPANOSAURUS ▶

The grasping hands, large, curved claws and prehensile tail of the Triassic reptile *Drepanosaurus*, known both from Italy and the USA, indicate proficient tree-climbing abilities. Perhaps this animal climbed in the hand-over-hand style used by chameleons and small primates today. The possibility also exists that it could climb upside-down.
• Digital | 2020

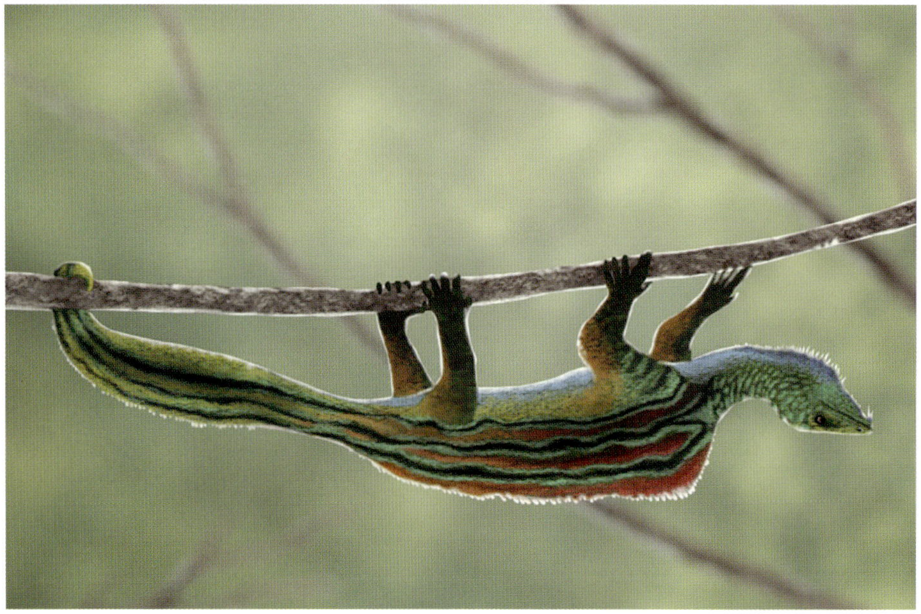

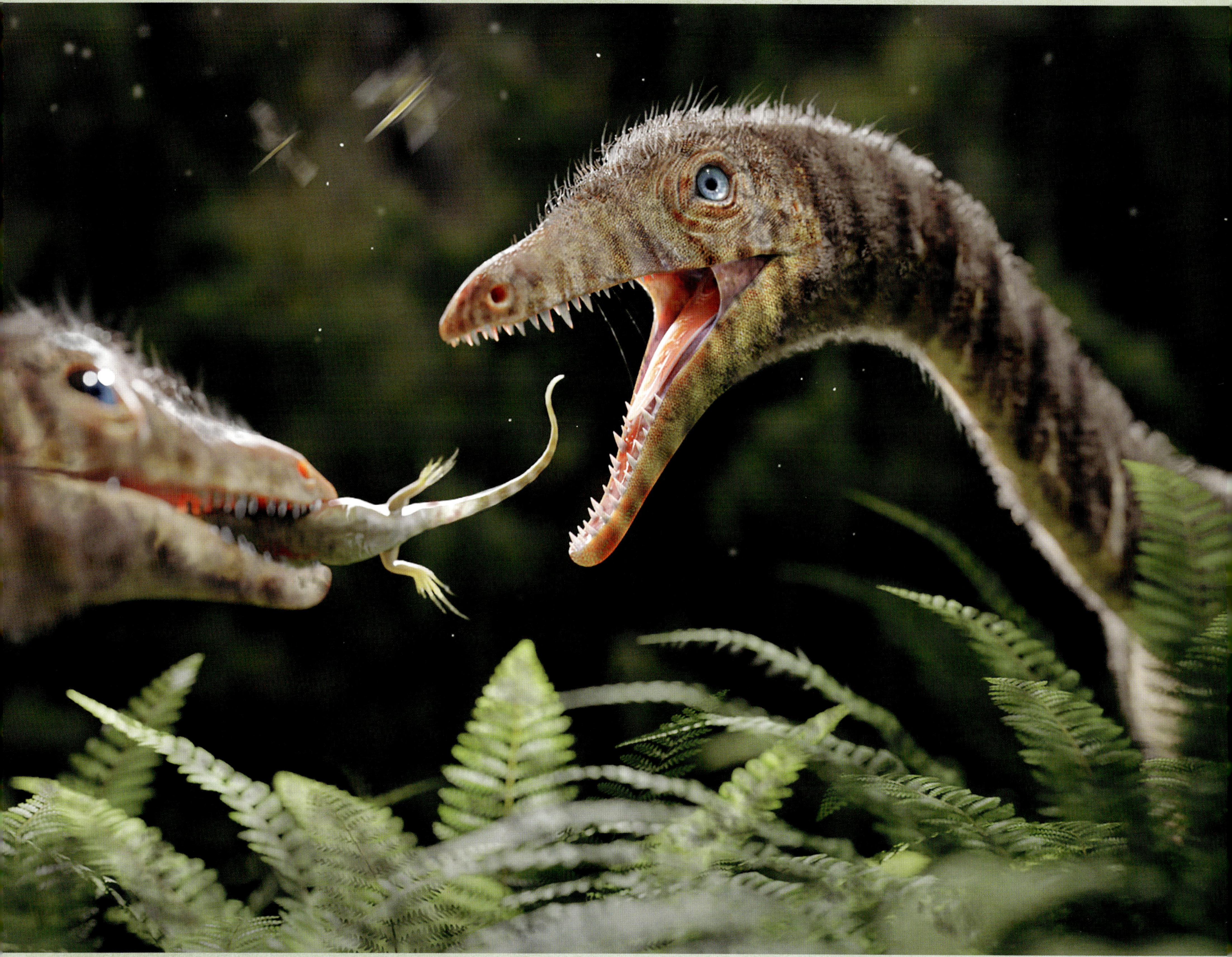

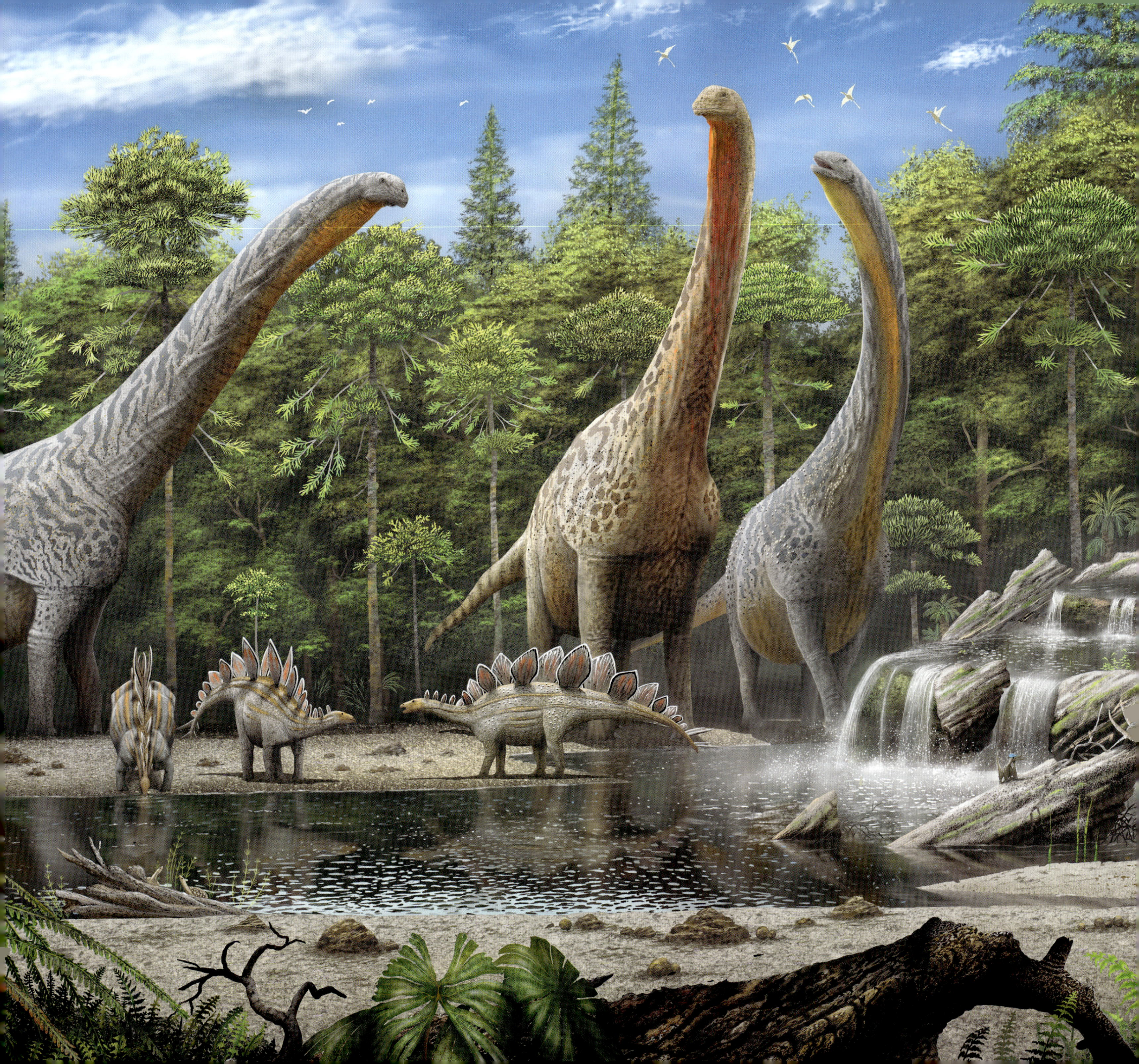

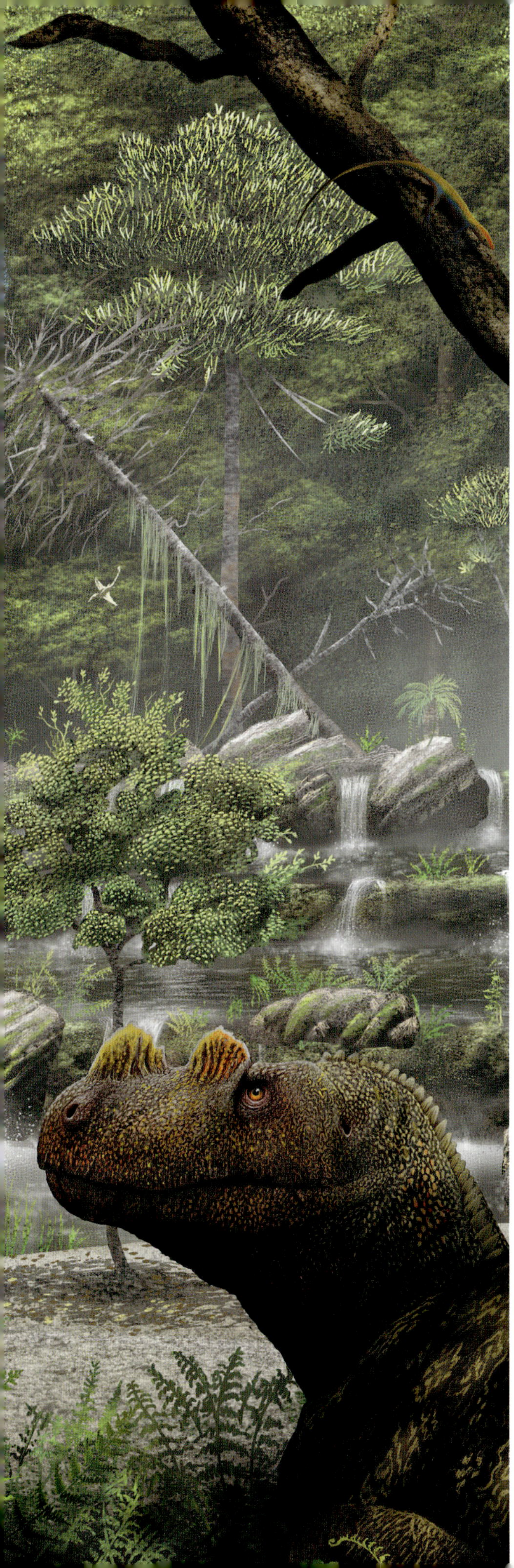

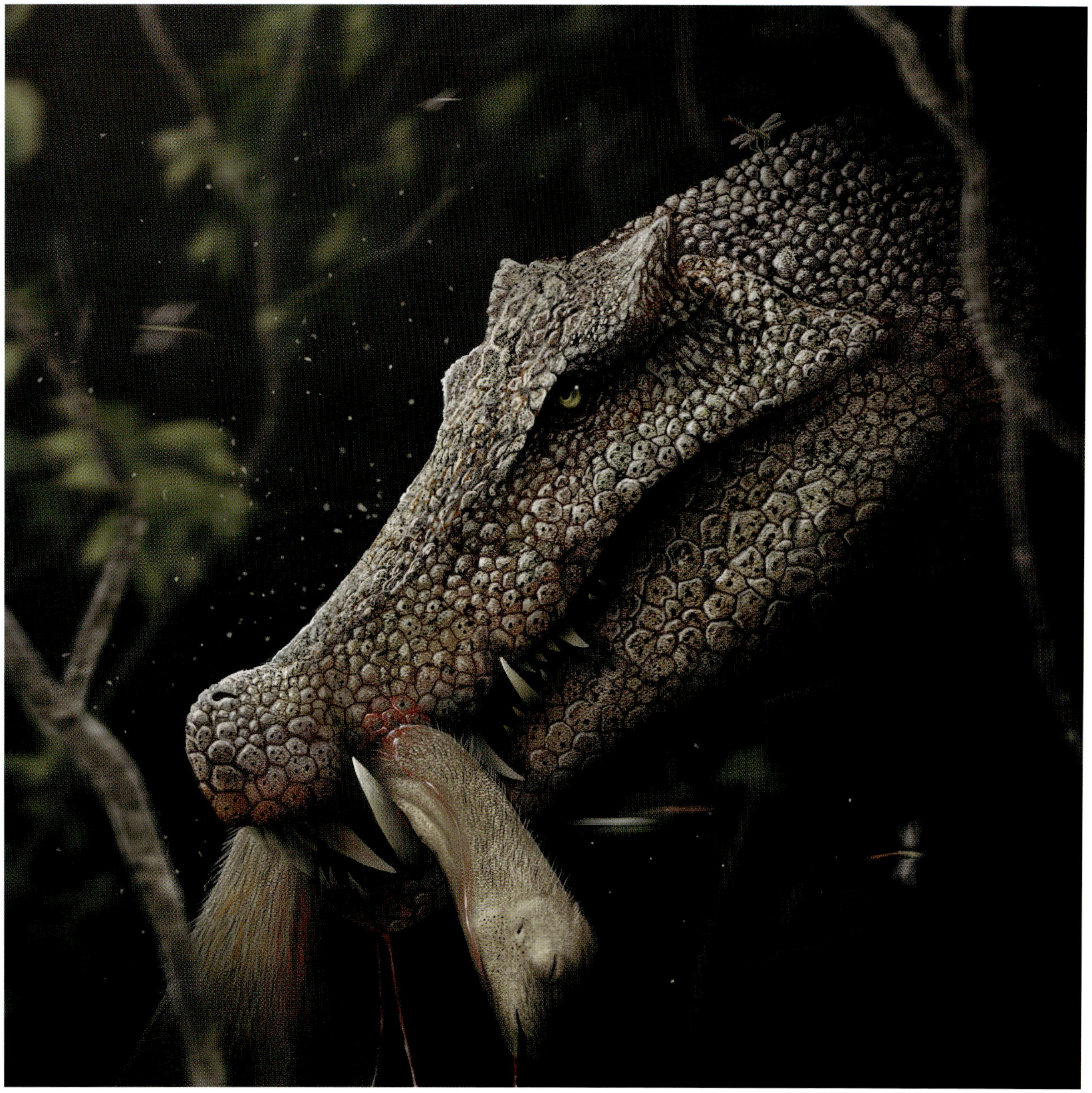

◄ WATER IS LIFE

This idyllic representation of Late Jurassic life shows dinosaurs, pterosaurs and plants of the USA's famous Morrison Formation. Giant sauropods like *Brachiosaurus*, the great *Stegosaurus* and the horned *Ceratosaurus* are among the Morrison's most familiar dinosaurs. Whilst an excellent piece of palaeoart, this scene could also be an exercise in the depiction of water, a subject many artists struggle to accurately portray.

• Digital | 2021

▲ *KAPROSUCHUS*

The spectacular tusk-like teeth and raised bony crests of *Kaprosuchus* – a Late Cretaceous crocodyliform from Niger, named in 2009 and nicknamed 'boar-croc' – guaranteed its 15 minutes of fame at the hands of the world's media. Initially portrayed as a long-limbed animal adapted for life on land, its flattened snout and elevated eye sockets are more consistent with an amphibious lifestyle.

• Digital | 2021

▼ *CRETOXYRHINA* HUNT

Sharks similar to the makos and great whites of today hunted the Cretaceous oceans too, and many of the surprising, sometimes spectacular, predatory feats of modern times surely occurred in the past. Here, the large mackerel shark *Cretoxyrhina* charges through shallow water in pursuit of small mosasaurs. This is a risky strategy, but tiger sharks and other modern species show that it sometimes pays off.
• Digital | 2024

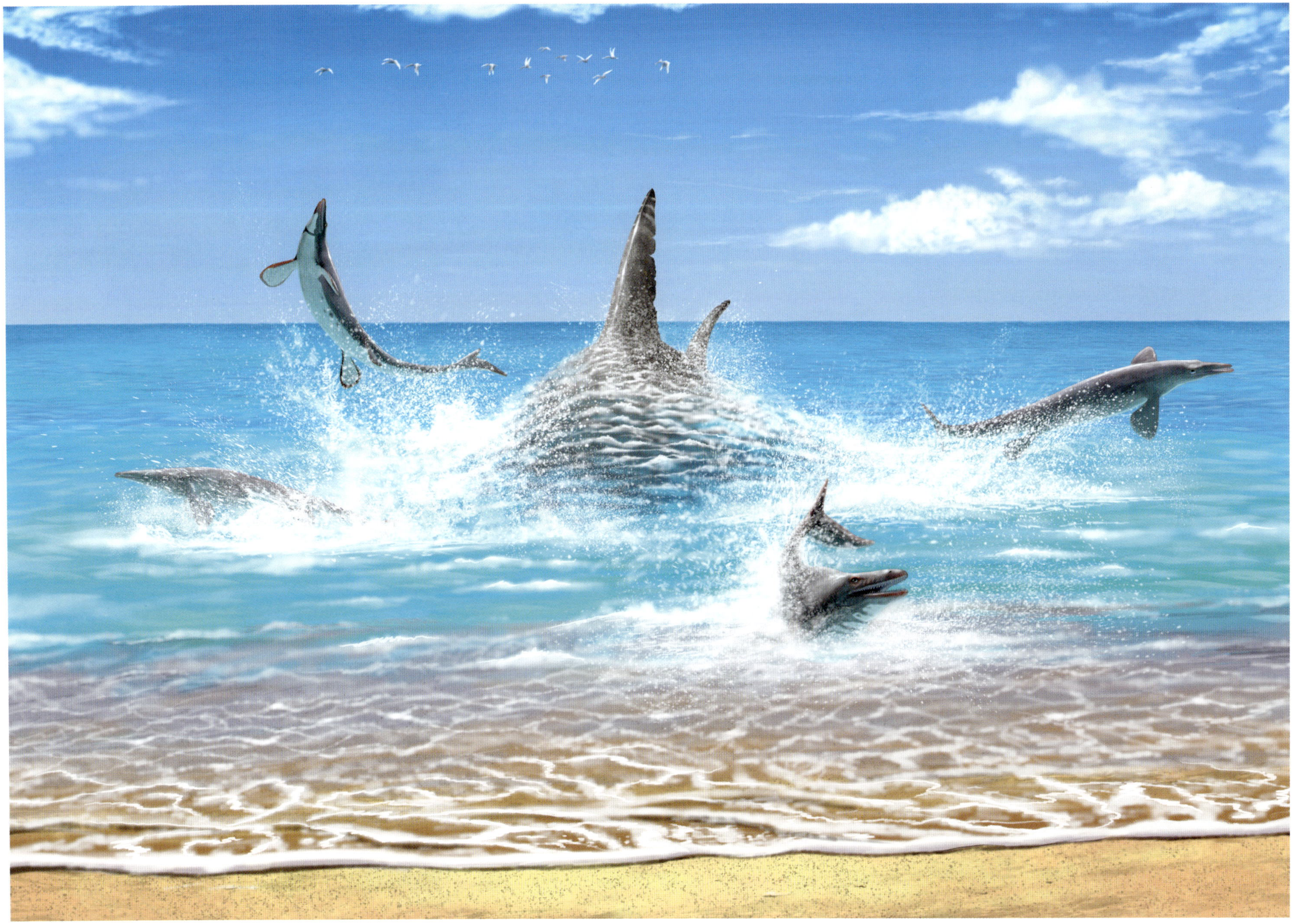

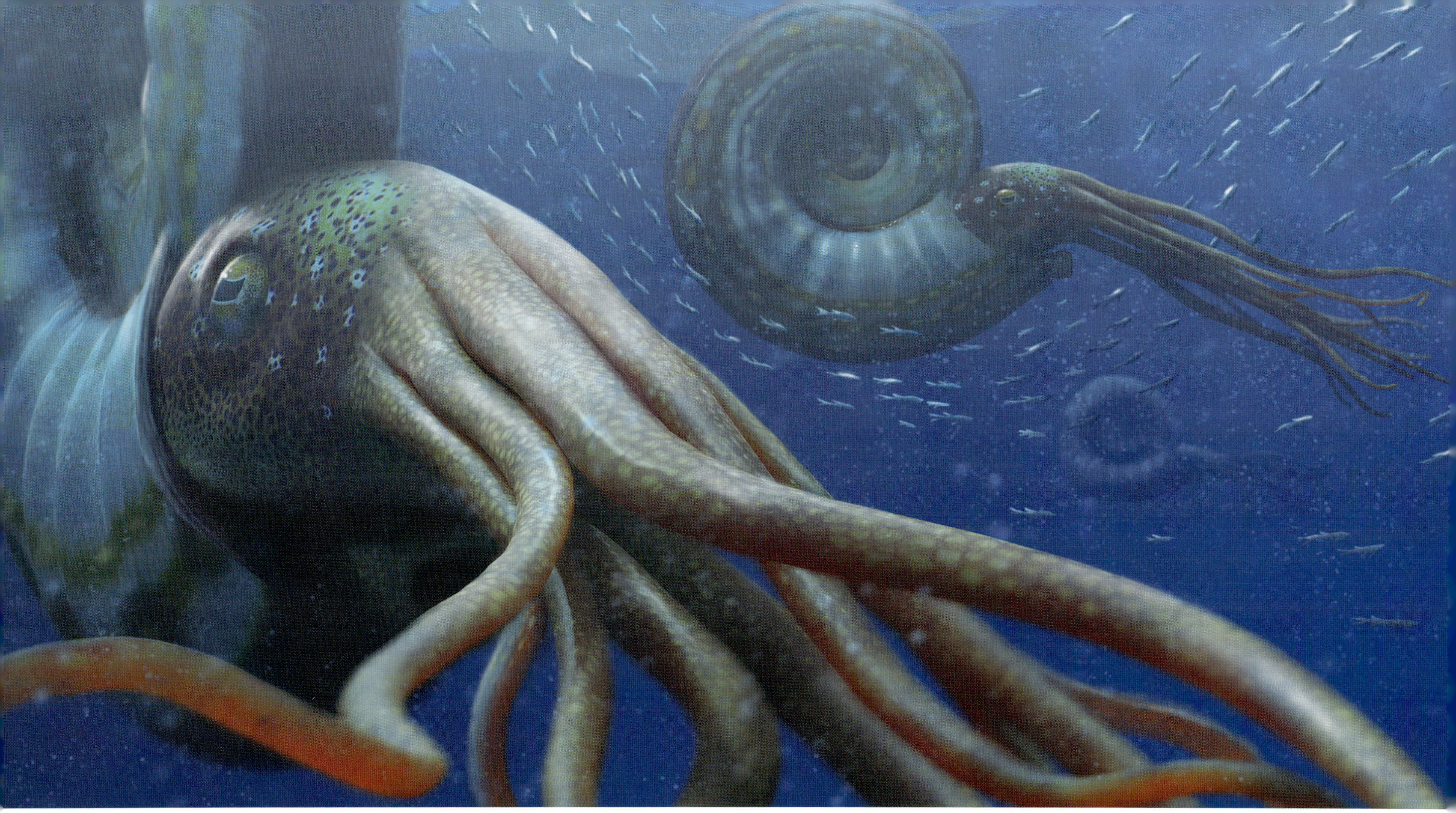

▲ *PARAPUZOSIA*

Large ammonites like the Late Cretaceous form *Parapuzosia* (its shell diameter is typically 60 cm [2ft]), shown here, could well have been squid-like predators, with big eyes and long, slender arms. Yes: *arms*. The term *tentacles* applies only to the paired, extra-long, normally concealed feeding appendages of squid and cuttlefish. The narrow, streamlined shape of *Parapuzosia*'s shell suggests an ability to move quickly.
• Digital | 2024

TULLIMONSTRUM ▶

The famous 'Tully monster' of Carboniferous Illinois (and thus alive about 310 million years ago) is an odd, streamlined animal, about 35cm (14in) long, equipped with a tail fin and an appendage that's tipped with serrated pincers or pseudo-jaws. Initially interpreted as a worm of some sort, it has more recently been interpreted as an archaic vertebrate, and more recently still as an indeterminate non-vertebrate animal.
• Digital | 2020

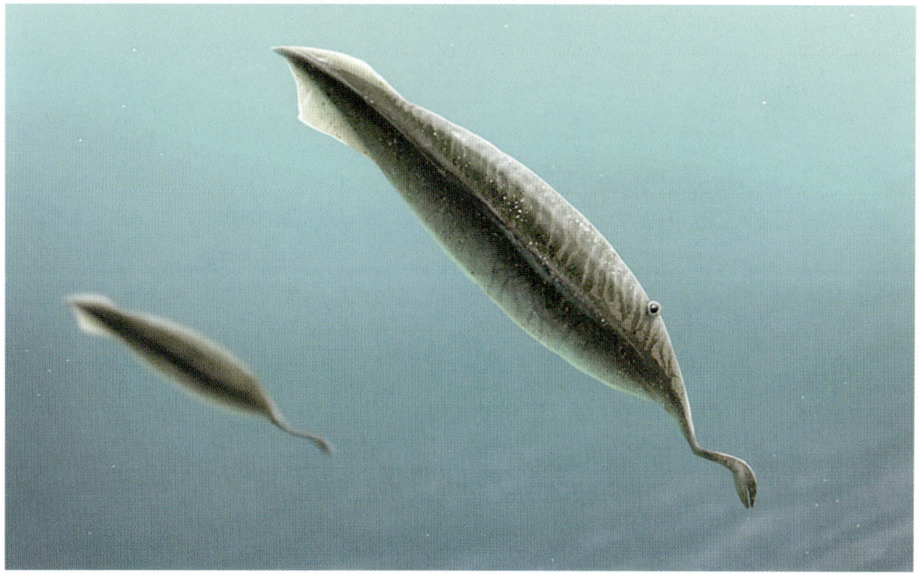

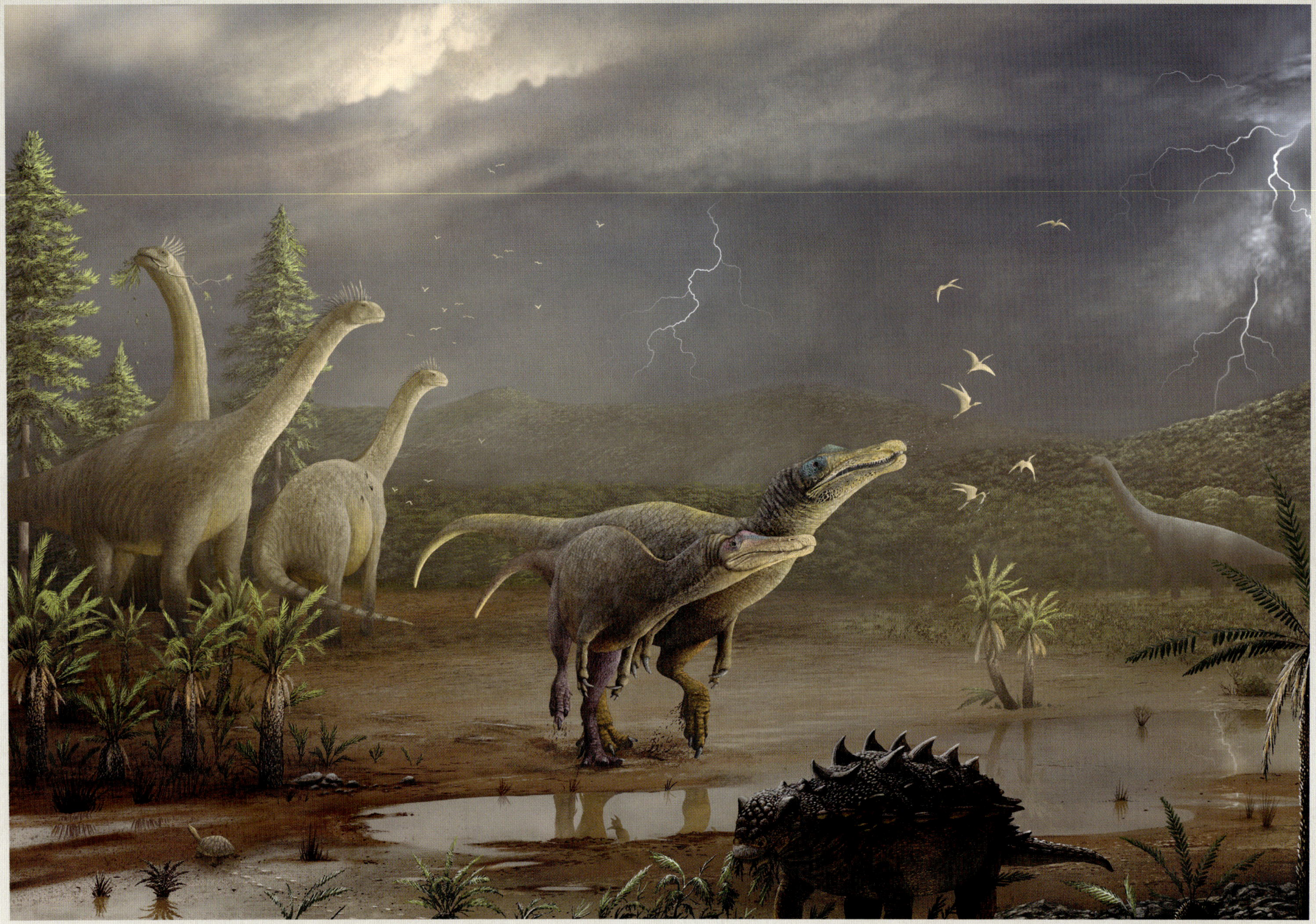

▲ SIERRA DE LA DEMANDA SCENE

This Early Cretaceous scene depicts the dinosaurs, pterosaurs, turtles and plants known from the Sierra de la Demanda Formation in Burgos, Spain. Polacanthine ankylosaurs, baryonychine spinosaurs and sauropods are known from this location, and indeed all three groups are characteristic of Early Cretaceous European faunas. The plants include conifers, cycads and ferns.
• Digital | 2022

HELL CREEK SCENE ▶

Most palaeoartists have to depict the environments and animals of the Late Cretaceous Hell Creek fauna of North America's West at some point in their careers! This scene shows an *Ankylosaurus* pair at the edge of a river, with azhdarchid pterosaurs and birds in the background. It is arguably the beautifully rendered plants, however, that command most attention here.
• Digital | 2020

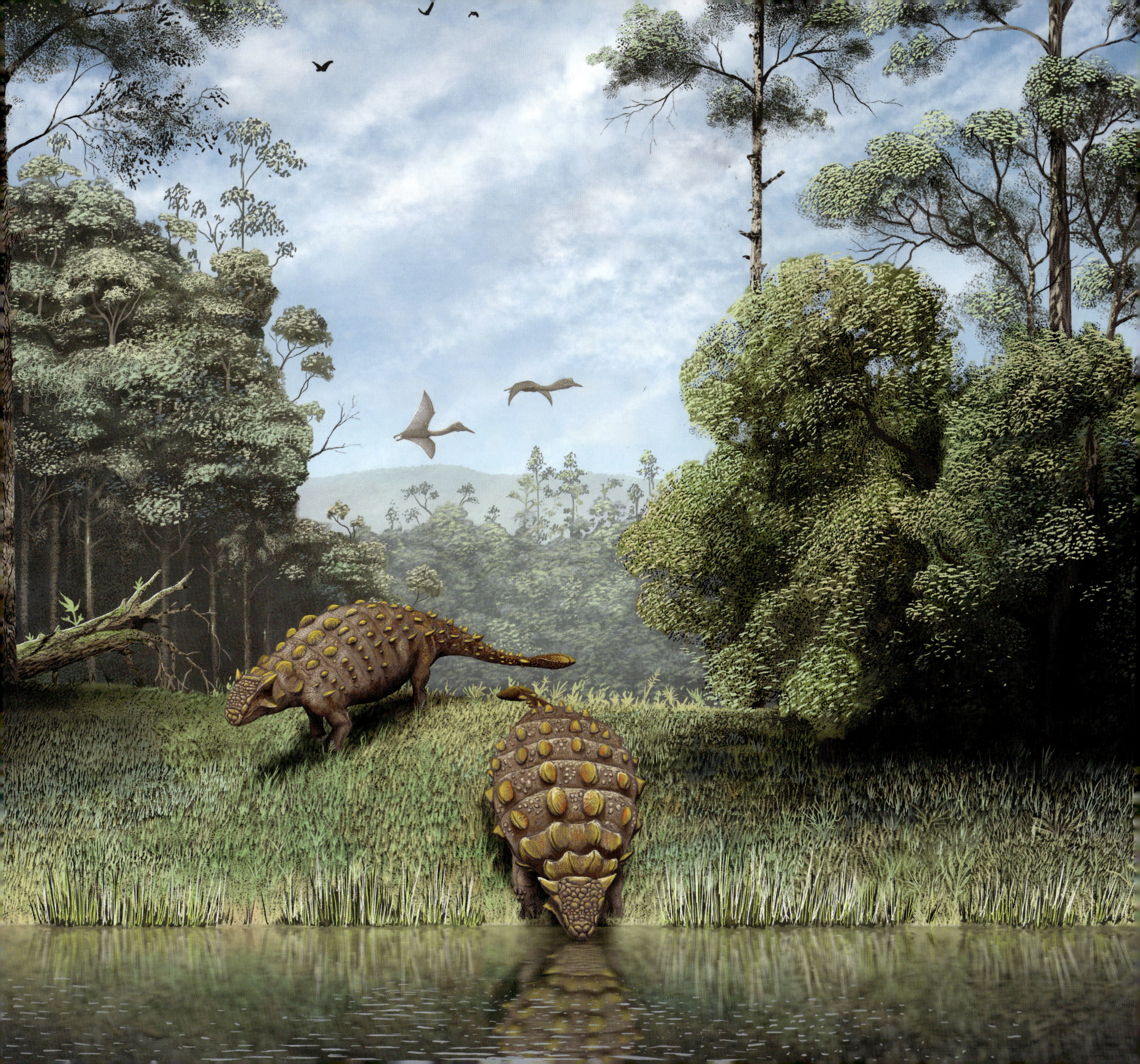

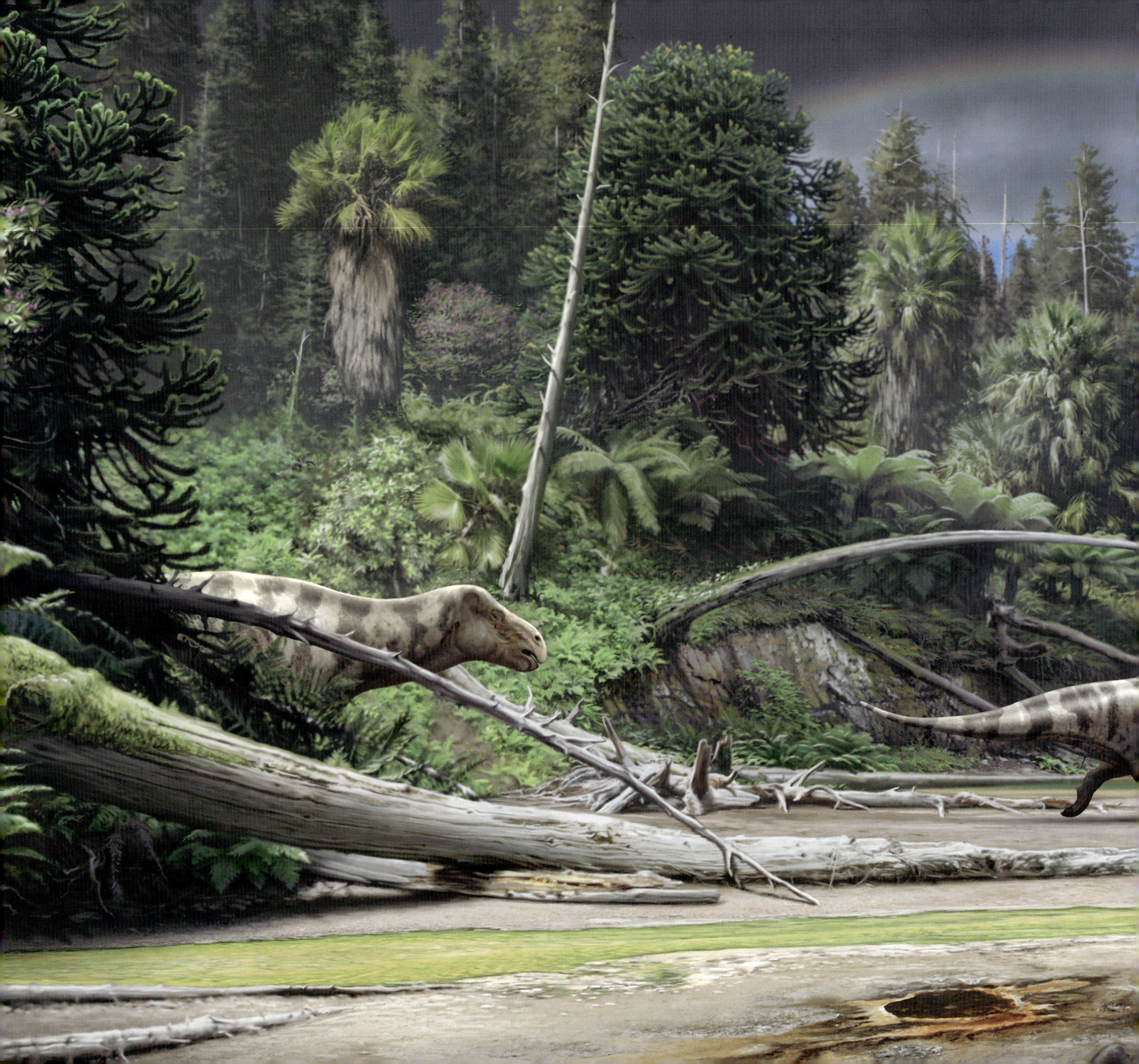

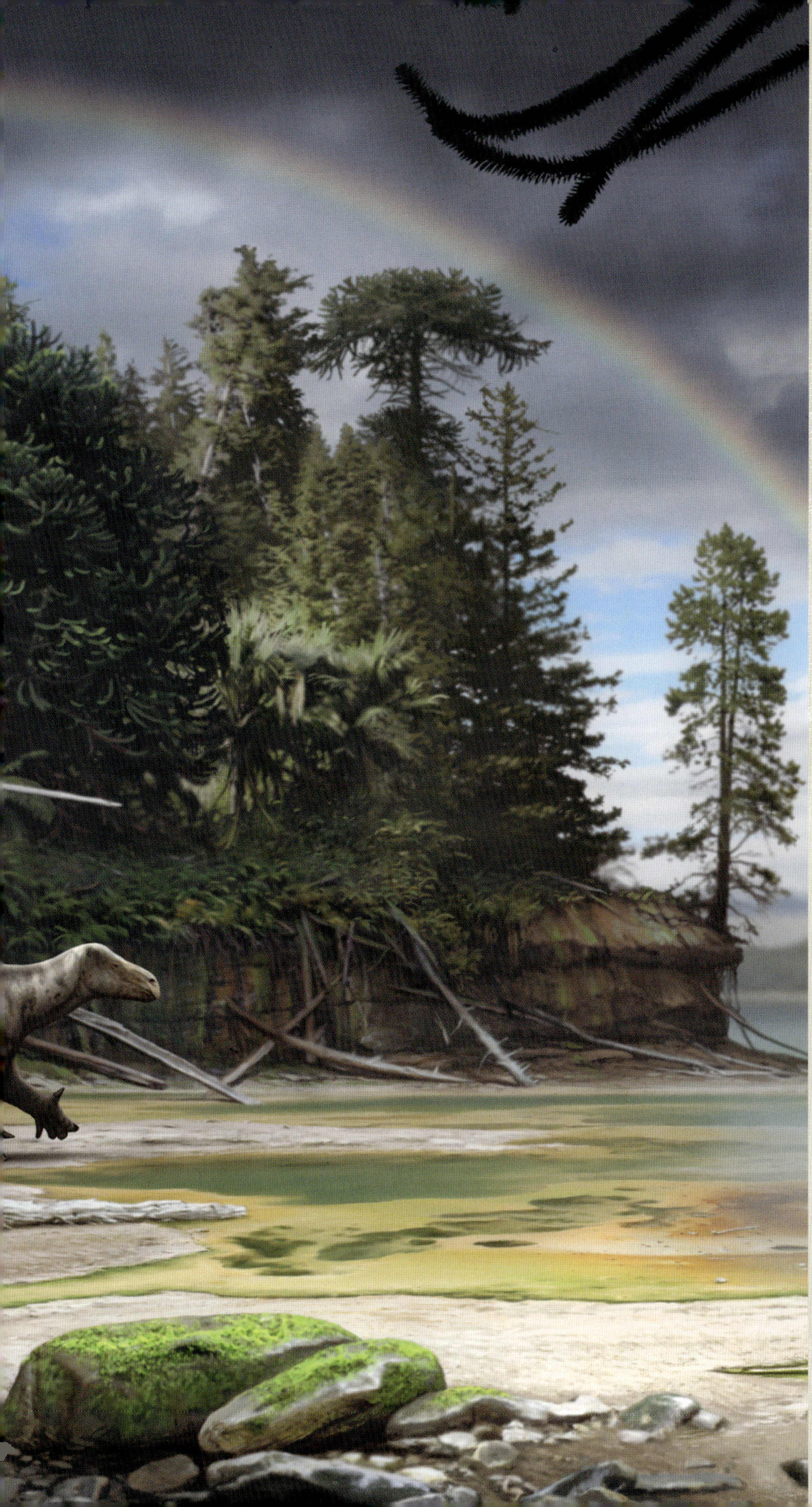

BOB NICHOLLS

ROBERT (BOB) NICHOLLS is a globally respected, multi award winning palaeo-reconstruction artist, with a reputation for producing unique, exciting and scientifically accurate artworks. Bob began drawing prehistoric animals from the moment he was old enough to hold a pencil, and today he has 25 years of professional palaeoart experience.

Bob's illustrations and models have appeared in more than 40 books and magazines published by the likes of *National Geographic*, Columbia University Press, Oxford University Press and *Nature*. His work is also exhibited in dozens of museums, universities and visitor attractions throughout the world, including the London Natural History Museum, The Chicago's Field Museum, the Children's Museum of Indianapolis, the Royal Mint and the *Walking with Dinosaurs* exhibition. Bob has also appeared on radio, podcasts and on TV many times. Between 2015 and 2024, Bob co-authored multiple scientific papers.

To reconstruct extinct animals, plants and environments, Bob undertakes a painstaking inside-out reconstruction process. This process ensures all of his artworks are accurate and completely unique. He uses various mediums, his portfolio containing traditional paintings, digital paintings, photographic composite illustrations, large-scale mural paintings, themed dioramas, digital sculptures and traditionally built models.

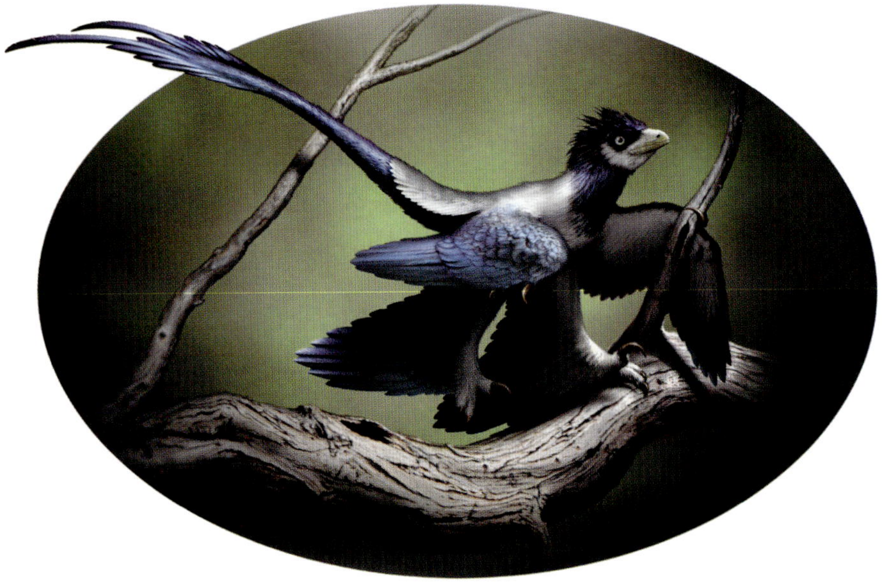

◄ ***WULONG***

Named in 2020 from China's Early Cretaceous Jiutofang Formation, *Wulong* is one of the microraptorines, a dinosaur group noted for their small size and long hindlimb feathers. *Microraptor* – named in 2001 – is the best-known member of this group. Bob's reconstruction reflects the results of a 2023 study which reported iridescent feathering on *Wulong*'s limbs and grey feathers on its body.
• Digital | 2023

▼ **THE DUET**

On the banks of an Early Cretaceous stream in north-east China, a male and female *Caudipteryx* bow to one another during a courtship ritual. Several features shown here have not conventionally been portrayed in reconstructions of this animal, including feathered toes, sexual dimorphism in wing patterning, and tight barring across the V-shaped tail fan.
• Digital | 2019

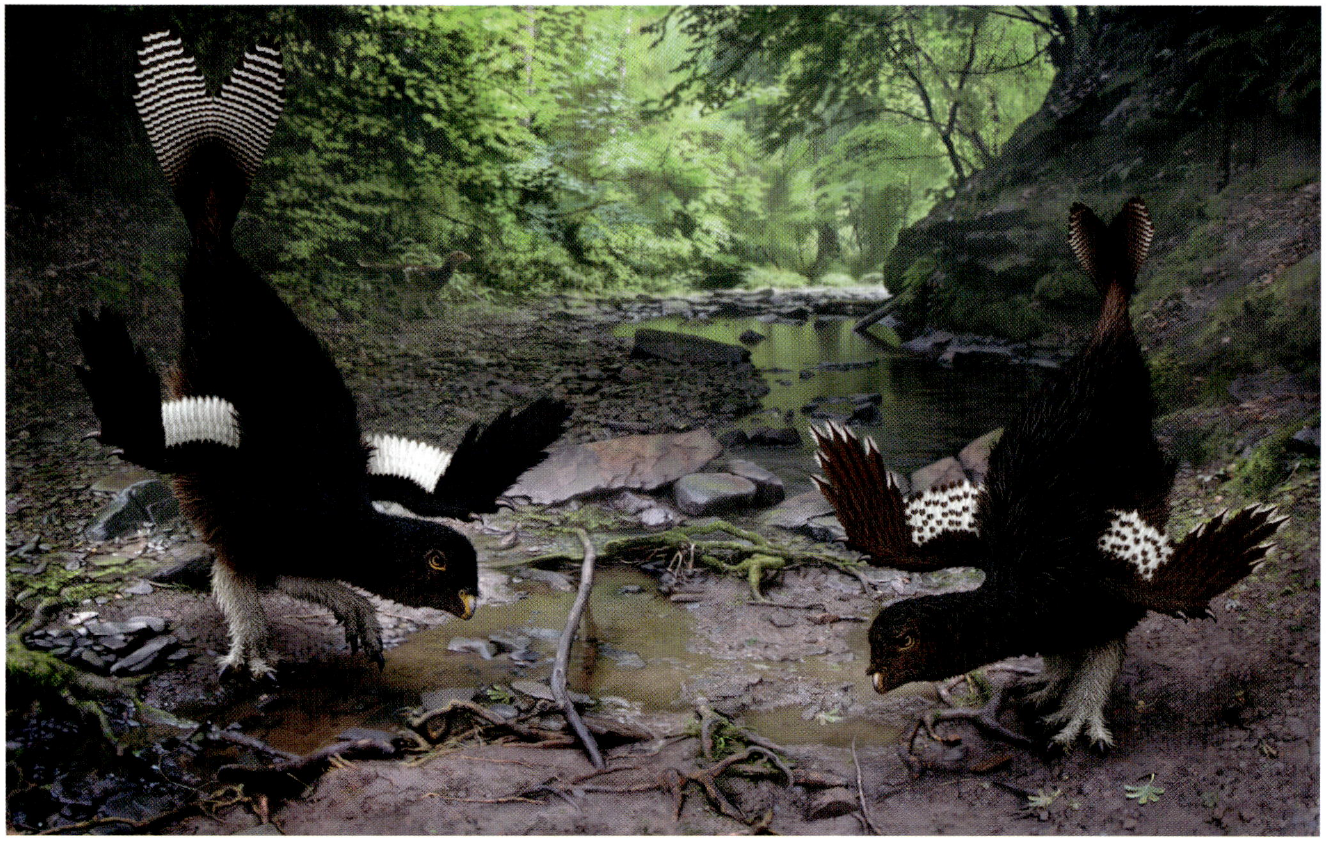

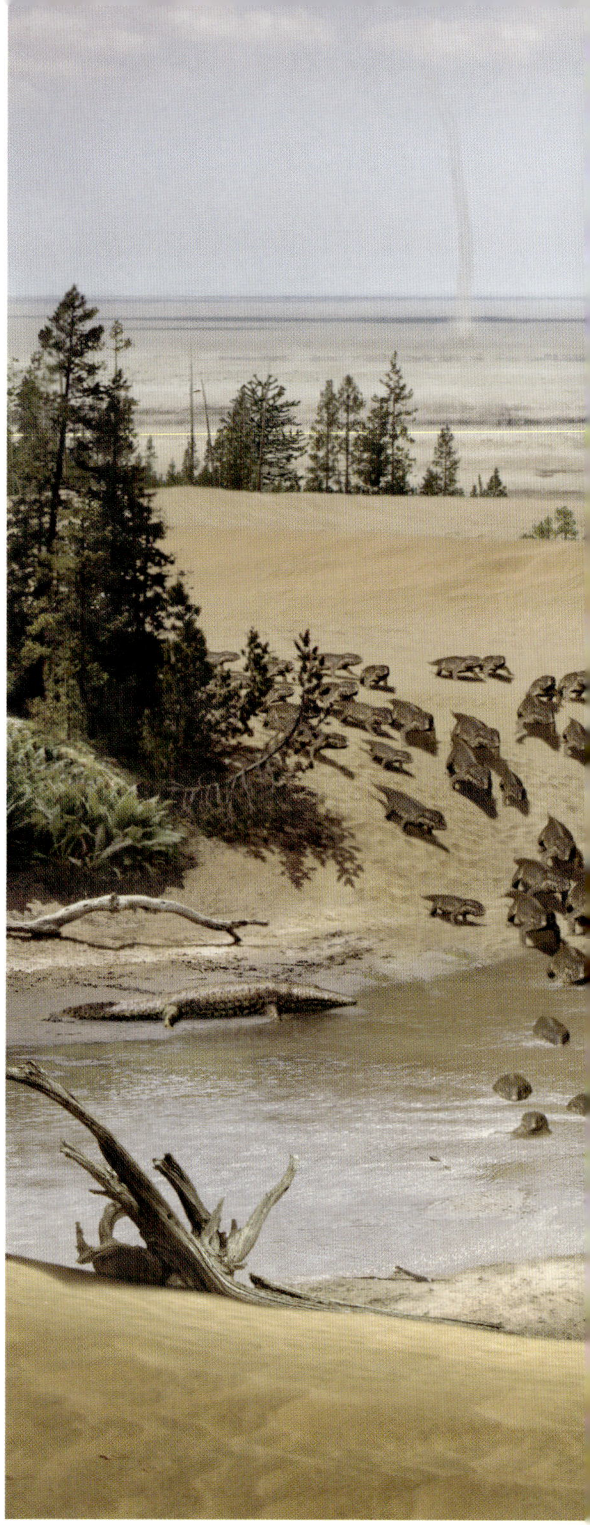

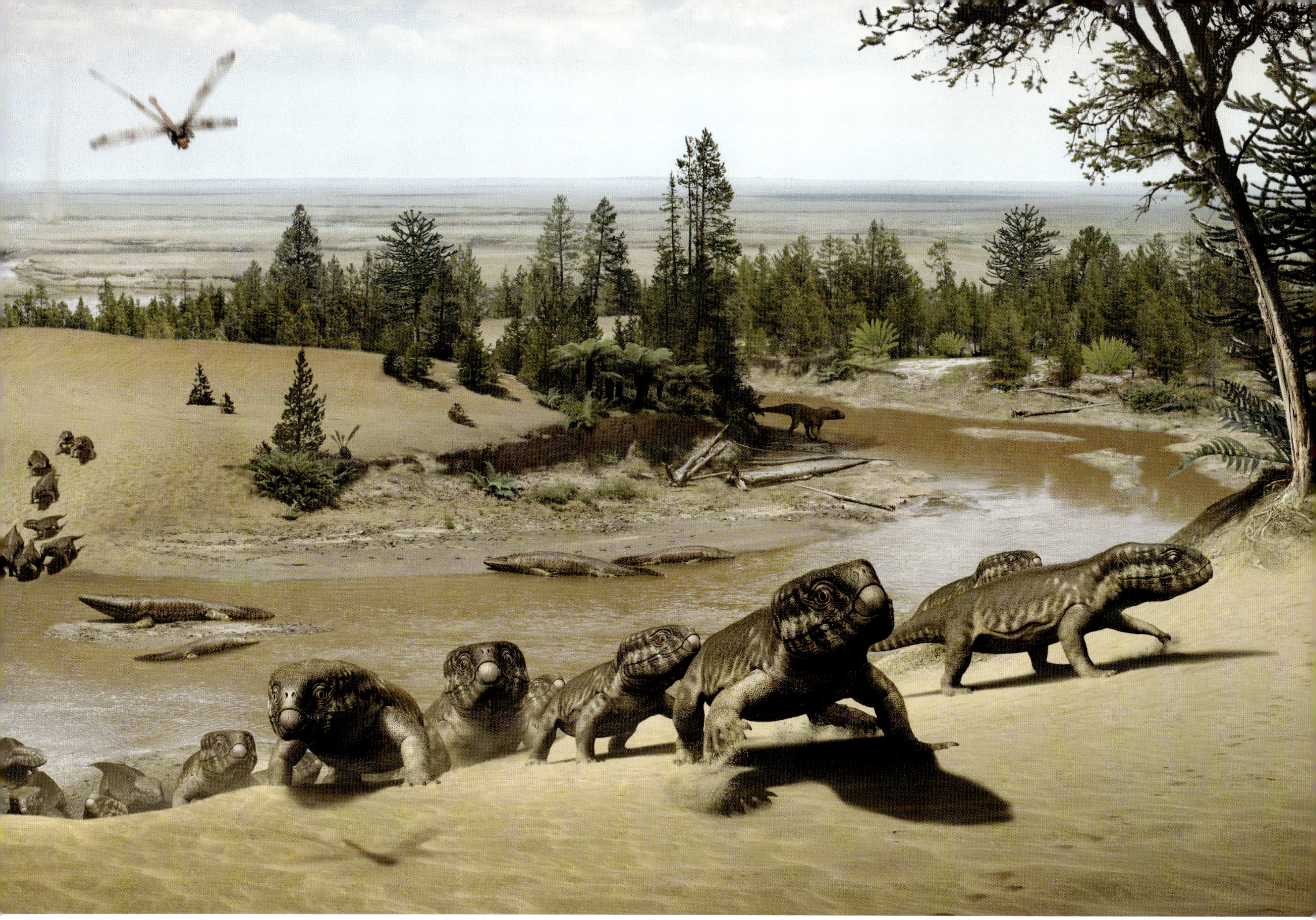

[Previous pages]

***IGUANODON* SCENE**

Geological data shows that the iconic Early Cretaceous ornithischian *Iguanodon* occurred in several western European environments, ranging from semi-deserts and coastal plains to dense woodland. Here, Bob has reconstructed a lavishly forested scene depicting numerous plant species of the time. This illustration was commissioned by the British Royal Mint to accompany a set of coins.

Used courtesy of the Royal Mint.

• Digital | 2020

▲ ***LANGERONYX* ON THE MOVE**

Between 250 and 210 million years ago during the Triassic, a group of reptiles termed rhynchosaurs occurred in terrestrial habitats worldwide. *Langeronyx*, shown here, is a British rhynchosaur. The discovery of specimens in close proximity suggests a degree of sociality. They perhaps moved in groups. Twinned, beak-like structures in the rhynchosaur upper jaw remain the source of argument, and artists have portrayed them in several ways, in this case as a single beak.

• Digital | 2016

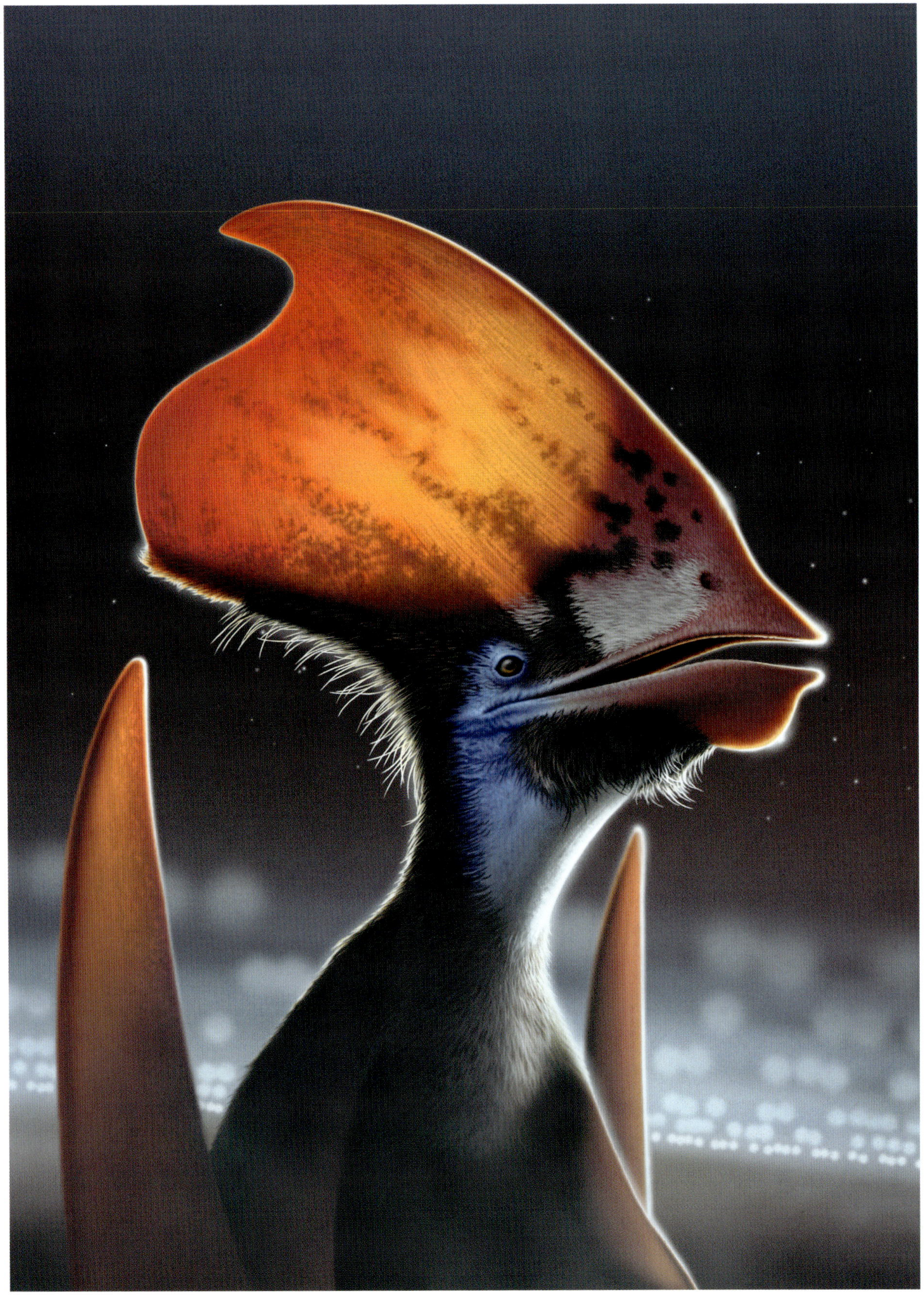

◂ *TUPANDACTYLUS*

The discovery of sail-crested tapejarid pterosaurs (like Brazil's *Tupandactylus*, shown here) caused a sensation in the 1990s. It was always suspected that these large crests were brightly coloured. A 2022 study found evidence for various pigments occurring across the crest, and in the feather-like filaments growing from the bony spar at the crest's trailing margin.

• Digital | 2022

SICK *T. REX* ▸

Various holes and hollows in the jaw bones of Sue the *T. rex* led experts to think that Sue had suffered from a mouth infection, as portrayed here. More recently, this idea has been contested, and the real cause behind Sue's disfigured jaws remains unknown. Whatever the truth, we can be confident that ailments of various sorts might have given a poorly tyrannosaur a grotesque appearance.

• Digital | 2021

▾ *PSITTACOSAURUS* PAIR

A now famous, spectacularly complete Chinese specimen of the ceratopsian *Psittacosaurus* preserves long filaments on the tail. Another surprising feature revealed by this fossil is that the area around the cloaca (the external opening for the gut, bladder and reproductive organs) is swollen and unusually dark. Perhaps this reveals an unusual role in behaviour, as hinted at here.

• Digital | 2021

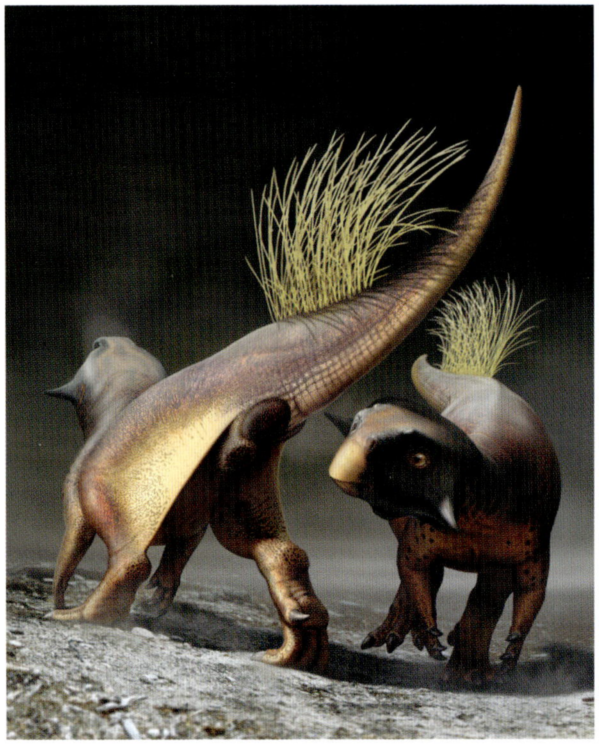

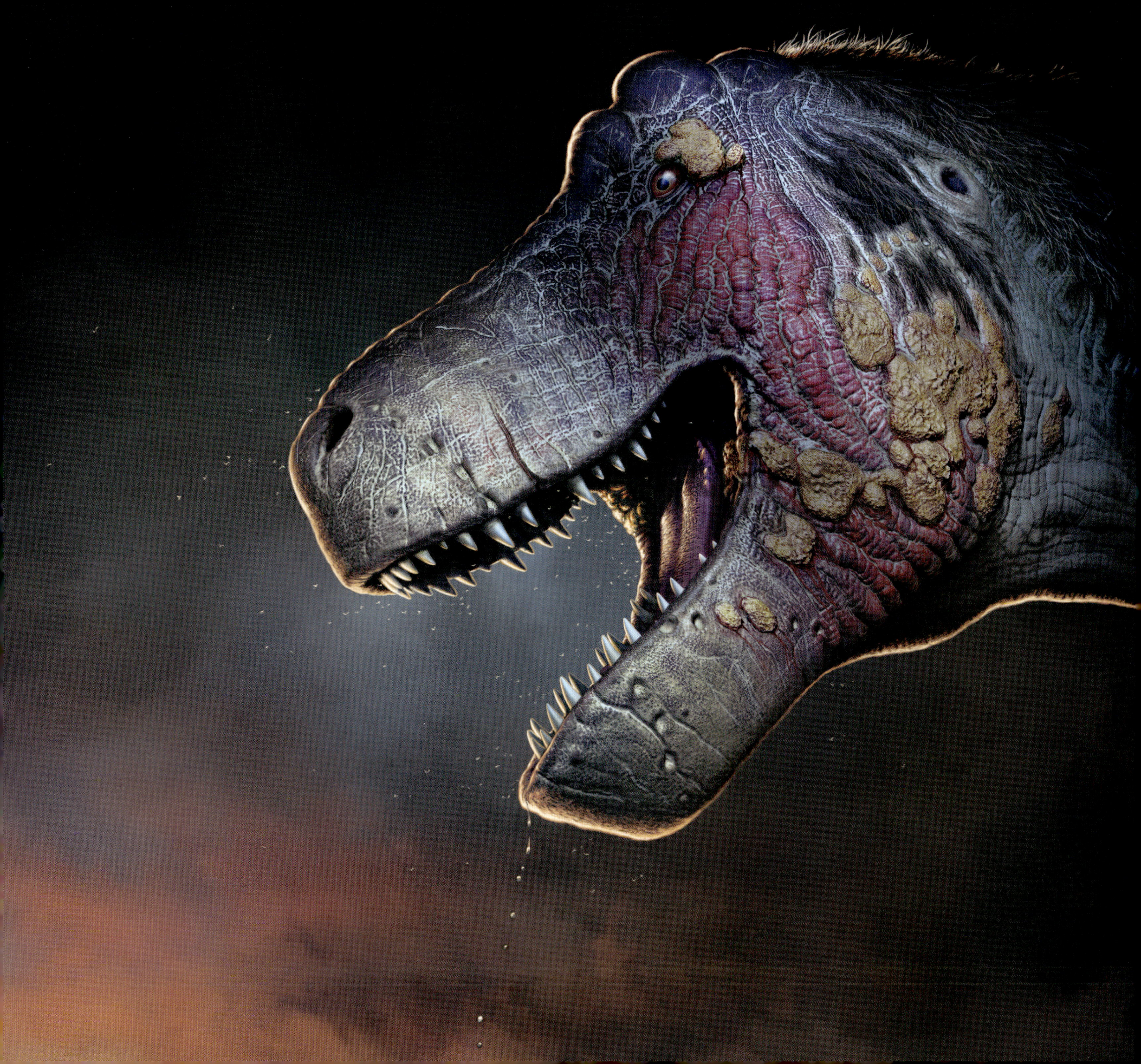

PLATECARPUS FLEEING *TYLOSAURUS* ▸

Numerous different groups of mosasaurs – the great swimming lizards of the Cretaceous – existed, differing substantially in proportions and the shapes of their skulls, flippers and tails. The two kinds shown here – giant, long-snouted *Tylosaurus* and small *Platecarpus* – belong to separate branches of the mosasaur family tree. Direct evidence that large mosasaurs preyed on their small cousins comes from stomach contents.

Used courtesy of the Children's Museum of Indianapolis

• Digital | 2021

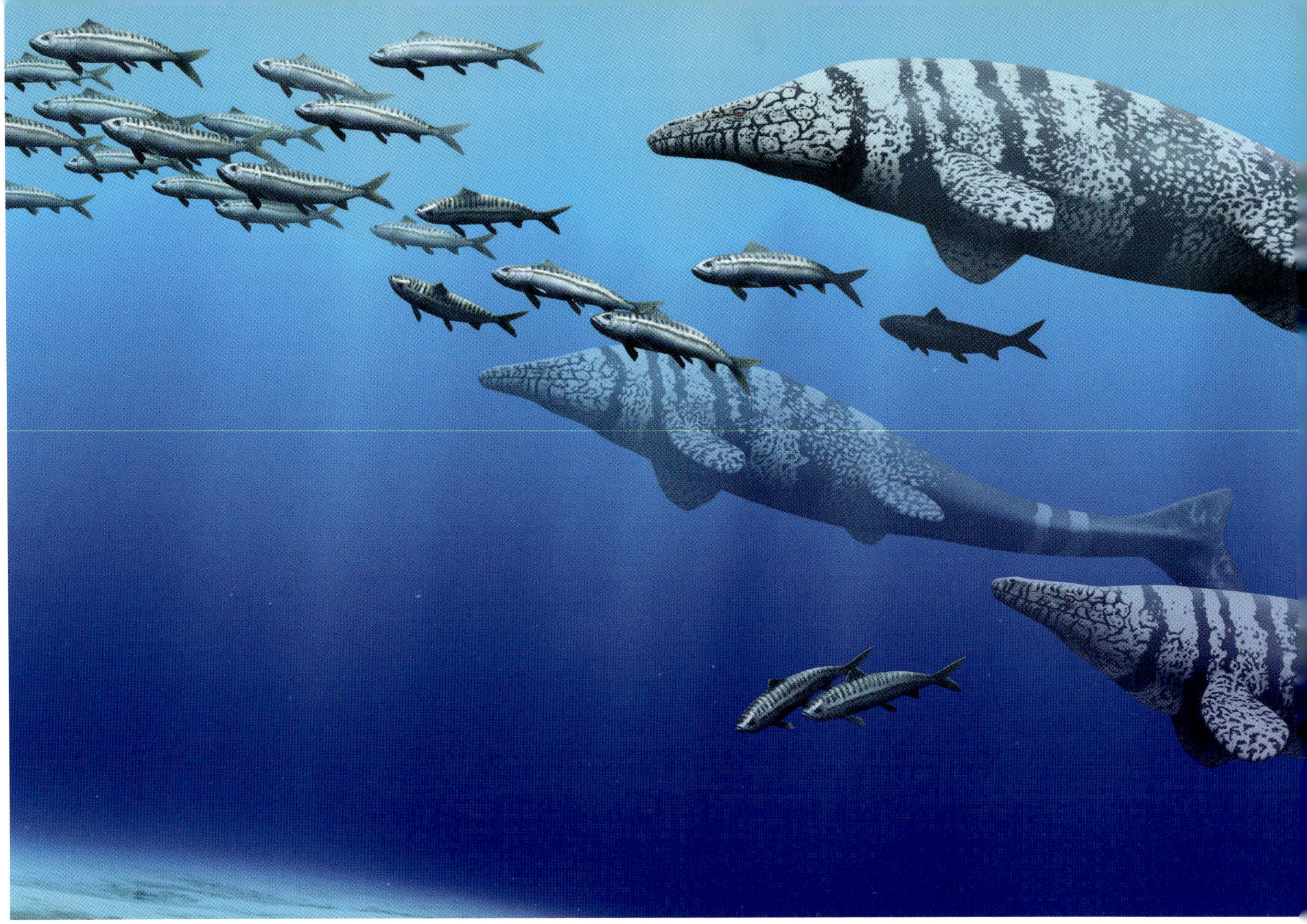

▾ ELASMOSAUR AND *PACHYRHIZODUS*

In the massive blue space of the open ocean, giant animals loom from the void in a way that makes their scale difficult to comprehend, and maybe elasmosaurid plesiosaurs used this effect to conceal their bulk. The large fish here is *Pachyrhizodus*, a widespread, tuna-shaped animal that reached 3m (10ft) and possessed a tall dorsal fin.

Used courtesy of the Children's Museum of Indianapolis.

• Digital | 2021

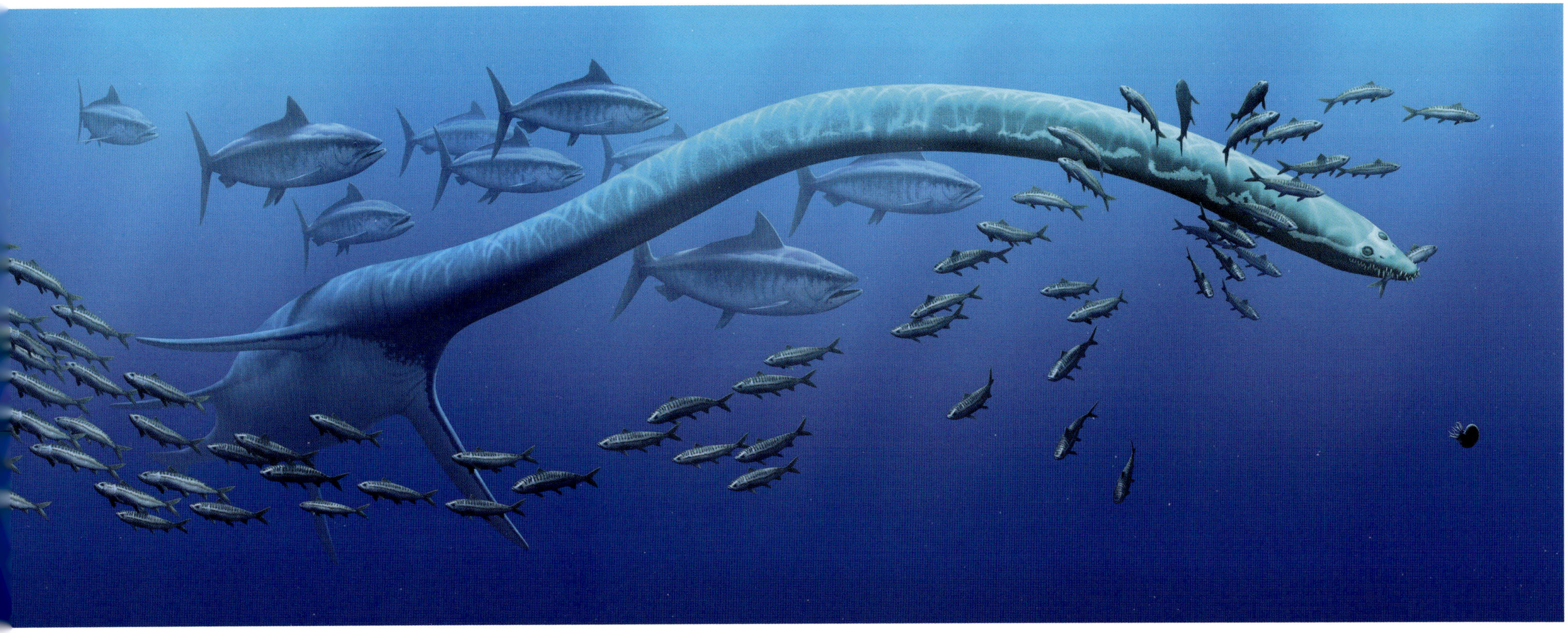

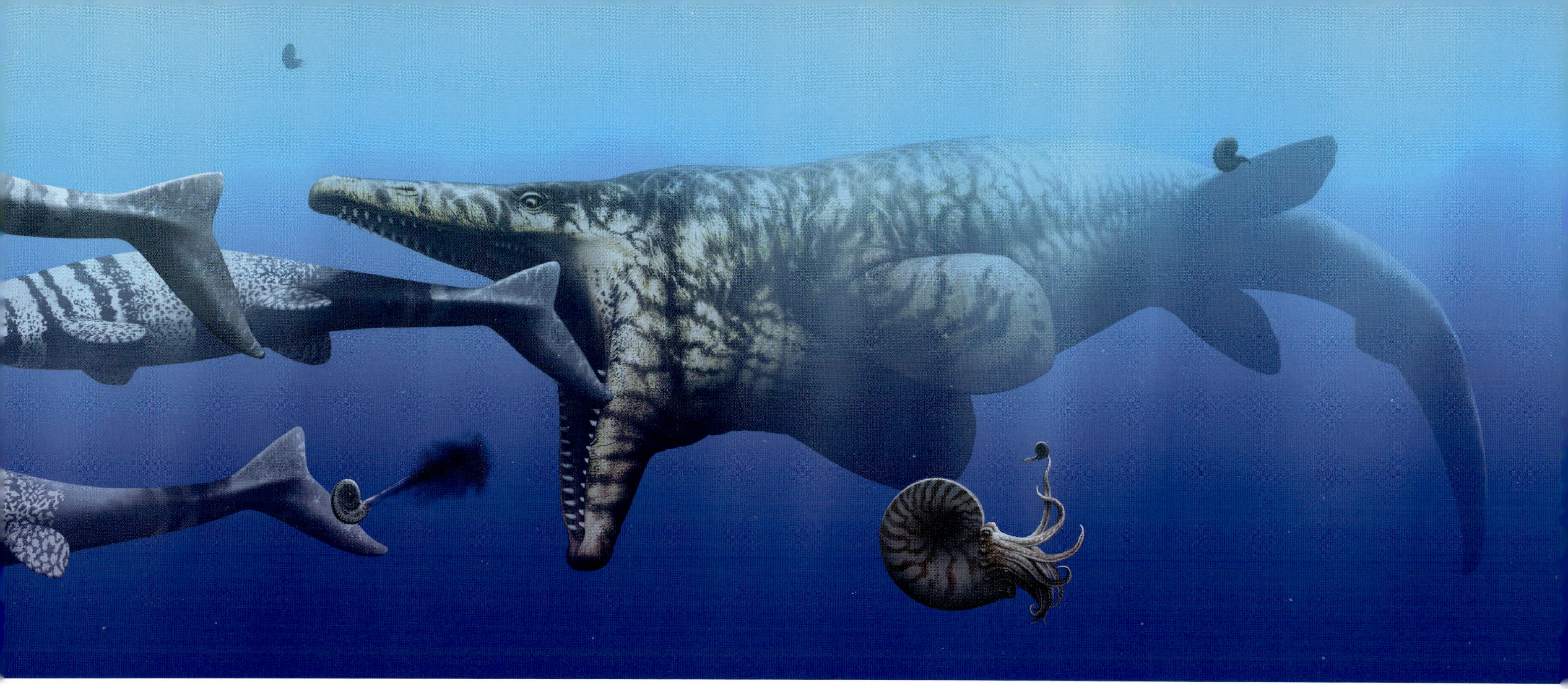

EDISCETUS ▶

A great surge in our knowledge of early whale evolution has occurred over the past few decades as new fossil species have been reported worldwide. This image shows the 30-million-year-old dolphin *Ediscetus*, named from the Oligocene of South Carolina, USA, in 2018. Like many archaic dolphins, *Ediscetus* had a long, slender rostrum and protruding, fang-like teeth.
• Digital | 2022

[Overleaf]

MEGALOSAURUS

In another lavish scene commissioned by Britain's Royal Mint, Bob has reconstructed the Middle Jurassic European landscape inhabited by the famous predatory dinosaur *Megalosaurus*. The original *Megalosaurus* fossils come from sediments laid down in a tropical coastal environment, perhaps with mangrove-type habitat nearby. Cycads, ferns, ginkgos and conifers are recorded here.
Used courtesy of the Royal Mint.
• Digital | 2020

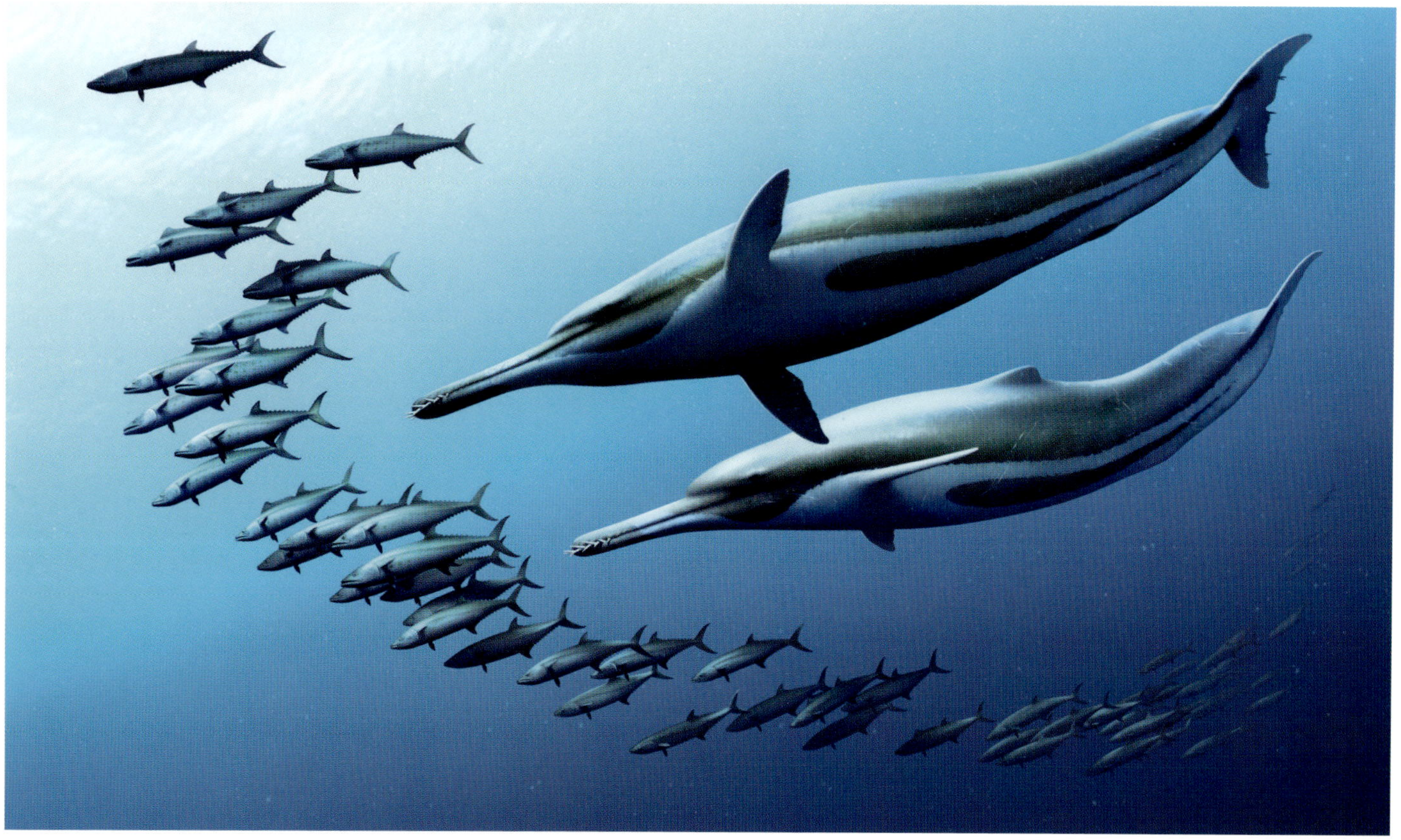

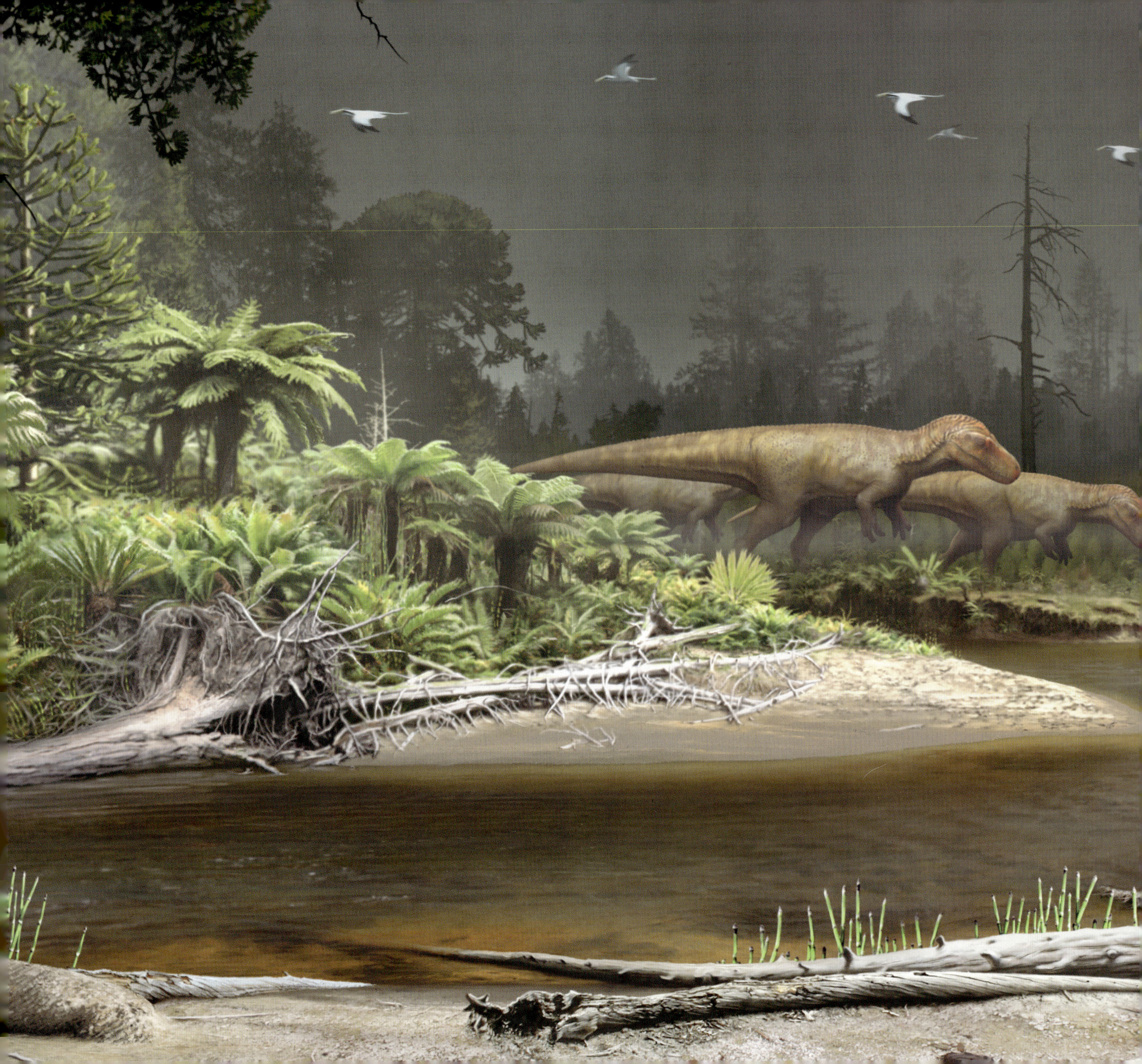

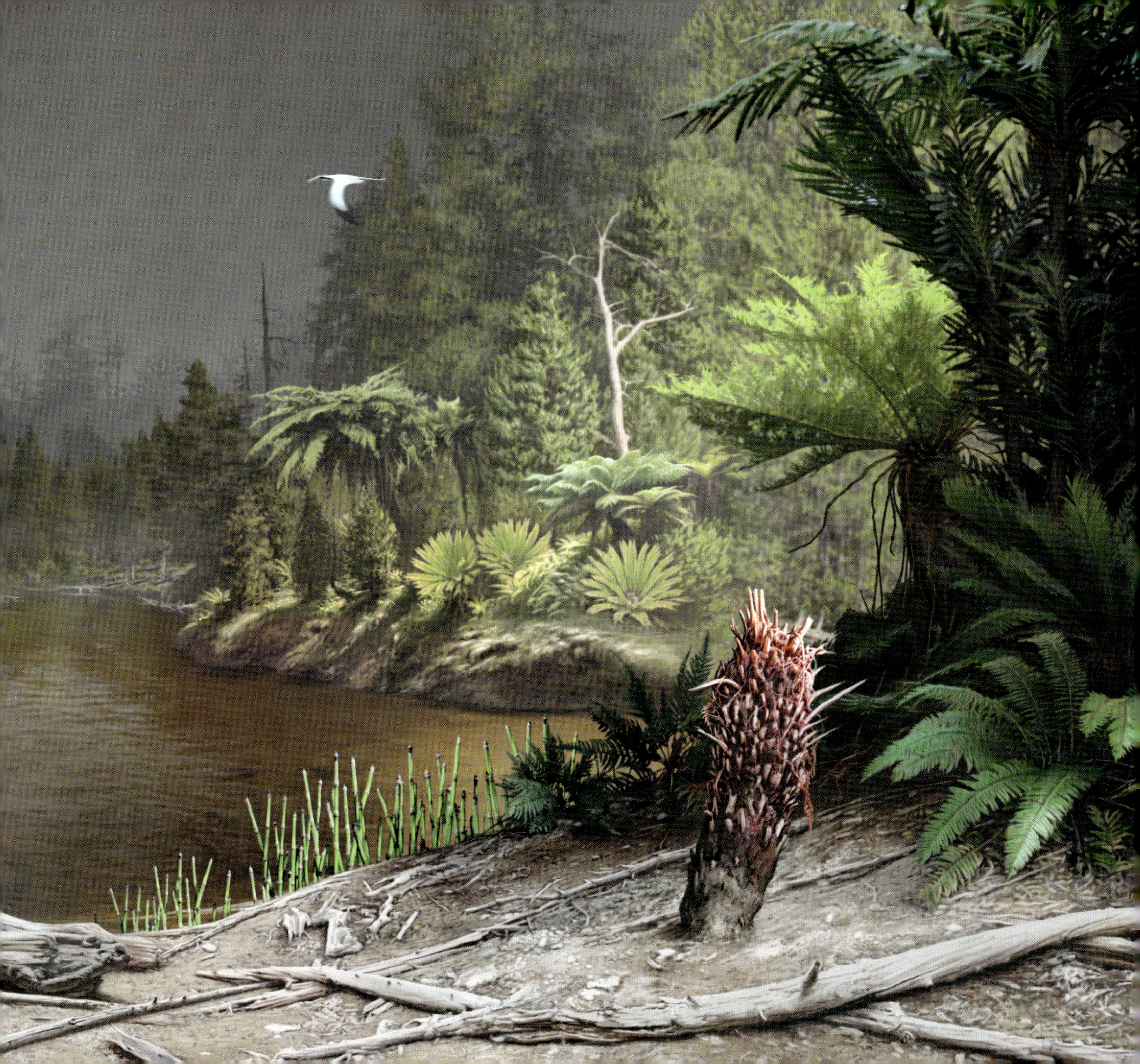

ACKNOWLEDGEMENTS

DARREN thanks, once again, the many friends and colleagues who, over the years, have provided so much insight on what palaeoart was, is, and can be, those who've purchased and praised the predecessor to this current volume, and the team who've helped bring *Mesozoic Art II* to fruition. Big love in particular to Steve, Emma, Jim, and the many artists and Bloomsbury staff we've worked with. Thanks to everyone on socials, Patreon and the TetZooniverse who've been supportive. Thanks to Chris Packham for swift work, and to Toni, Hel and Em for their tolerance of my Sisyphean workload.

STEVE would like to thank all the artists whose gorgeous work made *Mesozoic Art II* a joy to work on. Special thanks to Jim and Jenny at Bloomsbury for giving us a chance to do it all over again, and to Emma and all those anonymous members of staff at Bloomsbury who made producing and publishing this book such a smooth ride. And thanks, as ever, to Darren, who, despite everything, kept his side of the bargain. Working at home, alone, in your back room doubling as an office and studio, it's hard to remember sometimes just how much of a joint effort pulling together a book like this is. Cheers, everyone. Here's to volume III...

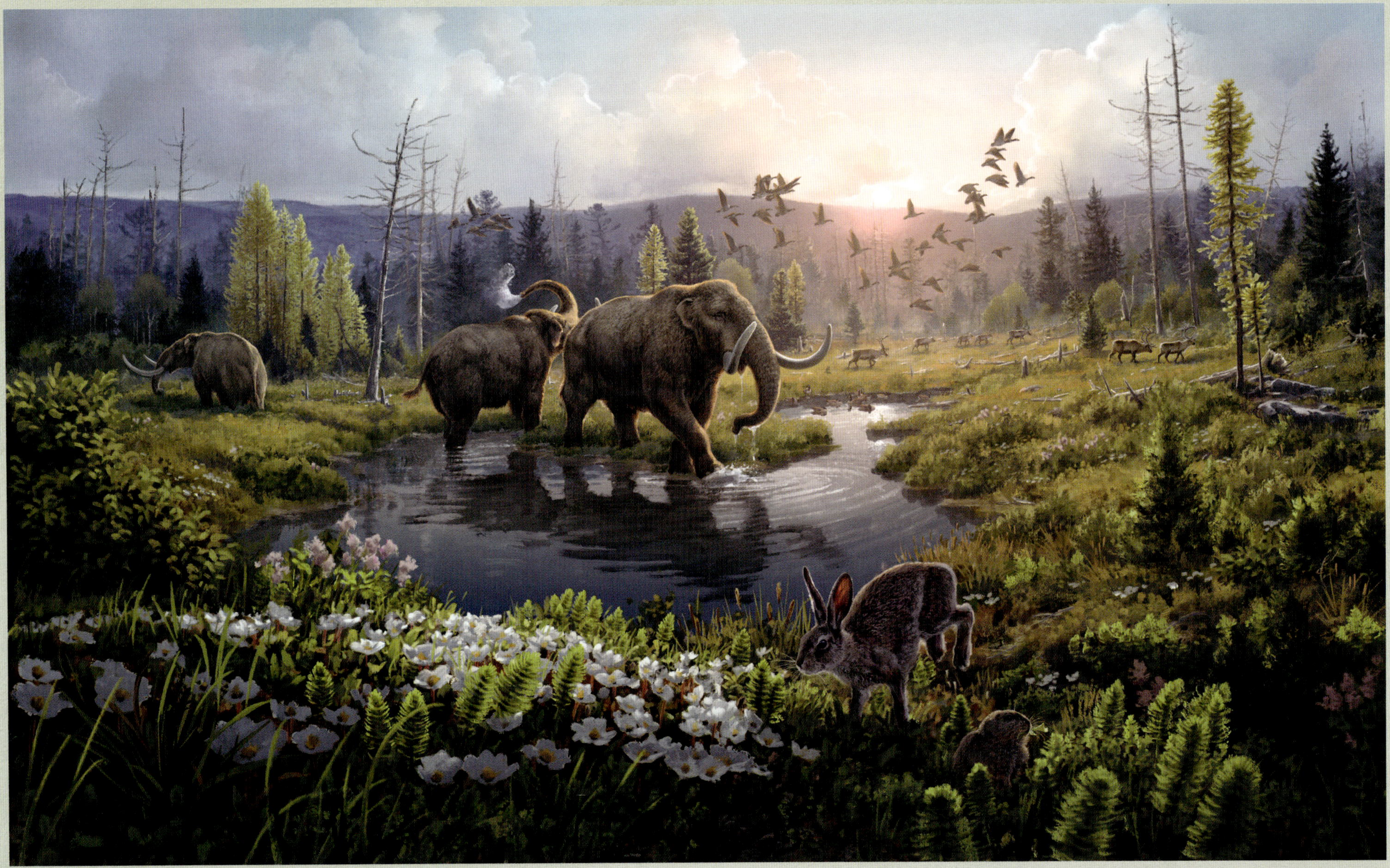

GREENLAND TWO MILLION YEARS AGO
Beth Zaiken • Digital | 2022

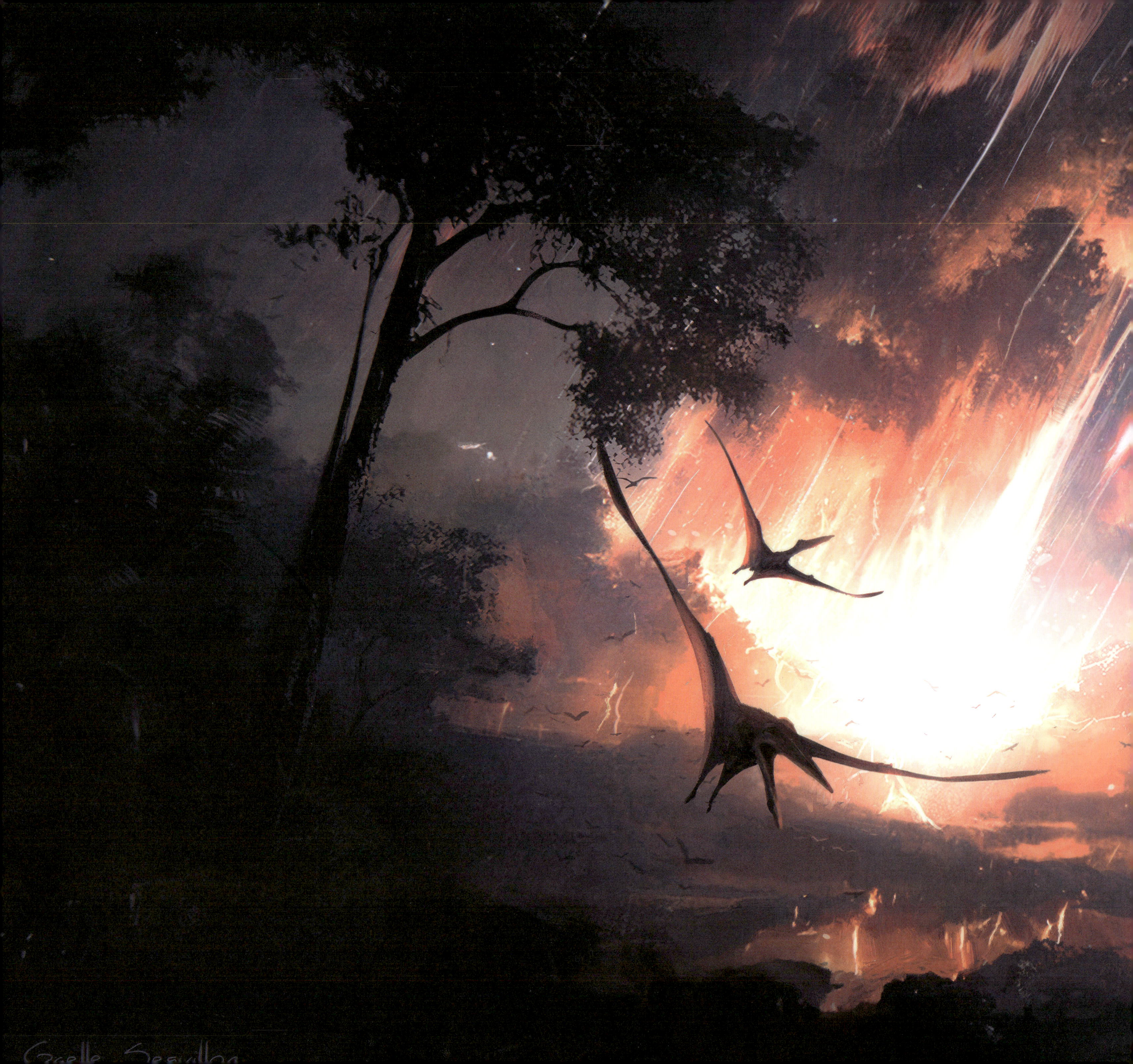